AF778255

GROUNDWATER INTENSIVE USE

SELECTED PAPERS ON HYDROGEOLOGY

7

INTERNATIONAL ASSOCIATION OF HYDROGEOLOGISTS

Groundwater Intensive Use

Selected Papers, SINEX, Valencia, Spain 10–14 December 2002

Edited by

A. Sahuquillo[(1)], J. Capilla[(1)], L. Martínez-Cortina[(2),(3)] and X. Sánchez-Vila[(4)]

With the collaboration of

R. Aliaga[(1),(5)]

[(1)] *Technical University of Valencia, Spain*
[(2)] *Marcelino Botin Foundation, Santander, Spain*
[(3)] *Spanish Association of Groundwater Users, Madrid, Spain*
[(4)] *Technical University of Catalonia, Barcelona, Spain*
[(5)] *Geological Survey of Spain, Madrid, Spain*

A.A. BALKEMA PUBLISHERS LEIDEN / LONDON / NEW YORK / PHILADELPHIA / SINGAPORE

Library of Congress Cataloging-in-Publication Data

Applied for

This book is a contribution to the implementation of the UNESCO IHP-VI programme 'Water Interactions: Systems at Risk and Social Challenges' 2002–2007.

Cover illustration: Groundwater discharge from the Oak Ridges Moraine Aquifer in Southern Ontario, Canada. Courtesy of Dr. Ken Howard.
Cover design: Gabor Lorinczy
Typesetting: Charon Tec Pvt. Ltd, Chennai, India
Printed in Great Britain by Anthony Rowe Ltd.

Copyright © 2005 Taylor & Francis Group plc. London, UK

All rights reserved. No part of this publication or the information contained herein may be reproduced, stored in a retrieval system, or transmitted in any form or by any means, electronic, mechanical, by photocopying, recording or otherwise, without written prior permission from the publishers.

Although all care is taken to ensure the integrity and quality of this publication and the information herein, no responsibility is assumed by the publishers nor the author for any damage to property or persons as a result of operation or use of this publication and/or the information contained herein.

Published by: A.A. Balkema Publishers, Leiden, The Netherlands, a member of Taylor & Francis Group plc.
www.balkema.nl, www.tandf.co.uk and www.CRCPRESS.com

ISBN 04 1536 444 2

Contents

IMPACTS

MANAGEMENT

Introduction

The use of groundwater is intensive when as a consequence of abstraction the flow pattern and the quality distribution are notably changed with respect to the natural situation. The consequences of intensive use include changes, even reversals, of hydraulic gradients, reduction of discharge through springs or into rivers and wetlands, increase of net recharge in areas previously fully saturated or subject to phreatic evapotranspiration, displacement of groundwater bodies of a given quality (salinity, chemical composition, presence of noxious, minor components such as Fe, As, F, ...), changes in flow conditions that lead to salts dissolution (gypsum, saline crusts, ...), and modification of redox potential and other chemical and biochemical circumstances that affect water-rock interaction (Fe and Mn presence, SO_4 and NO_3 reduction, dissolution of reduced sulphur compounds). Each one of these changes can be positive, negative or indifferent, depending on the point of view under which groundwater is evaluated, such as Nature conservation, stability of water quality and temperature, groundwater level depth, cost of water abstraction, compliance with existing legal rights, changes in groundwater storage, ... The term intensive use poses no previous judgement on the actual impacts. It only points out a situation regarding groundwater abstraction. The evaluation of the consequences of groundwater development is done later on, under the viewpoints of the technical, economic, social or political groups concerned, and using the objective quantifications resulting from a carefully oriented analysis of the current and the future situation. This is an improved way of thinking with respect to other conceptions that are often used in groundwater evaluation and management. The oldest of these conceptualisations is the *safe-yield* (annual volume of groundwater abstraction at a rate that does not produce undesirable effects), and the derived ones (*secure-yield*, *permanent-yield*). They intend to show how much water can be abstracted from an aquifer or aquifer system, considering some direct negative effects, but externalities (including effects on habitats) are not considered in full or at all. These conceptualisations represent mostly the approach of the water developer.

In the last two decades the designation of overexploitation has been widely introduced, mostly in non-hydrogeological fora, although it is now also used as a hydrogeological term. The meaning of overexploitation is poorly constrained and focuses on pointing out negative aspects (impacts) of aquifer development. As such, most people tend to think that overexploitation is the contrary to sustainable exploitation, and this is not necessarily the case. In practice, the very valuable and positive sides (benefits) of aquifer development, and of controlled intensive use, and the capacity for solving and securing freshwater supply problems not only for drinking purposes, but also for economic and social development of depressed areas, mostly through irrigation are often overlooked. The impact-oriented focus of the term overexploitation is mostly presented in a descriptive rather than in a quantitative form, and in its formulation the often favourable intrinsic characteristics of groundwater (delayed response, three-dimensional nature, long transient stages) are overlooked. The term has easily pervaded language and has been widely adopted by media. In some cases the lay man and the non-specialist decision-maker may erroneously conclude from media information and descriptive reports which tend to put the accent only in negative aspects, that groundwater development is essentially negative and full of problems, while reality may be very different and often much more complex.

Sustainability is currently a widely used term to guide human behaviour and social development. The idea of using resources in such a way that future generations can continue using them is appealing. This is also much in line with the concept of humans living in peace and harmony with the environment, now and in the future. It is a nice concept, but as generally happens to universal concepts, its widespread use wears and dilutes its meaning, and little by little it becomes poorly bounded. Sustainability does not strictly apply to a given local resource but to human needs, at least at a regional level, and considering the different and alternative sources. Evaluations often use a current scientific and technical framework, but the social, space and time frameworks are continuously changing, and the sustainability framework also changes. Knowledge refers to present time and the past. The future is the subject of forecast; this is uncertain for the short and medium-term and very uncertain for long-term situations. The consequence is that sustainability becomes a fuzzy concept. Decisions for the present must consider the future, but this future has a waning influence depending on the credibility of future scenarios and how much weight society and politions have decided to give to this scenarios. Then the analysis of intensive exploitation of groundwater is a correct framework to evaluate sustainability but does not define it. This explains the importance of considering the analysis of intensive exploitation as a needed basic step.

Personal, group and social feelings are important to qualify a given situation. This is fully acceptable if it is based on measured and quantifiable information, and the bases for interpretation are explained. But, aside from this, in most countries, and especially in arid and semiarid regions, groundwater evolution towards intensive development seems unstoppable given the clear immediate benefits it reports, the relatively very low investment needed, the easy drilling of wells and other works to win groundwater, and the widespread availability of pumps and drainage devices driven with relatively cheap energy. This is why the fast expanding groundwater development has been called by Prof. M.R. Llamas "the silent revolution". It goes bottom-up and in many aspects cannot be stopped under current institutional, economic, social and political circumstances. This is true even in countries where policies are imposed by the government (top–down policies). Regulations, institutions and users' participation in management to safeguard the benefits (sustainability) and to minimize and compensate social and environmental costs, in an integrated water resources framework, are urgently needed.

Under this perspective it is surprising that the recent (March 2003) Kyoto's 3rd World Water Forum Ministerial Declaration does not include any explicit mention to groundwater and its rapidly evolving role, in spite of the great effort of several organizations and the excellent promotion made in parallel sessions in Osaka by UNESCO-IHP, the International Association of Hydrogeologists and other leading groups. It seems clear that freshwater resource sustainability is still not recognised by politicians as a key issue. This is so in spite of the universal declarations on human freshwater needs and rights, of health urgent improvement goals and actions to foster economic development.

The present volume is a contribution to place groundwater use as an outstanding issue for humanity and for the environment, with clear advantages but also with drawbacks that may be serious if uncontrolled. To convey messages to policy-makers, politicians and mass media, the Geological Survey of Spain (IGME, Ministry of Science and Technology), the Marcelino Botin Foundation (FMB, private foundation, Spain) and the Regional Government of Valencia (GV, Spain) joined efforts to prepare and convene two main events. The venture was joined by the UNESCO's International Hydrological Programme (IHP-UNESCO, Paris), the International Atomic Energy Agency (IAEA-UN, Vienna), the International Water Resources Association (IWRA), the Food and Agriculture Organisation (FAO, Rome) and the National Ground Water Association (NGWA, USA).

The first event (WINEX) was a closed *Workshop on Intensive Use of Groundwater: Challenges and Opportunities*, held in November 2001, Madrid, and attended by 30 worldwide experts. A book with this same title was edited in September 2002 and published by Balkema (480 pages), in 2003. Twenty-two peer-reviewed papers on general topics about aquifer intensive use were gathered, forming the first book on the issue after the 1992 *IAH Selected Papers book on Aquifer Overexploitation*, printed by Heise (350 pages).

The second event (SINEX) was an open *Symposium on Intensive Use of Groundwater*, held in December 2002, Valencia, and attended by 120 world-wide experts. The goal was to widely discuss WINEX topics with oral presentations, posters sessions and round tables, and to prepare a declaration to be presented and discussed during the Kyoto-Osaka meeting (March 2003) of the 3rd World Water Forum.

Among the papers that were presented some of them contain new ideas, recent developments and case studies that deserve being published as a new IAH Selected Papers Issue. The aim is making them available not only to hydrogeologists but to all the multidisciplinary community interested in groundwater and freshwater resources, including managers, policy-makers and politicians. The task of selecting, preparing the reviews and editing the papers has been carried out by an active editorial team leaded by Dr. Andrés Sahuquillo (Technical University of Valencia), with the help of the IGME (Mr. Argimiro Huerga and the author of this presentation) and the collaboration of the FMB (Dr. M. Ramón Llamas).

The book contains contributions from North and South America, Europe (mostly Spain), South Asia and North and Subsaharan Africa. There are general contributions dealing with technical, economic, legal, administrative and political issues, which not only address economic and social development but also sustainability of ecosystems. These papers present different or complementary viewpoints with respect to the contents of the WINEX publication. A wide series of case studies are included, covering different issues from countries with a wide diversity of social circumstances. The papers cover from areas which are now trying to start development, to well developed ones, and deal with human supply, irrigation and mining problems. Some examples on natural noxious constituents in groundwater are also included. For sure the reader will find a so wide spectrum of viewpoints and cases to accommodate his/her own circumstances. As a last contribution the Valencia Declaration has been incorporated. This is a set of conclusions prepared and reviewed by the organisers, after giving the participants to the SINEX meeting the opportunity of contributing ideas and suggestions.

Both the WINEX and the SINEX, and especially this publication, are characteristic tasks of the IAH. Intensive use of groundwater is a key issue and a situation of paramount importance that affects and involves many hydrogeologists and groundwater concerned people in the world, and especially in arid and semiarid countries. Groundwater intensive use is and will produce large benefits, will help in poverty alleviation, and will be unstoppable. But groundwater intensive use will also be a future cause of concern, which needs good evaluation, correct decision-making and improved management. The IAH is attentive to these issues, tries to be continuously updated to serve its membership and society, will foster groundwater rational use, and will present new opportunities to hydrogeologists.

Emilio Custodio

President. International Association of Hydrogeologists

Professor. Technical University of Catalonia, Barcelona, Spain

Director. Geological Survey of Spain, Madrid

Foreword

Intensive use of groundwater is a recent phenomenon, less than half a century old in most places. Prof. Custodio gives a definition of this phenomenon in the Introduction of this book. This situation has occurred mainly in arid and semiarid countries, in some coastal zones and near to a few megacities. In most cases, almost all groundwater abstracted is used for irrigation.

This groundwater development has produced great socioeconomic benefits, mainly in developing countries. It has provided cheap drinking water that has helped to improve public health. The new irrigated lands have contributed to eradicate, or at least mitigate, malnourishment among those living in poverty. Millions of modest farmers with scarce public or governmental planning, assessment, financing and control have mainly performed this intensive groundwater development. In most countries the corresponding public water or irrigation agencies have been mainly devoted to design, build and operate large surface water irrigation systems.

In 1972, such an attitude by water decision-makers of strongly separating surface and groundwater projects, usually ignoring groundwater, was described by the American hydrologist, Raymond Nace, as *hydroschizophrenia*. This lack of interest for groundwater was caused by several motivations. One of them, but not the principal, was the lack of accuracy and scientific evidence people had of Groundwater physical and chemical Hydrology a century ago. In his book Applied Hydrogeology, Fetter (1994, page 523) quotes an Ohio court decision in the case Frazier *vs.* Brown (1861). The court noted: "Because the existence, origin, movement, and course of such waters, and the causes which govern and direct their movement are so secret, occult and concealed, an attempt to administer any set of legal rules in respect to them would be therefore, practically impossible". This was everywhere the prevalent legal attitude in most regions but the advances in Hydrogeology were progressively changing the judges' attitude. In the same book (page 524) Fetter notes another Ohio court decision, Cline *vs.* American Aggregates (1984), where the Justice wrote: "Scientific knowledge in the field of Hydrology has advanced in the past decade to the point that water table and sources are more readily discoverable. This knowledge can establish the cause and effect relationships of the tapping of the underground water to the existing water level. Thus liability can now be fairly adjudicated with these advances which were sorely lacking when this court decided Frazier more than a century ago". Unfortunately the awareness of this new situation is still lacking in many regions of this blue planet. For many people groundwater issues still continue to be something secret, more related to water witches than to engineers and scientists.

The spectacular worldwide increase in groundwater use, mainly for irrigation, has not been driven by the clear advances in Hydrogeology but mainly by economic reasons. The total direct cost of abstracting groundwater for irrigation in most cases is only a small fraction of the economic value of the crops obtained. This reduction in the cost of abstracting groundwater has been made possible by the technological advances in well drilling and in the devices to pump water from the wells. Groundwater cost for the farmers is often higher than the surface water price that they pay. Nevertheless, many farmers still choose to use groundwater because obtaining surface water (usually highly subsidized with public funds) is not easy, and the security groundwater offers against drought is much greater.

In some sites groundwater abstraction has caused different kinds of problems. As Custodio notes in his Introduction, most of these problems could be avoided or mitigated if the corresponding government agencies had been more active in assessing and controlling groundwater use by farmers. On the other hand surface water officials have frequently exaggerated such problems. This has created a pervasive *hydromyth* on the fragility of groundwater as a resource. The more serious problem of uncontrolled groundwater abstraction is usually quality degradation due to saline groundwater intrusion from the ocean or from other naturally saline groundwaters. Rarely is groundwater quality degradation related to its abstraction for economic uses, but to land use changes that do not take into account their impact on underlying aquifers. Other problems may be excessive drawdown of the groundwater levels, land subsidence, reduction of springs and baseflows, or degradation of groundwater dependent ecosystems.

Because of ignorance, or vested interests, or more frequently because of the low credibility of the water official warnings about the potential overexploitation threats, most farmers are not reducing their groundwater abstraction. They only consider the short-term benefits because these are tangible and have significantly contributed to their socio-economic improvement. There are practically no documented cases where intensive groundwater abstraction from medium or large size aquifers has caused serious social or economic problems similar to those caused by soil-waterlogging and salinization or by the people displaced or ousted by the construction of large dams. Intensive use of groundwater has really been a kind of *silent revolution* because it has been carried out without noise by millions of modest farmers, with practically no help from the conventional governmental agencies.

The mid or long-term sustainability of this intensive groundwater use may not be clear. In some cases, as happens currently in Spain, the farmer's lobby, helped by other lobbies, has been able to convince the government that the deficit due to real or pretended overexploited aquifers needs to be restored by importing water from other basins. They demand that the large hydraulic infrastructures necessary for such surface water diversions be financed with public funds. This solution is against the recent *European Union Water Framework Directive*, which tries to apply the principle of full cost recovery or *user pays*. On the other hand, this large surface water transfer (1 Mm^3/yr), although democratically approved by the Spanish Parliament, is fraught with strong opposition from different groups. Several demonstrations with over 300,000 participants each have already taken place. These clamorous demonstrations were for or against the water transfer depending on the region where they took place. In any case, this situation shows that water issues and more specifically interbasin water transfers are an emotional issue in Spain and in most arid or semiarid regions. Probably it will take a few years of negotiation and mediation before social conflicts of this type are settled and the social situation changes from confrontation to cooperation.

I expect that this book based on the papers presented in the *Valencia International Symposium on Intensive Use of Groundwater* will constitute a step forward in order to create a greater worldwide awareness on the relevance of groundwater in water resources policy.

Finally I would like to express the deep gratitude of the SINEX organizing group to the editors of this book.

M. Ramón Llamas

Emeritus Professor. Complutense University, Madrid, Spain

Sponsors

Sponsored by

Geological Survey of Spain
(Instituto Geológico y Minero de España, Madrid)

Marcelino Botin Foundation
(Fundación Marcelino Botín, Santander)

Regional Government of Valencia
(Generalitat Valenciana, Valencia)

Co-sponsors

FAO	Food and Agriculture Organization
IAH	International Association of Hydrogeologists
IAEA	International Atomic Energy Agency
IWRA	International Water Resources Association
NGWA	National Ground Water Association (USA)
UNESCO-IHP	United Nations Education and Science Organization-International Hydrological Programme

With the help of

FCIHS	Fundación Centro Internacional de Hidrología Subterránea, Barcelona, Spain
AIH-GE	Asociación Internacional de Hidrogeólogos – Grupo Español
CHJ	Confederación Hidrográfica del Júcar, Valencia, Spain

This book is a selection of papers presented at the Symposium held in Valencia, Spain, from 10 to 14 December 2002

ETHICAL, ECONOMICAL, AND LEGAL ASPECTS

Ethical issues in relation to intensive groundwater use [1]

M. Ramón Llamas [(1)] and Pedro Martínez-Santos [(2)]
Department of Geodynamics, Complutense University, Madrid, Spain

[(1)] *mrllamas@geo.ucm.es*
[(2)] *pemartin@geo.ucm.es*

ABSTRACT

Large engineering structures have been constructed since early civilisations to develop irrigation and urban water supply. These hydraulic structures and their operation contributed significantly to the building of the civil society: cooperation and not confrontation was necessary for the common benefit. These are the so-called hydraulic civilizations, like those developed in Egypt and Mesopotamia more than fifty centuries ago. Development of groundwater through wells and/or infiltration galleries was at a smaller scale and usually did not require important societal cooperation.

During the first two thirds of the 20th century, most of the large water developments were based on surface water structures (dams and canals). Most of them were designed, constructed and operated by government agencies and heavily subsidised with public money. Nevertheless, the second half of the 20th century might be characterised by a strong development of groundwater, mainly in arid and semiarid regions. Usually, many individual users with little or no government planning and control have performed this development. The growth in groundwater use has contributed significantly to provide food and potable water in arid and semiarid regions. This often spectacular increase in groundwater use has been mainly driven by economic reasons: the full direct cost of groundwater irrigation is usually a small fraction of the value of the crops obtained with such groundwater abstraction. Really this has been a *silent revolution*.

Nevertheless, mainly because of the lack of knowledge and planning, this *wildcat* groundwater development has caused significant problems in a few regions. Such problems are often exaggerated or unknown because the lack of hydrogeological experience among many water planners who are often surface water-engineers.

One usual false paradigm or *hydromyth* among water resource planners is that groundwater is an unreliable or fragile resource. For them: "almost always every water well becomes dry or brackish after a few years". Another *hydromyth* is that groundwater mining (or development of non-renewable groundwater resources) is always an unethical attitude because it is unsustainable and damages future generations. It will be shown that this general statement is wrong because it only presents a simplistic perspective of a rather complicated problem. There is not a blueprint. Each case is site-specific and all the factors (technological, economic, social, and ecological) should be assessed as accurately as possible, in order to make a scientifically sound and politically feasible decision. In summary, long-term groundwater mining may be ethical or unethical depending on the physical and socio-economic circumstances.

[1] This article is a modified and updated version of the eighth chapter of the UNESCO book: *Water and Ethics* (J. Dooge, ed., in preparation). This chapter was authored by the first author of this article. A preliminary CD ROM version of the UNESCO book was presented in the Third World Water Forum (Kyoto, March 2003).

Keywords: *groundwater, water ethics, intensive use, overexploitation, groundwater mining, sustainability.*

1 INTRODUCTION

Groundwater development has significantly increased during the second half of the last century in most semiarid or arid countries. This development has been mainly undertaken by a large number of small (private or public) developers and often the scientific or technological control of this development by the responsible Water Administration has been scarce. In contrast, the surface water projects developed during the same period are usually of larger dimension and have been designed, financed and constructed by Government Agencies which normally manage or control the operation of such irrigation or urban public water supply systems. This historical situation has often produced two effects: 1) most Water Administrations have limited understanding and poor data on the groundwater situation and value; 2) in some cases the lack of control on groundwater development has caused problems such as excessive depletion of the water level in wells, decrease of well yields, degradation of water quality, land subsidence or collapse, interference with streams and/or surface water bodies, ecological impact on wetlands or gallery forests.

These problems have been sometimes magnified or exaggerated by groups with lack of hydrogeological know-how, professional bias or vested interests. Because of this, in recent decades groundwater overexploitation has become a kind of *hydromyth* that has pervaded water resources literature. A usual axiom derived from this pervasive *hydromyth* is that groundwater is an unreliable and fragile resource that should only be developed if it is not possible to implement the conventional large surface water projects.

Another usual *hydromyth* is to consider that groundwater mining – i.e. the development of non-renewable groundwater resources – is always an *overexploitation*. The implication of this word is that groundwater mining goes against basic ecological and ethical principles.

2 SCOPE AND AIM

The aim of this article is to present a summary of: 1) the many and confusing meanings of the term *overexploitation* and the main factors of the possible adverse effects of groundwater development; 2) the criteria to diagnose aquifers prone to situations of over-use; 3) the strategies to prevent or correct the unwanted effects of groundwater development in *stressed or intensively used aquifers*.

A certain emphasis will be put on the ethical issues in relation to the use of non-renewable groundwater. Nevertheless, groundwater mining is only an end case in the Ethics of water resources use. Therefore, the general framework of this paper will be the technical and ethical issues related to the management of the so-called *stressed aquifers*. It might be better to call them *intensively used aquifers*. The proceedings of the *International Workshop and Symposium on Intensive Use of Groundwater* represent a significant contribution in order to bring more light on this topic (Llamas and Custodio, 2003).

But, what is a stressed aquifer? During the last decade the expression *water stressed regions* has become pervasive in the water resources literature. Usually this expression means that those regions are prone to suffer now or in the near future serious social and economic problems because of water scarcity. Some authors insist in the probable outbreak of violent conflicts, that is, water wars among water stressed regions. The usual threshold to consider a region under water stress is 1,000 m^3/yr per person, but some authors almost double this figure. If this ratio is only 500 m^3/yr per person, the country is considered in a situation of absolute water stress or water scarcity (Seckler *et al.*, 1998).

This simplistic approach of considering only the ratio between water resources and population has little practical application and is misleading. First of all most water problems are related to its quality and not to its relative abundance. As a matter of fact, a good number of regions – such Israel or several watersheds in Spain – with a ratio lower than 500 m^3/yr per person, are regions with a high economic and social standard of life.

United Nations (1997) in an *Assessment of Global Water Resources* has done a more realistic classification of countries according to their water stress. This assessment considers not only the ratio water/population but also the Gross National Product per capita. Other experts are beginning to use other more sophisticated indices in order to diagnose the current of future regions with water problems. There is not room now to present them.

The result of those analyses will probably show that a certain *water-stress* may be an incentive to promote the development of the region. In this case, it could be defined as a *eu-stress*, i.e. a good stress. For example, during the last decades in a good number of semi-arid or arid regions, tourist's activities or high value crops agriculture have been very intensive. The scarcity of precipitation has been fully compensated by the great amount of sun hours and the high radiation energy. The examples of these developments are the *sunny belt* in the USA and most of the European Mediterranean coast. The necessary water for these activities may have different origins. Groundwater is probably the greater and more frequent resource but also may be imported, recycled or even desalinated brackish or sea water. For example, within the year 2002 in Almería (Southern Spain) a rather large (42 Mm^3/yr) sea water desalting plant has begun to operate. This water will be used mainly to grow high value crops in greenhouses. The real cost of this desalted water will be less than US$ 0.60/m^3, but the price paid by farmers is about half of it because of the relevant public subsidies given by the EU Commission and the Spanish government for the plant construction.

3 THE MANIFOLD CONCEPTS OF OVEREXPLOITATION AND SUSTAINABILITY

The term *overexploitation* has been frequently used during the last three decades. Nevertheless, most authors agree in considering that the concept of aquifer overexploitation is one that is poorly defined and resists a useful and practical definition (Custodio, 1992, 1993, 2000a, 2000b, 2002; Foster, 1992; Llamas, 1992a, 1992b, 2001b, 2003; Collin and Margat, 1993; Adams and MacDonald, 1995; Sophocleous, 1997, 2000, 2003).

A number of terms related to overexploitation can be found in the water resources literature. Some examples are: safe yield, sustained yield, perennial yield, overdraft, groundwater mining, exploitation of fossil groundwater, optimal yield, and others (Fetter, 1994; Adams and MacDonald, 1995). In general, these terms have in common the idea of avoiding *undesirable effects* as a result of groundwater development. However, this *undesirability* depends mainly on the social perception of the issue. This social perception is more related to the legal, cultural and economic background of the region than to hydrogeological facts.

For example, in a relatively recent research study, financed by the European Union, called GRAPES (Acreman, 1999), three pilot catchments were analysed: the Pang in the UK, the Upper Guadiana in Spain and the Messara in Greece. The main social value in the Pang has been to preserve the amenity of the river, related to the conservation of its natural low flows. In the Messara the development of irrigation is the main objective and the disappearance of relevant wetlands has not been a social issue. In the Upper Guadiana the degradation of some important wetlands caused by groundwater abstraction for irrigation has caused a serious conflict between farmers and conservationists (Llamas *et al.*, 1996; Cruces *et al.*, 1997; Hernandez-Mora *et al.*, 2003).

The Spanish Water Code of 1985 does not mention specifically the concept of sustainability in water resources development but frequently indicates that this development has to be respectful to nature. Nevertheless, it basically considers an aquifer overexploited when the pumping is close or larger than the natural recharge. In other words, the Spanish regulations follow the common misconception of considering that the *safe yield* or *sustainable yield* is practically equal to the natural recharge (MIMAM, 1998).

This misconception, already shown by Theiss (1940), has been voiced by other American and Spanish hydrogeologists such as Bredehoeft *et al.* (1982), Llamas (1986), Bredehoeft (1997), and Sophocleous (1997, 2000). Bredehoeft *et al.* (1982, p. 53–56) describe the issue in the following way:

> "Water withdrawn artificially from an aquifer is derived from a decrease in storage in the aquifer, a reduction of the previous discharge from the aquifer, an increase in the recharge, or a combination of these changes. The decrease in the discharge plus the increase in recharge is termed *capture*. Capture may occur in the form of decreases in the groundwater discharge into streams, lakes, and the ocean, or from decreases in that component of evapotranspiration derived from the saturated zone. After a new artificial withdrawal from the aquifer has begun, the head of the aquifer will continue to decline until the new withdrawal is balanced by capture.
>
> In many circumstances the dynamics of the groundwater system are such that long periods of time are necessary before any kind of equilibrium conditions can develop".

As an example of the change in the social perception of water values it is interesting to remark that for Theiss (1940, p. 280) the water *was gained* by lowering the water table in areas of rejected recharge or where the recharge was *lost* through transpiration from *non-beneficial vegetation* (phreatophytes). In Theiss' times *wetlands were wastelands.*

Bredehoeft *et al.* (1982) present some theoretical examples to show that the time necessary to reach a new equilibrium or steady state between groundwater extraction and capture may take decades or centuries. Custodio (1992) and Sophocleous (2000) have also presented graphs to show the relationship between the size of the aquifer, its diffusivity, and the time necessary to reach a new steady state after the beginning of a groundwater withdrawal and obtain similar values to Bredehoeft *et al.*

On occasion of the preparation of the new Spanish Water Law of 1985, these misconceptions were also discussed before the Law was enacted and afterwards (Pulido *et al.*, 1989). Also two international Conferences on overexploitation were organised by Spanish hydrologists (cf. Custodio and Dijon, 1991; Simmers *et al.*, 1992) in order to contribute to dispel these misconceptions. Nevertheless, up to now the success of these activities has been rather limited as it can be seen reading the papers by Seckler *et al.* (1998), Postel (1999) or Cosgrove and Rijsberman (2000).

As was previously discussed, certain authors consider that *groundwater mining* is clearly against sustainable development and that this kind of *ecological sin* should be socially rejected and/or legally prohibited. Nevertheless, a good number of authors (Freeze and Cherry, 1979; Issar and Nativ, 1988; Llamas, 1992a, 2003, in press; Collin and Margat, 1993; Margat, 1994; Lloyd, 1997; Price, 2002; Abderrahman, 2003) indicate that, under certain circumstances, groundwater mining may be a reasonable option. As a matter of fact, groundwater mining is today practised in a good number of regions (Issar and Nativ, 1988; Custodio, 1993, 2000a, 2002; Zwingle, 1993; Bemblidia *et al.*, 1996; Amin, 2002). Fossil groundwater has no intrinsic value if left in the ground except as a potential resource for future generations, but are such future generations going to need it more than present ones? As it was previously told, it is hoped that the recent activities in order to bring more data and transparency on this issue (see the Valencia Declaration in this book, and Llamas and Custodio, 2003) will mean a clear step forward for a better water policy.

4 COSTS OR PROBLEMS IN INTENSIVELY DEVELOPED AQUIFERS

4.1 *Indicators of overexploitation*

Adams and MacDonald (1995) noted that, in general, overexploitation is only diagnosed *a posteriori*. They tried in their report and in other subsequent papers to present a method to analyse *a priori* the susceptibility of an aquifer to become stressed (or overexploited). They consider three main effects or indicators: a) decline in water levels; b) deterioration of water quality; and c) land subsidence. In this paper two other relevant effects are considered: d) the hydrological interference with streams and lakes;

e) the ecological impact on aquatic ecosystems fed by groundwater.

Before describing these five indicators, it is relevant to mention that these indicators are sometimes wrongly used. This is either because of lack of hydrogeological knowledge or because certain lobbies may have an interest in expanding the *hydromyth* of the unreliability (or fragility) of groundwater development in order to promote the construction of large hydraulic works.

4.2 *Groundwater-level depletion*

It has not been unusual – like in the Spanish 1985 Water Law – to define overexploitation as the situation when groundwater withdrawal exceeds or is close to the natural recharge of an aquifer. The observation of a trend of continuous significant decline of the levels in water wells during one or two decades is frequently considered as a clear indication of imbalance between abstraction and recharge. This is a simplistic approach that might be a long way from the real situation as it has been shown previously, with reference to the papers of several authors, mainly Bredehoeft *et al.* (1982), and Custodio (1992).

When a well field is operated, even if the general input is much greater than pumping, a transient state will always occur before the water levels in wells stabilise. The duration of the transient state depends mainly on aquifer characteristics such as size and hydraulic diffusivity, degree of stratification and heterogeneity. On the other hand, the natural recharge of an aquifer in semiarid and arid climates does not have a linear relationship with precipitation. In dry years recharge might be negligible or even negative due to evapotranspiration or evaporation from the watertable. Significant recharge may only occur once every one or more decades. Therefore the water table depletion trend during a long dry spell – when the recharge is almost nil and the pumping is high – might not be representative of a long-term situation.

Groundwater depletion caused by deep wells can cause the drying up of shallow wells or *khanats* (infiltration galleries) located in the area of influence of the deep wells. This may cause social problems in regions where many farmers can not afford to drill new wells or the Water Authorities are not able to demand the just compensation in water or money to the poor farmers. Some situations of this type are described in the books by Moench (1999, 2003) and Burke and Moench (2000). There is no doubt that here there is an ethical issue of equity, but probably in most cases the solution is not to forbid groundwater abstraction, but to try overcoming a poverty atmosphere that facilitates a situation of weakness in the legal situation or in the enforcement of the existing regulations. Mukherji (2003) in her analysis of some Eastern Indian states considers that unjust situation caused by the drilling of deep tube well by the rich farmers that dried up the shallow well of the poor farmers had been corrected. The poor farmers improved their agricultural results by buying groundwater from the rich farmers. After a few years they were able to drill their own deep tube wells. This is an example of the social transition facilitated by groundwater irrigation.

4.3 *Degradation of groundwater quality*

Groundwater abstraction can cause, directly or indirectly, changes in groundwater quality. The intrusion into a freshwater aquifer of low quality surface water or groundwater because of the change in the hydraulic gradient due to groundwater abstraction is a frequent cause of quality degradation. Saline intrusion may be an important concern for the development of aquifers adjacent saline water bodies. This is a typical problem in many coastal regions of semiarid or arid regions. The relevance of the saline water intrusion not only depends on the amount of the abstraction, in relation to the natural groundwater recharge, but also on the well field location and design, and on the geometry and hydrogeological parameters of the pumped aquifer. In many cases the existing problems are due to uncontrolled and unplanned groundwater development and not to excessive pumping (cf. Custodio and Bruggeman, 1982).

The degradation of groundwater quality may not be related at all to excessive abstraction of groundwater in relation to average natural recharge. Other causes may be responsible, such as return flow from surface water irrigation, leakage from urban sewers, infiltration ponds for wastewater, septic tanks, urban solid waste landfills, abandoned wells, mine tailings, and many other activities not related to groundwater development (Barraqué, 1997; Foster *et al.*, 1998). Also a temporary situation, such as a serious drought, can contribute to the degradation of groundwater quality (López Geta and De la Orden, 2003).

According to the European Commission, groundwater pollution is the most serious problem of the EU water resources policy. The *Programme for the Integrated Management and Protection of Groundwater* (Official Journal of the EU, 25 November 1996) has been designed to deal with this problem, although it is still too early to assess the practical effectiveness of this EU Programme. Its first positive impact is the greater attention paid to groundwater in the new *European Union Water Framework Directive* (WFD) enacted in December 2000. Nevertheless, it is now recognised that the WFD is not specific enough to cope with the European groundwater pollution problems. As a consequence a *daughter* groundwater Directive is being prepared.

An awareness of the crucial importance of preventing groundwater pollution in order to avoid a future water crisis exists only in a few countries. The old proverb: *out of sight out of mind*, is very apt in this case. A strong educational effort is necessary in order not to bequeath next generations some of our better aquifers almost irreversibly polluted (Llamas, 1991; Custodio, 2000a). This is the real problem in most countries: humid, arid and semiarid. The depletion of groundwater storage (classical misconception of overexploitation) is not generally a problem as serious as groundwater quality degradation and may often be solved without great difficulty, e.g. if water-use efficiency is improved.

One might think that the problem of groundwater quality degradation is mainly an issue in humid and industrialised regions. This does not seem to be the general situation. For instance, Salameh (1996) in his study of Jordan water resources says: "it is not water quantity, but its worsening quality that will bring us to our knees". And Jordan is one of the countries with least amount of renewable water resources per capita (about 160 m^3/yr per person) (Gleick, 1993, p. 131; Bemblidia *et al.*, 1996).

It is significant that the book *Groundwater Protection* published by the Conservation Foundation (1987) shows no significant concern about groundwater overdraft or overexploitation in the USA. Its main interest is to mitigate groundwater pollution. This aspect has been also emphasised in the *Workshop and Symposium on Intensive Use of Groundwater* (see Llamas and Custodio, 2003, and the Valencia Declaration in this book).

4.4 *Susceptibility to subsidence*

Sedimentary formations are deposited at low density and large porosity. As subsequent layers are deposited the overburden compresses the underlying strata. The overburden is in static equilibrium with the intergranular stress and the pore water pressure. This equilibrium is quickly reached in coarse-granular layers, but in fine-grained layers with low permeability, it may take a long time. The effect of this process is the natural progressive consolidation of sediments.

When an aquifer is pumped, the water pore pressure is decreased and the aquifer solid matrix undergoes a greater mechanical stress. This greater stress may produce compaction of the existing fine-grained sediments (aquitards) if the stress due to the decrease in water pore pressure is greater than the so-called *preconsolidation* stress. This situation has occurred in some aquifers formed by young sediments, such as those in Mexico City, Venice, Bangkok and others (Poland, 1985). A brief update of the situation on the spectacular case of Mexico City subsidence can be read in Ragone *et al.* (2003).

Caves and other types of empty spaces may exist under the water table in karstic aquifers. When the water table is naturally depleted the mechanical stability of the *roof* of such empty spaces may be lost and the roof of the cave collapses. This is a natural process that gives rise to the classical *dolines* and *poljes* in the karstic landscape. When the water table depletion or oscillation is increased by groundwater abstraction, the frequency of karstic collapses can be also increased. The accurate prediction of such collapses is not easy (LaMoreaux and Newton, 1992).

In both cases the amount of subsidence or the probability of collapses is related to the decrease in pore water pressure which is related to the amount of groundwater withdrawal. Nevertheless, the influence of other geotechnical factors may be more relevant that the amount of water abstracted in relation to the renewable groundwater resources of the aquifer.

4.5 *Interference with surface water*

Some anthropogenic activities may have a significant impact on the catchment hydrologic cycle, as was already stated by Theiss (1940) and Bredehoeft *et al.* (1982). For example in the Upper Guadiana catchment in Spain (Cruces *et al.*, 1997; Hernandez-Mora *et al.*, 2003), a serious water table depletion (about 30–40 m) has decreased the evapotranspiration from the water table and wetlands between 100 and 200 Mm^3/yr. This depletion has degraded several important wetlands but has increased significantly the renewable water resources that can be used for irrigation, which were estimated between 300 and 400 Mm^3/yr under non-disturbed situation.

The artificial depletion of the water table can also change dramatically aquifer-streams relationship. *Gaining rivers* fed by aquifers may become dry except during storms or humid periods when they may become *losing rivers*, an important source of recharge to the aquifer. Nevertheless, this *new water budget* may present legal problems if the downstream water users have previous water rights (Sophocleous, 2000).

4.6 *Ecological impacts*

Ecological – real or pretended – impacts are becoming an important new constraint in groundwater development in some countries (Llamas, 1992b; Acreman and Adams, 1998; Fornés and Llamas, 1999; Custodio, 2000b; Sophocleous, 2000, 2003). These impacts are mainly caused by water table depletion. This can induce different effects such as: 1) decreasing or drying up of springs or low flow of streams; 2) diminution of soil humidity to an extent in which phreatophitic vegetation cannot survive; 3) changes in microclimates because of the decrease in evapotranspiration. In some cases, the ecological impact of such changes is obvious. For instance, if the water table that was previously at land surface and it is lowered by more than 10 meters during more than twenty years, it is obvious that the peatland or riparian forests that might exist on that aquifer are not going to survive. But if the water table is depleted only during one or two years and not more than one or two meters, probably it cannot be assured that the ecological impact will be irreversible. Quantitative and detailed studies on this type of problems are still rather scarce.

Although sometimes labelled a groundwater development enthusiast, the first author of this paper has long since been an advocate for responsible groundwater use. He was a pioneer in calling the attention on possible conflicts between groundwater development and wetlands conservation (Llamas 1988, 1992b). In fact, his advocacy for the conservation of the Doñana National Park in Spain against a large governmental groundwater irrigation project, was the main cause of the cancellation of a contract that he had signed with the Direction General for the Environment of the European Union Commission. This topic was the subject of questions by several EU Members of Parliament in 1993 (1 April and 14 June), and 1994 (1 January), and a response by the Director General of the Environment on 11 October 1994.

5 STRATEGIES AND CRITERIA TO VALUE THE BENEFITS AND TO MITIGATE THE PROBLEMS OF GROUNDWATER INTENSIVE USE

5.1 *Introduction*

In this section seven criteria or strategies are presented in order to assess not only the benefits that groundwater development can induce, but also the assessment and mitigation of the corresponding potential impacts or problems. The main aim of this paper is to analyse the Ethics of groundwater mining or intensive use. Such analysis demands a more general framework.

Like the one in preparation by UNESCO (see Dooge *et al.*, 2003). Perhaps the main moral of this paper is that a *stressed aquifer system* can become an *eu-stressed aquifer system* if the criteria described hereafter are applied. In other words, the groundwater intensive use can often be beneficial if such development is well designed and controlled.

5.2 *Diagnostic method*

As previously mentioned, Adams and MacDonald (1995) proposed a method to make an *a priori* diagnosis of aquifer susceptibility to overexploitation effects. The method established three levels of susceptibility, which relate to groundwater level decline, saline intrusion and subsidence. The ecological impacts and the influence of groundwater abstraction in surface water bodies or stream are not graded. The technique involves assigning numerical values to the contributing factors and then summing them up to give an overall grade or susceptibility to the particular impact under consideration. Only relative values are used in the final designation (high, medium, and low) due to the high parameter variability at individual locations. According to these authors, as only relative values are used in the grading, this diagnostic method should only be used with great caution for interregional comparisons. There is a need to recognise the great diversity of hydrogeological and socio-economic situations which makes it almost impossible to apply the same groundwater management tools everywhere: *there is not a blueprint.*

5.3 *Management of uncertainties*

A generally accepted principle is that *prevention is better than cure*. But this version of the precautionary principle should be applied with considerable prudence. In general, groundwater development should not be rejected or seriously constrained if it is well planned and controlled. During recent decades, notable socio-economic benefits have derived from groundwater withdrawal, particularly in developing countries. It has provided affordable potable and irrigation water, thus improving public health and significantly contributing to alleviate malnutrition and famine. Moreover, groundwater development is usually the cheapest and fastest way to achieve some of the UN Millennium Declaration goals of halving by the year 2015 the proportion of people under the poverty threshold (US$ 1/yr per person), without affordable drinking water, and malnourished (United Nations, 2000a, 2000b). It is estimated that about one billion human beings are now in this situation. The *UN World Summit of Sustainable Development of Johannesburg* 2002 endorsed these goals (United Nations, 2002). The *World Conference on Freshwater* (Bonn, Germany, 2001) also endorsed the same goals (German Federal Government, 2001).

An important first step when trying to manage a resource in the face of uncertainties is to assess the seriousness and type of the assumed problem. Often the adverse effects of overexploitation may be misunderstood or exaggerated. This is often the case in relation to the interpretation of a long (e.g. 10 years) water level decline as an indication of a groundwater abstraction higher than the average renewable resources. As previously several times explained, such a decline may be due to: 1) a dry spell; 2) a transient situation; or 3) scarce or incorrect data about streamflow, groundwater levels, climatic conditions, groundwater abstractions and natural recharge. The two last factors are usually difficult to determine in arid and semi-arid countries.

Frequently it will be necessary to ask for more funds in order to obtain better data in regard to quantity or quality of information. Nevertheless, the natural recharge in semi-arid regions will only be accurately known after a good number of years of good climatic and hydrological data have been collected. One should avoid transferring to the public a sense of accuracy that is really only illusory. In most cases the required funds to obtain the necessary data to dispel the main uncertainties on groundwater issues are only a quite small fraction of the public funds spent in the same countries for designing and constructing large surface water infrastructures. Therefore, it is mainly a problem of professional bias or of institutional inertia and not a problem of lack of funds.

The use of numerical models to analyse groundwater flow and management might be useful. Such models should employed to perform sensitivity analysis of the plausible variations of the stochastic and deterministic parameters, including those related to social sciences, such as the possible future scenarios of the irrigated agriculture in the next decades.

Uncertainty about water resources is usually no higher when dealing with groundwater than when dealing with surface water or other water policy-related problems. A good example of such uncertainties is related to the general exaggeration that is associated with the prediction of future water demands. Gleick (1998) has analysed the progressive decline in estimates of future water demands, according to different authors. These have decreased from 7,000 km^3/yr about 20 years ago, to less than 4,000 km^3/yr in one of the latest predictions issued by United Nations (Shiklomanov, 1997). Even this last prediction is probably exaggerated. For example, it estimates a 20% growth in North America's water demand. However, the U.S. Geological Survey (Solley *et al.*, 1998) has indicated a steady decline in total water uses in the USA over the past two decades, while during that same period population and standard of life have continued to grow. Wood (1999) considers that this decline may be due to the pressure of conservation groups that have demanded a more efficient use of water.

In summary, professional hydrogeologists should transfer the awareness of these uncertainties to decision-makers and the general public. This transfer must be done with prudence and honesty in order to avoid loss of credibility of the scientific community either in the short term (by giving the impression of lack of knowledge) or in the medium term (because of the failure of the announced predictions). The frequent and widely voiced *gloom and doom* pessimistic predictions done by certain individuals and/or institutions about the depletion of natural resources or the population explosion have usually not been realised. For example, Dyson (1996) shows how the predictions done along the last three decades by the *pessimistic neo-malthusians* have not been realised. On the other hand, quite recently, according to Pearce (1999), it seems that the focus about population explosion is misplaced and next century may have to worry about falling birth rates, not rising ones. Nevertheless, the fashion of *preaching environmental scares* (The Economist, 1998) seems to be so deeply rooted in most media (Lomborg, 2001) that it will not be an easy task to bring objectivity and transparency in this field.

5.4 *Abstraction of fossil groundwater*

In most countries it is considered that groundwater abstraction should not exceed the renewable resources. In other countries – mainly in the most arid ones – it might be considered that groundwater mining is an acceptable policy, as long as available data assure that the groundwater development can be economically maintained for a long time; for example, more than fifty years, and that the potential ecological costs and socio-economic benefits have been adequately evaluated (Llamas *et al.*, 1992).

Nevertheless, some authors consider this option as unsustainable development or a dishonest attitude with respect to future generations. What Lazarus (1997, p. 22) proposed for South Africa could also be the policy in many other countries: "in essence, current thinking in the sector is that strategies need to be developed to ensure that groundwater resources are utilised within their capacity of renewal. It is recognised however that quantification of sustainable use levels requires extensive research".

In contrast, few authors speak of the frequent unsustainability of most dams in arid regions. Bemblidia *et al.* (1996, p. 20) consider that the *useful life* of most dams in the North African Mediterranean countries use to be between 40 and 200 years because of their silting.

Lloyd (1997) states that the frequently encountered view that the water policy of arid zone countries should be developed in relation to renewable water resources is unrealistic and fallacious. Ethics of long-term water resources sustainability must be considered with ever improving technology. With careful management many arid countries will be able to utilise resources beyond the foreseeable future without major restructuring.

In Saudi Arabia, according to Dabbagh and Abderrahman (1997), the main aquifers (within the first 300 m of depth) contain huge amount of fresh fossil water – a minimum of 2,000 km^3 – that is 10,000 to 30,000 years old. It is considered that these fossil aquifers can supply useful water for a minimum period of 150 years. Current abstraction seems to be around 15–20 km^3/yr. During a couple of decades the Saudi government has pumped several km^3/yr of non-renewable groundwater to grow low cost crops (mainly cereals) which were also heavily subsidised. The official aim of such activity was to help to transform nomadic groups into farmers. Apparently such *overdraft* has been a social success. Now the amount of groundwater abstraction has been dramatically reduced and the farmer nomads have become high-tech farmers growing cash crops. Another example is the situation of the Nubian sandstone aquifer located below the Western desert of Egypt. According to Idriss and Nour (1990), the fresh groundwater reserves are higher than 200 km^3 and the maximum pumping projected is lower than 1 km^3/yr. Probably similar situations do exist in Libya and Algeria. Price (2002) reminds that "sustainability should not be seen as an end in itself", while Abderrahman (2003) states that in certain arid climates (such as Saudi Arabia), it is neither practical nor feasible to reduce the exploitation of non-renewable groundwater resources to the recharge values. Other examples of mining groundwater can be found in Llamas and Custodio (2003).

It is not easy to achieve a virtuous middle way. As Collin and Margat (1993) state: "we move rapidly from one extreme to the other, and the tempting solutions put forward by zealots calling for *Malthusian underexploitation of groundwater* could prove just as damaging to the development of society as certain types of excessive *pumping*". The number of analyses that consider jointly the hydrogeological, environmental, social and economical aspects in the intensively used aquifers is rather scarce. This is a gap that should be bridged as soon as possible.

5.5 *Apportioning available groundwater*

The distribution of the estimated available groundwater renewable resources or fossil groundwater among the potential or actual users may be a source of conflict between persons, institutions or regions. Again it has be repeated that there is no universal solution or blueprint. Each case may be different according to the cultural, political and legal background of the region. Nevertheless, it may be useful to try to achieve some kind of universal agreement on the ethical principles that should rule water distribution and management. The initiative of the International Association of Hydrologists to create a Working Group to analyse the problems in internationally shared aquifers may become a positive step forward. Nevertheless, one should not forget the slow path of the global legislation on International Rivers and also the almost universal lack of control of governments on the groundwater abstraction in their own countries. If this problem is not previously solved the enforcement of international agreements on shared aquifers will be difficult.

5.6 *Mitigating ecological impacts*

The ecological cost of groundwater development should be compared with the socio-economic benefits produced (Barbier *et al.*, 1997; National Research Council, 1997; Custodio, 2000b). The evaluation of the ecological impacts is highly dependent on the social perception of ecological values in the corresponding region. This social perception is changing rapidly in most countries. For example, the recent Framework Directive on Water of the European Union pays great attention to monitoring and conservation of aquatic ecosystems and especially to wetlands. In arid and semi-arid regions, wetlands or oases are usually rare and related to groundwater discharge zones. The development of groundwater for irrigation or other uses may often have a significant negative impact on the hydrological functioning of wetlands or oases (Fornés and Llamas, 1999). These impacts should be properly evaluated by decision-makers.

The social relevance of the conflict between nature conservation and groundwater development and its solution will be different from country to country and also changes with time. However, considering the relevance of this issue, the team of a research project on groundwater, funded by the Spanish Foundation M. Botin, submitted through the Spanish Government, a proposal of Resolution to the *8th Meeting* (Valencia, Spain, 18–26 November 2002) *of the Conference of the Contracting Parties (COP8) to the Convention on Wetlands* (Ramsar, Iran, 1971). This proposal was approved with some minor changes as Resolution No. 40: *Guidelines for rendering the use of groundwater compatible with the conservation of wetlands* (see Llamas, 2003). It is expected that this resolution will contribute to avoid or, at least, mitigate the current conflicts between groundwater development for utilitarian uses and groundwater as a support of ecosystems.

5.7 *Socio-economic issues*

Groundwater development has produced great economic and social benefits in many respects during the latter half of the last century. For example, the intensive use of groundwater for irrigation has contributed significantly to alleviate the problem of hunger or famines and of potable water supply to cities and rural areas. Although in some cases groundwater development has induced some of the problems previously described in small aquifer systems of areas such as Yemen (Moench, 2003), this author is not aware of any case of a large aquifer (e.g. with a surface greater than 1,000 km^2) in which intensive groundwater development has caused social disturbances (see De Villiers, 2001, p. 282–283).

In contrast, serious social problems are well known because of the construction of dams (e.g. Narmada Valley, in India), soil water logging and salinisation caused by excessive surface water irrigation and/or poor drainage (e.g. San Joaquin Valley in California, and Punjab plain in Pakistan) or surface water diversions (e.g. Aral Sea disaster).

Economic studies analysing in detail intensively developed aquifers cases are still rare. In his economic analysis of overexploitation, Young (1992) defined it as a "failure to achieve maximum economic returns of the resource". Nevertheless, the estimation of the real economic cost of the different factors is a difficult and controversial matter. Therefore the final solution of conflicts related to overexploitation will not be only dictated by economic rules; socio-political motivations may play the leading role.

Different scenarios can be presented in relation to the economy of over-used aquifers. Among them, two theoretically extreme situations are: unrestricted (free) development against controlled development (Custodio and Gurgui, 1989; Foster, 1992).

Another issue to be considered is the almost universal policy of public *perverse subsidies* for water supply, mainly for irrigation (Llamas, 2003, in press). According to Myers and Kent (1998) these subsidies are those which are noxious both for the economy and the environment. In most cases, the water users only pay a small fraction of the real cost of the water supplied. This is especially true in surface water for irrigation. Water policy all over the world has, during the past decades, focused on the management of the supply and not on the management of the demand. This has induced an almost universal wasteful use of water.

In most groundwater developments the situation may be quite different. The owners of the water wells usually pay for the wells' construction, maintenance and operation. But they do not usually pay the external costs caused by the impacts of the groundwater abstraction.

The great socio-economic benefits produced by groundwater developments are rarely documented. According to Dhawan (1995), research in India indicates that yields in areas irrigated with groundwater are one third to one half higher than those in areas irrigated with surface resources. In a previous report Dains and Pawar (1987) estimated that as much as 70–80% of India's agricultural output might be groundwater dependent. More recently, the Indian Water Resources Society (1999) has published, among others, the following significant data:

a) Groundwater is contributing at present 50% of irrigation surface, 80% of water for domestic use in rural areas, and 50% of water in urban and industrial areas.
b) Groundwater abstraction structures had increased from 4 million in 1951 to nearly 17 million in 1997.
c) In the same period groundwater irrigated area had increased from 6 to 26 million ha.

d) It is estimated that this rapid pace of development is likely to continue and will reach 64 million ha in the year 2007.

By indirect calculation it may be estimated that in India the average amount of water applied in surface water irrigation is around 16,000 m^3/ha/yr; in groundwater irrigation this ratio is only 4,000 m^3/ha/yr. In other words, it seems that in India (as an average) the economic yield in irrigation and by m^3 is from 5 or 10 times higher when groundwater is used than when irrigation is made with surface water [2].

Llamas (2000), analysed the relevance of groundwater to mitigate drought effects in Spain. Corominas (1999) and Hernández-Mora *et al.* (2001) published an assessment of irrigated agriculture in Andalusia (Spain). It is a well-documented and transparent analysis. Some significant data from this study are:

a) Out of 800,000 ha currently under irrigation, 75% use surface water and 25% groundwater.
b) Average water applied is 4,000 m^3/ha/yr in groundwater irrigation and 7,500 m^3/ha/yr in surface water (including the losses in surface water transport from the dam or river to the field).
c) The average economic yield per hectare is more than three times greater in groundwater irrigated areas than in surface water irrigated areas.
d) The economic yield by m^3 used is five times higher in groundwater than in surface water. And the jobs provided by volume unit of groundwater irrigation were more than three times greater that the similar jobs provided by surface water irrigation.

This analysis of Andalusia (Spain) has been extended to other regions of Spain (Hernández-Mora and Llamas, 2001) with different climatic and social conditions, and the results are similar.

It would be highly desirable that similar studies be done in many other arid and semi-arid countries. Also it would be appropriate to include this type of analysis in the framework suggested in the report of the World Commission on Dams (WCD, 2000) in order to study the economic and social feasibility of new dams. Nevertheless, it seems necessary to state that the ideas on the role of groundwater in water resources policy presented in the WCD's report follow the pervasive *hydromyth* of groundwater fragility.

A positive new situation worldwide is that in several international conferences and reports the practical importance of corruption and bribery in the water management has been emphasised. This was an issue that previously was very rarely mentioned in such conferences or in international reports. Perhaps the most important document on this topic is the book by OECD (2000) dealing with the *Convention on Combating Bribery of Foreign Public Officials in International Business Transactions.* This book was specifically quoted in the Report of the World Commission on Dams (WCD, 2000, p. 186–187 and 249). During the *Fresh Water Conference of Bonn* (December 2001) the debate on corruption was a frequent topic (German Federal Government, 2001). This author has also mentioned the practical relevance of corruption as a frequent driving force in water policy in several papers (Llamas and Delli Priscoli, 2000; Llamas, 2001a, 2001b; Delli Priscoli and Llamas, 2001). Postel (1999, p. 229) presents an interesting view of the role of consulting firms, construction companies and politicians in promoting large surface water irrigation projects, which is transcribed below:

> "The rules, especially for large schemes, often look something like this: a politician seizes the potential for bolstering political support by proposing an irrigation project in a strategic location. Engineering firms lobby the decision-makers in order to raise their chances of winning the project's construction contract. The politician, in collusion with colleagues who also want to please their constituents with pork-barrel projects, sees to it that the nation's taxpayers foot most of the

[2] Most of these data are ten or more years old. The spectacular increase of groundwater use has continued during the last two decades. A good number of papers have been published in the last years. Among them, see: Deb Roy and Shah, 2003; Moench, 2003; Mukherji, 2003. The current groundwater irrigated surface in India seems to be larger than 35 million ha.

bill. The farmers themselves pay only a small fraction of the project's cost; in return, they support the politician in the next election. With prices kept artificially low, farmers have little incentive to use water efficiently. The irrigation system never becomes financially self-sustaining because the meager fees collected from the farmers do not cover the system's operation and maintenance costs, much less its capital costs. National or state irrigation agencies keep the projects running with taxpayer funds. If budgets become tight and maintenance work is neglected, the systems fall into disrepair. Gradually, agricultural output and benefits begin to decline. Either the government saves the project by allocating more taxpayer money for expensive rehabilitation work, or the system deteriorates until farmers abandon it. Alternatively, international donors come to the rescue with funds provided by taxpayers in wealthier countries."

The most relevant socio-economic fact in relation to groundwater development in arid and semiarid countries is that the full direct cost (not including externalities) of abstracting groundwater for irrigation is only a small fraction of the value of the crops obtained. This fact has been the driving force of the practically universal *silent revolution* caused by the intensive use of groundwater. Table 1 (from Llamas and Martinez Cortina, 2003) shows a rough estimate of these values in most not very poor countries where marketing of agricultural products is not irrelevant. The values in the table have to be reduced in poor countries, e.g. where the Gross National Product per person and day is lower than one or two US$. In this case the average lower value of crops may be only 200 or 300 US$ per hectare and year.

Table 1. Rough estimates of direct cost[1] of groundwater irrigation in relation to crop value (Llamas and Martínez Cortina, 2003).

Type of aquifer (cost of water)	GA: Good aquifer[2] (US$ 0.01/m^3) PA: Poor or deep aquifer[3] (US$ 0.10/m^3)	GA	GA	PA	PA	GA	GA	PA	PA
Type of crops (water consumption)	LW: Low water-consuming[4] (1,500 m^3/ha/yr) HW: High water-consuming[5] (15,000 m^3/ha/yr)	LW	HW	LW	HW	LW	HW	LW	HW
Typical cost of irrigation water (US$/ha/yr)	Combining type of aquifer and type of crops	15	150	150	1,500	15	150	150	1,500
Crop value	HV: High-value crops[6] (US$ 50,000/ha/yr) LV: Low-value crops[7] (US$ 500/ha/yr)	HV	HV	HV	HV	LV	LV	LV	LV
Cost of water/ Crop value	**Combining type of aquifer, type of crops and crop value**	**0.03%**	**0.3%**	**0.3%**	**3%**	**3%**	**30%**	**30%**	**300%**

[1] Direct cost means that externalities are not included.
[2] This group mainly includes alluvial and karstic aquifers, and the water depth is not significant (e.g. <50 m).
[3] This group mainly includes low permeability aquifers, and the water depth is significant (e.g. >100 m).
[4] Typical low water-consuming crops are: cereals, vineyards, olive-trees.
[5] Typical high water-consuming crops are: sugar cane, bananas, rice.
[6] Typical high-value crops are: tomato, cucumber, bananas, oranges, flowers.
[7] Typical low-value crops are: cereals, rice, alfalfa, cotton.

Some conclusions that come to mind after reading this table include:

- In good aquifers all types of crops are usually economically feasible, except those that have low value and are high water consuming.
- In low permeability aquifers or in aquifers where the pumping depth is significant, only crops that are low water consuming and/or have high value are economically feasible.
- In summary, the economic feasibility of groundwater irrigation is mainly conditioned by the crop value. High value crops can usually be developed even using low permeability or deep aquifers. In most normal aquifers the cost of abstracting groundwater is usually affordable to grow any type of crops. These data explain why, in the last decades, the farmers in most arid and semi-arid countries have intensively developed groundwater irrigation.

5.8 *Stakeholder participation*

There exists a general consensus that, in order to avoid conflicts and to move from confrontation to cooperation, water development projects require the participation of the social groups affected by the projects, the stakeholders. The participation should begin in the early stages of the project and should be, as much as possible, bottom-up and not top-down. The first question is to define who the stakeholders are; the second, how, when and where they should intervene in the decision-making processes.

The Spanish experience, in trying to implement groundwater management as a public dominion, indicates clearly that the active collaboration of Groundwater Users Associations is a key element (Aragonés *et al.*, 1996; Hernández-Mora and Llamas, 2000, 2001). However, the process implementation demands sometime to change from the old to the new paradigms (Lopez Gunn and Llamas, 2001).

In July 2001 the Spanish Parliament approved the Law of the National Water Plan. This Plan includes several provisions that if really enforced, will change the current almost chaotic situation of groundwater development in Spain. Perhaps the most important article of this Law is the one which strictly demands the set up of Groundwater User Associations in every intensively developed aquifer and a thorough hydrological assessment of every aquifer which may be supposed to receive a surface water transfer from other catchments. The National Water Plan Law also states that an intense and broad Water Education Programme has to be implemented. The Spanish experience about the enforcement of this regulation is still short but the attitude of the official of the Water Agencies does not allow for a very optimistic outlook about the final result.

Obviously, there is not a universal solution. For example, in some arid and semi-arid developing countries, when dealing with correction of ecological impacts of overexploitation, the influence of conservationist groups will probably be weak compared to the influence of farmers associations or urban water supply companies.

The necessary participation of the stakeholders demands that they are aware of how the issue at hand will affect them directly or indirectly, and also a basic knowledge of the hydrogeological concepts involved in aquifer development. Probably in most countries there exist a good number of *hydromyths* or obsolete paradigms about the origin, movement and pollution potential of groundwater. In any stressed aquifer it is essential to organise different types of educational activities aimed at different groups: from school students and teachers to officials of Water Administrations, as described by McClurg and Sudman (2003).

6 CONCLUSIONS

6.1 *Various factors have made possible the significant increase of groundwater development over the second half of the 20th century, particularly in arid and semi-arid regions*

a) Technological: invention of the multistage pump, improvements in drilling methods and in the advance of the scientific knowledge on occurrence, movement and exploration of groundwater.

b) Economic: the real cost of groundwater is usually low in relation to the direct economic benefits obtained from its use. The *in situ* or social value of groundwater is rarely estimated.
c) Institutional: groundwater development can easily be carried out by individual farmers, industries or small municipalities, without financial or technical assistance from Government Water Authorities. It does not require significant financial investments or public subsidies like surface water projects typically do.

6.2 *The socio-economic benefits of groundwater development have been significant*

Groundwater is an important source of potable drinking water. World-wide 50% of municipal water supplies comes from groundwater. In some regions the proportion is much higher. In general, groundwater is particularly important as a source of drinking water for rural and dispersed population.

70% of all groundwater withdrawals world-wide are used for irrigation, particularly in arid or semi-arid regions. Irrigation with groundwater has been crucial to increase food production at a greater rate than population growth.

Irrigated agriculture using groundwater is often much more efficient than irrigation using surface water. This is mainly because groundwater irrigation farmers typically assume all abstraction costs (financial, maintenance and operation) and produce high value crops because they have a greater security in their investment, as groundwater usually is not affected by droughts.

6.3 *Groundwater administration*

In most countries, groundwater development has not been adequately planned, financed, or controlled by the existing Water Authorities. Historically, officers of these agencies have been trained to manage surface water systems and lack adequate hydrogeological training. The result use to be a bias toward surface water management and the frequent mismanagement of groundwater sources.

Groundwater management presents particular challenges given the great number of users on a single aquifer system. Coordination among the thousands of stakeholders that generally exist on an aquifer of medium or large size is scarce. Various reasons can account for this: 1) the coordination was not really necessary at the beginning of the development; 2) the usual tendency among farmers to individualism; and 3) the lack of willingness to promote such coordination by the Water Authorities.

6.4 *Emerging problems in groundwater developments*

In certain regions unplanned and uncontrolled development has caused problems, which can be classified in five groups:

1) *Excessive drawdown of the water level in wells*, which increases costs by requiring more pumping energy. Some shallow wells may become dry. Nevertheless, there is no documented case of a medium or large aquifer which has been physically emptied.
2) *Degradation of water quality* because of different factors such as point pollution coming from the surface or from saline groundwater intrusion from adjacent aquifers. Pollution coming from the surface is generally not caused by groundwater use but by inadequate land use planning. In most countries, groundwater pollution or degradation is the main threat to achieve a sustainable water resources management.
3) *Land subsidence or collapse* may be induced by groundwater abstraction but it is more related to the geotechnical properties of the terrain and to the location of the well fields than to the amount of groundwater withdrawal.
4) *Impact on surface water bodies and in the water cycle of the whole basin*. In some rivers intensive groundwater pumping has caused significant changes in their hydrological regime with the consequent legal problems when the water of such river was previously allocated to other users. Nevertheless, the total renewable water resources in the basin can be significantly increased because of the augmentation of the natural recharge.

5) *Impact on wetlands and other aquatic ecosystems*. Relatively small (e.g. 2 m) but long-term (e.g. 10 yr) depletion of the water table often causes dramatic changes in wetlands, springs and riparian forests. These impacts have only been a cause of concern during the last three of four decades and almost exclusively in industrialised countries.

6.5 *Five ethical issues in groundwater use*

Five ethical issues are considered relevant in trying to achieve sustainable or reasonable groundwater use:

1) *Perverse subsidies to surface water projects.* The hidden or open subsidies that have traditionally been a part of large hydraulic works projects for surface water irrigation, are probably the main cause of the pervasive neglect of groundwater problems among water managers and decision-makers. Surface water for irrigation is usually given almost free to the farmers; and its wasteful use is the general rule.

   Progressive application of the *user pays* or *full cost recovery* principle would probably make most of the large hydraulic projects economically unsound. As a result, a more comprehensive look at water planning and management would be necessary and adequate attention to groundwater planning, control and management would probably follow.
2) *Public, private, or common groundwater ownership.*
   Some authors consider that the legal declaration of groundwater as a public domain is a *conditio sine qua non* to perform a sustainable or acceptable groundwater management. This assumption is far from evident. For many decades groundwater has been a public domain in a good number of countries. Nevertheless, sustainable groundwater management continues to be a significant challenge in many of those countries. Highly centralised management of groundwater resources is not the solution but to promote solidarity in the use of groundwater as a *common good.* Groundwater management should be in the hands of the stakeholders of the aquifer, under the supervision of the corresponding Water Authority. The stakeholders' participation has to be promoted bottom-up and not top-down.
3) *Lack of hydrogeological knowledge and/or education.*
   Adequate information is a prerequisite to succeed in groundwater management. It has to be a continuous process in which technology and education improve solidarity and participation to the stakeholders and a more efficient use of the resource.
4) *Transparency in groundwater related data.*
   Good and reliable information is crucial to facilitate cooperation among aquifer stakeholders. All stakeholders should have easy access to good and reliable data on abstractions, water quality, aquifer water levels. Current information technology allows information to be made available to an unlimited number of users easily and economically. Nevertheless, in a good number of countries it will be necessary to change the traditional attitude of water agencies of not facilitating the easy access to water data to the general public.
5) *The ethics of pumping non-renewable groundwater resources (groundwater mining).*
   Some arid regions have very small amounts of renewable water resources but huge amounts of fresh groundwater reserves, like for example the existing reserves under most of the Sahara desert. In such situations, groundwater mining may be a reasonable action if various conditions are met: 1) the amount of groundwater reserves can be estimated with acceptable accuracy; 2) the rate of reserves depletion can be guaranteed for a long period, e.g. from 50 to 100 years; 3) the environmental impacts of such groundwater withdrawals are properly assessed and considered clearly less significant than the socio-economic benefits from groundwater mining; and 4) solutions are envisaged for the time when the groundwater is fully depleted. Selbourne (2000), chairman of the Water Resources Committee of the World Commission of the Ethics of Science, seems to agree with this approach.

REFERENCES

Abderrahman, W.A. (2003). Should intensive use of non-renewable groundwater resources always be rejected? In: M.R. Llamas and E. Custodio (eds.), *Intensive Use of Groundwater. Challenges and opportunities*. Balkema Publishers. Lisse, The Netherlands: 191–206.

Acreman, M.C. (compiler) (1999). *Guidelines for the sustainable management of groundwater-fed catchments in Europe*. Report of the Groundwater and River Resources Action Programme on a European Scale (GRAPES). Institute of Hydrology. Wallingford, UK. 82 pp.

Adams, B. and MacDonald, A. (1995). *Overexploited aquifers. Final Report*. British Geological Survey. Technical Report WC/95/3. 53 pp.

Amin, I.E. (2002). *Groundwater mining in the Tripoli area, Libya. 2002* Denver Annual Meeting. Geological Society of America. Colorado, USA.

Aragonés, J.M.; Codina, J. and Llamas, M.R. (1996). Importancia de las Comunidades de Usuarios de Aguas Subterráneas (CUAS). *Revista de Obras Públicas*, 3355 (June): 77–78.

Barbier, E.B.; Acreman, M. and Knowler, D. (1997). *Economic evaluation of wetlands: a guide for policy makers and planners*. Ramsar Convention Bureau. Gland, Switzerland. 127 pp.

Barraqué, B. (1997). Groundwater management in Europe; regulatory, organizational and institutional change. *Proceeding of the International Workshop: how to cope with degrading groundwater quality in Europe*. Stockholm, Sweden, 21–22 October 1997. 16 pp (preprint).

Bemblidia, M.; Margat, J.; Vallée, D. and Glass, B. (1996). *Water in the Mediterranean Region*. Blue Plan for the Mediterranean. Regional Activity Centre. Sophia-Antipolis, France. 91 pp.

Bredehoeft, J.D. (1997). Safe yield and the water budget myth. *Ground Water*, 35(6): 929.

Bredehoeft, J.D.; Papadopoulos, S.S. and Cooper, H.H. (1982). The water budget myth (scientific basis of water management). *Studies in Geophysics*. National Academy of Sciences: 51–57.

Burke, J.J. and Moench, M. (2000). *Groundwater and Society: Resources, Tensions and Opportunities*. United Nations Publication. Sales No. E.99.II.A.1. 170 pp.

Collin, J.J. and Margat, J. (1993). Overexploitation of water resources: overreaction or an economic reality? *Hydroplus*, 36: 26–37.

Conservation Foundation (1987). *Groundwater Protection*. Conservation Foundation. Washington DC, USA. 240 pp.

Corominas, J. (1999). El papel de las aguas subterráneas en los regadíos. In: J. Samper and M.R. Llamas (eds.), *Actas de las Jornadas sobre las Aguas Subterráneas en el Libro Blanco del Agua en España*. Asociación Internacional de Hidrogeólogos–Grupo Español: 65–79.

Cosgrove, W.J. and Rijsberman, F.R. (2000). *World Water Vision*. Earthscan Publications Ltd. London, UK. 108 pp.

Cruces, J.; Casado, M.; Llamas, M.R.; De la Hera, A. and Martínez Cortina, L. (1997). El desarrollo sostenible en la Cuenca Alta del Guadiana: aspectos hidrológicos. *Revista de Obras Públicas*, 3362 (February): 7–18.

Custodio, E. (1992). Hydrogeological and hydrochemical aspects of aquifer overexploitation. In: I. Simmers, F. Villarroya and L.F. Rebollo (eds.), *Selected Papers on overexploitation*. International Association of Hydrogeologists. Selected Papers, Vol. 3. Heise, Hannover, Germany: 3–28.

Custodio, E. (1993). Aquifer intensive exploitation and overexploitation with respect to sustainable development. *Proceedings of the International Conference on Environmental Pollution*. European Centre for Pollution Research. Vol. 2: 509–516.

Custodio, E. (2000a). *The complex concept of groundwater overexploitation* . Papeles del Proyecto Aguas Subterráneas. No. A1. Fundación Marcelino Botín. Santander, Spain. 58 pp.

Custodio, E. (2000b). *Groundwater-dependent wetlands*. Acta Geológica Hungarica. Budapest, Hungary. 43(2): 173–202.

Custodio, E. (2002). Aquifer overexploitation: what does it mean? *Hydrogeology Journal*, 10: 254–277.

Custodio, E. and Bruggeman, G.E. (1982). *Groundwater problems in coastal areas*. Studies and Reports in Hydrology, No. 45. UNESCO. Paris, France. 650 pp.

Custodio, E. and Dijon, R. (1991). Groundwater overexploitation in developing countries. *Report of an UN Interregional Workshop*. United Nations. INT/90/R43. 116 pp.

Custodio, E. and Gurgui, A. (eds.). (1989). Groundwater Economics. *Selected Paper from a UN Symposium held in Barcelona*, Spain. Elsevier. Amsterdam, The Netherlands. 625 pp.

Dabbagh, A.E. and Abderrahman, W.A. (1997). Management of groundwater resources under various irrigation water use scenarios in Saudi Arabia. *The Arabian Journal for Science and Engineering*, 22(IC): 47–64.

Dains, S.R. and Pawar, J.R. (1987). *Economic return to irrigation in India*. Report prepared by SDR Research Group Inc. for the U.S. Agency for International Development. New Delhi, India.

De Villiers, Marq (2001). *Water: the fate of our more precious resource*. Mariner Books. Houghton Miffing Co. Boston-New York, USA. 352 pp.

Deb Roy, A. and Shah, T. (2003). Socio-ecology of groundwater irrigation in India. In: M.R. Llamas and E. Custodio (eds.), *Intensive Use of Groundwater. Challenges and opportunities*. Balkema Publishers. Lisse, The Netherlands: 307–336.

Delli Priscoli, J. and Llamas, M.R. (2001). International perspective in ethical dilemmas in the water industry. In: C.K. Davis and R.E. McGinn (eds.), *Navigation Rough Waters*. American Water Works Association. Denver, Colorado, USA: 41–64.

Dhawan, B.D. (1995). *Groundwater depletion, land degradation and irrigated agriculture in India*. Commonwealth Publisher. New Delhi, India.

Dooge, J.C.I.; Delli Priscolli, J. and Llamas, M.R. (2003). Overview. In: J. Dooge (ed.), *Water and Ethics*. UNESCO. Preliminary presentation in CD ROM in the Third World Water Forum (Kyoto, Japan, March 2003).

Dyson, T. (1996). *Population and Food*. Rootledge. London, UK. 220 pp.

Fetter, P. (1994). *Applied Hydrogeology* (3rd edition). Macmillan, New York, USA. 691 pp.

Fornés, J. and Llamas, M.R. (1999). Conflicts between groundwater abstraction for irrigation and wetlands conservation: achieving sustainable development in La Mancha Húmeda Biosphere Reserve (Spain). In: C. Griebler *et al.* (eds.), *Groundwater Ecology. A Tool for Management of Water Resources*. European Commission. Environmental and Climate Programme: 227–236.

Foster, S.S.D. (1992). Unsustainable development and irrational exploitation of groundwater resources in developing nations. An overview. In: I. Simmers, F. Villarroya and L.F. Rebollo. (eds.), *Selected Papers on overexploitation*. International Association of Hydrogeologists. Selected Papers, Vol. 3. Heise, Hannover, Germany: 321–336.

Foster, S.; Lawrence, A. and Morris, B. (1998). Groundwater in Urban Development, *World Bank Technical Paper*, nº 390. 55 pp.

Freeze, R.A. and Cherry, J.A. (1979). *Groundwater*. Prentice-Hall. Ins. Englewood Cliffs. New Jersey, USA. 604 pp.

German Federal Government (2001). *Ministerial Declaration. The Bonn Keys. Bonn Recommendation for Action*. International Conference on Fresh Water. Bonn, Germany, 3–7 December 2001. 20 pp.

Gleick, P.H. (1993). *Water in Crisis*. Oxford University Press, UK. 473 pp.

Gleick, P.H. (1998). *The World's Water. The biennial report on freshwater resources*. Island Press. Washington DC, USA. 308 pp.

Hernández-Mora, N. and Llamas, M.R. (2000). *The role of user groups in Spain: participation and conflict in groundwater management*. CD ROM of the X World Water Congress. Melbourne, Australia, 12–16 March 2000. International Association of Water Resources. 9 pp.

Hernández-Mora, N. and Llamas, M.R. (eds.) (2001). *La Economía del Agua Subterránea y su Gestion Colectiva*. Fundación Marcelino Botín and Ediciones Mundi-Prensa. Madrid, Spain. 549 pp.

Hernández-Mora, N.; Llamas, M.R. and Martínez Cortina, L. (2001). Misconceptions in Aquifer Overexploitation. Implications for Water Policy in Southern Europe. In: C. Dosi (ed.), *Agricultural Use of Groundwater. Towards Integration between Agricultural Policy and Water Resources Management*. Kluwer Academic Publishers: 107–125.

Hernández-Mora, N.; Martínez Cortina, L. and Fornés, J. (2003). Intensive Groundwater Use in Spain. In: M.R. Llamas and E. Custodio (eds.), *Intensive Use of Groundwater. Challenges and opportunities*. Balkema Publishers. Lisse, The Netherlands: 387–414.

Idriss, H. and Nour, S. (1990). Present groundwater status in Egypt and environmental impacts. *Environmental Geology and Water Sciences*, 16(3): 171–177.

Indian Water Resources Society (1999). *Water: Vision 2050*. New Delhi, India. 74 pp.

Issar, A.S. and Nativ, R. (1988). Water beneath the desert: keys to the past, a resource for the present. *Episodes*, 11(4): 256–262.

LaMoreaux, P.E. and Newton, J.G. (1992). Environmental effects of overexploitation in a karst terrain. In: I. Simmers, F. Villarroya and L.F. Rebollo (eds.), *Selected Papers on overexploitation*. International Association of Hydrogeologists. Selected Papers, Vol. 3. Heise, Hannover, Germany: 107–113.

Lazarus, P. (1997). *Towards a regulatory framework for the management of groundwater in South Africa*. Draft prepared for the Directorate of Geohydrology. South Africa. 67 pp.

Llamas, M.R. (1986). Aguas Subterráneas e Ingeniería Civil. *La Voz del Colegiado*, No. 172 (March–April). Colegio de Ingenieros de Caminos, Canales y Puertos. Madrid, Spain: 12–21.

Llamas, M.R. (1988). Conflicts between wetland conservation and groundwater exploitation: two case histories in Spain. *Environmental Geology and Water Sciences*, 11(3): 241–251.

Llamas, M.R. (1991). The future of groundwater: a forecast of its exploitation and quality compared with past exploitation. In: *XXI Journées de l'Hydraulique. Les Eaux Souterraines et la Gestion des Eaux*. Sophia-Antipolis, 29–31 January 1991: IV.2.1– 8.

Llamas, M.R. (1992a). La surexploitation des aquifères: aspects techniques et institutionnels. *Hydrogeologie*, 4: 139–144. Orleans.

Llamas, M.R. (1992b). Wetlands: an important issue in Hydrogeology. In: I. Simmers, F. Villarroya and L.F. Rebollo (eds.), *Selected Papers on overexploitation*. International Association of Hydrogeologists. Selected Papers, Vol. 3. Heise, Hannover, Germany: 69–86.

Llamas, M.R. (2000). Some lessons learnt during the drought of 1991–1995 in Spain. In: Vogt and Somma (eds.), *Drought and drought mitigation in Europe*. Kluwer Academic Publisher: 253–264.

Llamas, M.R. (2001a). Considerations on Ethical issues in relation to groundwater development and/or mining. UNESCO Conference on international aquifers systems in arid zones; managing non-renewable resources. Tripoli, Lybia, 20–24 November 1999. *Technical Documents in Hydrology*, V IHP, No. 42. UNESCO. París, France: 467–480.

Llamas, M.R. (2001b). *Cuestiones éticas en relación con la gestión del agua en España*. Discurso de Ingreso en la Real Academia de Doctores. Madrid, Spain. 85 pp.

Llamas, M.R. (2003). *El Proyecto Aguas Subterraneas: resumen, resultados y conclusiones*. Papeles del Proyecto Aguas Subterraneas, No. 13. Fundación Marcelino Botín. Spain. 101 pp. (It can be downloaded from www.fundacionmbotin.org).

Llamas, M.R. (in press). *Groundwater and Human Development*. International Congress of the International Association of Hydrogeologists. Mar del Plata, Argentina. October 2002. Balkema Publishers. The Netherlands.

Llamas, M.R. and Custodio, E. (eds.) (2003). *Intensive Use of Groundwater. Challenges and Opportunities*. Balkema Publishers. The Netherlands. 478 pp.

Llamas, M.R. and Delli Priscoli, J. (2000). *Water and Ethics*. Papeles del Proyecto Aguas Subterráneas, No. A5. Fundación Marcelino Botín. Madrid, Spain: 56–99.

Llamas, M.R. and Martinez Cortina, L. (2003). *Overview of groundwater management for the Middle East and the Mediterranean Region*. CD-ROM corresponding to the Middle East and Mediterranean Region Day. Third World Water Forum. Kyoto, Japan, 20 March 2003. World Bank. (A written article in preparation: The Silent Revolution caused by Intensive Use of Groundwater).

Llamas, M.R.; Back, W. and Margat, J. (1992). Groundwater use: equilibrium between social benefits and potential environmental costs. *Applied Hydrogeology*, 1(2): 3–14. Heise, Verlag.

Llamas, M.R.; Casado, M.; De la Hera, A.; Cruces, J. and Martínez Cortina, L. (1996). El desarrollo sostenible de la cuenca alta del río Guadiana: aspectos socio-económicos y ecológicos. *Revista Técnica de Medio Ambiente (RETEMA)*. Madrid, Spain. September–October: 66–74.

Lloyd, J.W. (1997). The future use of aquifers in water resources management in arid areas. *The Arabian Journal for Science and Engineering*, 22(IC): 33–45.

Lomborg, B. (2001). *The skeptical environmentalist. Measuring the Real State of the World*. Cambridge University Press. 515 pp.

López Geta, J.A. and De la Orden, J.A. (2003). Drought as a catalyser of intensive groundwater use. In: M.R. Llamas and E. Custodio (eds.), *Intensive Use of Groundwater. Challenges and Opportunities*. Balkema Publishers. The Netherlands: 177–190.

López Gunn, E. and Llamas, M.R. (2001). New and old paradigms in Spain's Water Policy. In: Water security in the Third Millenium: Mediterranean countries towards a regional vision. *UNESCO Science for Peace Series*, Vol. 9: 271–293.

Margat, J. (1994). Groundwater operations and management. *Groundwater Ecology*. Academic Press: 505–522.

McClurg, S. and Sudman, R.S. (2003). Public and stakeholder education to improve groundwater. In: M.R. Llamas and E. Custodio (eds.), *Intensive Use of Groundwater. Challenges and Opportunities*. Balkema Publishers. The Netherlands: 271–286.

Ministerio de Medio Ambiente (MIMAM) (1998). *Programa de ordenación de acuíferos sobreexplo-tados/salinizados*. Serie Monografías. Secretaría de Estado de Aguas y Costas. Madrid, Spain. 66 pp.

Moench, M. (1999). Addressing constraints in complex systems. Meeting the water management needs of South Asia in the 21st Century. In: M. Moench *et al.* (eds.), *Rethinking the Mosaic*. Institute for Social and Environmental Transition. Boulder, Colorado, USA: 1–56.

Moench, M. (2003). Groundwater and Poverty: exploring the connections. In: M.R. Llamas and E. Custodio (eds.), *Intensive Use of Groundwater. Challenges and Opportunities*. Balkema Publishers. The Netherlands: 441–456.

Myers, N, and Kent, J. (1998). *Perverse subsidies: their nature, scale and impacts*. International Institute for Sustainable Development. Winnipeg, Canada. 210 pp.

Mukherji, A. (2003). *Groundwater development and agrarian change in Eastern India*. Comment No. 9 in IWMI-TATA Water Policy Program. (www.iwmi.org/iwmi-tata). 11 pp.

National Research Council (1997), *Valuing Ground Water*. National Academy Press. Washington DC, USA. 189 pp.

OECD (Organisation for Economic Cooperation and Development) (2000). *No longer business as usual: fighting bribery and corruption*. Paris, France. 276 pp.

Pearce, F. (1999). Counting down: focus about population explosion is probably misplaced, say demographers. Next century may have to worry about falling birth rates, not rising ones. *New Scientist*. 2 October: 20–21.

Poland, J.F. (1985). Guidebook to studies in land subsidence due to groundwater withdrawal. *Studies and Reports in Hydrology*, No. 40. UNESCO. Paris, France. 350 pp.

Postel, S. (1999). *The Pillar of Sand*. W.W. Norton and Co. New York, USA. 313 pp.

Price, M. (2002). Who needs sustainability? In: K.M. Hiscock, M.O. Rivett and R.M. Davison (eds.), *Sustainable Groundwater Development*. Balkema Publishers. Lisse, The Netherlands: 191–207.

Pulido, A.; Castillo, A. and Padilla, A. (eds.) (1989). *La sobreexplotación de acuíferos*. Instituto Tecnológico GeoMinero de España. Madrid, Spain. 687 pp.

Ragone, S.; Rivera, A.; Vecchioli, J.; Goodwin, C.; Marín, L.E. and Escolero, O.A. (2003). Intensive use of groundwater in North America. In: M.R. Llamas and E. Custodio (eds.), *Intensive Use of Groundwater. Challenges and Opportunities*. Balkema Publishers. The Netherlands: 287–306.

Salameh, E. (1996). *Water quality degradation in Jordan*. Royal Society for the conservation of Nature. Amman, Jordan. 179 pp.

Seckler, D.; Amarashinge, U.; Molden, D.; de Silva, R. and Barker, R. (1998). *World Water Demand and Supply, 1990 to 2025, Scenarios and Issues*. Research Report 19. International Water Management Institute. Colombo, Sri Lanka. 42 pp.

Selborne, L. (2000). *The ethics of freshwater use: a survey*. COMEST, UNESCO. Paris, France. 58 pp.

Shiklomanov, I. (1997). Comprehensive assessment of the fresh water resources of the world, Report E/CN

17/1997/9. World Meteorological Organisation. 88 pp.

Simmers, I.; Villarroya, F. and Rebollo, L.F. (eds.) (1992). *Selected papers on overexploitation*. International Association of Hydrogeologists. Selected Papers, Vol. 3. Heise, Hannover, Germany. 392 pp.

Solley, W.B.; Pierce, R.R. and Perlman, H.A. (1998). *Estimate of water use in the United States*. U.S. Geological Survey, Circular 1200. 71 pp.

Sophocleous, M. (1997). Managing water resources systems: why *safe yield* is not sustainable. *Ground Water*, 35(4): 361.

Sophocleous, M. (2000). From safe yield to sustainable development of water resources. The Kansas Experience. *Journal of Hydrology*, 235: 27–43.

Sophocleous, M. (2003). Environmental implications of intensive groundwater use with special regard to streams and wetlands. In: M.R. Llamas and E. Custodio (eds.), *Intensive Use of Groundwater. Challenges and Opportunities.* Balkema Publishers. The Netherlands: 93–112.

The Economist (1998). *Environmental Scares*. 20 December 1998: 21–23.

Theiss, C.J. (1940). The source of water derived from wells. Essential factors controlling the response of an aquifer to development. *Civil Engineering*, 10: 277–280.

United Nations (1997). *Comprehensive assessment of freshwaters of the world.* Report of the Secretary General. Commission on Sustainable Development. 7–15 April 1997. E/Cn/17/1997/9. 35 pp.

United Nations (2000a). *United Nations Millenium Declaration*. A/RES/55/2.

United Nations (2000b). We, the people: the role of the United Nations in the twenty-first century. Report of the Secretariat General. A/54/2000.

United Nations (2002). The Johannesburg Declaration on sustainable development. World Summit of Sustainable Development. Johannesburg, South Africa. 26 August–4 September 2002. A/conf. 199/L.6.

WCD (World Commission on Dams) (2000). *Dams and developments. A new frame for decision-making*. Earthscan. 404 pp.

Wood, W.W. (1999). Water use and consumption: what are the realities? *Ground Water*, 37(3): 321–322.

Young, R.A. (1992). Managing aquifer overexploitation. Economics and policies. In: I. Simmers, F. Villarroya and L.F. Rebollo (eds.), *Selected Papers on Overexploitation*. International Association of Hydrogeologists. Selected Papers, Vol. 3, Heise, Hannover, Germany: 199–222.

Zwingle, E. (1993). Ogallala aquifer: wellspring of the High Plains. *National Geographic*, March: 80–109.

Governing the groundwater economy: comparative analysis of national institutions and policies in South Asia, China and Mexico

Tushaar Shah
International Water Management Institute (IWMI), India
t.shah@cgiar.org

ABSTRACT

In many parts of the world, especially in South Asia, the size of the groundwater economy has rapidly grown during the past 5 decades, and is growing still. Elsewhere in Asia – Sri Lanka, Vietnam, Laos, Thailand – and in Maghreb countries, groundwater use in agriculture has begun to grow during the past decade and is likely to peak in the coming 10 years. Global concerns with growing groundwater use in agriculture have focused mostly on its sustainability, quality degradation and adverse impacts on environment and ecological flows. Direct regulation of groundwater draft through stringent laws, regulatory frameworks and aggressive water pricing has been strongly advocated. However, despite the consensus for need to move in these directions, many governments have dragged their feet in operationalizing direct regulation. Where governments have taken pro-active stance, as in Mexico and to lesser extent, China, the impacts are variable. Governing groundwater economies is proving intractable; and responses to intensive groundwater use vary widely across nations. This paper attempts to understand why. It also argues that particularly in Asia, direct regulation of groundwater use may remain a pipe dream for a long time to come; and for effective governance of the groundwater economy, there is need to invent a wider toolkit including direct and indirect instruments of management – that can be adapted to peculiar contexts of the groundwater economy in different countries.

Keywords: *groundwater, governance, poverty, sustainability, institutions, policy.*

1 THE COMMON CHALLENGE

Regions of Asia where food security and rural livelihoods have come to depend precariously on intensive use of groundwater in agriculture have expanded at a frightening pace, especially after 1970. Recent IWMI analyses suggest that 1970 was probably the watershed year: prior to that, steady rise in food production depended squarely on growth in surface irrigation. Since then, however, South Asia and North China have experienced a massive groundwater boom; and by the early 1990s, groundwater irrigation had overtaken all other sources in explaining the total area irrigated as well as in contribution to farm output and incomes (Deb Roy and Shah, 2001). An extraordinary aspect of this boom is its quiet, furtive character: governments in many Asian countries remain unaware of this wildfire growth of wells and tubewells fueling right under their noses. As a result, by the time resource managers begin to size up the challenge facing them, they find they are fighting a losing battle. Many regions of Asia are now discovering that the groundwater boom comes with a price tag in the form of groundwater depletion, pollution and quality deterioration (because of fluoride, arsenic, nitrates, etc.) raising serious concerns about the future sustainability of such intensive groundwater irrigation. What these need direly is a practical strategy of managing this runaway growth in the bubble of the groundwater economy before it bursts, causing misery all around. The phrase *groundwater governance* has come into currency primarily in the wake of the recognition that managing groundwater has come to involve dealing with not just physical and ecological processes but a complex of socio-economic and institutional relationships with far reaching impacts on society.

Drawing from the experience of Western USA, Europe and Australia, international thinking on the ways forward on improving groundwater governance has veered towards a complex of stylized prescriptions: a) countries should get an appropriate legal and regulatory framework for groundwater appropriation and use; b) a new system of groundwater rights should replace the present open-access regime; c) groundwater should be treated as an economic good and priced to reflect its scarcity value; d) an institutional structure created for development of the resource should be transformed into one appropriate for resource management; and e) policies should be redefined and adjusted to the new priority of sustainable management.

Against these stylized prescriptions, we find vast variations in the way nations actually respond to the problems of groundwater stress. In this paper, we try to understand why. We do this by developing a comparative analysis of institutions and policies for groundwater management in South Asia, China and Mexico. Our purpose is to explore to what extent the responses to groundwater overdevelopment in these regions conforms to these stylized prescriptions, and why. In Section 2, we draw a broad comparison between the South Asian and Chinese situations because we find marked similarities in these in several respects. In Section 3, we develop the Mexican case study. In Section 4, we offer some general conclusions and argue for a more nuanced understanding of the context within which groundwater economies operate in different parts of the world and which shapes their strategies of governing its appropriation and use.

2 COMPARING GROUNDWATER INSTITUTIONS AND POLICIES IN SOUTH ASIA AND NORTH CHINA

2.1 *History and context*

South Asia and North China have important similarities in terms of very high population densities, small land holdings, and predominance of groundwater. While South Asia's irrigation history goes back to the millennia, North China's goes back to all of 50 years. However, when it comes to history of groundwater irrigation, unprecedented expansion in it after 1970 was spurred in both the regions by nearly the same compact of factors, *viz*., intensification of farming and the propagation of seed-fertiliser technologies, reduced and undependable surface water supplies, role of groundwater in mitigating the impacts of drought, and early encouragement from public policy makers to groundwater irrigation. In both the regions, well-densities increased in spurts during drought periods, and with active support from the state (Ronghan, 2000, p. 83). Booming groundwater-based irrigated agriculture in both the regions is facing imminent threat of decline as a result of resource depletion or salinization caused by constant overdraft. Secondary salinization has yet not emerged as a critical problem in large areas of North China, especially in the Western Parts (Zhang and Zhang, 2001; Kendy *et al*., 2002) as it has in Pakistan Punjab and Sindh and Indian Punjab and Haryana. However, groundwater depletion and secular decline in water table, high fluoride content, rising energy costs of pumping, problems of land subsidence, falling well-yields and high rate of failure of wells are problems common to both the regions.

2.2 *Organization of village groundwater economies*

However, there are notable differences in other respects, mostly in the institutional fabric of the two regions. First, it is the sheer numbers; China has an estimated 3.5 million agricultural tubewells, mostly in the North China Plains (NCP) that extract an estimated 75 km^3/yr of groundwater; in comparison, South Asia had 19 million agricultural wells and tubewells during mid 1990s, which may well have increased to 23–25 million now, extracting some 210–230 km^3/yr of groundwater. Then, the structure of land and water rights and groundwater institutions are different. Throughout South Asia, overwhelming majority of groundwater wells and pumps are privately owned by farmers. Rights to groundwater are not separately specified, and are treated as an easement attached to land; as such, land owners act as if they have unrestricted ownership rights on groundwater. However, exercising this right requires a well and a pump; and many small farmers' holdings are too small to make a mechanized well viable. Moreover, because of high level of land fragmentation, even those who own wells cannot irrigate all their fragments by their own wells. A major institutional response to this problem is the emergence of pervasive, local, fragmented pump irrigation markets which have helped smooth out these rough edges of the groundwater economy in a South Asian village and have expanded access to groundwater irrigation to the resource poor.

This institution of South Asian groundwater markets – or, to be precise, pump rental markets– has been extensively studied in South during recent years (Kolavalli and Chicoine, 1989; Janakarajan, 1992; Shah, 1993; Palmer-Jones, 1994; Meinzen-Dick and Sullins, 1994; Saleth, 1994; Strosser and Kuper, 1994). Like markets in general, these create new wealth and help alleviate rural poverty; but these also make the organization of the groundwater economy chaotic maze of intense, criss-crossing interaction amongst pump owners and water buyers without any mediating influence. In many regions of South Asia, farmers invest in tubewells and pumps primarily for selling water for profit; even when the primary motive is irrigating own land, pump owners can still earn significant supplemental income from selling water as a side activity (Kolavalli and Chicoine, 1989). As water markets mature, intense competition amongst water sellers confers benefits to buyers in terms of lower price and better service; but it also encourages huge overlap in command areas of private tubewells and excess pumping capacity.

In a typical South Asian village, the groundwater economy is completely untrammeled by any regulatory authority or mediating agency. It permits no role for even the village level governance structures (such as India's *Gram Panchayats* or Pakistan's *numberdars*). No norms are effectively in place for siting and licensing of groundwater wells. The only government agency the pump owners have any interaction with is the electricity utility, and some times public sector banking institutions. India's National Bank for Agriculture and Rural Development, which refinances bank loans for groundwater structures, stipulated some siting norms; however, these are extensively violated (Shah, 1993). In groundwater depletion areas such as North Gujarat region in Western India, where capital investments as well as risks in making successful tubewells are high, farmers come together to form co-operative tubewell organizations (Shah and Bhattacharya, 1998); however, their prime aim is to secure irrigation for their members, and play no role whatever in guiding or managing the overall groundwater socio-ecology of the village. All in all, in South Asia's chaotic village groundwater economies, all formal and informal institutions function with the sole aim of maximizing present wealth creation from groundwater irrigation.

In a typical North China village, however, nearly the opposite was the case until early 1980s. Before the sweeping agrarian reforms initiated by the Deng administration in 1983, irrigation organization in the Chinese country-side was uniform and orderly in comparison. Collectives were responsible for making and maintaining tubewells as well as pumps and distribution systems. This does not necessarily mean that they were efficient in techno-economic terms; however, it did mean the presence of an over-arching governance mechanism at the level of the collective and above that oversaw the working of the irrigation economy. With the onset of reforms and the household responsibility system, we find a wide variety of institutional arrangements have now come into play in the Chinese countryside (Xiang *et al.*, 2000). Table 1 outlines a range of institutional arrangements we came across for tubewell management in 9 villages of Hanan and Hebei province in course of fieldwork during 2002. Where water tables are high and the cost of making tubewells low – as, for example, in the lower Hanan province – it is common for pre-existing shallow tubewells to be owned and maintained by Village Committees from agricultural taxes.[1] Where new Shallow Tube Wells (STW) need to be built, Village Committees still do so, especially if they have buoyant tax revenue; else, they invite private farmers to build and operate tubewells under formal contracts which vary from very simple to quite complex.

Regardless of whether STW are collectively managed or contractor-managed, pumps and ground pipes are generally owned by farmers or borrowed from friends or relatives. Unlike in South Asia, where the ownership of a pump in many regions is not only source of significant extra income but, as some social scientists claim, also of social status and political power (Wood, 1995; Dubash, 2002), in STW areas in North China, pump ownership yields the owner neither profit, nor power nor status. The chief reason is the relatively low real cost of machine capital in China in comparison to the rest of the developing world.

In deep tubewell areas of Hebei and Shandong provinces, village irrigation organization undergoes marked change; tubewells are bigger and fewer, each serving a larger command. Some Village Committees here also build and operate Deep Tubewells (DTW) which are a much costlier affair compared to STW. Here tubewells going to the depth of 350 m or so, motor-pumps generally of 28 kW and 1,000–1,500 m of buried pipeline network – which comprises a tubewell assembly may entail an investment of Y (Chinese yuan) 250,000–300,000 (US$ 30,000–36,000)[2] apiece. Each tubewell here commands 600–1,000 mu (40–67 ha)[3] and is beyond the reach of any individual farmer unless he fancies himself as a water entrepreneur. It is common for deep tubewells to be established, funded, and owned by the Village Development Committee; but its operation is commonly contracted out. A variety of contracting arrangements seem to be in vogue, each presenting interesting alternatives in the design of incentives. Regardless of whether they are privately or collectively managed, in deep tubewell areas of NCP, such as in most of Hebei province, irrigation is a far more expensive proposition than in the STW areas. In these villages, everyone is a water buyer and pays a water rate that is generally linked to energy use. In this increasingly complex maze of irrigation institutions in North China, private irrigation service providers are emerging as key players. Their margins vary and do not seem to show any particular pattern but did not seem to us to contain high monopoly premia, except in deep tubewell villages in Hebei.

Unlike in South Asia, where local water markets find their own prices, in North China, the Village Leaders or Party Leaders often fix or have a say in deciding the margins to be charged by contractors which hovered around Y 0.2–0.35/kWh (US$ 0.02–0.04/kWh) in the villages where we worked. Even where private irrigation providers fixed prices, their gross margins tend to be in this range. There is hardly any competition in the sense it operates in South Asian groundwater markets mostly because there is little overlap in the command areas served by different tubewells; and it is common for contractors and private sellers to collude in setting a common price under the watchful guidance of the Village Committee. The net result is that the ratio of irrigation fee to energy – cost which in India may be as high as 2.5 or more – seldom exceeds 1.5 in China. In villages with alternative irrigation sources or in years of good rainfall, tubewell irrigation is sparingly used and contractors make

[1] Chinese farmers pay several types of land tax – notably, crop tax, education tax, water use tax, electrician tax, etc. In Xiaotan Village, Yanjin county in Henan, for instance, half of the annual tax collection of Y 20,000 (US$ 2,400) is turned into the township government and half is retained by the village leader. Of the village share of Y 10,000 (US$ 1,200), some 15% was earmarked for maintenance of tubewells.

[2] Y 1 = US$ 0.12095 (July 1, 2003).

[3] 1 mu = 0.0667 ha.

Table 1. Variety of institutional arrangements for groundwater irrigation in 9 villages of Henan and Hebei Provinces.

Village, County, Province	Pumping water level	Shallow tubewell	Motor pump	Deep tubewell	Transformer	Collection of electricity fee
Xiaotan, Yanjin county, Henan	7–8 m	BOM by Village Committee	Farmers own and share pumps	Nil	O&M by Township Electricity Bureau	TEB's electrician collects based on meter reading on each motor pump
Guantun,Yanjin county, Henan	10 m	BOM by Village Committee	Farmers own and share pumps	Nil	O&M by Township Electricity Bureau	TEB's electrician collects based on meter reading on each motor pump
Xijie, Yanjin county, Henan	17 m	STW maintenance by Village Committee; but O&M of STW and pump by contractor who charges Y 1/kWh against Y 0.7/kWh to be paid to electrician		Nil	O&M by Township Electricity Bureau	TEB Electrician collects irrigation fee at the rate of Y 1/kWh and pays the contractor Y 0.3/kWh as his margin
Zhao Zhuang, Ci county, Hebei	26–30 m	12 contractors operate collective STW; 12 private service providers; 5 share-holder service providers		Nil	4 managed by private contractors; 2 by Village Committee	STW operators collect Y 1/kWh from irrigators against electricity cost of Y 0.565/kWh
Dong wan gnu, Ci county, Hebei	5 m	Farmer-contractors operate STW and maintain them		Nil	Owned and managed by Village Committee	Contractors charge Y 0.82/kWh from irrigators and pay Y 0.565/kWh to electrician
Shi cun Ying, Ci county, Hebei	220 m	Private and stake-holder group owned and managed deep tubewells with pumps and buried pipe networks			Collectively owned by Village Committee	Tubewell owners charge Y 10/hour from irrigators and pay 0.565/kWh
Yao Zhung Zi, Chang Zhou county, Hebei	220 m	15 STW managed by Village Committee	Farmers use own or borrowed pumps	6 DTW owned and managed by the Village Committee	Collectively owned by Village Committee	DTW operators employed by VC collect Y 16/hour from irrigators and pay Y 0.45/kWh to electrician
Xi Huayuan, Chang Zhao county, Hebei	250 m	VC owns and operates 6 DTW and private farmers operate 4, all with 28 kW pumps; the former operate as a utility, the latter as a business			Four private DTW owners own a transformer each	Private DTW owners charge Y 1.1/kWh and pay Y 0.48/kWh; VC DTW charges Y 0.65/kWh and pays Y 0.55/kWh to TEB
Xi Tun Zi, Chang Zhou county, Hebei	250 m	VC collected Y 200/mu to build 7 collective DTW at a cost of Y 620,000; each has a 400 m deep DTW, 30 kW pump and 1,200–1,500 m of underground pipeline network.			Collectively owned and managed by VC	VC employed DTW operators charge Y 15/hour and pay Y 0.48/kWh to electrician

little money. All in all, compared to South Asia, village governance institutions are stronger in China in that they enjoy and use greater authority in village affairs, and therefore have a pervasive influence not only on the irrigation organization but on the entire village economy and society. One aspect of this is the larger political system through which state authority percolates down; but another aspect – which must promote some measure of responsiveness in the governance structure to people's aspirations has also to do with the fact that governance structures are supported from locally generated resources, an aspect completely absent in South Asian village.

Compared to the South Asian farmer, who virtually pays no direct taxes, the Chinese farmer is heavily taxed. A major reason for the heavy taxation is the burden of the salary of the local government officials. Every village has a Village Leader and a Village Communist Party Leader. The former is elected by a Village Committee of 7 elected members. The Party Leader for each village is selected by the Township level party leadership from all party members in the village. The party leader is all powerful in village affairs; if there is a dispute between the Village Committee and the Party, there are negotiations to settle the differences, but ultimately what the party leader says goes. Party members, and particularly the party leader are selected supposedly for their social concern and awareness and a strong *extension motive* (McClelland, 1985). This helps somewhat in keeping the institution from becoming oppressive and hegemonic. The party leader as well as the village leader and members of the village committee are salaried officials. In Henan villages we covered, the party leader gets Y 140 (US$ 17) per month; while others get Y 120 (US$ 14.5) per month. These salaries have to come from land tax. Land tax is also used to sustain the Township government which lays claim to upto 50% of all land tax collections at the village level. Considering there is hardly any subsidization of agriculture, there is probably a heavy transfer of wealth from agriculture to industry in China, a situation akin to Japan's under Meiji restoration period (Mellor, 1995).

This parallel structure of local government and party organization may not be perfect or even ideal; however, it ensures the presence in the average Chinese village of state authority which is largely or completely absent in the South Asian village. In an Indian village, farmers can chase away an Electricity Board meter reader with impunity; and in Pakistan, Water and Power Development Authority has to use the military to take meter readings on electric tubewells (Sunday Times, 2002); but in China, many rules of the game get formulated as well as enforced by the village level governance structures. The *soft state* is as evident in the South Asian village as the *hard state* is in a Chinese village. This authority structure is at the heart of the way irrigation institutions function in China. This village level governance structure with an all-encompassing mandate in the village life can underpin a national and regional structure of water governance in ways that would be impossible in South Asian country side.

All in all, as outlined in Table 2, the organization of village groundwater economy in North China differs in several material ways from that in South Asia in that: a) the Village Committee and the village leader play a significant mediating and regulatory role in shaping the irrigation economy in North China whereas throughout South Asia, the village groundwater economy operates in a laissez faire style; b) monopoly premia are non-existent or marginal on pump rental markets in NCP where as they are significant in South Asia; in the NCP, monopoly rents emerge with the rise of private water sellers in DTW areas; c) since STW as well as DTW in the NCP have no overlapping command areas, the opportunities for *competitive deepening* and destructive chasing of falling water tables encountered in South Asia is absent in the NCP; this advantageous feature is likely to stay as long as the village committee and the village leader play an influential role; d) finally, effective cost of groundwater irrigation tend to rise as one moves from STW areas to DTW areas in a manner that broadly reflects the social cost of groundwater as shown in the Figure 1.

2.3 *Direct and indirect cost of groundwater irrigation*

Neither in South Asia nor in North China is groundwater itself priced on the margin. Under the new Chinese Water Law, farmers are required

Table 2. Comparing features of Village Groundwater Economies in South Asia and North China.

	South Asia	North China
Ownership of tubewells	Private	Collective, contracted
Ownership of pumps	Overwhelmingly private	Mostly private; some collective
Do all farmers own pumps?	No	No
Do tubewell command areas overlap?	Yes, extensively	No, rarely
Do pump owners compete to increase water sales?	Yes, because of active markets in pump irrigation service with powerful productivity and equity impacts	No, because tubewells are sited to serve specified command areas
Are water prices fixed by the operation of the market?	Yes, entirely; there is no regulation whatever of the way fragmented, local pump irrigation markets function	No, it is guided by Village Committee and Village leader; usually it is fixed on energy-cost plus basis
Is water selling viewed as a source of significant income?	Yes, especially in Eastern India, Nepal terai and Bangladesh	No, except in DTW areas where farmers make heavy investments
Irrigation cost as a proportion of total value of output?	20–25% for water buyers	3–5% for water buyers

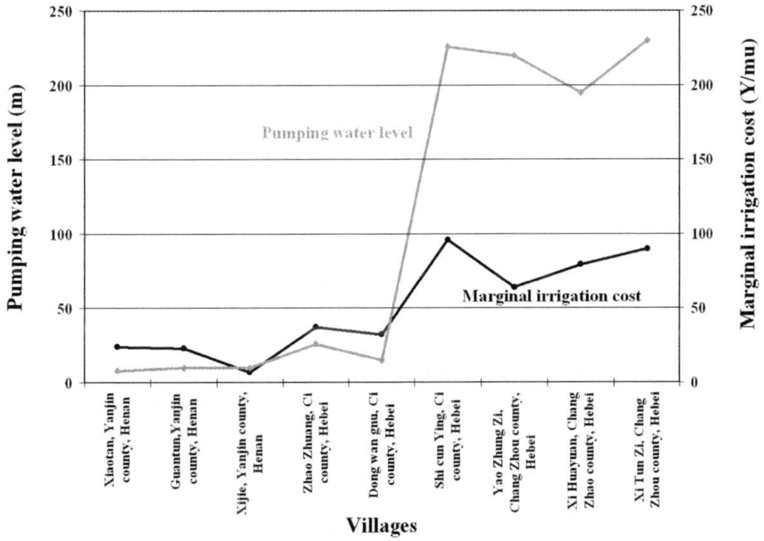

Figure 1. Impact of groundwater depletion on marginal irrigation cost. Based on fieldwork by author during 2001. 1 Y/mu = 1.8133 US$/ha (July 1, 2003)

to obtain a *permit* for which they have to pay a fixed fee. This was nowhere in effect; even if it were, it would not determine the marginal cost of groundwater use. What does however affect the marginal cost of groundwater use is the cost of energy used for pumping. In this, there are major differences between South Asia and North China. Energy costs of groundwater in South Asia seldom fully reflect the scarcity value of groundwater or energy. For instance, the cost of 1 m^3 of groundwater purchased by a small farmer is around Rs (Indian rupees) 4 (US$ 0.08)[4] in Eastern Uttar Pradesh or North Bihar in India where it is abundantly

[4] Rs 1 = US$ 0.02161 (July 1, 2003).

available;[5] but it is less than Rs 2 (US$ 0.04)[6] in North Gujarat where it is mined from 240 m or more. In Bangladesh, where groundwater is abundant and can be pumped from 3 m below ground, irrigating a hectare of paddy with purchased groundwater costs a high Taka (Bangladeshi Taka) 6,000 (approx. US$ 100)[7] (Mainuddin, pers. comm.) which drives many small holders to manual irrigation; but in Tamilnadu, where almost all groundwater presently being used is mined, irrigating a hectare of paddy with purchased groundwater costs less than Rs 1,500 (US$ 32).

The main reason why groundwater irrigation costs do not reflect its scarcity in India is the distorted electricity pricing policies pursued by Indian state governments. Collecting electricity charges from millions of farmers scattered over a huge country side has been a nightmare for South Asian countries like India and Pakistan. The logistical difficulty and economic costs of metering electricity used by tubewells has been found so high that most Indian states have done away with metering and instead charge a flat tariff based on horse power rating of the pumps (see Shah, 1993; Shah *et al.*, 2002a). Pakistan too tried flat tariff for nearly a decade before reverting to metering in 2000. In India, there is growing opposition to flat tariff in part because it is believed to induce inefficient use of power and groundwater but in part also because flat tariff has been used by populist politicians to subsidize tubewell irrigation. Electricity subsidy is thought to be the prime reason why many State Electricity Boards in India are on the verge of bankruptcy. Despite this state of affairs, many still argue that reintroduction of metering may not be a practical idea in the Indian context unless innovative technologies and/or institutional arrangements for collecting electricity charges can be used to reduce the transaction costs of metering and charge collection (Godbole, 2002; Shah *et al.*, 2002a). All in all, the management of this *energy-irrigation nexus* in South Asia is central to the governance of the region's groundwater as well as energy economies. Now that many Indian states are waking up to this stark reality, their attempts to go back to metered electricity are frustrated by two blocks: first, sustained opposition from electric tubewell owning class; and second, the formidable logistical problems and high transaction costs of managing metered electricity supply to farmers.

Surprisingly, the electricity-irrigation nexus in ways widely discussed in South Asia is not a subject of discussion in China at all. Indeed, researchers and technocrats with whom we raised the topic had difficulty in understanding why the two need to be co-managed at all. The Chinese electricity supply industry operates on two principles: a) of total cost-recovery in generation, transmission and distribution at each level with some minor cross-subsidization across user groups and areas; and b) each user pays in proportion to his use. Unlike in much South Asia where farmers pay for power either nothing or much less than domestic and industrial consumers, agricultural electricity use in many parts of North China attracts the highest charge per unit, followed by household users and then industries. Operation and maintenance of local power infrastructure is the responsibility of local units, the Village Committee at the village level, the Township Electricity Bureau at the township level, and the County Electricity Bureau at the county level. Equally, the responsibility of collecting electricity charges too is vested in local units in ways that ensures that the power used at each level is paid for in full. At the village level, this implies that the sum of power use recorded in the meters attached to all irrigation pumps has to tally with the power supply recorded at the transformer for any given period. The unit or person charged with the fee collection responsibility has to pay the Township

[5] Water purchased from 5 HP diesel pump with an hourly discharge of 12,000 L costs Rs 50 (US$ 1.08) in most parts of Eastern India.

[6] Water purchased from a 75 HP electric pump with an hourly discharge of 55,000 L costs Rs 90 (US$ 1.94) in North Gujarat.

[7] Taka 1 = US$ 0.01765 (July 1, 2003).

Electricity Bureau for power use recorded at the transformer level. To allow for normal line losses, 10% allowance is given by the Township Electricity Bureau to the village unit.[8] Under a new Network Reform program initiated by the national government with the objective of improving power supply infrastructure, village electricians in many areas of NCP have organized to provide improved services to their customers. However, these too levy a service charge for attending each request for help to cover their cost or transport on motor cycle. The village electrician, who generally enjoys the support of the party leader, is feared; and the new service orientation is designed partly to project the electrician as the friend of the people. The hypothesis that with better quality of power and support service, farmers would be willing to pay a high price for power is best exemplified in Henan where at Y 0.7/kWh (US$ 0.08/kWh; Rs 4/kWh) farmers pay a higher electricity rate compared to all categories of users in India and Pakistan, as also compared to the local diesel price at Y 2.1/L (US$ 0.25/L).

The village electrician, Network Reform Program, Township Electricity Bureau, the incentive payments, and new service organization are – all elements of the Chinese strategy that has turned the energy-irrigation nexus into a positive ratchet. In India, there has been some discussion about the level of incentive needed to make privatization of electricity retailing attractive at the village level. The village electrician in Hanan and Hebei is able to deliver on fairly modest reward of Y 200/month (US$ 24/month) which is equivalent to half the value of wheat produced on a mu (or 1/30 of the value of output on a hectare of land). For this rather modest profit, the village electrician undertakes to make good to the Township Electricity Bureau full amount on line and commercial losses in excess of 10% of the power consumption recorded on the transformers; if he can manage to keep losses to less than 10%, he can keep 40% of the value of power saved.

All in all, the Chinese have all along had the solution to the energy-irrigation nexus that has befuddled South Asia for nearly two decades. In the way the Chinese collect metered electricity charges, it is well nigh impossible to make financial losses since these are firmly passed on downstream from one level to the level down below. Take for example the malpractice common in South Asia of end-users tampering with meters or bribing the meter-reader to under-report actual consumption. In the Chinese system, it is very unlikely that such malpractice can occur on a large scale since the village electrician is faced with serious personal loss if he fails to collect from the farmers electricity charges for at least 90% of power consumed as reported at the transformer meter. And since malpractice by a farmer directly hits other farmers in the village, there likely exist strong peer control over such practices. There are similar incentive-control mechanisms at the level of the Township Electricity Bureau as well so that major malpractices at the transformer level would likely be detected and curbed early.

Would transposing the Chinese institutional design for consumption based pricing of electricity and water work in South Asia? After all it should be simple to put the meter reader of a state electricity board on a salary plus performance-linked incentive or disincentive; and equally, to put the canal guard too on a similar system of performance-linked reward system with minor adjustments in physical infrastructure. Our assessment is that it would not work because of the break-down of local authority structures in South Asia. The primary reason why the metering system works in China is that, in order to perform their tasks effectively, the electrician can invoke the authority of the State

[8] The village electrician's reward system encourages him to exert pressures to achieve greater efficiency by cutting line losses. In Dong Wang nu village in Ci county, the village committee's single large transformer which served both domestic and agricultural connections caused heavy line losses at 22–25%. Once the Network Reform Program began, he pressurized the VC to sell the old transformer to the Township Electricity Bureau and raise Y 10,000 US$ 1,200 (partly by collecting a levy of Y 25 US$ 3 per family, and partly by a contribution from the Village Development Fund) to get two new transformers, one for domestic connections and the other for pumps. Since then, power losses have fallen to the permissible 12% here.

through the Village Committee, the Village Leader and, above all, the Village Party leader. And since the Chinese are used to taking this authority seriously, the electricians too invoke a measure of fear and compliance. The ease with which an electrician in a Chinese village can recover the difference between power fee deficit by levying a cess on all users is suggestive of the authority these vicariously enjoy.

2.4 *The organization and reach of the Water Bureaucracy*

Never in the 2,500 year history of South Asia have ordinary citizens been subjected to a unified system of governance for a sustained period of time. A major reason probably is that except for brief periods – when regents like Asoka, Harshawardhan, Akbar unified vast territories – what are now India, Pakistan, Bangladesh, Sri Lanka and Nepal were ruled over by numerous kings through feudal chiefs and overlords constantly engaged in internecine strife. These regions came under unified administration only during the Colonial period which created a bureaucracy as an institution of governance. Until then, each South Asian village was pretty much a republic.

In contrast, for most parts of over 2,000 years, right until 1911, China has been a unified, tightly-governed state that ensured respect for law and the authority of the state. Since the time of Qinshi Huangdi *circa* BC 250, China's first Emperor who unified numerous feuding kingdoms into an efficient and organized state, China's political system and governance institutions seem to have changed very little. The Chinese state Huangdi built has survived, in its essentials, to date with a single currency, nationalized land and natural resources, standardized weights and measures, a single script with 3,000 characters. In a brief reign of 11 years, Huangdi also produced homogeneity in people's thought by destroying all books apart from legalist works and rallied society around the common goal of creating a *rich and powerful country*. Despite numerous efforts to recreate pre-Huangdi kingdoms, China retained, until well into the 20th century, the tradition of a unified state with, uniform penal code, the Legalist political system and a vast, centralized bureaucracy with a formidable reach and ambit, which are evident even today.

The organization of the south Asian water administration is thin, fragmented, top-heavy, bureaucratic and in general ill-equipped to manage a sector that is rapidly growing in size and complexity. Take for instance India; in a typical Indian block (or *taluka* or *tehsil*) that covers some 100 villages and a population of over 200,000 people, the total number of government officials (excluding the lowest rung, such as canal *chawkidaars* and public tubewell operators) working on water probably does not exceed 10; and for a district, which may have 18–30 such blocks and a population of 2–3 million, this number is probably around 100. Moreover, these are vertically organized into line departments such as – irrigation, groundwater, water supply and sanitation – which hardly interact with each other. Canal irrigation departments are commonly the largest; whereas groundwater departments are either absent (as, for instance in Gujarat, where the Gujarat Water Resources Development Corporation doubles up as one) or thinly staffed. Each department functions as a bureau, often pursuing a mandate that has long become irrelevant. For instance, the groundwater departments in most Indian states still believe further development of groundwater to be their key mandate; these are nowhere close to making a transition from the *resource development mode* to *resource management mode*. Likewise, once the construction of new projects gets over, canal irrigation bureaucracies too feel unable to move into the new role of system management and service delivery. Their ambit of operation is linked to administrative units – such as districts and *talukas* – rather than a river basin or sub-basin. Finally, the water bureaucracies in much of South Asia have increasingly become a drag on the society; over 90% of their budgets get used up by salaries and establishment costs; and the heavily subsidized water fees – of which only a small fraction is actually collected – can hardly meet even a part of this salary and establishment cost, leave alone contribute to infrastructure maintenance and improvement.

As shown in Figure 2, Chinese water administration differs from South Asian in at least in

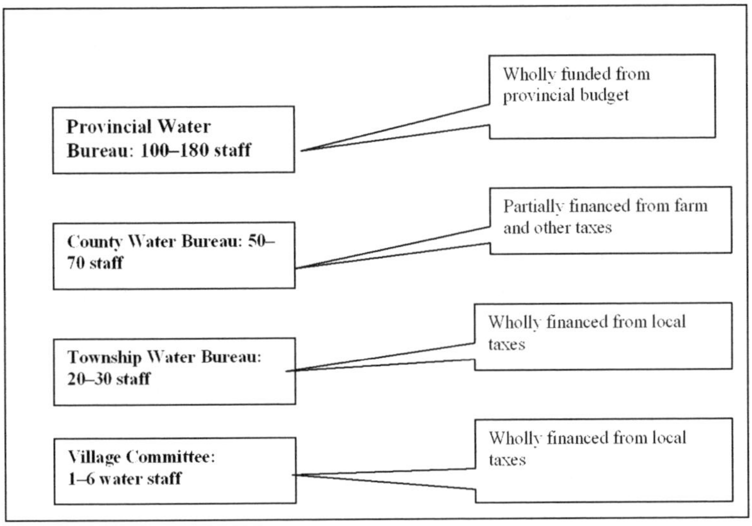

Figure 2. Structure of Chinese Water Administration and its funding.

two respects: it has much greater presence at the grass-roots; and increasingly, it is paying for itself through service fees. China has a nested hierarchy of water institutions at each level, controlled mostly by the government at that level but within an overall policy influence of the Ministry of Water Resource (MWR) (Wang and Huang, 2002). Like South Asian bureaucrats, Chinese water bureaucrats too have a resource development rather than a resource management mindset. However, there are indications that water management concerns are increasingly coming to a head, especially in provinces like Hebei where groundwater scarcity, depletion and quality deterioration are emerging as paramount concerns.

Compared to South Asia's fragmented water administrations, China's Water Bureaus represent an effective first step towards integrating and unifying water management tasks at local levels. Until a decade ago, water management in a typical county in China was fragmented as in a South Asian local administrative territory (such as a Tehsil or a district). A serious and avoidable – water crisis in Shenzhen in 1991 led its municipal administration to create a unified Water Affairs Bureau which manages water source development and construction, flood control and drainage infrastructure, urban water supply, water saving programs all – under one umbrella. The Water Affairs Bureau model caught on; and by May 1999, 160 counties had Water Affairs Bureaus. In May 2000, Shanghai brought in even more functions under a new Shanghai Water Resources Bureau (Wang and Huang, 2002).

Water Bureau's are substantial outfits even at the county level (equivalent to 2–3 *talukas*/blocks/*tehsils*/*thana* in South Asia). In Ci county in Hebei province, which has 19 townships and 390 villages under it, the County Water Bureau staff is only 60 and a typical township water bureau employs 20–30 officials; however the entire hierarchy of water bureaus in the county employs some 560 people. Hebei province, for instance, has 9 city level water bureaus and 200 water bureaus of counties like Ci which manage water resources in 4,000 small townships and villages. Thus, when all levels are taken together, the Water Bureau structure in a province may employ several thousand officials. Where as the Provincial Water Bureau is fully supported by the state budget, the county water bureau has to raise a portion of its own budget and the township water bureau is wholly self-financed. Thus, Ci county, for example, has an annual budget of Y 30 million (US$ 3.6 million); in this Y 10 million (US$ 1.2 million) is contributed by the national government under drought mitigation program; however, the balance of Y 20 million (US$ 2.4 million) has to be raised by the County Water Bureau from farmers taxes and local incomes. Kendy *et al.* (2002) note, in their study of groundwater institutions and policies in Luancheng county near Beijing that "fee revenues are sufficient to fund the County Water Affairs Bureau, but not to finance water conservation county-wide". Changing incentives facing bureaucrats is an economy-wide phenomenon that seems designed to transform China's bureaucrats into entrepreneurs (Rozelle *et al.*, 2000)

The Water Bureau structure is apparently undergoing a strategic transformation under the 1998 reform; indeed, recently, in some provinces these are renamed Water Management and Service Bureaus and are strongly encouraged to adopt a business ethic rather than regulatory-bureaucratic approach, and generate resources locally by selling services. True, this may be easier said than done, especially since revenue yielding water infrastructural assets are commonly held by provincial bureaus or the national government. Even so, many researchers believe that unified water resources management under the overall leadership of the much-restructured MWR is gradually becoming a reality in China (Wang and Huang, 2002). From the groundwater perspective, another major 1998 reform was to remove groundwater management from Ministry of Geological and Mineral Resources to MWR, a more logical home.

The Chinese bureaucracy – of the government as well as the party – in general has been a subject of much criticism by western scholars and researchers. However, the potentially powerful role of an effective bureaucracy in governance of scarce natural resources such as water has in general been underestimated. In India, for example, the Supreme Court announced two far-reaching environmental decisions in the span of a decade: in the first, it enjoined the Forest Department to bring illegal felling of trees in reserved forest areas forthwith; and the Forest Department which has a large bureaucracy with significant presence at the local levels effectively implemented the Supreme Court's injunction throughout Indian country-side; and deforestation of reserved forest was significantly reduced. In 1996, alarmed by widespread groundwater depletion, the Supreme Court, in an equally momentous judgment, empowered the Central Groundwater Board of India as the Central Groundwater Authority charged with the task of controlling groundwater depletion forthwith. Six years later, nothing has changed; beyond launching a limited regulatory program in the Union Territory of Delhi, the Central Groundwater Authority has been totally unequal to the task because it has no operational bureaucracy comparable to the Forest Department (see also, Down to Earth, 2002). The Groundwater Board has been used to its traditional role of groundwater monitoring, which it has been performing with the help of a thin force of scientific staff at the state level.

2.5 *Groundwater Law, Policy and their implementation*

In the context of growing scarcity, the task of managing water resources is becoming complex entailing numerous tasks at the ground level such as, for instance: "1) the need to register users and control free riders; 2) [building] the technical capacity to deliver agreed upon discharges at different points on the network; 3) the establishment of a process of collective decision making where groups of users are federated in higher hierarchical levels, with corresponding representatives; 4) the definition of partnership between users and irrigation officials, where service fee contributes to payment of field staff; 5) a legal framework to support this new institutional setting; 6) a strong commitment from the administration and politicians." (Barker and Molle, 2002, p. 21). Barker and Molle also argue that in the Asian context, "the growing importance of common pool groundwater resources add greatly to the complexity of the problem". Doing this will require resource management and regulatory institutions with wide reach.

South Asian countries are at ground zero in all these. None of them has in place a system of registering water users nor a law or a legal framework. In India as well as Pakistan, draft groundwater bills have been making rounds for several years; but there is no will to make them into a law because of doubts about their enforceability (Van Steenbergen and Oliemans, 2002). China has more of the necessary conditions in place to make a beginning. Starting with the epoch-making 1988 National Water Law, which defined a new legal and policy framework for water management, China has enacted 3 more laws and issued some 30 water management regulations during the 1990s (Wang and Huang, 2002). A slew of new laws is in the making. The 1998 reforms, which marked a further transition from a planned economy to a *socialist market economy*, pressures have been created for water bureaus at various levels to increase efficiency, reduce staff and generate resources through service provision. In 1992, when the Communist Party voted in favour of transition to a socialist market economy, the MWR proposed a strategic framework for water conservancy reform that focused on five key areas: water investment system, water asset management, water price and charge collection, water legislation and regulation; and water services provision (Wang and Huang, 2002). Institutional reform in China's water sector has relentlessly pushed this five-point agenda in recent years.

As of now, however, there is little evidence that this is having much effect on the ground. Chinese as well as Western observers and researchers are critical of the ineffectual role of the Chinese water bureaucracy in managing groundwater depletion in North China. Several reasons explain this: first, the Chinese bureaucracy has for long been fed on the developmental rhetoric of *protecting people against floods and droughts* (Boxer 2001, p. 337); moreover, rising from the farmers' ranks, the local bureaucracy empathizes more with farmers' needs to eke out a livelihood than the objective of long term environmental sustainability. Then, there are also informal kinship ties and network – and cultural institutions such as *quanxi* – which create a gulf between macro-level policy making and micro-level implementation. Finally, in their exhortations, even national and provincial leaders betray this ambivalence between protecting livelihoods and food security on the one hand and mitigating groundwater degradation. This ambivalence deepens as we move from national to provincial and county levels.

In course of our fieldwork, we found, however, that the water administration was more concerned about sustainability in water-stressed regions than where the water situation is more comfortable. For instance, in Hebei, regulating groundwater over-draft – and in general, managing scarce water efficiently – has become an important goal of administrative action, and initiatives are designed at all levels to focus on demand as well as supply side issues. It starts from the top; for example, the Ministry of Agriculture (China), is working on a national policy to wean farmers away from wheat and rice and encourage them to grow water-saving high value crops. Water Affairs Bureaus came up in water scarce North China faster than elsewhere; for instance, in Fuoyang river basin, a sub-basin of Yellow, 49% of counties established water affairs bureaus by 1999 where as only 7% of the counties did so at the national level (Wang and Huang, 2002). Similarly, although licensing is provided for all tubewells and water users by the National Water Law of 1988, it is enforced more vigorously and exhaustively in provinces like Hebei than in relatively less water-stressed provinces like Hanan.[9] Licensing of industrial and municipal tubewells is already in vogue in many provinces especially in North China. Farmers are to be licensed individually; however, in practice, only villages are licensed for irrigation tubewells; the next step is to issue tubewell license. In Luancheng county in Hebei, while existing tubewell owners are still outside

[9] Wang and Huang (2002) however cite a 1995 Report on Implementation Situation of Water Withdrawal Permit System by MWR which asserts that 95% of users (barring domestic water users who are exempt) had applied for permit by July 1995. If this is true, it is a major step forward in resource management since it automatically created a registry of water users, and brought these within the ambit of the resource management agency. Apparently, the performance of water resource fee by water affairs bureaus in urban areas too is quite satisfactory; however, in 1993, the farmers were specifically exempted from the fee for a period of 5 years by the Central Government to alleviate their burden; and the exemption continues todate. Wang and Huang (2002) suggested that water bureaus in Hebei were to begin collecting water fees from farmers irrigating with groundwater in 2000; however, in course of our fieldwork in Hebei in mid-2002, we found no sign of farmers paying any water resource fee in Hebei.

the ambit of the permit system, farmers who drill new wells are obliged to obtain permits through which the spacing between wells is regulated (Kendy *et al.*, 2002).[10]

Under the 1989 water law, borewell drilling contractors too are supposed to be licensed; but these are not covered by the permit system in most of the Hanan province. In Hebei, however, licensing of groundwater structures is taken far more seriously. In several villages we visited, we found that all the tubewells were individually licensed for 5 years. Drilling companies were licensed too; bigger ones were licensed by the Province Water Bureau and are allowed to operate anywhere in the province. Local contractors are licensed by Township Water Bureau and can operate only within the township area. There seemed to be no unlicensed drilling units in this township. Since 1999, the licensing drive has become more serious; besides domestic users, all other water diverters are to get a license which costs only Y 4 (US$ 0.50). In some other counties of Hebei, we still found permits obtained at the village level; and the license drive is yet to begin with individual farmers in full earnest. But there were indications that this would gradually happen. Some county courts in Hebei have already constituted Water Law Teams whose job is to enforce the National Water Law. If a farmer has not obtained a license, the Bureau staff persuade him to get one; however, they can not impose a penalty on him; they can merely lodge a court complaint; and only the court can punish unlicensed diversion of water.

In Chang Zhou county of Hanan, which exemplifies the groundwater crisis of North China plains somewhat in the manner of the Mehsana district in North Gujarat, the County Water Bureau is pursuing a 5 point strategy combining demand management as well as supply augmentation to counter it: a) promoting water saving technologies; b) discouraging water intensive crops and promoting water saving, high value ones; c) water import; d) limiting the number of tubewells; and e) limiting the draft from each tubewell. There is some progress, especially on promoting water saving technologies. Buried pipes and over-ground pipes now cover 70% of the farm lands in the county. Little progress is made in encouraging a shift from food grain crops to high value water saving crops; however, rice which was once a significant crop is no longer permitted in Chang Zhou county or elsewhere in North China. On water imports, major progress will occur only when the mega water transfer project from Yangtzee to Yellow gets completed; but on a smaller scale, some water imports are already taking place from Shandong province. While little progress is seen in restricting the number of tubewells for agricultural use, industrial use of groundwater is much more tightly regulated in Chang Zhou county now than ever before. Each industrial unit is provided a licensed draft limit up to which groundwater withdrawn is charged at the rate of Y 1/m^3 (US$ 0.12/m^3); beyond that, the rate goes up steeply to Y 5/m^3 (US$ 0.60/m^3).

There is growing though scattered evidence of successes in groundwater demand management in North China. In Luancheng county in Hebei, Kendy *et al.* (2002) note that a cost-share program of water saving – in which the provincial, prefecture and county water bureaus share 30% each while the farmer contributes 10% – resulted in shift from flood irrigation to sprinkler sets serving 2,900 ha, drip irrigation to 20 ha and buried pipe networks to 6,700 ha. Similarly, a panel of UN experts studying basin management in Huiahe river basin east of Yellow noted that "with the same irrigated area and water consumption, the grain output [in the basin] almost doubled from 1980 and 1993, increasing from 40.4 to 73.6 million tons. This may point to a significant improvement in agricultural and irrigation practices over a short space of time, but is probably due largely to the uncontrolled expansion of groundwater irrigation to supplement existing surface schemes" (UNDP, 2000, p. 8). And later, the report says: "While the predominant approach has been supply oriented, demand management has made its mark. In irrigation, efficient water use is an important programme

[10] However, a little later in the same paper, the authors suggest that even new tubewells easily escape the permit system because their owners use private drillers in preference to the Water Bureau since the former are cheaper, demand no labour contribution and evade the permit system (Kendy *et al.*, 2002: 15).

that is reported to have had significant impact during the past 15 years." (UNDP, 2000, p. 9).

In South Asia, such demand management initiatives by local governments or water bureaucracies are rarely to be found even in areas like Mehsana in North Gujarat, Ramnathapuram in Tamilnadu, or Kolar in Karnataka, examples of Indian districts suffering extreme groundwater stress. Here, to start with there is no legal or regulatory framework under which groundwater use can be regulated; even if there were one, there is no administrative structure that might enforce it. In any case, there is no water administration at the district or *taluka* (block) level that might develop and implement anything like the strategy that the Chang Zhou water bureau has come up with. Above all, even at the higher levels of the bureaucracy and political leadership, there is no recognition of environmental sustainability as an important policy variable; the focus of attention is still on how best to protect livelihoods. It is not surprising that in most Indian states, electricity supplied to farmers by state-controlled power utilities tend to become cheaper, not costlier, as one moves from groundwater abundant to groundwater-depleted areas.

2.6 *Larger context: Political, Economic and Institutional*

The strongest factor that may help China act decisively to manage its groundwater socio-ecology for environmental sustainability is that it has become a growth economy. Pressure on groundwater use in agriculture tends to decline as economies industrialize; no better example can be found to illustrate this than Hong Kong, where groundwater use – which was never very intensive – has almost stopped with the decline of agriculture and most of the workforce getting absorbed in secondary and tertiary sectors of the economy (Chen, 2001). Hong Kong also exemplifies how water scarcity is seldom absolute; and as economies grow, public water systems are able to invest in multiple avenues of balancing water demand and supply. At a Gross Domestic Product (GDP) of US$ 3,600 per capita in Purchasing Power Parity (PPP) terms, the Chinese economy's capacity to absorb surplus rural labour is huge. Already, the work permit system which for long has kept China's rural labour force confined to the country-side is gradually relaxing; and migration to regions such as the Pearl River delta that are emerging as China's economic powerhouses is beginning to ease population pressure on land. The major agrarian challenge that China will face in the coming two decades is of producing enough food for its growing population so that its food deficits do not grow so large as to destabilize global food markets (Brown, 1995). Long term forecasts made by China IWRH[11] suggest that by 2050, 60% of China's population of 1.55 billion will be urban (compared to 25% now); irrigation water use will increase moderately from 391 km^3 to 399.1 km^3 due to growing use of water saving technologies;[12] industrial water use will grow manifold but water use per unit (specific product) of GDP will fall drastically from 100 to 200 m^3 to 20. China will face an acutely negative water balance until 2020 but demand supply balance will be restored by 2050. In this scenario, China's industrial growth rate of 6–7% per year will play a pivotal role in ultimately overcoming the water scarcity. (Zhang and Zhang, 2001).[13]

[11] Institute of Water Resources and Hydrology.

[12] According to Zhang and Zhang (2001: 236), between 1980 and 1993, water use in agriculture in the four major provinces in Yellow, Huai and Hai river basins in China fell by up to 6 km^3.

[13] That these projections are credible is indicated by the experience of Taiwan over the past 50 years. In most respects similar to mainland China, Taiwan has been an economic powerhouse and its economy has been transformed from a predominantly rural-agricultural to urban-industrial between 1960–1990. Taiwan has history of advanced irrigation; even in 1895, 57% of its farm land was irrigated. Irrigated farming got a boost in Taiwan during the Japanese occupation during 896–1947 when irrigated area increased to 570,000 ha. This trend continued upto early 1962 when irrigation peaked at 676,000 ha. With rapid industrialization, however, share of agriculture in GDP began falling rapidly and reduced to around 3%; moreover, although nearly 800,000 families farm land, only 13% are full time farmers; irrigated area fell 44% between 1962 and 1998 to 381,000 ha. With soaring incomes, people's food habits changed; and per caput rice consumption fell from 134 kg/yr in 1974 to 59 kg/yr in 1996 which also helped release water from agricultural uses (Sakthivadivel *et al.*, 2001).

South Asian economies are lagging substantially behind China on the economic growth trajectory; India, Pakistan, Bangladesh and Nepal are still far more rural, poor and agricultural. Despite free movement of labour between rural and urban sectors, population pressure on land and water in agriculture is high and will likely stay so for the next 20 years. In acting resolutely to make agricultural use of groundwater sustainable, the prime barrier in these countries is not so much food security but livelihood security. India is sitting atop a huge buffer-stock of 60 million tons of food grains; and it is expecting a big harvest of 210 million tons in 2002. If anything, India's growing mountains of food stocks are becoming an embarrassment for the government because they feed rodents more than poor people who do not have the purchasing power to buy it. Acting decisively to curtail groundwater use in South Asian agriculture may invite stiff popular resistance if it is seen to hit the incomes and livelihoods of rural poor households who depend more than India's better off farmers on groundwater to protect and improve their crops.

Major technological advances being available – especially, genetically modified (GM) crops – are far more effective in addressing the problem of food insecurity than improving livelihoods and incomes of poor people. China, which has embraced the use of bio-technology in agriculture enthusiastically, seems well on its way to enhancing its food production substantially in the next decade. However, South Asian economies will take longer periods of broad-based economic growth and transformation in order to shift sizeable chunks of South Asia's rural, agrarian population to urban industrial and tertiary sectors. In the interim, South Asian governments will tend to be lukewarm to any water management strategy that promotes environmental sustainability by putting rural livelihoods at risk.

Ironically, the nature of the political system many South Asian countries have may encourage their leadership to deal with the environmental challenge facing their water sectors with kid gloves. The parliamentary democracy in India, Nepal, Bangladesh, Sri Lanka and Pakistan seems to be at the heart of their *soft states*. The vote-bank politics here inspire a populist myopia amongst the political class that makes it difficult for them to take a hard stand on national issues. The inability of South Asian states to act on some fundamental aspects of water governance such as pricing water to at least meet operation and maintenance costs (O&M), reforming electricity pricing to save power sectors from ruin as well as aligning the cost of lifting groundwater to its scarcity value on the margin, instituting simple regulatory measures such as registration of wells and basic well-siting norms suggest their incapacity to choose what is rational and sensible over what is populist, and panders to the vote banks that can sometimes destabilize popularly elected governments.

The Chinese state, in contrast has been a *hard state* that has systematically transferred wealth from agriculture to build its industrial economy. There are several indications to this: at the county level, leave alone subsidies, farmers pay more for water and electricity compared to industrial users; agricultural taxes in China are a significant portion of the value of land, whereas agricultural land as well as income taxes are either absent (as in India) or levied at trivial rates (as in Pakistan). Since the 1960s public investments in agriculture has declined rapidly; for instance, the share of local governments in infrastructural investment (such as irrigation projects) has fallen rapidly from 63% in 1978 to 47% in 1985 (Wang and Huang, 2002).

2.7 *Summary*

Will national water administrations in Asia be in a position to act swiftly and decisively to protect their groundwater socio-ecologies? In our analysis so far, we argue that China has in place more of the socio-economic and institutional preconditions needed to make direct as well as indirect management work on the ground (see Table 3). In particular: a) even as these are weakening, China's village, township and county level governance structures play a more proactive and effective executive and regulatory role than comparable local governance structures in South Asia; b) China has in place, for well over a decade, a Water Law and a water permit system that are already enforced on industrial and municipal users where as South Asian countries

Table 3. Comparing water institutions and policies in South Asia and China: summary.

	South Asia	China
1. Does the village government have significant regulatory role?	No, except in Baluchistan	Yes
2. Are there significant taxes on agriculture? Are these collected?	No	Yes
3. Is there a system of registering and licensing groundwater structures? Is it enforced?	No	Yes; but not enforced strictly
4. Nature of the water bureaucracy?	Fragmented; thin presence	Less fragmented; but more presence
5. Water as an economic good; does water command an economic price?	No; most users pay a tax	Yes; most users pay a water price
6. Does the water administration have capability to enforce broad-spectrum measures?	No	Yes; rice cultivation in NCP completely eliminated
7. Are there institutional limits to *competitive deepening of tubewells*?	Only indirect; unenforced	Avoided easily, even with privatization
8. Adoption of water saving methods and technologies	Very limited	Extensive and growing
9 Macro-economic safety valves: Is there scope for shift of population from farm to off-farm livelihoods?	No; except in small pockets	Yes, with the work permit system liberalized.
10 Institutional reform: is the focus just on cost recovery or productivity and environment sustainability?	Focus on cost recovery through Irrigation Management Transfer (IMT)	Chinese water admin. in a *franchise mode* rather than IMT

are still debating a water law; c) the Chinese water administration is better integrated and has a greater and more effective grassroots presence and reach compared to South Asia where water administrations are fragmented, and have thin or no presence at the local level; d) the Chinese are much closer to transforming water into an economic good than South Asians; a large proportion of Chinese water use – domestic, industrial, agricultural – is paid for based on consumption or its surrogate; most South Asian water charge is aimed at recovery of O&M, and is collected as a *tax* rather than as a *price*; e) in search of viability for its water infrastructure and institutions, China is transforming its water bureaucracy into a business-oriented service provider which is likely to place water productivity at the centre stage; South Asia, in contrast, is trying to turn over irrigation management to water user organizations; such institutional reform may achieve better cost recovery but it is unlikely to mount effective regulation aimed at sustainable use; f) finally, with work permit system being liberalized, a rapidly industrializing China is likely to witness massive population shifts from water-stressed North to wealthy South and East, especially the Pearl River delta enjoying economic boom; China's industrial growth presents it with a safety valve to take population pressure off its irrigated land; and its challenge of producing enough food is easier to meet than of creating millions of rural livelihoods, which is South Asia's central concern.

To South Asian policy makers, the lessons China's experience offers are four: a) local resource management or community rule-making are unlikely to offer effective solutions to unsustainable groundwater use in the South Asian rural context where food and livelihood security are uppermost concerns of water users; effective regulatory frameworks and vigorous demand management require strong authority structures at micro, meso and macro levels; b) making a national water policy or groundwater law has no meaning unless it is underpinned at meso and local levels by institutional structures to implement these; c) the first essential step South Asian countries need to take in order to manage water better as an economic good is to start charging a price for it rather than a tax; to do this, two things seem essential: first, focus needs to expand from infrastructure creation to service provision and resource management; second, ways need to be explored to drastically reduce transaction costs of consumption linked pricing; in doing both these, the Chinese experience is valuable; and d) finally, in the medium to long term, a big part of the solution to the upcoming groundwater crisis is economic growth and urbanization, and shifting people from farm to off-farm livelihoods.

3 MEXICO: AGGRESSIVE REFORMS IN THE GROUNDWATER ECONOMY

Like India and China, Mexico too suffers from chronic imbalance of population and water availability in different regions. Arid and semi-arid areas of Mexico account for 76% of the population, 90% of the irrigated area, and 70% of the industries but these receive only 20% of Mexico's total precipitation (Barker *et al.*, 2000). As a result groundwater depletion is rampant in North, North Western areas and in the Mexico Valley. States like Sinaloa, Sonora, Guanajuato, Coahuila and Tamaulipas are water short but have intensive agriculture; Chiapas, Tabasco, Campeche, Yucatan and Quintana Roo are water abundant but have the bulk of Mexico's poverty. In the former, which constitute Mexico's food baskets, dealing with groundwater depletion is a critical policy issue that Mexico's water reforms have tried to grapple with.

Mexico's irrigation reforms – of which groundwater reforms are an integral part – are a product of its agrarian history and the larger program of restructuring the economy that began during the early1980s. The agrarian structure we find in Mexico today can be traced back to the series of peasant uprisings that culminated in the 1915 revolution and ensuing 1917 constitution. The far reaching land reforms – driven by the principle *land belongs to those who work it* that were ushered in by the 1930s but that in fact took decades to consummate, declared the Mexican state as the custodian of all land and broke up large feudal estates into 100–800 ha holdings. Two different forms of land rights followed – *pequeña propiedad* (*small* private property) and the *ejido* (or agrarian collective). The former had unattenuated ownership rights over land; the *ejidatarios* (or *ejido* members) got a legal identity but had only usufruct on land; they could use and inherit land but not mortgage or sell it. Up to 1983, 25,589 *ejidos* were formed.

Mexico enacted its first Irrigation Law in 1926; this was replaced by a Federal Water Law in 1972. But it was the Law of the Nation's Waters of 1992 combined with an amendment to article 27 of the constitution in the same year that became a watershed in Mexican agrarian as well as water reforms. Up until 1989, all irrigation was managed by the Ministry of Agriculture and Hydraulic Resources; and like in India, the government policy towards agriculture and irrigation was guided by the socialist thinking of a welfare state. The reform process pursued four fundamental and far reaching aims: a) make water infrastructure self-financing by withdrawing the government from its management; b) improve the efficiency of water use by establishing tradable private rights on water as well as by involving users in managing water infrastructure; c) restrict and even reduce groundwater depletion by the CNA (*Comisión Nacional del Agua*) operationalizing the authority to issue rights (concessions) to draw groundwater and by enforcing the concessions; d) achieve basin level optimality in water use through basin level co-ordinating mechanisms. Did Mexico's reform process achieve all these aims? A discussion of this question is presented elsewhere (Shah *et al.*, 2002b); here we focus on how far have Mexico's water reforms helped achieve sustainable management of its groundwater economy.

Before 1992, groundwater rights in Mexico were tightly linked to land rights, much like in Asia today (Wester *et al.*, 1999). There was some discussion of creating private water rights separate from land rights during the 1980s itself; and a National Registry of Water Rights was created well before the sweeping reforms in water sector took place in 1992. In 1989, the National Water Commission (or CNA), was created as the first step to separating the management of water from that of the agrarian economy, recognizing the declining role of agriculture in Mexican economy and growing non-agricultural demand for water.

The new Law of the Nation's Waters aimed to: a) "provide for administrative modernization, planning and programming" in the water resources sector; and b) "reinforce a more efficient and rational use of natural resources". The National Water Registry was charged with the responsibility to maintain a national register of newly created private property rights in water. The design manual of the CNA provided that no user could impound or divert more than 1,080 m^3/yr of water except by obtaining a *concession* from the CNA. In sum, all water used for purposes other than domestic personal use, had to be *titled*.

Thus, Mexico has sought to create tradable private property rights in water by: a) first, declaring water as national property, thereby severing the linkage between land rights and water rights; b) allowing existing users to get their use *regularized* by obtaining a concession from the CNA; c) by setting up a structure for enforcing the concessions; and d) by levying a volumetric water fee from concession holders (barring irrigators) which would help generate resources to maintain water infrastructure. Under the new Water Law, all diversions of water other than for direct personal use are allowed only through concessions. Even sand-mining in river beds – these are considered Federal property – requires a concession. Concessions for different users, uses and sources are for different periods and specified volumes. The Law enjoins the concession holders to abstain from over-stepping the agreed volumes, to establish mechanisms to measure volumes used and report these periodically to the CNA.

What has been the outcome and impact of this rights reform? Mixed, as of now. Large water users, especially industrial and commercial establishments have been quick to secure proper concessions and pay water fee to the CNA. This has been a significant source of revenue for the CNA. Surface irrigation associations (Water User Associations or WUAs) are few, organized and therefore easy to bring within the purview of the concessions; and since each WUA holds a concession on behalf of its members, it is administratively simple to formalize their water rights. Municipal Councils similarly are to obtain concessions that cover all users within their ambit. By and large, municipal diversion has conformed to the volumes they are entitled; however, municipal Water Boards have regularly defaulted on the payment of water fees to the CNA which recently had to write off Mex$ (Mexican peso) 72 billion (US$ 6.9 billion)[14] owed by them to it by way of accumulated water fees. One expectation was that the new system of rights would stimulate an active market in water; however, this expectation has been largely belied because "water rights are not rigidly enforced and legal processes to redress grievances are difficult, costly and drawn out" (Scott *et al.*, 2001).

The real difficulty has been with water rights of numerous agricultural users who account for over 80% of the water use and seem to be at the heart of the matter. In particular, there are three problems: a) getting agricultural users to get *regularized* by obtaining a concession; b) coping with the administrative workload involved in processing applications for concessions and issuing them; and c) enforcing the terms of the concession. Even amongst agricultural users, tubewell irrigators have responded to the Law quite well. Most tubewell irrigators we interviewed, on private farms as well as in *ejidos* held – a concession or had already applied for one. One reason perhaps is that tubewells in Mexico are quite large, by Asian standards. A typical tubewell in Guanajuato goes to a depth of 150–250 m, has a lift of 60–90 m, and has a 75–150 HP motor-pump and a 6″ outlet pipe yielding 30–60 L/s. Thus, a typical tubewell may have a command area of 40–80 ha; only large private farmers have individual tube wells; most *ejidatarios* share tubewells through informal *well societies* similar to the tubewell partnerships and companies found in North Gujarat (Shah and Bhattacharya, 1998).

[14] Mex$ 1 = US$ 0.09587 (July 1, 2003)

Another reason why tubewell owners keenly seek *regularization* by securing concessions is that they are linked to the formal economy through their dependence on the Federal Electricity Commission for power supply. The Federal Electricity Commission would require a concession before issuing an electricity connection for a new tubewell. Then, there is also an incentive for existing tubewell owners. Power supply to agricultural users in Mexico is subsidized; farmers pay around Mex$ 0.23–0.28/kWh (US$ 0.02–0.03/kWh) against the average power tariff of Mex$ 0.55–0.65/kWh (US$ 0.05–0.06/kWh). And although the CNA and the federal government have yet not used that stick, they have certainly issued threats that tubewells without concessions would attract commercial power tariff, while *concessioned* tubewells will keep enjoying subsidized tariff. This is a major factor; an average tubewell in Mexico probably uses 50,000–80,000 kWh of power in a year; and access to power subsidy at current rates would mean a saving of Mex$ 12,000–18000/yr (US$ 1,100–1,700/yr) in their electricity bill, high enough to make it worth getting the concession.

However, it is one thing to issue a concession to a tubewell; it is quite another to specify its volumetric water right and yet another to limit its pumping to the volume specified. The *concession* in itself is nothing more than the registration of a well, which is easily done from the records of the Federal Electricity Commission in the Mexican context where all groundwater pumping is done by electric pumps. The creation of a water right lies in entitling each concessioned tubewell to a particular volume of extraction. We found, however, that the volumes entitled are based on a combination of the current use implicit in the yield of the well and the area owned. Thus, groundwater concessions merely regularize the status quo and do not aim to curtail present levels of groundwater use, except through ban on new tubewells which can be more efficiently imposed simply by putting a cap on new agricultural power connections.

Monitoring the actual extraction and enforcing it to *entitled volumes* has proved impossible even in a small state like Guanajuato where agricultural tubewells are all of 15,000 in number. The CNA has legal powers to undertake surprise inspections and monitor water use under concessions. However, it has only 2 field teams in Guanajuato; and if these were to make a single inspection visit to each irrigation well, it would take several tens of years to complete one round. Now, the state CNA has got 7 brigades of 2 members each against a request for 20 brigades. This is better, but is still much less than what is needed to begin to monitor actual groundwater extraction. In law, concessions are supposed to forfeit if the concessioned volumes are not used by the holder; however, this provision can be enforced only if there is regular monitoring of water use by concession holders. This is proving well nigh impossible; and there is already talk of extending the ambit of the *environmental police force* –already created at the Federal level primarily to enforce industrial pollution – to cover groundwater extraction.

Compared to tubewells, a far trickier animal is the *bordo*, a small tank-like water harvesting and storage structure – and *presas that* are somewhat larger – which have been proliferating in uplands of Mexico at a frightening pace.[15] *Bordos* and *presas* too are growing especially in up-land areas with intensive livestock farming for meat or dairying. In Guanajuato alone, around 200 large *presas* are organized as *Unidades de Riego*[16] – nominally controlled by the state agriculture department but are in fact farmer controlled and managed as much as smaller *bordos* and *presas* are. If the tubewells listed as *unidades* (because they have received some government assistance for drilling, etc.) are included, together, these

[15] IWMI estimated their number at 29,000 in late 1990s (Scott and Flores-López, submitted). The State Water Commission believed that although *bordos* are traditional structures, a large majority of these came up during the past 10 years as popular response to growing water scarcity. Local farmers we interviewed supported the view that a majority of *bordos* found today are less than 8–10 years old.

[16] These are used by rainfed farmers essentially to get one irrigation to establish the rainfed sorghum crop.

informal water structures irrigate more land in Guanajuato than all the WUAs do together. Under the new Water Law, each of these structures needs a concession; but most, as yet, do not have them. *Bordos* and *presas* present a catch-22 situation for the Mexican experiment in creating private water rights: if their owners persistently avoid applying for concessions, the intent of the Water Law will be frustrated in substantial ways. However, if they begin applying for concessions in large numbers, it may raise important issues of administrative logistics as also of equity and integrated river basin management that Mexico is trying to achieve.

In the hilly upland areas of Mexico, and the catchment areas of major river basins, *bordos* have emerged as the backbone of a rainfed crop-livestock farming system. Conditions in these hilly upland areas are worse than in the plains. The new Water Law, under which all water bodies are required to be concessioned by the CNA has created enormous confusion for owners of the *bordos* which store 5,000 to 50,000 m^3 of water. Besides finding it pointless, upland farmers we interviewed were worried about the hassle and transaction costs which are out of all proportion to the value of *bordos*. Concessions have set into motion a new race for privatizing the rain water run off. Another major concern was also about how the Water Law hits the poor in the remote areas particularly hard. The government keeps issuing ordinances and new time limits for compliance. But people in the remote areas do not even know about these for months and get left out. In the meanwhile, the smart and aggressive use these proactively to entrench and strengthen their positions by legalizing them. They have found that getting concessions is an easy way of establishing private rights over what was so far open access run off.

This kind of mass manipulation also occurred in the *bajío* areas, the low lands of south-central Guanajuato for their intensive groundwater use in agriculture. Here, groundwater depletion is a 50 year old problem; the first ban on new groundwater structures was announced way back in 1948; and since then, 14 such bans have been issued. However, every announcement of an imminent ban – or injunction to regularize existing tubewells such as in 1995 – here stimulated a flurry of tubewell making activity in the hope that if made before the deadline, they would get regularized. Indeed intended bans and injunctions for regularization can be counted as one of the chief reasons for the run away rise in tubewell density in central Guanajuato. One such injunction was issued without a time limit in 1995; another with a time limit in 1996; and yet one more was issued on February 2002 with time limit up to September 2002. Farmers also used other ways to manipulate the concession-grants. Many made new wells in the name of *repositioning*. Francisco García, a senior CNA official lamented that in 2001 against 250 applications for repositioning wells, 1,000 new wells were made commonly with power connections drawn from a concessioned transformer.

The CNA's decision to form and support COTAS (Aquifer Management Councils) was born out of the recognition that concessions and private water rights by themselves would be of little help in getting the water users in the *informal sector* to play the ball-game of sustainable water management, and that new mechanisms and structures need to be experimented with to engage this vital sector in implementing the spirit of the Water Law. To their protagonists in the CNA, COTAS were government promoted NGOs fashioned as user organizations; and Guanajuato, where these early experiments first began under the leadership of Governor (now President) Fox, continues to lead Mexico's COTAS experiment to date. Of the 47 COTAS in Mexico, 14 are in Guanajuato, one for each of the 14 aquifers delineated in the state. Now COTAS have been adopted as a national model and the CNA is promoting them in the rest of the country. However, federal COTAS differ from Guanajuato COTAS (Technical Councils for Water Management) in that the latter are termed water management *councils* that sound more inclusive where as the federally promoted COTAS (Technical Committees for Groundwater Management) seem limited in their scope. Guanajuato COTAS concern themselves with managing all water resource; COTAS in other states focus squarely on groundwater. Guanajuato COTAS are also supported more liberally with state financial support; each is provided a rented office, a car and salaries for a Manager, a technician and an administrative assistant.

Federal COTAS have far more meager support from the CNA. Everywhere, however, COTAS have key design features that are common: their operational domain is defined by an aquifer boundary, which clearly gives primacy to their groundwater management role; they are all designed as representational non-profits; registered as a Civil Association, each has a general assembly, an elected board and a small hired staff. Recently, all the Guanajuato COTAS were federated in to a State Water Management Council with a representational structure akin to a COTAS. The Office of the Guanajuato Water Resources Council (CEH) is the organization that represents all water users in the state. In its evolutionary process, the State Council first brought the 14 COTAS together in this representational structure; but its ultimate goal is to bring all water users/stakeholders into the forum. They already have 6 representatives of surface irrigators now, 4 from two important irrigation districts of the state and 2 more to represent the 200 odd *Unidades de Riego*.

The idea of COTAS is bold; and the expectations from these structures high. A COTAS is expected "to be an IWRM[17] promoter in the state bringing together different actors and stakeholders to protect the water resources in quantity and quality". The State Water Commission of Guanajuato (CEAG) expects that a COTAS should become a local water management organization, and will mature to a stage where it becomes a rallying point for all water users; that as they get formally recognized by the Water Law (which for the present they are not), they will come up with and implement practical water management and conservation actions and policies; they will mediate water conflicts; enforce/implement national water policy on the ground level (Sandoval, 2002). A common expectation is also that the COTAS – particularly, their state-level federation – will become a powerful instrument of implementing the Law of the Nation's Waters; they will interact with authorities and water regulatory agencies and provide decisive inputs on the creation, establishment, control and changes in water management plans. Above all, COTAS are expected to mediate between the state and the federal water authority and water users they represent. This is why COTAS were designed as representational organizations.[18] The sub-text in all this is that with their closer grassroots presence, COTAS will do what the CNA can not: restrict groundwater extraction by enforcing the Water Law.

Will Mexico's COTAS fulfill these multifarious, often conflicting expectations? It is early days to say; COTAS even in Guanajuato, the state that pioneered them, are all of 4 years old; and according to Francisco García, Deputy Director of Water Administration, CNA, Guanajuato, COTAS will take time to become effective. "After all, Texas took 16 years to constitute its first aquifer management organisation through a state assembly decree, and 5 more years to actually put it on the ground. Mexico's COTAS need to be given time to congeal and find their feet". Guanajuato's COTAS have until 2004 to find their feet; after that, the financial support from the State's Water Commission will cease; and COTAS left without alternative sources of funds will have to liquidate their operating systems, and will in effect cease to exist.

Stuck in such a situation, the normal propensity of a member organization would be to turn to its members for sustenance; it would begin providing services that its members value and

---

[17] Integrated Water Resources Management.

[18] In a typical COTAS in Guanajuato, accordingly, the general assembly elects a 10 member board that has President, Treasurer, General Secretary, one representative each from agricultural users, public service (which includes domestic and municipal users) and industrial users. A back up candidate is elected for each of these which takes the total board size to 10. The General Assembly generally meets twice every year; the board meets every month. There are few women on elected boards; however, 5 are hired as employees by different COTAS. In keeping with their broader, more ambitious mandate of IWRM, the Guanajuato COTAS drew representatives from various stakeholder groups although their domain was defined by the aquifer boundaries. In these 14 COTAS, 12 of the 29 elected representatives of farmers represent surface water irrigators where as 17 represent groundwater users. In theory, only concession holders are official members of the general assembly of a COTAS; however, in practice, the State Water Commission staff ends up spending a great deal of effort in getting all users to participate.

in turn expect them to contribute fees for such support. This is what Guanajuato's *módulos* (WUAs) do; for instance, as a member organization, the Irapuato *módulo* offers its members better irrigation service and has jacked water fees five times in five years, partly to fund its own growth and partly to improve the services. A fundamental design flaw in COTAS may well be its concept itself: it is not allowed to provide what a majority of its members value most, *viz.*, unrestrained access to groundwater; and its members are reluctant to want to pay it membership fees for enforcing the Water Law on them which its creators think is the mandate of the COTAS. It is not surprising then that industrial players – whose water use was closely regulated even before the new Law – have been quick to take to the COTAS and even dominate them; but the farmers, the prime target of the Water Law's groundwater provisions, have been staying away from the COTAS.

As a result, COTAS are ploughing along without a strong sense of direction. Most have no notion of formal membership. With its 20,000 concession holders, Guanajuato's 14 COTAS should each have 1,000–15,000 members with full user participation;[19] but their general assembly meetings often have a few dozen participants. The Ocampo COTAS, one of the few which offers formal membership has less than 100 members of the several hundred groundwater concession holders (Ocampo is not an important groundwater irrigation region). COTAS are little known amongst common people; and their presence on the ground is thin or non-existent. Some 45 farmers we interviewed in various parts of Guanajuato these included all types, small holders as well as large farmers, men and women, a few young and mostly old farmers – all were uniformly blank on COTAS. Most COTAS boards were elected by a general assembly attended by a small fraction (often 5–10%) of total concession holders.[20] Partly for that reason, the office bearers of COTAS enjoy little regard and allegiance of the wider public and citizenry, neither do they seem under pressure to respond to an aggregate of member priorities from COTAS. Many elected office bearers of COTAS seek to pursue their own ideals or have their own passions and bees in bonnets, and drive their COTAS in that direction rather than working on aggregated priorities of members, as would be the vogue in a responsive member organization (Shah, 1996). For instance, the president of a predominantly agricultural COTAS has been able to focus all its work on issues related to industrial water use that he feels strongly about although most concession holders in the COTAS domain are farmers.

A major reason for member apathy is that the high-ground assumed by the COTAS leadership often fails to connect with the here-and-now priorities of its members. In Jaral de Berrios, one of the best performing COTAS according to Guanajuato Water Commission, only a couple of dozen farmers, all above 70 years of age participated in a council meeting to strategize for groundwater management.[21] Concerned about

[19] As a matter of fact, the State Water Commission (CEAG) has tried to break out of the norm that only concession holders can be COTAS members; it has been trying to broaden participation into COTAS affairs from wider cross-section of the citizenry.

[20] For instance, in the 4-year old Silao and Romita COTAS which also covers the town of Guanajuato, the general assembly included 2,049 well owners, 93% of them agricultural, should reflect farmer concerns. However, a general assembly attended by some 100 members elected a manager of the General Motors as the President, a manager of the Leon airport as the Treasurer.

[21] Jaral de Berrios aquifer, shared by Guanajuato and the neighbouring state of San Luis Potosí, faces critical problems of over draft but has only a few hundred wells; the Guanajuato side in fact has only 334 wells, 89% of them agricultural. Only 200 of these are concessioned yet; others have applied some years ago and were waiting to receive their concessions. These large tube wells, going to 250–280 m and using pumps of 75–100 HP produce discharges of 25–40 L/s which they deliver into a large tank and from thence, water conveyed by buried pipes to different fields. Typically, a well irrigates 30–35 ha and if all users use drip irrigation, the area irrigated can go up to 55–60 ha. However, only 3% of the tubewell irrigated area uses drip; and farmers feel reluctant to *technify* their irrigation systems because of the dismal after-sales support of irrigation equipment companies. Many would also expect government support to install such technologies. Wells are private as well as group-owned, the latter common among *ejidatarios*.

the bleak agricultural future of the region, the COTAS president, a large private land holder, delivered an impassioned speech advocating the need to restrict groundwater use by regulating the area under tube well irrigation, and presented an elegant formula to link total groundwater draft to the previous year's rainfall. The old farmers in the audience were unmoved; one 75 year old *ejidatario* got up and said: "my farming is already down to 2 ha; how much more do you expect me to cut?" Another rose and said: "it took me 6 years after making an application to get my concession; and by then, my tubewell needed to be deepened and I was ready for a new concession. Can't the COTAS help us cut through this maze?" Yet another farmer described how he put his life's savings in an expensive drip irrigation system, which failed and irrevocably damaged his well and pump, due to lack of technical support.

Many people we met thought COTAS have done well as *talking shops*. Numbers support this suggestion; during the first half of 2001, all 14 COTAS together organized some 30 meetings in Guanajuato. But the participation often tends to be thin; indeed, an important criterion the State Water Commission (CEAG) uses in judging the appeal and robustness of a COTAS is the level and extent of participation in its meetings. Some COTAS seem to have served well as a public platform for raising and debating water issues. In Silao-Romita, the major purpose the COTAS has served so far is in mobilizing the community to protest the transport of groundwater from its aquifer to Leon, the largest industrial city of Guanajuato, a transaction wholly valid within the Law of the Nation's Waters.[22] Many COTAS are deeply into research, training and capacity building as a core activity, somewhat unusual for a member organization which exists to provide services that members would pay for. Valuable as these service may generally be, it is doubtful if COTAS members will be willing to pay for them, and that at a rate that would enable the COTAS to survive. As a result, managers and staff of the Guanajuato COTAS – arguably, the stakeholder group most concerned about the future survival of the COTAS – are exploring strategies of resource generation from sources other than their members. Many will continue to look up to the CNA and state governments for continued support. Laguna Seca COTAS has been planning research collaboration with the Ecology Institute and State Department of Agriculture which the CNA will probably support.[23] The Silao-Romita COTAS hopes that a research collaboration with Guanajuato University's Agricultural Research Station on researching water saving irrigation technology will help it generate resources. The Ocampo COTAS has planned a range of services and activities to generate resources from members: such as registering *bordos*, tanks, wells, etc. with the CNA so that members have secure rights; training and technical support in irrigation, *bordo* construction; help members deal with the CNA's titling process and concessions; help in installing small treatment plants at the community level. However, with its current human resource base, it is open to question whether the Ocampo COTAS can fructify this plan, or even a fraction of it.

In sum, the present role and future direction of the COTAS are unclear to say the least. The CNA expects them to implement the Water Law, in particular, help contain groundwater extractions to concessioned limits, and help curb illegal well-drilling. Ambitious COTAS presidents, such as in

[22] Leon has its own groundwater concessions; however, in order to meet its growing municipal demand, Leon municipality (SAPAL) purchased groundwater rights from 3 well owners amounting to about 600,000 m^3/yr for Mex$ 6 million (US$ 575,000), at Mex$ 10/m^3 (US$ 0.96/m^3). They now intend to set up a 30 km pipeline to transport water to Leon. The Silao-Romita COTAS is averse to the idea primarily because it is not sure that the Leon municipality can be restrained to pump only the concessioned quota. It is difficult, if not impossible, to continuously monitor and account for the pumpage from each well; they suspect that once it is allowed to pump, Leon can pump the aquifer dry.

[23] For a member organization to police and spy over its own members would be a curious role indeed. This is what Laguna Seca has been doing. This COTAS has already put its head in the lion's mouth and declared to the CNA several farmers making illegal wells clandestinely. CNA can never find these out; but being closer to the field of action, COTAS can; and Laguna Seca COTAS tipped off the CNA while the clandestine wells were being made.

Jaral de Berrios, want to transform the COTAS into a strong water user organization that can mediate between the users and the authorities. Many COTAS managers view their role as one of promoting IWRM. There is no indication yet that COTAS are ready to play any of these roles. However, what they have been doing may not be without value. Many COTAS have been monitoring water levels; most have been carrying out water education campaigns. They have served as forums in which users can participate in discussing their water problems. And others have been trying to promote technification. At least in one COTAS, farmers shifted wholesale from cultivation of wheat to barely which uses less water.[24] In any case, regulating agricultural use of water, especially of groundwater, is a challenge that has nowhere been met fully; and perhaps, the CNA will be well placed to support COTAS for a long time to come with full recognition that they will not be able to achieve the ulterior goal behind the CNA support, *viz.,* to help CNA implement the provisions of the Water Law on the informal water sector. Considering that the 14 Guanajuato COTAS have cost the state Water Commission (CEAG) less than US$ 2.5 million to support for five years,[25] one can easily argue that the capacity building and attitudinal impact COTAS can produce through targeted research and public education activity may justify such investment in view of growing importance of water in Mexico's evolution. An early vision of the COTAS was that they would foster self-policing by users themselves taking the responsibility of self-monitoring his extraction to the agreed volume. Even though idealized, some believe that such a scheme can work in Mexico aided by European style *water notaries* that might be used to certify the actual extraction. We believe many conditions will need to be fulfilled before such a schema might work reasonably well; one of these is high quality public education on groundwater issues. And COTAS are certainly equipped to deliver this.

4 CONCLUSION

In this paper, we have reported results of field research on groundwater management institutions and policies in three regions of the world where agriculture, food and livelihoods depend heavily on intensive use of groundwater which is becoming increasingly unsustainable. Our purpose was to review institutions and policies in place to promote sustainability and draw lessons from comparative analysis. Table 4 summarizes our key conclusions from such a comparative analysis. Our overriding impression is that South Asian countries have not even begun to address the problem in any serious manner; China has but will take time before its initiatives bear fruit. Mexico has gone by far the furthest in creating a legal and property rights structure that might be drawing a leaf from an institutional economics text book. Interestingly, we find no evidence that these have helped Mexico move towards sustainability; and that Mexico's efforts need to produce better results before they can be held out as a model that other groundwater-using countries can follow. However, our comparative analysis does suggest the outline of a framework that tells us what might work where.

How countries respond to the challenge of sustainable management of their groundwater

[24] This is a great achievement; however, besides the educational effort of COTAS, the key catalyst to this change has been the establishment of a Corona beer plant in its territory that created a steady, remunerative market for barley far better than wheat. This suggests that in the complex business of groundwater regulation, an ounce of positive incentive may do the work of a ton of regulatory effort.

[25] The cost of catalyzing and sustaining COTAS in Guanajuato has been met by the State Water Commission (CEAG) through providing them budgetary support as follows:

Year	Support to 14 Guanajuato COTAS
1998	US$ 153,471
1999	US$ 459,184
2000	US$ 607,142
2001	US$ 510,204
2002	US$ 766,490

Table 4. Groundwater Governance: comparative analysis of institutions and policies in South Asia, China, and Mexico.

	South Asia	China	Mexico
1. Government share in GW provision to agriculture	Miniscule; < 0.01%	No	No
2. State provision of GW to urban settlements	Significant	Significant	Significant
3. State participation in GW monitoring	Yes	Yes	Yes
4. Incentives to private investment in groundwater development	Significant in India and Sri Lanka, often perverse; discontinued in Pakistan, Nepal, Bangladesh	None or insignificant	None
5. Incentives to operating costs	Huge in India; less in other countries	Nil or insignificant	Yes, energy subsidies
6. Targeted disincentives in capital or operating costs	None	None	None
7. Registration of GW structures	No	No	Yes
8. Permits to abstract groundwater	No	Yes, but mostly to villages, municipalities and industries	Yes, but water quantities unenforceable
9. Promotion of water saving technologies	Ineffective	Yes, strong	Some
10. Promotion of small-scale water harvesting and recharge works	Strong in Western India; but growing elsewhere in India	South-North water transfers	Yes, in highlands where *bordos* are the mainstay of livestock farmers

economies depends on a constellation of factors that defines the peculiar context of each country. This constellation of factors differs vastly across regions and countries; and these differences have decisive impact on whether an approach that has worked in one country will work in another with a different context. As a simple illustration of this point, Table 5 sets out some key variables that define the organization of the groundwater economy in six different countries which make intensive use of groundwater in agriculture. The USA uses around 100 km^3 of groundwater for irrigation; but to manage its economy, it has to monitor and regulate only around 200,000 pumping plants, each producing around 500,000 m^3/yr of groundwater. Mexico is in the same league as the USA. India uses 150 km^3; but to manage this groundwater economy, it has to manage the owners of over 20 million small wells, each producing an average of 8,000 m^3/yr of water. Clearly, the task of USA groundwater managers is enormously simpler compared to their Indian counterparts. With just 95,000 agricultural tubewells, the task of governing Mexico's groundwater economy is even simpler.[26]

The nature of the political system also matters. Iran has been able to impose a complete

[26] However, in actuality, even in Mexico, enforcing concessions on agricultural tubewells has proved almost impossible. Similar is the experience even in Spain where with the passing of the 1985 Water Law, it was declared as of *public ownership*. This represented a fundamental change in relation to water rights. Yet this drastic change, compounded by

Table 5. Structure of national groundwater economies.

Country	Annual groundwater use (km^3)	No of groundwater structures (million)	Extraction/structure (m^3/yr)	% of population dependent on groundwater
India	150	19	7,900	55–60
Pakistan-Punjab	45	0.5	90,000	60–65
China	75	3.5	21,500	22–25
Iran	29	0.5	58,000	12–18
Mexico	29	0.07	414,285	5–6
USA	100	0.2	500,000	< 1–2

ban on sinking of new tubewells throughout its central plains that encompass 2/3 of the entire country (Hekmat, 2002).

But as we noted, Mexico has been trying to ban new tubewells in its *bajio* for 50 years, and has yet not succeeded. China has a large number of tubewells scattered over a huge country-side; yet chances are that over the coming decade, it will be able not only bring these within the ambit of its permit system but also succeed in influencing their operation. Doing something like this in India or Pakistan will remain unrealistic for a long time to come because of their political structures and systems.

Besides what is feasible and practical, there is also the question of social impacts of approaches adopted. In Mexico and the USA, where a miniscule proportion of people depends on groundwater for livelihoods, governments may easily adopt a tough regulatory posture. In South Asia, where over half of the total population may directly or indirectly depends on groundwater use for their livelihood, it is not surprising that political and administrative leadership is reluctant to even talk about regulating groundwater use, leave alone acting on it. In point of fact even in China, where political resistance from farmers is not an overriding issue, and Mexico where irrigator class is small enough to be ignored, governments have steered clear of tough regulatory measures.

Table 6 lists a tentative set of *contingencies* that seem to influence the way different countries respond to groundwater over-development. Countries where public systems will aggressively manage the groundwater economy by proactively intervening in demand and well as supply side will have some of all of the context factors in the middle column aligned in an enabling mode, as outlined in the right-hand column. Where some or all of the context factors operate in a disabling mode, public intervention will tend to be absent, or half-hearted or even perverse; here, proactive response to groundwater depletion will commonly be in the form of projects to enhance supply rather than containing demand. This is perhaps why no amount of opposition from within or outside will deflect China from its mega-project for South-North water transfer; and no matter how much scholars emphasize the upstream-downstream externalities of decentralized water harvesting and recharge, governments and communities in Western and Southern India will for long pursue these proactively as a strategy of sustaining groundwater irrigation and a more equitable allocation of a basin's water between catchment areas and downstream irrigation commands.

lack of knowledge and a poor information campaign (in relation to the legal changes and to groundwater use) has led to many situations of *hydrologic disobedience* in relation to water rights and abstraction in almost every stressed aquifer. Indeed the question remains as to what came first, hydrologic disobedience or stressed aquifers. A typical example of this situation is the Upper Guadiana basin (López Gunn and Llamas, 1999).

Table 6. Overall context and national strategies for groundwater management.

Under-managed resource with accent on supply-side measures	*Larger Social and Political Context*	Conducive to demand and supply-side management
Weak and unwilling to implement hard measures (India, Sri Lanka, Pakistan, Nepal)	*Political System: central and local authority structures*	Capable of tough measures (e.g., Iran, China, Pakistan under early years of military rule)
Numerous small users (South Asia, North China plains)	*Organization of the groundwater Economy*	Few large users (as in USA, Mexico, Iran)
Agricultural contribution to GDP > 30–50%; population dependent on farming > 50% (South Asia)	*Stage of economic development*	Agricultural contribution to GDP < 10%; population dependent on farming < 20% (USA, Mexico, Spain)
High (South Asia)	*Relative significance of groundwater economy to national and household food and livelihoods security*	Low (USA, Mexico, Spain)
Water rights as an easement of land ownership (Asia)	*Structure of property rights on land and water*	Water rights independent of land rights (Mexico's concessions)
Low (South Asia)	*Experience and effectiveness with using law to regulate people's behaviour*	High (Europe, USA)
High (India, Iran)	*Perverse incentives in GW irrigation (energy subsidies; tubewell subsidies)*	Low (China, Pakistan, Mexico)
Low (South Asia)	*Economics of groundwater irrigation: benefit-cost ratio*	High (as in North China, Mexico)
Low (South Asia)	*Capacity, reach and effectiveness of water bureaucracy*	High (China, Mexico)

REFERENCES

Barker, R. and Molle, F. (2002). *Perspectives on Asian Irrigation.* Paper presented at the Conference on Asian Irrigation in Transition-Responding to the Challenges Ahead. Asian Institute of Technology. Bangkok, Thailand, 22–23 April 2002.

Barker, R.; Scott, C.A.; de Fraiture, C. and Amarasinghe, U. (2000). Global water shortages and the challenge facing Mexico. *International Journal of Water Resources Development*, 16 (4): 525–542.

Boxer, B. (2001). Contradictions and Challenges in China's Water Policy Development. *Water International*, 26 (3): 335–341.

Brown, L. R. (1995). *Who will feed China? Wake-up call for a small planet.* WW Norton and Co. New York, USA.

Chen, Y.D. (2001). Sustainable Development and Management of Water Resources for Urban Water Supply in Hong Kong. *Water International*, 26 (1): 119–128.

Deb Roy, A. and Shah, T. (2001). *Socio-ecology of Groundwater Irrigation in India.* IWMI-Tata Water Policy Research Program. Anand, India.

Down to Earth (2002). *Ground Reality*, 11 (1): 38. May 31.

Dubash, N. (2002). *Tubewell Capitalism: Groundwater Development and Agrarian Change in Gujarat.* Oxford University Press. New Delhi, India.

Godbole, M. (2002). Electricity Regulatory Commissions: the jury is still out. *Economic and Political Weekly*, Vol. XXXVII (23): 2195–2201, June 8–14.

Hekmat, A. (2002). *Overexploitation of groundwater in Iran: need for an integrated Water Policy.* Paper for the IWMI-ICAR-Colombo Plan sponsored Policy Dialogue on: Forward-Thinking Policies for Groundwater Management: Energy, Water Resources, and Economic Approaches. Organized at India International Center. September 2–6. New Delhi, India.

Janakarajan, S. (1992). Inter-linked Transactions and Markets for Water in the Agrarian Economy of a Tamilnadu Village. In S. Subramanian (ed.), *Themes in Development Economics: Essays in honour of Malcolm Adiseshiah.* Oxford University Press. Delhi, India.

Kendy, E.; Steenhuis, T. and Liu, C. (2002). *Policies Drain the North China Plain: Hydrologic Impacts of Institutional Changes Affecting Luancheng County, Hebei Province, 1949–2000.* Unpublished draft. Department of Biological and Environmental Engineering, Riley Robb Hall, Cornell University, Ithaca, New York, USA.

Kolavalli, S. and Chicoine, D.L. (1989). Groundwater Markets in Gujarat, India. *Water Resources Development*, 5 (1).

López Gunn, E. and Llamas, M.R. (1999). *New and old paradigms in Spain's Water Policy*. Paper presented in Forum of the UNESCO International School of Science for Peace on: Water security in the Third Millenium. Mediterranean countries as a case. April 12–15. Villa Olmo, Como, Italy.

McClelland, D. (1985). *Achieving Society*. Replica Books. Boston, USA.

Meinzen-Dick, R. and Sullins, M. (1994). *Water Markets in Pakistan: Participation and Productivity*. International Food Policy Research Institute. Washington, USA (unpublished).

Mellor, J.W. (ed.) (1995). *Agriculture on the road to industrialization*. Johns Hopkins Press. Baltimore, USA.

Palmer-Jones, R. (1994). Groundwater Markets in South Asia: A Discussion of Theory and Evidence. In M. Moench (ed.), *Selling Water: Conceptual and Policy Debates over Groundwater Markets in India*. VIKSAT-Pacific Institute-Natural Heritage Institute.

Ronghan, H. (2000). Development of Groundwater for Agriculture in the Lower Yellow River Alluvial Basin. In H. Ronghan, C. Lingen and J.E. Nickum (eds.), *Case Studies in Conjunctive Use: The North China Plain*.

Rozelle, S.; Park, A.; Huang, J. and Jin, H. (2000). Bureaucrat to Entrepreneur: the changing role of the State in China's grain economy. *Economic Development and Cultural Change*, 48: 227–252.

Sakthivadivel, R.; Aloysius, N. and Matsuno, Y. (2001). *Assessment of Performance and Impact of Irrigation and Water Resources Systems in Taiwan and Sri Lanka*. International Water Management Institute. Working paper 31. Colombo, Sri Lanka.

Saleth, M.R. (1994). Groundwater Markets in India: A Legal and Institutional Perspective. *The Indian Economic Review*, Vol. XXXIX (2): 157–176.

Sandoval, R. (2002). Los COTAS de Guanajuato en el contexto del manejo del agua en Mexico. *Aqua Forum*, 26: 3–8.

Scott, C.A.; Silva-Ochoa, P.; Florencio-Cruz, V. and Wester, P. (2001). Competition for water in the Lerma-Chapala basin. In A. Hansen and M. van Afferden (eds.), *The Lerma-Chapala Watershed: Evaluation and Management*. Kluwer Academic/Plenum Publishers.

Shah, T. (1993). *Groundwater Markets and Irrigation Development: Political Economy and Practical Policy*. Oxford University Press. Bombay, India.

Shah, T. (1996). *Catalyzing Co-operation: Design of Self-governing Organisations*. SAGE Publishing Company. New Delhi, India.

Shah, T. and Bhattacharya, S. (1998). *Groundwater organization in North Gujarat: Tubewell Companies of Mehsana*. Anand: Policy School. Working Paper 3.

Shah, T.; Scott, C.; Kishore, A.; Sharma, A. and Deb Roy, A. (2002a). *Energy Irrigation Nexus in South Asia: Agrarian Prosperity with Power Sector Viability*. IWMI-Tata Water Policy Program. Anand, India (unpublished).

Shah, T.; Scott, C. and Bucheler, S. (2002b). *Mexico's Experience with Water Sector Reforms: Lessons for India's New Water Policy*. IWMI-Tata Water Policy Program. Anand, India (unpublished).

Strosser, P. and Kuper, M. (1994). *Water markets in the Fordwah/Eastern Sadiquia area: an answer to perceived deficiencies in canal water supplies?* International Irrigation Management Institute. Working Paper 10. Colombo, Sri Lanka.

Sunday Times (2002). April 7, p. 7.

UNDP (2000). Perspectives on Integrated Basin Management for a Policy Study on the Huaihe River Basin, China. United Nations Development Programme. New York, USA (unpublished draft).

Van Steenbergen, F. and Oliemans, W. (2002). *A review of policies in groundwater management in Pakistan 1950–2000*. *Water Policy*, 4: 323–344.

Wang, J. and Huang, J. (2002). *Water institutional and management system at National and River Basin level in China*. Center for Chinese Agricultural Policy. Beijing, China (internal paper).

Wester, P.; Pimentel, B.M. and Scott, C. (1999). *Institutional responses to groundwater depletion: the Aquifer Management Councils in the State of Guanajuato*. Presented at the International Symposium on Integrated Water Management in Agriculture, Gómez Palacio, Mexico (June 16–18). International Water Management Institute. Colombo, Sri Lanka (internal draft).

Wood, G. (1995). *Private provision after public neglect: opting out with pumpsets in North Bihar*. Centre for Development Studies. University of Bath, UK.

Xiang, Q.; Huang, J. and Wang, J. (2000). *Property right innovation and cropping pattern adjustment: case study of groundwater irrigation system in Hebei, China*. Center for Chinese Agricultural Policy. Beijing, China (draft).

Zhang, Q. and Zhang, H. (2001). Five current important water issues for sustainable development in China. *Water International*, 26 (2): 231–238.

Socio-Ecology of groundwater irrigation in South Asia: an overview of issues and evidence

Aditi Mukherji[(1)] and Tushaar Shah[(2)]
International Water Management Institute (IWMI), India
[(1)] *a.mukherjee@cgiar.org*
[(2)] *t.shah@cgiar.org*

ABSTRACT

Rapid growth of groundwater irrigation in South Asia has been at the heart of its recent agrarian growth. Here, over the past 50 years, public investments and donor funds have been showered over surface irrigation, but the bulk of its irrigation and agrarian growth have been delivered by millions of small pumps and wells financed mostly through private farmer investments. Not only has area under groundwater irrigation grown in leaps and bounds, but portion contributed by groundwater irrigation to total agricultural production is very nearly twice than that of surface irrigation. At the same time, groundwater development has been spatially dispersed and even whereas canal irrigation projects have created small islands of affluence leaving large catchments areas poor and deprived. It is not surprising then that providing access to groundwater irrigation through pump subsidies or public tubewell programs has been at the centre-stage of poverty reduction programs in South Asia. But with growing problems of resource depletion and deterioration, Asia's groundwater socio-ecology is under siege. Much concern about the problems of groundwater depletion, pollution and quality deterioration is fueled by worries about their environmental consequences. These are indeed serious; however, equally serious are their consequences for the sustenance of agrarian economies and millions of rural livelihoods that have come to precariously depend upon groundwater irrigation, particularly in India, Pakistan, and Bangladesh. Therefore, the issue is: how long can this good run continue without any mechanism for governing this colossus? What kind of governing structures and mechanisms might help bring a modicum of order in the functioning of this booming but anarchic economy? In this quest for better governance, need to understand the spatial variation within South Asia itself is of great importance. Recognizing the need for a refined understanding of groundwater socio-ecology of south Asia, the International Water Management Institute (IWMI) undertook a groundwater survey in four South Asian countries, *viz*. India, Bangladesh, Pakistan and Nepal plains. This paper presents the findings of the IWMI cross country groundwater survey.

Keywords: *groundwater, irrigation, South Asia, agriculture, water markets.*

1 INTRODUCTION

Groundwater has become the mainstay of irrigated agriculture in South Asian countries of India, Pakistan, Bangladesh and Nepal *Terai*.[1] Throughout South Asia, the history of protective well irrigation goes back to the millennia. However, intensive groundwater irrigation of the scale we encounter today is a relatively recent phenomenon, say of the last 30 years or so. In India, groundwater irrigated

[1] *Terai* refers to the narrow strip of foothill plains of Nepal adjoining India. It is the homeland to the bulk of Nepali population.

area increased from 13 million ha in 1970–1973 to 57 million ha in 1998–1999 (GOI – Agricultural Census, – several years). Similarly, here number of water extraction mechanisms (WEMs) rose from less than one million in 1960 to almost 26–28 million[2] in 2002. In Pakistan Punjab, number of mechanized wells and tubewells increased from barely a few thousand in 1960 to 0.5 million today (Figure 1). Bangladesh saw an increase in number of deep and shallow tubewells from mere 93,000 in 1982–1983 to almost 0.8 million in 1999–2000 (Figure 2).

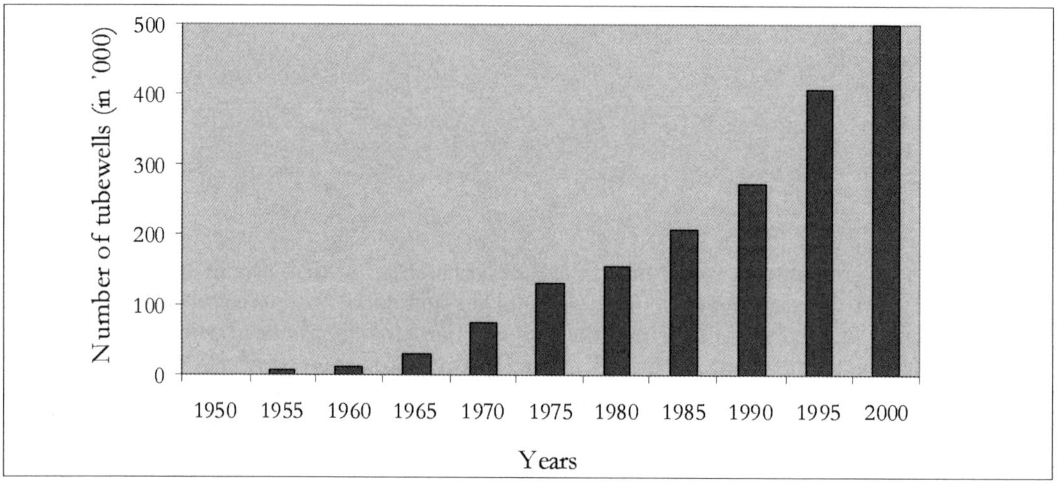

Figure 1. Growth of private tubewells in Punjab Province of Pakistan, 1950–2000.
Source: PPSGDP (2001)

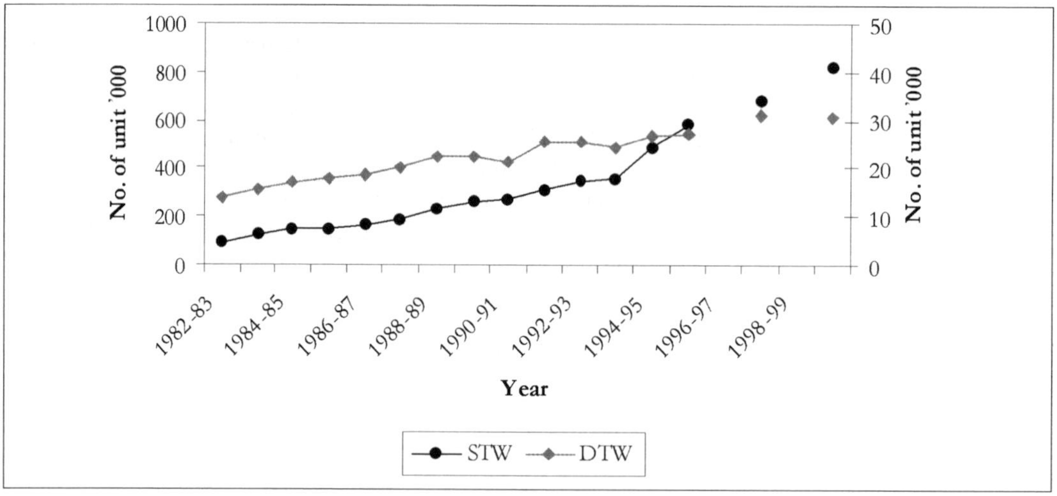

Figure 2. Tubewell development in Bangladesh, 1982–2000.
Source: National Minor Irrigation Census of Bangladesh (GOB, 1995–1996 and 1999–2000)

[2] There are no firm and accurate estimates of total number of groundwater structures in India. Minor Irrigation Census of GOI (1993) estimates the total number of groundwater structures to be 10 million. However, this estimate excludes four major states of India, *viz*. Gujarat, Maharashtra, Tamil Nadu and Karnataka. These states alone had almost 5 million structures in 1986–1987 (GOI–MI Census, 1986–1987). Based on average annual compound growth rates of groundwater structures between 1986–1987 to 1993–1994, we estimated that the total number of structures were anything between 19 to 20 million in 1993 and now (2002) it would be almost 26–28 million, given that every year 0.75 to 1 million structures are added.

With increase in number of groundwater based WEMs, the share of groundwater in total agricultural productivity has gone up several folds in much of South Asia. New research suggests that in the recent decades, of the agricultural productivity of a *representative* hectare, the portion contributed by groundwater irrigation is almost 35% more than that contributed by surface water irrigation (Deb Roy and Shah, 2003). There are various reasons for this. Groundwater is produced at the point of use, needing little transport; it offers individual farmers irrigation *on demand*, and because it entails significant incremental costs for lift, farmers tend to economize on its use and therefore maximize application efficiency. Due to reliability of supply, groundwater irrigation encourages complimentary investments in fertilizers, pesticides and high yielding varieties (HYV), leading to higher yield (Kanhert and Levine, 1989). Evidence from India suggests that output/ha from groundwater irrigated area is 1.2 to 3 times higher than that of surface water irrigated areas (Dhawan, 1989, p. 167). Similar evidences were documented in a number of early studies in Pakistan (Meinzen-Dick, 1996) and in Gujarat and Eastern Uttar Pradesh (Shah, 1993).

2 DATA AND COVERAGE

At the outset, it was decided to cover every part of South Asia through primary questionnaire surveys. For this purpose, our study countries were divided into a number of grids, based on latitude and longitude. From the approximate center of each grid, a village was chosen and questionnaire was administered to 12 to 20 randomly chosen pump owners. In India and Pakistan, we concentrated only on those farmers who own mechanized WEMs, while in Bangladesh and Nepal we also included farmers owning non-mechanized WEMs. Figure 3 shows the survey plan for India, Pakistan and Bangladesh and Table 1 gives some details about the sample size of farmers, sample size of number of wells and tubewells covered under the survey, etc. Two questionnaires were framed (one for tubewell owner, another for the village) and pre tested in all the countries. Thereafter we developed final questionnaires, which were administered by various agencies in India (NGOs and data collection agencies), by IWMI office at Pakistan, by Bangladesh Agricultural Research Institute (BARI) in Bangladesh, and IWMI Nepal office and a firm of consultants called TECHDA in Nepal. Visual Basic based software was specially developed for data feeding and analysis. This paper presents the analysis of data so generated. For purpose of refined analysis, we further divided all the countries (except Nepal) into different regions based on agro-climatic, geohydrological and socio-economic factors. Table 2 provides a brief summary of regions and their principal characteristics.

3 RESEARCH QUESTIONS

The survey attempted to shed new understanding on the following aspects of groundwater based irrigated economy in South Asia.

1) Significance and size of groundwater irrigation in village economies in terms of agricultural economy and its environmental dimensions.
2) Pace and pattern of growth of groundwater structures.

Table 1. Sample size.

Country	Number of villages	Number of tubewell owners surveyed	Number of WEMs owned by sample farmers
India	149	1,508	1,936
Pakistan	79	0,525	0,736
Bangladesh	40	0,245	0,298
Nepal	24	0,351	0,380
Total	**292**	**2,629**	**3,350**

Source: Primary survey conducted by IWMI in 2002.

Table 2. Regionalization and sample area characteristics.

Sl. No	Country/Region	Number of villages	Coverage	General Characteristics
1	North Western India	16	Punjab, Haryana and Western Uttar Pradesh	Most developed agricultural states in India, dense canal networks, intensive groundwater use, salinity is a major problem, good alluvial aquifers.
2	Eastern India	21	West Bengal, Bihar, Coastal Orissa and Eastern Uttar Pradesh	Huge groundwater potential but not enough use, high growth potential, annual flooding a major problem, low to medium agricultural productivities, very good alluvial aquifers.
3	Central Tribal India	22	Chhatisgarh, Jharkhand, 2, 4 and 5 villages from Gujarat, Madhya Pradesh and Orissa respectively	Most backward of all regions in India, inhabited by tribals. Forested and undulating terrain, agriculture not developed. Low to very low productivities. High precipitation and runoff.
4	Central and Western India	55	Rajasthan, Maharashtra, non tribal districts of Madhya Pradesh and Gujarat	Water scarce region of India. Low rainfall, varying aquifers, from over exploited alluvial aquifer in North Gujarat to hard rock geology and shallow aquifers in Maharashtra. Very intensive use of groundwater. Groundwater depletion and fluoride a major threat.
5	Interior South India	25	Interior districts of Karnataka, Tamil Nadu and Andhra Pradesh	Hard rock aquifer, heavy dependence on groundwater irrigation mostly through shallow dug wells. Depletion is a major problem.
6	Coastal South India	10	Coastal districts of Andhra Pradesh, Tamil Nadu and Kerala	High rainfall, good alluvial aquifers, intensively cultivated with 3 crops of paddy, agriculturally dynamic area within Southern states.
7	Pakistan Punjab	36	Punjab province in Pakistan	Semi arid region, but very extensive canal network, home to 80% of Pakistan's population and the bread basket of the country, conjunctive use and secondary salinization are major issues.
8	Pakistan Sindh	21	Sindh province in Pakistan	Arid to semi arid with only 50% of its area under cultivation. Unlike Punjab, this region does not have extensive canal network. The main crops are wheat, cotton and sugarcane.
9	Pakistan NWFP	22	NWFP in Pakistan	Sparsely populated area with less than 30% of its geographical area under cultivation. Receives rainfall in winter in addition to monsoon rain. The main crops are wheat, vegetables and sugarcane.
10	North Western Bangladesh	11	Former (or greater) districts of Dinajpur, Rangpur, Bogra, Pabna and Rajshahi	Agriculturally dynamic part of Bangladesh, produces more than half of country's total food grain output, very intensive cultivation and high tubewell densities.
11	Rest of Bangladesh	29	Rest of Bangladesh	Not as developed as North Western region, flooding and water logging are major problems, though groundwater potential is very high, not much is put to use due to constraints.
12	Nepal *Terai*	24	26 districts in Nepal *Terai*	Foothills of Himalayas, this narrow stretch adjoining India is the main agricultural belt in the country. High water table and water logging are major problems, coupled with generally low productivities and fragmented land sizes.

Source: Primary survey conducted by IWMI in 2002.

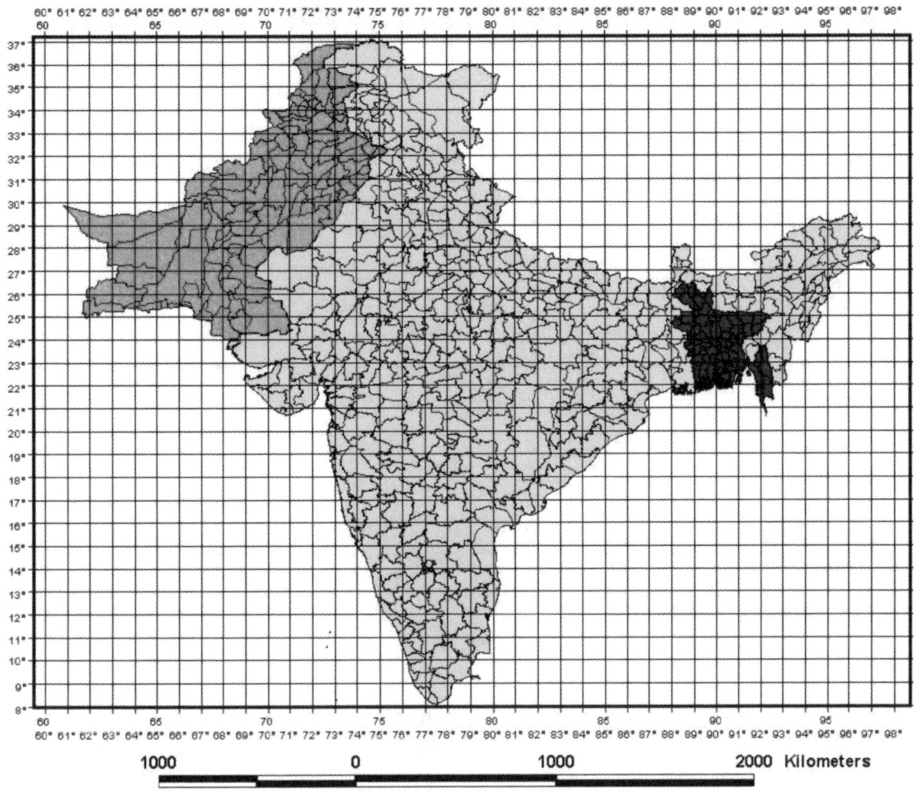

Figure 3. Survey plan for India, Pakistan and Bangladesh.[3]

3) Profile of the well owners and its equity and poverty reduction implications.
4) Technological configuration of the groundwater irrigation economy.
5) Well owners' farming operations.
6) Pump irrigation markets: what are the economic institutions that have emerged in the wake of the emergence of the groundwater economy? How do these vary across space and time and why?

This paper is accordingly divided into six sections (4 to 9), each section highlighting one of the research questions. In addition to the primary data collected, findings are also substantiated from other secondary sources, such as Minor Irrigation Censuses of India and Bangladesh, and relevant secondary literature.

4 THE SOUTH ASIAN GROUNDWATER ECONOMY

4.1 *Size of the South Asian groundwater economy*

There is growing evidence that groundwater has become a very important source of irrigation in South Asia. Due to its inherent characteristics such as supply *on demand* and decentralized development, groundwater productivity is much higher than that of surface water productivity. These two together, *viz.* increasing area under groundwater irrigation, coupled with higher productivity in groundwater irrigated regions has resulted in increasing share of groundwater in generating agricultural wealth in all of South Asia. This growing importance of groundwater

[3] One village at the centre of each grid was selected for the survey. In India, we excluded states of Jammu and Kashmir, Himachal Pradesh and all the northeastern states, and in Pakistan few grids adjoining Afghanistan were excluded due to logistical problems.

Table 3. Overall profile of groundwater irrigation in villages surveyed.

Sl. No	Country/Region	Number of sample villages	GWI area to NCA (%)	GWI area to NIA (%)	Canal irrigated area to NIA (%)	Tubewell density (per 100 ha of NCA)	Number of diesel pumps	Number of electric pumps
1	North Western India	16	82	90	3	26	2,652	1,682
2	Eastern India	21	24	54	7.5	13	483	74
3	Central Tribal India	22	26	62	1.6	6	157	165
4	Central and Western India	55	25	87	2.2	12	728	4,146
5	Interior South India	25	13	58	10.5	13	126	3,304
6	Coastal South India	10	20	33	26	9	841	920
7	Pakistan Punjab	36	14	19	0.3	8	1,534	200
8	Pakistan Sindh	21	7	15	42	3	109	13
9	Pakistan NWFP	22	6	13	53	1	36	36
10	North Western Bangladesh	11	79	97	0	21	263	54
11	Rest of Bangladesh	29	15	36	4.3	6	270	121
12	Nepal *Terai*	24	32	51	45	9	309	7

Source: Primary survey conducted by IWMI in 2002

in agrarian economies is clearly brought forth by our survey data. Of the 292 villages surveyed, 132 villages or 45% reported high to very high contribution of groundwater to the total agricultural production in the village. This figure varied from as high as 100% in North Western Bangladesh to only 18% in villages of central tribal belt of India. Another way to capture the significance of groundwater in the village economy would be to see the proportion of groundwater irrigated (GWI) area to net cropped area (NCA) and net irrigated area (NIA). This varies considerably across South Asia, from as low as 13% of NIA in NWFP in Pakistan to as high as 90% and 87% in Northwestern India and Western and Central India and almost 97% in Northwestern Bangladesh. The rather low GWI to NIA figure in Pakistan is due to conjunctive use of surface and groundwater. In Nepal *Terai* region more than 50% of the net area irrigated is accounted for by groundwater irrigation and another 15% by conjunctive use of canal and groundwater irrigation, while the balance 35% is accounted for by canal irrigation. Tubewell density (number of tubewell/100 ha of NCA) is another indicator of level of groundwater development and its significance in an area. Northwestern India and Northwestern-Bangladesh have very high tubewell densities of 26 and 21 tubewells per 100 ha of cultivated area, while NWFP in Pakistan has less than one tubewell per 100 ha. Table 3 presents the overall and region wise profile of groundwater irrigation in 292 surveyed villages across South Asia. As can be seen from the table, there are wide variations across regions, and even within regions themselves. That groundwater structures and therefore groundwater irrigated areas have proliferated across South Asia is not hard to explain. While much development of the surface water over the past 50 years has been guided more by geo-hydrology (or supply factors) than demand for irrigation, all of groundwater development has been demand induced and made by farmers themselves, as and when they felt the need for assured irrigation. Figure 4 shows the well density in each of the regions.

Given the growing importance of groundwater economy, it is imperative that we estimate an approximate size of the groundwater economy. One practical way of estimating the scale and size of groundwater economy is to measure the economic value of groundwater production. Most of South Asia has active market in pump

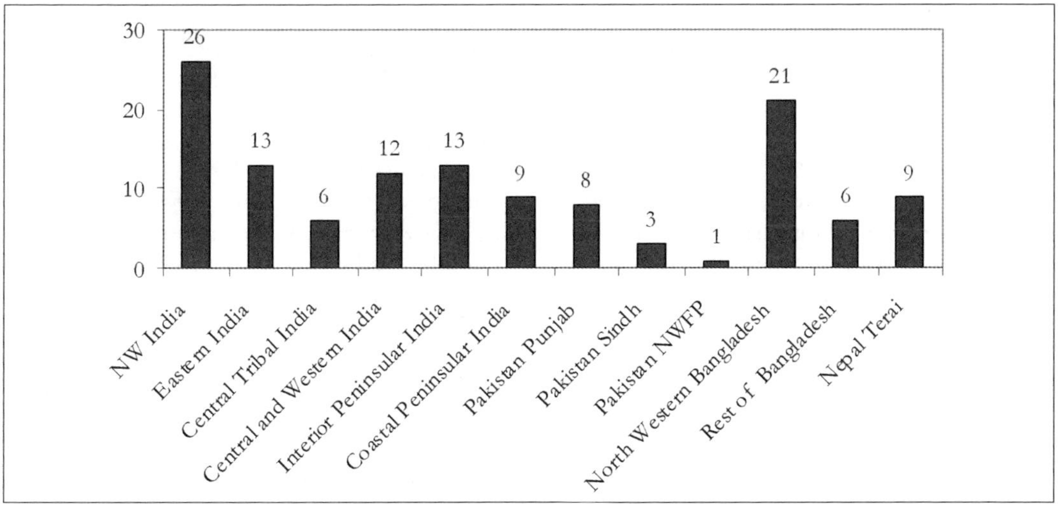

Figure 4. Pump density (numbers/100 ha of NCA) in different regions of South Asia.
Source: Primary survey conducted by IWMI in 2002.

Table 4. The size of agricultural groundwater economy of India, Pakistan and Bangladesh.

		India	Pakistan[4]	Bangladesh	Nepal *Terai*
A	Total number of groundwater structures (million)[5]	26	0.5	0.8	0.06
B	Average output of groundwater structures (m^3/h)[6]	25	100	30	30
C	Average hours of operation/well/yr[7]	330	1,090	1,300	205
D	Price at which pump irrigation from standard size pump sells (US$/h)	1	2	1.5	1.5
E	Estimated groundwater used (km^3) {(A*B*C)/10^9}	215	54.5	31.2	0.37
F	Imputed value of groundwater used/yr in billions US$ (E/B*D) or {(D*C*A)/1000}	8.6	1.1	1.6	0.02

Source: Given in footnotes

irrigation services in which tubewell owners sell groundwater to their neighbours at a price that exceeds their marginal cost of pumping. This price offers a market valuation of groundwater use in irrigation (Shah *et al.*, 2001). Table 4 constructs a profile of groundwater economy in three South Asian countries (India, Pakistan and Bangladesh) using such valuation and suggests

[4] Estimate for Pakistan includes only that of Pakistan Punjab which has almost 90% of groundwater structures in the country.

[5] Total number of groundwater structures have been estimated for India based on GOI–MI Census (1987 and 1993), for Bangladesh based on GOB MI Census (1996–97) and that of Pakistan based on estimates provided by Punjab Private Sector Groundwater Development Project (PPSGDP, 2000).

[6] Average output of groundwater structures (m^3/h) will depend among other things on average HP of pumps and depth to water table. In Pakistan, average HP is almost two or three times that of India and Pakistan Punjab has high water tables due to canal recharge. In Bangladesh, though pump HP is comparable with that of India, water table is very near the surface and WEMs pump water from an average depth of 5–8 m in most places. So, average output of WEMs in Bangladesh has been assumed to be marginally higher than that of India, and that of Pakistan is 4 times that of India.

[7] Average hours of operation (h/well/yr) and price at which water sells are based on primary data generated through survey conducted by IWMI, 2002.

that groundwater irrigation in South Asia may well be of the tune of US$ 10–11 billion/yr.

4.2 *Environmental dimensions of groundwater development in South Asia*

No resource is an unmixed boon, neither is groundwater. Groundwater supports a booming agricultural economy in South Asia. But in many parts of this region are under serious threat of resource depletion and degradation. The rate at which groundwater is being drawn exceeds annual recharge in many parts like North Gujarat. In Pakistan Punjab, while estimated groundwater recharge has remained stagnant, there has been a secular growth in the number of tubewells. Groundwater depletion has major environmental consequences, but it has far equally serious economic consequences, especially for the poor and marginal farmers in South Asia. While, the marginal and small farmers own and operate only 15% of India's total agricultural land, they are saddled with 73% of wells and tubewells that are abandoned due to various reasons, such as lowering of water tables or high energy costs. Figure 5 shows Lorenz curve of distribution of area operated by farmers (according to land size) and distribution of abandoned wells. Thus while the bottom 40% of farmers operate less than 10% of the cultivated land, they own more than 60% of the WEMs that have been abandoned. However, degree of WEMs abandonment depends on a number of factors which vary from one region to the other. Based on our primary survey, we have estimated the total number of wells and tubewells that have been abandoned by sample farmers. This (Figure 6) seems to be the highest in Western India and interior peninsular India, characterized by very intensive groundwater use and low rainfall, leading to groundwater withdrawal exceeding groundwater recharge.

Coupled with the problem of resource depletion is that of resource pollution. Much of coastal Gujarat and Tamil Nadu have witnessed widespread problems of saline ingress up to 5–10 km inland, rendering several thousand hectares of agricultural land useless. Fluoride contamination of groundwater too has emerged as an important environmental problem in much of Gujarat, so much so that drinking water has been rendered unfit for consumption. This in turn has led to some amount of local entrepreneurship in the form of setting up Reverse Osmosis (RO) plants for supplying drinking water to urban areas. Again, not all can afford treated drinking water. While fluoride plagues Western India, arsenic has emerged as a menace in Gangetic plains of West Bengal and Bangladesh. Potentially far more vicious is the problem of secondary salinity of groundwater and soils arising from poor drainage of canal irrigated areas. Punjab province has about 25% of its irrigated area severely

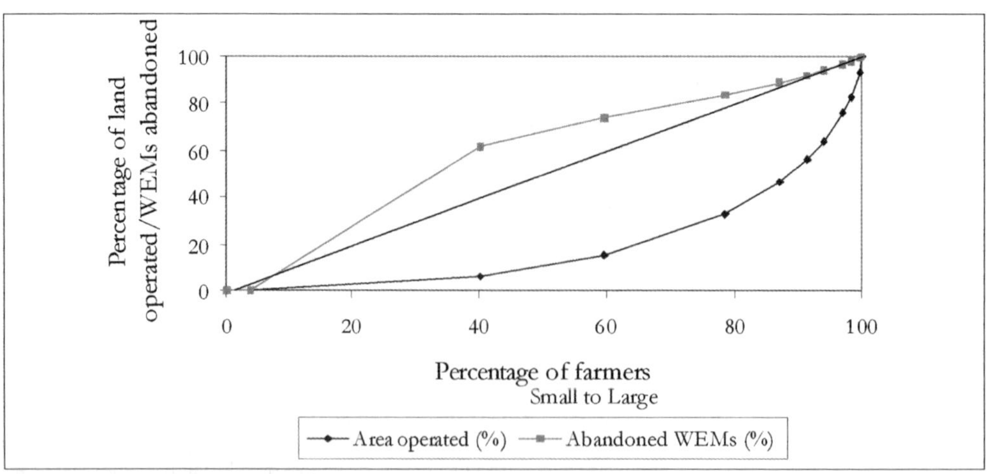

Figure 5. Lorenz curve showing area operated and WEMs abandoned, India.
Source: Based on Minor Irrigation Census (GOI, 1993).

waterlogged and every year around 40,000 ha of farm land is being abandoned within the Indus basin alone, due to secondary salinity (WAPDA, 1989). In Pakistan Punjab, around 30% of the irrigated area has groundwater tables within 1.5 m of the soil surface right after monsoon. In India, on the other hand, some 22,000 villages have water tables (Figure 7) below 50 m (GOI–MI

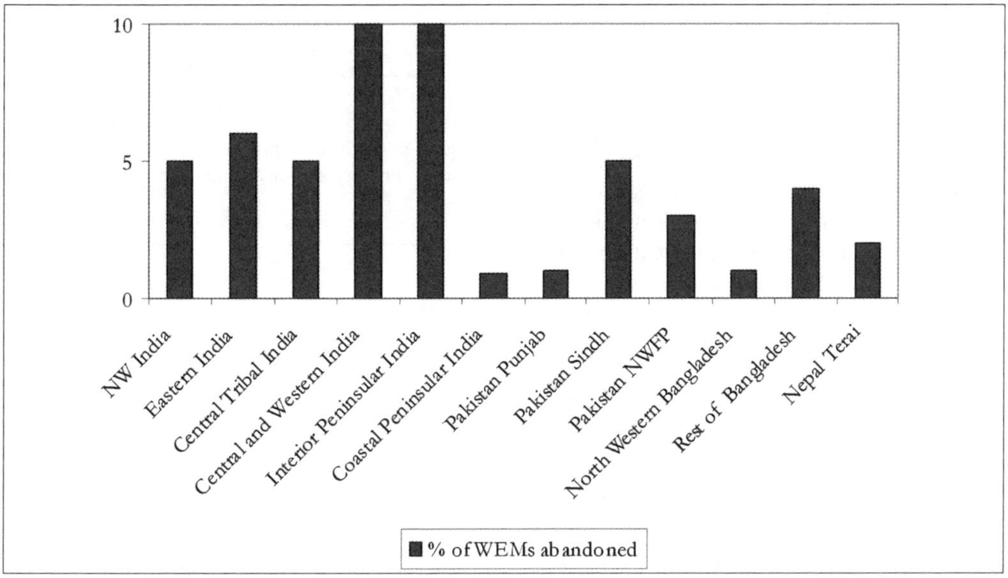

Figure 6. Percentage of WEMs abandoned due to various reasons.
Source: Primary survey conducted by IWMI in 2002.

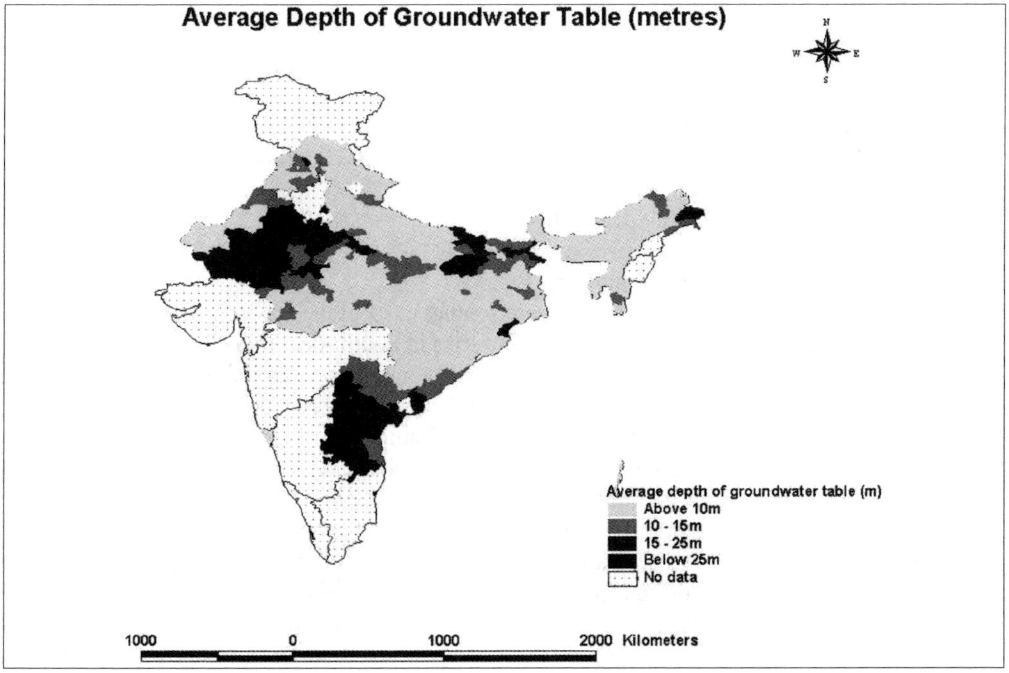

Figure 7. Average depth of groundwater table in India.
Source: Minor Irrigation Census (GOI, 1993).

Census–, 1993). Thus, South Asia is plagued by twin problems of lowering of groundwater tables on the one hand and water logging on the other.

5 PACE AND PATTERN OF GROWTH IN GROUNDWATER STRUCTURES

Perhaps the most intriguing aspect of the Asian groundwater economy has been its rapid pace of growth within the last two or three decades. Well irrigation has been known to South Asian farmers for ages, but until the Green Revolution, it was used solely for protective irrigation. However, Green Revolution and adoption of HYV seeds changed the rules of the game. Modern varieties of seeds were designed to be extremely sensitive to water stress and given the high input costs and higher returns, irrigation and fertilizer became the most important determinants of crop yields. Modern development of groundwater irrigation started in much of South Asia through government initiatives in the form of construction of public tubewells. By the early 1960s, Pakistan had embarked upon a large programme of deep tubewell development, the SCARPs, primarily to combat water logging. Over the last 30 years, some 12,500 deep tubewells were installed in canal commands in Pakistan. India too witnessed a similar public funded beginning of groundwater irrigation. In Uttar Pradesh, for example, public tube wells were installed way back in the 1930s. In Bangladesh, state funded BADC was instrumental in installing deep tubewells all across the country. However, over time, all these government funded schemes under performed, so much so that all these countries have started programme of management transfer of public tubewells. The public tubewells did have a desired impact it essentially created a *demonstration effect* that later led to large private investments in groundwater irrigation.

However, the real boom in groundwater irrigation started after Green Revolution. In India, number of tubewells rose from less than one million in 1960s to around 26 million in 2002. Most of the new tubewells were added in the 1970s and 1980s in Punjab, Haryana and other agriculturally dynamic states and later in the 1980s to 1990s in much of Eastern India and

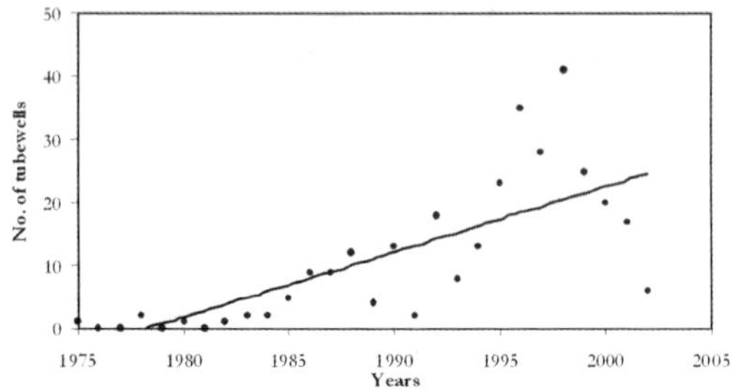

Figure 8. Growth of private tubewells in Bangladesh.
Source: Primary survey conducted by IWMI in 2002.

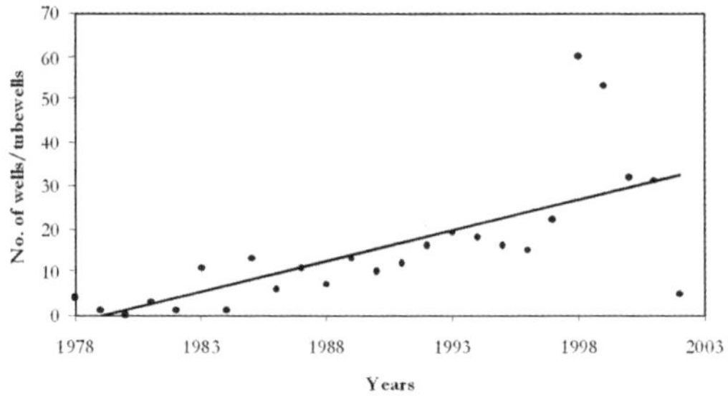

Figure 9. Growth of private tubewells in Nepal.
Source: Primary survey conducted by IWMI in 2002.

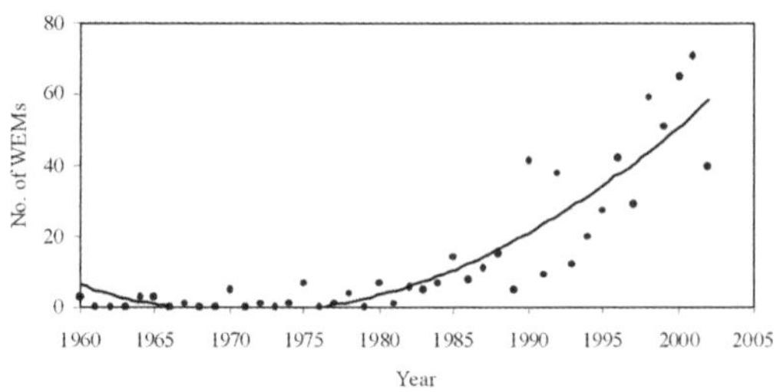

Figure 10. Growth of private tubewells in Pakistan.

other parts of the country. Among the four countries in South Asia, India has seen the fastest rise in the number of groundwater structures during the last two-three decades. Based on year of mechanization, we plotted graphs depicting the growth in number of private tubewells in India, Pakistan, Bangladesh and Nepal. Similarly, we also plotted growth of private tubewells in India in a GIS based map showing our sample grids. Figures 8–11 show the pace of growth of WEMs in whole of South Asia. There has been a linear and secular growth in groundwater structures in Bangladesh and Nepal, while much of India has actually witnessed an exponential growth in

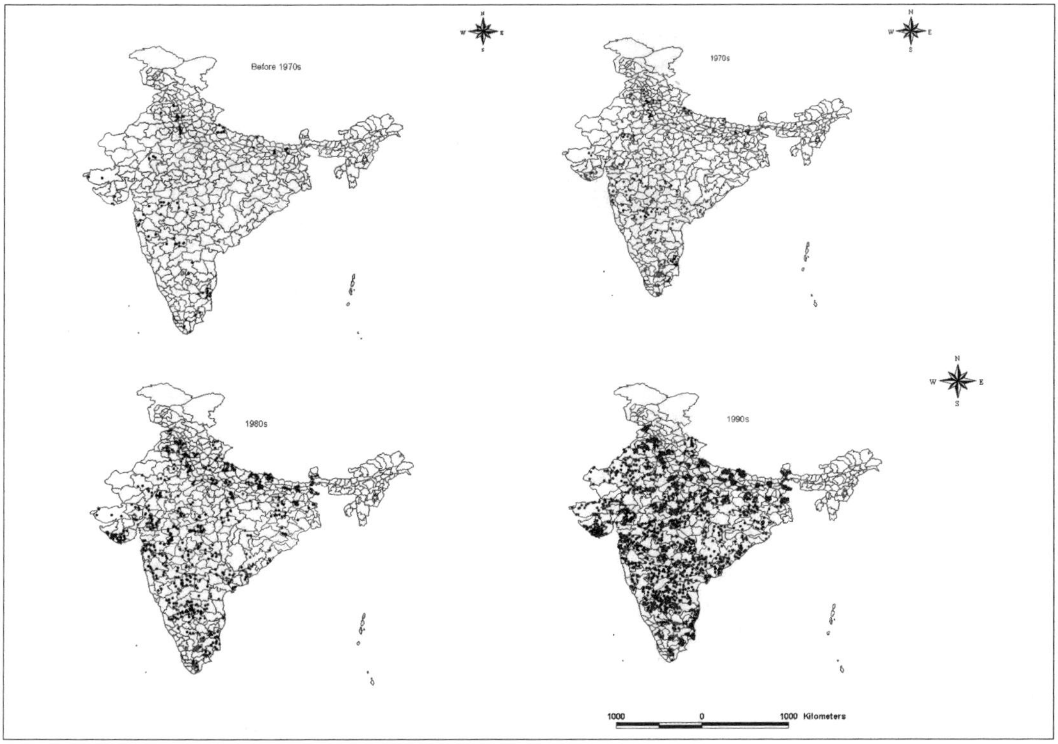

Figure 11. Pace of growth in mechanized WEMs in sample villages, India.

WEMs during 1970s to 1990s. Similarly, Figure 11 shows that the decade of 1990s can be justly called to be the decade of *pump explosion* in India just as the 1970s was the decade of *population explosion*.

6 PROFILE OF THE WELLS AND WELL OWNERS

6.1 *Who owns wells?*

Groundwater has become the lifeblood of irrigated agriculture in South Asia and creates more wealth and any other source of irrigation. But the pertinent question is: who owns and controls access to this very precious resource? Answer to this question will have wide ranging social and economic implications. Previous research shows that access to groundwater irrigation is less skewed against the small and marginal farmers when compared to access to land. In India, while 78% of operational holdings are small and marginal farms (of less than 2 ha), they operate only 32% of the area. However, they constitute 38% of net irrigated area by wells and account for 45% of WEMs fitted with electric pumps (GOI – Agricultural Census, 1991). Thus, in relation to the amount of land they cultivate, the poor are better represented in ownership of groundwater assets. No wonder, then, in developing countries of Africa and Asia, groundwater development has become the central element of livelihood creation programs for the poor (Kanhert and Levine, 1989; Shah, 1993).

Our primary survey results yield comparable results. Figure 12 shows the ownership of WEMs according to farm size. Marginal and small farmers own up to 70% of groundwater structures in Bangladesh, 60% in Nepal and 50% in Eastern India. Except for tribal India and Western India, in all the other regions, small and marginal farmers own at least 30% of the groundwater structures that we surveyed. In Punjab and Haryana, small and marginal farmers operate only 12% of the total land (GOI – Agricultural Census–, 1991) but they own almost 30% of WEMs. Tribal

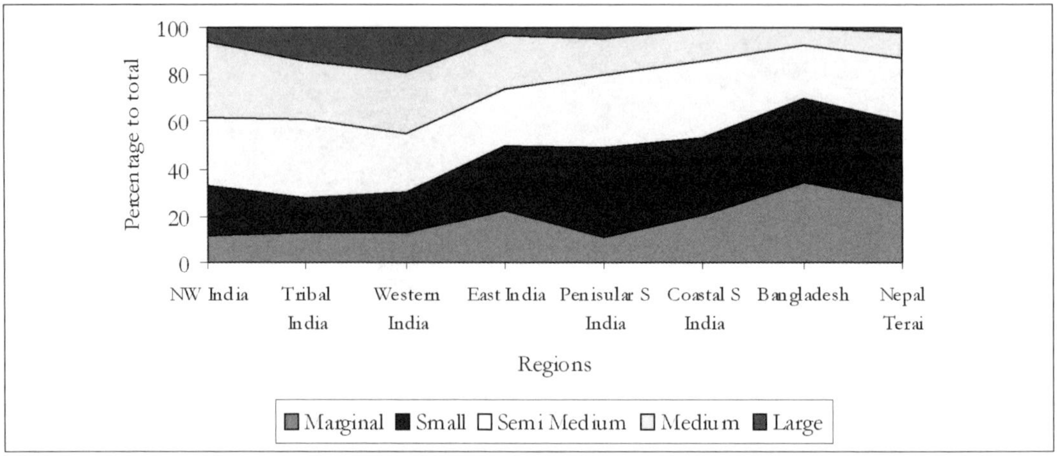

Figure 12. Ownership of WEMs according to farm size.
Source: Primary survey conducted by IWMI in 2002.

India is an exception. For one, here medium and large farmers own 72% of WEMs. This region, as already mentioned, is one of the poorest in the country. According to the MI census, only 7% of Scheduled Tribes own WEMs, while they constitute almost 15% of India's total population. In Western India too, almost 70% of WEMs are owned by medium and large farmers. Western India is characterized by rapid groundwater depletion. Thus, we find more skewed distribution of groundwater assets in regions that are either agriculturally backward, or regions that have experienced rapid groundwater depletion.

6.2 *Characteristics of the well owners*

Though small and marginal farmers have better representation in well ownership, the chunk of the resource still lies with the medium and large farmers. This seems inevitable given the bulky investment required for mechanized groundwater irrigation in the first place and it is further aggravated due to problems related to groundwater depletion, where the poor and marginal farmers are the first to loose their access to groundwater. Nevertheless, within the small and marginal farmers, those who own or have access to groundwater are better off than those who have neither. Past research has shown that cropping intensity and productivity is higher in groundwater irrigated areas. This holds not only for South Asia, but also for China (see Shah *et al.*, 2001) and Spain (see Hernández-Mora *et al.*, 2003). However, since our present study concentrated solely on tubewell owners, we have no direct way of ascertaining if they do achieve higher cropping intensities or crop yields than surface water irrigators. We shall therefore, in this Section limit ourselves to presenting basic characteristics of well owners we surveyed (Table 5). Quite expectedly, there are wide variations across and within regions, in terms of average land size, number of parcels owned, average irrigated area, source wise cultivated area, etc. These in turn have important policy implications. For example, steep increase in electricity tariff has the potential of crippling the entire groundwater based economy in much of Western and Southern India, while it will have very limited effect in Eastern India, because electric pumps form a miniscule part of the total number of WEMs in this part.

Average landholding varies widely from one region to the other, ranging from as large as 22 ha in Sindh in Pakistan to 1.6 ha in Bangladesh. Similarly, while 96% of land holdings are cultivated in Eastern India, only 60% or so are cultivated in Sindh province, leaving the rest as fallow. Eastern India and Nepal have very fragmented land holdings; on an average one tubewell owner owns 6 to 7 parcels of land. Tiny stamp sized land holdings was thought to be chief impediment of developing groundwater in Eastern India, because it was contended that WEMs needed lumpy investments and that small farmers would

Table 5. Characteristics of well owners.

Characteristics of sample well owners	Pak. Punjab	Sindh	NWFP	Punjab, Haryana, West UP	Eastern India	Tribal India	Western India	Interior South India	Coastal South India	Bangladesh	Nepal *Terai*
1. Average farm size for sample well owners (ha)	7.4	21.94	13.49	4.2	3.0	5.2	6.6	3.62	2.43	1.9	2.39
2. Average cultivated area for sample well owners (ha)	6.84	13.19	13.4	4.1	2.9	4.8	6.3	3.33	2.36	1.6	2.37
3. Average number of parcels per sample well owner	2	2	1	2.6	6.6	3.3	2.5	2.5	2.1	7	5
4. Average irrigated area per sample well owner (ha)	6.24	11.77	12.62	3.8	2.3	3.1	4.0	2.5	2.36	1.3	1.73
5. Rainfed area (ha)	0.59	2.94	0.79	0.66	0.34	1.64	2.82	1.13	0.11	0.3	0.61
6. By only canals (ha)	0.02	1.98	0	0.45	0.07	0.10	0	0.05	0.03	0.01	0.14
7. By only wells (ha)	1.65	3.63	8.55	2.34	1.9	2.83	3.23	1.87	1.49	0.98	1.44
8. By wells and canals (ha)	4.43	6.08	3.98	0.9	0.06	0.06	0.44	0.37	0.41	0.05	0.12
10. By other (ha)	0.14	0	0.17	0.02	0.06	0.11	0.6	3.74	0.32	0	0.01
11. Cropping Intensity (%)[8]	178	154	132	175	171	155	143	126	146	188	187

Source: Primary survey conducted by IWMI in 2002.

not be able to make good of it (Boyce, 1987). However, through the emergence of vibrant groundwater markets, small and marginal farmers could sell their excess water to other farmers and thereby justify big investments in WEMs. At least 60% of the total cultivated area is irrigated in case of well owners. Area irrigated as a percentage of cropped area varies between 100% in coastal peninsular India to 65% in tribal India. Quite predictably, most of the area irrigated is under groundwater irrigation, which invariably accounts for at least 61% in Northwestern India to as high as 83% in Nepal *Terai*. However, Pakistan Punjab and Sindh (not NWFP though) are an exception in that conjunctive use of canal and tubewell water is more important as a source of irrigation than tubewell water alone. On an average, all tubewell owners take two crops in a year and in some cases even three crops, such as three crops of paddy in coastal Andhra Pradesh. Consequently, cropping intensities are very high, ranging from 126% in interior peninsular India to 188% in Bangladesh.

7 TECHNOLOGICAL CONFIGURATION OF GROUNDWATER ECONOMY

South Asia has come a long way from the days of manual or animal driven WEMs. Intensive input based Green Revolution has had its impact on changing the technological configuration of these regions groundwater economy. Traditional ways of protective irrigation gave way to mechanized WEMs in the form of diesel engines and electric pumps. Even the configuration of groundwater structures changed-from shallow dug wells to bore-cum-dugwells to shallow and deep tubewells. Changes in technical configuration of WEMs are a reflection of number of factors, such as demand for irrigation, type of cropping pattern, groundwater level and rate of extraction, etc.

[8] Cropping intensity has been calculated on the basis of area cultivated once, twice and thrice.

Table 6. Distribution by type of groundwater structure.

	Pak. Punjab	Sindh	NWFP	Punjab, Haryana, West UP	Eastern India	Tribal India	Western India	Interior South India	Coastal South India	Bangladesh	Nepal *Terai*
1. Dug and dug-bores as % of total wells	26	39	56	63	17	61	66	31	42	0	0.5
2. Borewells, STWs and DTWs as % of total wells	74	61	44	37	83	39	34	69	58	100	99.5

Source: Primary survey conducted by IWMI in 2002.

In the sections that follow, we shall describe some of the important technical parameters, such as type of well, motive power of WEMs, depth of well and water table, average HP of pumps and mode of water distribution.

7.1 *Well types and its configuration*

The MI Census of India classifies groundwater structures into three major types, *viz.* dugwells, shallow tubewells and deep tubewells. We add another category to this; *viz.* dug cum bore wells, where a dug well has been converted to a bore or a tubewell. Depending on the nature of aquifer and its depth, the technological configuration of groundwater structures changes. Generally speaking, dugwells are amenable to shallow and poor yielding aquifers, while shallow tubewells are constructed in areas of shallow and alluvial aquifers. Deep tubewells are rather rare because they need huge investments. For India as a whole, 63% of the total groundwater structures are shallow tubewells, while another 35% are dugwells and the rest 2% are deep tubewells (GOI–MI Census, 1993). In Bangladesh, among mechanized groundwater structures, 97% are shallow tubewells and only 3% are deep tubewells (GOB–MI Census, 1999–2000). Table 6 shows the distribution of different types of wells among our sample farmers.

7.2 *Type of water lifting device and motive power*

WEMs can be either powered by electric or diesel. In most of South Asia, diesel operated pumps are more common than electric operated ones. This is because in all the South Asian countries, electricity is highly subsidized making most electricity boards unviable in the long run. This seriously limits their capacity to provide rural electricity, which anyway is down the lost of priorities of policy makers in these countries. As a result, more and more well irrigators use diesel pump and are quite susceptible to changes in diesel prices from time to time. In India for example, the whole of Eastern India has been progressively de-electrified due to apathy of State Electricity Boards. Figure 13 shows the distribution of electric and diesel pumps in India (based on MI census), which we call as the *energy-divide* in India. Western India, South Interior India and South Coastal India have more of electricity operated pumps, the figures being 85%, 94% and 94% respectively. Electricity tariff is very low, in fact, in Tamil Nadu state, electricity is supplied free of cost to the farmers and this has deleterious effect on groundwater tables. In tribal India as well, the ratio of electric to diesel pumps is 3:1 and this ratio is 1:1 in Northwestern India. In Punjab and Haryana, electricity situation has become so uncertain that almost all well owners keep a spare diesel pump as a standby to be used in case there is no electricity. NWFP in Pakistan too has more electric operated pumps than diesel ones. At the other end of the continuum are all the other regions of South Asia, where diesel pumps are more important than electric pumps (Figure 13). Pump HP too varies across regions and ranges between 2 to 10 HP in case of electric pumps and 4 to 16 in case of diesel pumps. Table 7 summarizes our

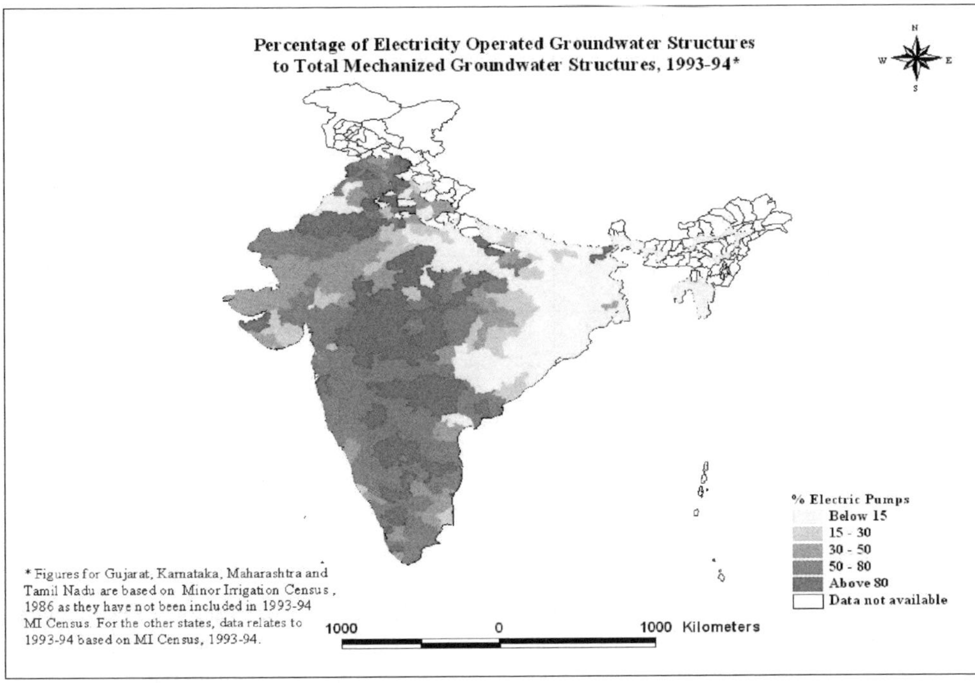

Figure 13. Energy divide in India. *Source*: MI Census (GOI, 1986–87 and 1993)

Table 7. Lifting device and motive power.

	Pak. Punjab	Sindh	NWFP	Punjab, Haryana, West UP	Eastern India	Tribal India	Western India	Interior South India	Coastal South India	Bangladesh	Nepal *Terai*
1. Percentage of electric WEMs	18	10	56	51	8	63	85	94	95	27	3
2. Percentage of diesel WEMs	82	90	44	49	92	37	15	6	5	63	79 [9]
3. Average HP of electric pumps	10	10	10	7	6	5	7	5	5	5	2
4. Average HP of diesel pumps	16	16	16	8	5	4	5	6	4	8	7

Source: Primary survey conducted by IWMI in 2002.

findings on water lifting devices and pump HP among surveyed farmers.

7.3 *Water distribution technologies*

There are essentially two mechanisms of water distribution from the groundwater structure to the farmers' field. The first is through channels, either unlined (*kutchha*) or lined (*pucca*). The second is piped distribution. Piped distribution can be of many types, either through over ground rubber pipes (in some places aluminum and even canvass pipes are used) or through underground pipelines. Underground pipelines

[9] In Nepal, manual WEMs constitute almost 18% of total wells and tubewells.

Table 8. Summary of distribution systems.

	Pak. Punjab	Sindh	NWFP	Punjab, Haryana, West UP	Eastern India	Tribal India	Western India	Interior South India	Coastal South India	Bangladesh	Nepal *Terai*
1. Percentage of wells with piped distribution	0.22	0	3.13	6.5	35	65	66	63	35	4	6.5
2. Average length of pipes used/ well (m)	366	0	1,100	600	200	150	500	260	220	300	109
3. Percentage of wells with lined channels	6.45	0	1.56	14.1	30.1	19	15	26	27	91	89
4. Average length of lined channel/well (m)	539	0	180	184	315	84	100	155	80	260	175
5. Percentage of wells using unlined channels	93	100	92.19	46	15.1	5.2	16	9	40	5	3.5
6. Average length of unlined channels/well	627	950	583	300	270	100	400	50	100	50	276

Source: Primary survey conducted by IWMI in 2002.

are uncommon, except for government managed deep tubewells and in water scarce Gujarat. Because opportunity cost of groundwater irrigation is high in Gujarat, most farmers have installed intricate network of underground pipelines for water conveyance and actually transport water to as far as 2 km from their own field to water buyers' fields. In West Bengal and Bihar, farmers have devised different mechanism for water conveyance. In this part, cost of boring a shallow tubewell is less and has stagnated at the level of Ind Rs[10] 2,000 (US$ 43) since 1970s, showing that the real cost of boring has gone down several times (Ballabh *et al.*, 2002). Land holdings too are highly fragmented. Therefore in order to overcome the problem of conveying water to far flung fields, farmers have invested in shallow bores in almost all the parcels and in one diesel pump, which they carry from field to field and irrigate. Water markets have proliferated in this part too, mostly because land holdings are small and tubewell owners have excess of water over and above what they need for their own fields, making it a necessity for them to sell of their surplus water to their neighbors. However, unlike Gujarat, where water *per se* is sold to the buyers (i.e., the buyers receive water in their fields directly from the seller in lieu of certain water charges), in Eastern India, rental market for irrigation equipment is more common than outright water sale.

Based on our primary survey we found that piped water distribution is relatively uncommon, except in regions where groundwater loss is high either due to high evaporation rates such as in Gujarat, Maharashtra and Rajasthan or where water conveyance is a problem due to undulating terrain as in tribal India. Besides, in order to increase water use efficiency and decrease seepage loss in alluvial plains of Bihar, rubber pipes are used to transport water from one field to the other. In fact in Bihar plains, rental market has also emerged for rubber pipes. On the other hand, unlined field channels are the most common way of conveying water in much of South Asia, followed by lined channels. Table 8 gives more details.

8 WELL OWNERS FARMING OPERATIONS

8.1 *Cropping pattern of sample villages and well owners*

In much of South Asia, food crops such as wheat and paddy are the most important crops, followed

[10] Indian rupee. Ind Rs 1 = US$ 0.02161 (July 1, 2003).

Table 9. Groundwater irrigated crops in South Asia.

Country/Region	GCA of sample farmers (ha)	Three groundwater irrigated crops (name) occupying maximum GCA and percentage area to GCA under each of them.		
		Crop 1	Crop 2	Crop 3
North Western India	1,020	Wheat, 41%	Paddy, 32%	Sugarcane, 11%
Eastern India	553	Paddy, 43%	Wheat, 33%	Sugarcane, 10%
Central Tribal India	872	Wheat, 38%	Oilseeds, 22%	Paddy, 8%
Central and Western India	4,061	Wheat, 28%	Oilseeds, 26%	Coarse Cereals, 10%
Interior South India	893	Paddy, 28%	Coarse Cereals, 14 %	Sugarcane, 14%
Coastal South India	280	Paddy, 53%	Sugarcane, 21%	Cotton, 8%
Pakistan Punjab	3,880	Wheat, 43 %	Cotton, 14 %	Fodder, 11%
Pakistan Sindh	955	Wheat, 47%	Cotton, 17%	Vegetables, 12%
Pakistan NWFP	1,148	Wheat, 40%	Coarse Cereals, 29%	Vegetables, 17%
Bangladesh	844	*Boro* Paddy, 78%	Wheat, 7%	Vegetables, 6%
Nepal *Terai*	1,086	Summer Paddy, 43%	Wheat, 29%	Coarse Cereals, 7%

Source: Primary survey conducted by IWMI in 2002.

by coarse cereals. Thus wheat occupies the largest cropped area in our sample villages in all of Pakistan and Indian Punjab and Haryana, while paddy is the main crop in Eastern India, Tribal India, Coastal South India, Bangladesh and Nepal *Terai*. In Western India and Interior South India, coarse cereals occupy largest area. Similar cropping pattern is also seen among well owners in these regions.

As already mentioned, well owners achieve very high cropping intensities, ranging from 126% in Tribal India to as high as 188% in Bangladesh. This is made possible through groundwater irrigation, especially in winter and summer months. In Bangladesh, for example, *aman* paddy (paddy grown during monsoon) is almost totally rainfed crop, while *boro* paddy (summer rice) is essentially irrigated with tubewell water. Our survey shows that in almost 72% of the villages, groundwater is the main source of irrigation for *boro* paddy in Bangladesh. In Nepal, winter wheat is the main groundwater irrigated crop, while wet paddy grown during monsoon is predominantly rainfed. In Pakistan Punjab and Sindh, however, conjunctive use is the most important source of irrigation for wheat, with 42% and 64% of sample villages reporting this as the main source of irrigation. In NWFP in Pakistan, wheat is irrigated either with canal water (22% of villages report it as the main source) or groundwater (27% villages report this as the main source). Table 9 summarizes cropping pattern of sample well owners in different regions of South Asia.

Food grains, such as wheat, paddy and coarse cereals are the main irrigated crops in whole of South Asia. In addition, sugarcane and vegetables occupy large parts of land irrigated by groundwater. Though, area under sugarcane and vegetables are low in absolute terms, these crops need more frequent irrigation and therefore hours of irrigation per hectare is much higher than say wheat. Most of irrigated paddy summer paddy needing intensive irrigation. We can therefore

Table 10. Impact of type of WEMs (electric/diesel) on hours of pumping and crop yields for paddy in Eastern, Tribal and Northern India.

Equation 1: Number of hours of operation of well/tubewells for paddy irrigation as a function of electric or diesel pumps (dummy = 0 for diesel and 1 for electric)

Hours of operation = 194.659* + 146.987D* ... (Equation 1)
$R^2 = 0.029$, N = 360, * means significant at 1% level of significance

Equation 2: Yield of paddy (100 kg/ha) as a function of electric or diesel pumps (dummy = 0 for diesel and 1 for electric)

Yield of paddy = 40.110* – 8.702D ... (Equation 2)
$R^2 = 0.04$, N = 360, * means significant at 1% level of significance

Source: Primary survey conducted by IWMI in 2002.

expect very high pumping requirements (h/ha) for paddy. Analysis on cropping pattern will be incomplete without an analysis of water requirement of irrigated crops and their yields.

8.2 *Hours of pumping and crop yields*

Assured irrigation leads to a shift in cropping pattern away from subsistence crops to high yielding varieties and water intensive crops that are remunerative on the one hand and very sensitive to water requirements on the other. In India, for example, area under coarse cereals and oilseeds went down in most parts of the country, while area under HYV of wheat, paddy and sugarcane went up after Green Revolution (Bhalla and Singh, 2001). Tubewell irrigation in areas of good (both quality and quantity) groundwater availability is amenable to cultivation of water intensive crops, because of its few inherent characteristics, such as control of the farmer over his source of water. Hours of pumping depend on crop requirements and we hypothesize that it also depends on the source of energy. Thus, while a diesel tubewell operator tries to economize on water use by irrigating as efficiently as possible, an electric tubewell owner facing flat tariff has no incentive to do so and he over irrigates.

Shah *et al.* (2002) talk about the energy irrigation nexus and one of their hypotheses is that electric pumps operate for a significantly higher number of hours than their diesel counterparts, though there is not much difference in yields. Our finding seems to confirm this hypothesis. We regressed hours of pumping for wet paddy cultivation in Eastern, Tribal and Northern India against type of motive power (diesel and electric), using dummy variable. The result shows that an average diesel pump gives 195 hours of irrigation to paddy, while an average electric pump owner irrigates his paddy for an additional 147 hours, though their yields are not significantly different (Table 10: Equations 1 and 2). Electric pumps also operate for 31% more hours for irrigating one hectare of wheat and though wheat yields are higher for those irrigated with electric tubewells (at 10% level of significance), it is only marginally so (Table 11: Equations 3 and 4). The R^2 values are predictably very low, one does not really expect merely type of water extracting device to explain differences in crop yields. This therefore presents an instance of over irrigation prevalent in much of South Asia and Shah *et al.*'s (2002) argument for fine tuning electricity supply according to seasonal crop water requirement merits attention. Based on our survey data, we present an estimate of crop yields and hours of pumping (both diesel and electric) required for wheat and paddy the two most important irrigated crops across regions (Table 12).

Table 12 brings out the salient features of two main groundwater irrigated crops, *viz.* wheat and paddy. Wheat yields are more or less comparable across various regions, though it is still Pakistan and Indian Punjab that achieve the highest yields in the sub-continent. However, because in Pakistan Punjab, farmers depend

Table 11. Impact of type of WEMs (electric/diesel) on hours of pumping and crop yields for wheat in Eastern and Northern India.

Model 1: Number of hours of operation of well/tubewells for wheat irrigation as a function of electric or diesel pumps (dummy = 0 for diesel and 1 for electric)

Hours of operation = 70.998* + 22.124D**... (Equation 3)
$R^2 = 0.028$, N = 279, * means significant at 1% level of significance
** means significant at 5% level of significance

Model 2: Yield of wheat (100 kg/ha) as a function of electric or diesel pumps (dummy = 0 for diesel and 1 for electric)

Yield of wheat = 24.933* + 3.171D*** ... (Equation 4)
$R^2 = 0.011$, N = 279, * means significant at 1% level of significance
*** means significant at 10% level of significance

Source: Primary survey conducted by IWMI in 2002.

Table 12. Yield and hours of groundwater irrigation supplied for cultivating wheat and paddy in South Asia.

Region/Country	Wheat		Paddy	
	Yield (kg/ha)	Hours of pumping/ha	Yield (kg/ha)	Hours of pumping/ha
North Western India	3,600	91	3,700	352
Eastern India	2,600	58	5,200	321
Central Tribal India	3,100	67	2,800	104
Central and Western India	3,200	144	Not important as a crop	Not calculated
Interior South India	Not important as a crop	Not calculated	4,400	542
Coastal South India	Not important as a crop	Not calculated	4,700	607
Pakistan Punjab[11]	3,500	44	3,100	168
Pakistan Sindh[11]	3,100	69	3,300	85
Pakistan NWFP	2,700	58	Not important as a crop	Not calculated
Bangladesh[12]	2,900	20	5,100	571
Nepal *Terai*[13]	2,300	54	3,400	57

Source: Primary survey conducted by IWMI in 2002.

[11] Conjunctive use is more important in Pakistan Punjab and Sindh, which partly explains lower pumping hours for wheat and paddy as compared to Indian Punjab which has similar cropping pattern.

[12] Data for paddy in Bangladesh here refers to only *boro* paddy, as this is the most important irrigated crop in the country, while *aman* paddy is mostly rainfed. But data from Eastern India paddy includes both monsoon paddy (rainfed and occasionally 1 or 2 irrigations) and *boro* paddy in West Bengal. This explains the lower pumping hours reported for paddy in Eastern India *vis-à-vis* Bangladesh.

[13] Paddy in Nepal *Terai* refers to rainy season paddy. This crop is mostly rainfed, only 1 or 2 irrigations are given from tubewell, which explains the lower pumping h/ha.

more on conjunctive use of canal and groundwater, total number of hours of pumping for wheat is lower in Pakistan Punjab than Indian Punjab. Growing wheat is highly water intensive in Western India and very much less so in Eastern India and Bangladesh. In Bangladesh for example, on an average 2,800 kg can be obtained per hectare of land using merely 20 hours of pumped groundwater. This is because wheat only need on an average 2 irrigations towards maturity stage, while in the initial stages, soil moisture fulfills water requirement for wheat in Bangladesh. On the other extreme, growing *boro* paddy is a very water consuming activity in Bangladesh. The *boro* season starts towards end of December and the crop is harvested between April to June, depending on whether it is a long or a short growing variety. Thus, *boro* cultivation coincides with high evapotranspiration and paddy being highly sensitive to water stress, yields can reduce drastically if water requirements of the crop are not met. Paddy cultivation is also water intensive in Coastal Peninsula and Interior Peninsula of India. In these areas, it is a common practice to take 2–3 crops of paddy and East and West Godavari districts in the state of Andhra Pradesh are known for the huge paddy harvests that they take every year. However, this booming groundwater based paddy cultivation is artificially supported by high electricity subsidies. In fact, in the Indian state of Tamil Nadu, electricity for agricultural purposes is supplied free of cost, and in Andhra Pradesh and Karnataka, electricity tariffs are very low. These raises questions about long term sustainability of groundwater based cropping systems, which if not rectified soon, will spell disaster for millions of poor farmers in decades to come. Punjab and Haryana are already facing serious problems of soil salinity and groundwater depletion and governments in these states are trying to provide incentives to the farmer to shift away from water intensive rice-wheat cropping systems.

9 WATER MARKETS IN SOUTH ASIA

The role of irrigation from groundwater has been crucial in the growth of agricultural production in South Asia and many scholars claim that most of it is allocated through groundwater markets (Shah, 1993; Palmer Jones, 1994). Evidence of water markets are found in all of South Asian countries of India, Pakistan and Bangladesh and in different geo-hydrological regimes. Shah (1997) puts forward a tentative hypothesis regarding agrarian transformation through accumulation of machine capital, particularly, pump capital and emergence of groundwater markets in much of South Asia. He presents a simplified, stage wise, linear model of agrarian transformation in the context of Eastern India much along the contours suggested by Rostow. He proposes that under *initial* conditions of low groundwater development, share of land in agricultural value added is very high, thereby precluding the poor and landless farmers from benefiting substantially from agriculture. This stage is also characterized by dependence of human and animal labour and nature of asset lease market is crude and rudimentary. The next phase is characterized by accumulation of machine capital (particularly pump sets) by rural elite. Water markets or lease market for other assets (except land) does not develop fully at this stage, because large farmers use all the water that their machines pump for self cultivation. However, in the later stages, even the small and medium farmers start investing in machine capital in general and pump capital in particular, water markets develop because they can not use all water on their own small fields. Close on the heel follows rental markets of other agricultural equipments such as tractors, threshers, etc. Consequently, the share of land goes down, while that of machine capital and labour goes up and rural economy stands transformed. For example, share of landowner reduces from 50% in backward regions such as Baliya, Gazipur and Azampur in Uttar Pradesh to only 25% in groundwater back vibrant economies in North Gujarat. Similarly, in the former case, the land owner does not share any cost of cultivation, while in the later, tenant and landlord share costs. Similar evidences are also got from Bangladesh (Crow, 1999). To this, we add another stage, the fifth and possibly the last stage, when more and more farmers accumulate pump capital and therefore the need to buy water from other farmers' declines. Thus water markets thin out.

Alternatively, heavy accumulation of pump capital might also lead to depletion of groundwater resources, thereby making water markets unsustainable. Perhaps, this is what we note in much of South India, where it seems that only 9% to 10% of our sample farmers reported selling water. This is in sharp contrast to figures quoted by Janakarajan[14] (1994, p. 52) a decade ago when he reports that almost 21% of the well owners were involved in water sales in Vaigai Basin in Tamil Nadu.

In our present survey, we asked a few questions regarding water selling and buying operations of well owners and price paid for bought water. Table 13 summarizes those findings.

Water prices vary according to regions, crops and whether it is an electric or a diesel pump. Invariably, water is cheaper from an electric WEM because of lower fuel costs, while it is much higher for diesel WEMs. There are generally three modes of payment for purchased water, *viz*. hourly rates (Rs or Taka/h), area based rates (Rs or Taka/ha) and crop share. Mode of payment too varies from one crop to the other. For example, in West Bengal and Bangladesh, for *boro* paddy, the most prevalent mode of payment till recently was 1 to 2 *maunds*[15] of paddy per *bigha*[16] (300–600 kg/ha), which comes out to be roughly 25% of total output. From institutional economics point of view, it can be explained that because *boro* paddy is a highly water intensive crop and chances of crop failure are very high, while market prices was good till early 1990s, the water seller and water buyer share risk of crop failure through this arrangement. Share payment provided incentive to the water seller to provide timely water and it helps the buyer because his risk is shared. However, he might have to shell

Table 13. Dynamics of water markets in South Asia.

	Pak. Punjab	Sindh	NWFP	Punjab, Haryana, West UP	Eastern India	Tribal India	Western India	Interior South India	Coastal South India	Bangladesh	Nepal *Terai*
1. Percentage of well owners selling pump irrigation	33	29	19	24	46	2.5	10	6	9	88	62
2. Average number of buyers/seller	3	3	8	4	5	4	3	3	1	11	4.5
3. Average annual hours of sale/seller	317	106	237	127	150	90	680	340	72	634	98
4. average size of buyer area served per seller (ha)	10	6	8	5	3.7	0.5	4	1.3	0.53	3	3
5. Percentage of well owners who bought pump irrigation	5	0	0	11	6.6	Negligible	2	1	0	36	2
6. Average number of hours bought/yr	133	NA	NA	53	85	NA	98	35	NA	140	31
7. Average area irrigated with purchased pump irrigation (ha)	6	NA	NA	2	1.1	NA	1	2	NA	0.6	0.67

Source: Primary survey conducted by IWMI in 2002.
NA: Not available

[14] Janakarajan (1994) presents the case of groundwater transactions in Vaigai Basin of Tamil Nadu. He bases his findings on field work conducted in 1990–91, exactly 11 years before our present survey.

[15] *Maund* in West Bengal and Bangladesh context is equal to 40 kg.

[16] 7.47 *bighas* is equal to 1 ha in most of West Bengal.

out a larger share (in terms of monetary equivalence), since *boro* prices are high. However, in recent years, we observe a change in this mode of payment. For one, *boro* prices plummeted in the early 1990s in most of Bangladesh (Palmer Jones, 1999) as well as in West Bengal. Consequently, from the tubewell owners' perspective, share payment became less lucrative than before and most of them shifted to area based payment. The prevailing rate in Bangladesh is Taka[17] 2,950/ha (US$ 52/ha) for *boro*, provided the buyer provides diesel. When diesel costs are included, this figure becomes Taka 5,800–6,000/ha (US$ 100–105/ha). In West Bengal, the prevailing rate for *boro* paddy is around Ind Rs 4,000/ha (US$ 85/ha). However, in case of vegetables, water rates are charged on an hourly basis, ranging from Ind Rs 40–60 (US$ 0.85–1.30) for diesel tubewells and Ind Rs 20–25 (US$ 0.45–0.55) for electric tubewells. Among all the modes of payment, hourly payment seems to be the most popular mode in most of South Asia. Table 14 summarizes the level of water charges prevalent in different areas.

Debating the desirability of water markets, researchers have time and again revisited the question of equity implications of water markets. In doing so, some scholars like Janakarajan (1994) and Adnan (1999) working on two entirely different geo-hydrological regimes of Deccan plateau and Bangladesh plains respectively have contended that groundwater markets are a

Table 14. Water charges for bought groundwater in South Asia.

Region/Country	Ind Rs or Pak Rs[18] or Taka or Nep Rs[19] /h	Ind Rs or Pak Rs or Taka or Nep Rs/h currency/ha	Share of crop
North Western India	Ind Rs 15–25/h for electric tubewells, Ind Rs 35–50/h for diesel tubewells	Rs 2,000/ha for 5 watering to wheat crop.	15% of paddy, only 2 reported case out of 55
Eastern India	Ind Rs 15–25/h for electric, but Ind Rs 35–60/h for diesel. Rates are higher for vegetables than for wheat	Varies between Ind Rs 3,500–5,000/ha for *boro* paddy	25% of *boro* paddy, very rare arrangement
Central Tribal India	Ind Rs 15–25/h for electric tubewells	None	None
Central and Western India	Ind Rs 20–40/h	None reported	1/3rd to 1/4th
Interior South India	Ind Rs 15–30/h	None reported	1/4th
Coastal South India	Ind Rs 20–30/h	Ind Rs 2,500/ha for paddy	1/4th
Pakistan Punjab	Pak Rs 35–60/h for electric, Pak Rs 65–95/h for diesel, and Pak Rs 110–220/h for tractor operated	None reported	None reported
Pakistan Sindh	Pak Rs 60/h for electric, Pak Rs 70/h for diesel	None reported	None reported
Pakistan NWFP	Pak Rs 35–50/h for electric, Pak Rs 80–100/h for diesel and Pak Rs 200/h for tractor driven	None reported	None reported
Bangladesh	Taka 35–40/h	Taka 5,500–7,000/ha including price of diesel which the buyer has to bear	1/3rd to 1/4th share
Nepal *Terai*	Nep Rs 80–100/h on an average	None reported	None reported

Source: Primary survey conducted by IWMI in 2002.

[17] Bangladeshi taka. Taka 1 = US$ 0.01765 (July 1, 2003).
[18] Pakistan rupee. Pak Rs 1 = US$ 0.01797 (July 1, 2003).
[19] Nepalese rupee. Nep Rs 1 = US$ 0.01398 (July 1, 2003).

vehicle for "few farmers (to) emerge with power to exercise control over this precious resource and extract surplus" (Janakarajan, 1994, p. 45). But a host of other scholars (Shah and Raju, 1988; Shah, 1993; Fujita and Hossain, 1995; Palmer Jones, 2001; Dubash, 2002) forcefully argue that water markets have been very crucial in alleviating rural poverty. Palmer Jones goes on to say that water markets "... are far more important institutions for human welfare than the *community irrigation*, *community forestry*, or even ... the group savings and credit examples ..." (Palmer Jones, 2001, p. 3). Equity implications are not always clear, but, what is more important: poverty alleviation or equity? Perhaps in the (very) short run a dominant class of rich and middle peasantry does accumulate monopolistic surplus through water selling, but in the long run the poor and the marginal farmers do benefit substantially since pump led machine capital reduces the share of land in total agricultural output and they (who have less or no land) get better value for their labour.

Water markets have both positive and negative externality in many pockets of South Asia. Strosser and Meinzen-Dick (1994) contend that increased pumping and sale of groundwater is beneficial in lowering water tables in much of salinity and water logging affected canal command in Pakistan. The same holds true to much of Nepal *Terai* and low flood plains of West Bengal and Bangladesh. However, in regions where groundwater resource base is already scarce, water markets could hasten resource depletion, as possibly might have happened in North Gujarat and part of Interior South India. However, in order to ascertain both the claims, one needs to undertake in depth study of water markets in different agro-ecological regions, which this study does not quite do.

10 CONCLUSIONS AND POLICY IMPLICATIONS

Growing contribution of groundwater in South Asia's agricultural economy remains both underestimated and understudied. However, now there is an increasing awareness about the important role that groundwater has played in fostering food sufficiency in much of this poverty stricken belt of the world. At the same time, there is a realization that much of this precious resource stands the chance of rapid and irreversible exploitation in many parts of South Asia. Thus, South Asia is plagued by twin problems, one of resource depletion and deterioration in the Western and South India and parts of Indian and Pakistan Punjab (affected by salinity) and that of concentrated rural poverty ironically concentrated in the most water abundant parts of South Asia, *viz*. Eastern India, Bangladesh and Nepal *Terai*.

Groundwater irrigation has been at the heart of agrarian transformation in South Asia. On the plus side, groundwater has ensured food sufficiency and provided livelihoods to millions of people. On the downside, it has created chronic problems of resources depletion and degradation. However, a major pitfall of this over simplified characterization of South Asia's groundwater socio-ecology is that it hides more than it reveals. This is because the South Asian region is immensely complex in terms of agro-ecological, socio-economic and institutional characteristics. This is compounded by the fact that South Asia is undergoing a rapid transition in almost all spheres of the economy, including the agrarian economy. Thus the first step towards understanding this colossal economy is to undertake a spatially disaggregated, refined and nuanced analysis of it. Having done that, the next step will be to suggest appropriate groundwater management policies suited to this region and its uniqueness.

Broadly speaking, groundwater management poses entirely different challenges in the water abundant eastern parts such as Indian states of West Bengal and Bihar, Bangladesh and Nepal, and water scarce parts such as Western India. In eastern parts, extreme poverty persists amidst abundance of groundwater. In such areas, groundwater irrigation can be geared towards poverty alleviation through proper agricultural, credit and energy support policies. On the other hand, hard rock peninsular and Western India poses serious groundwater depletion problems leading to further immiserization of the poor. In this context, different sets of policies aimed at supply augmentation as well as demand management needs to be pursued.

Uniqueness of the South Asian situation precludes it from emulating water management practices of the developed countries like USA and Australia. Thus, instead of *importing* water solutions from the west, South Asia needs to develop its local *home grown* solutions. This process has already begun. In parts of South Asia where groundwater is under the worst threat of depletion there is a growing groundswell of popular action in rainwater harvesting and local groundwater recharge. At the frontline of this movement are regions like Rajasthan and Gujarat in India where untold havoc and misery are a certain outcome if the groundwater bubble were to burst. However, so far, *home grown* solutions have focused more on supply augmentation. Sustainable groundwater management will remain elusive unless demand management mechanisms are developed. It seems unlikely that there will be groundswell of popular movement to facilitate demand management given the current scenario of crucial dependence on groundwater resource. Thus, indirect management options such as proactively managing the energy irrigation nexus and tinkering with the food policy might offer hitherto untried mechanisms for optimal management of this precious resource.

Groundwater, it is said, will be the enduring test of this generation's capability in water and land management and nowhere will this be to put to harder test than in South Asia, which uses the largest chunk of world's groundwater for agriculture.

REFERENCES

Adnan, S. (1999). Agrarian structure and agricultural growth trends in Bangladesh: the political economy of technological change and policy interventions. In: B. Rogaly, B. Harriss-White and S. Bose (eds.), *Sonar Bangla? Agricultural Growth and Agrarian Change in West Bengal and Bangladesh.* Sage Publications. New Delhi, India.

Ballabh V.; Chowdhary, K.; Pandey, S. and Mishra, S. (2002). *Groundwater development and agricultural production: a comparative study of Eastern Uttar Pradesh, Bihar and West Bengal.* IWMI-Tata Water Policy Research Program. Anand, India.

Bhalla, G.S. and Singh G. (2001). *Indian Agriculture: four decades of development.* Sage Publications, New Delhi, India.

Boyce, J.K. (1987). *Agrarian impasse in Bengal: agricultural growth in Bangladesh and West Bengal 1949–1980.* Oxford University Press. New York, USA.

Crow, B. (1999). Why is agricultural growth uneven? Class and agrarian structure in Bangladesh. In: B. Rogaly, B. Harriss-White and S. Bose (eds.), *Sonar Bangla? Agricultural Growth and Agrarian Change in West Bengal and Bangladesh.* Sage Publications. New Delhi, India.

Deb Roy, A. and Shah, T. (2003). Socio-ecology of groundwater irrigation in India. In: M.R. Llamas and E. Custodio (eds.), *Intensive Use of Groundwater: Challenges and Opportunities.* Balkema Publishers. The Netherlands. Pp. 307–335.

Dhawan, B.D. (1989). *Studies in irrigation and water management.* Commonwealth Publishers. New Delhi Publishers. New Delhi, India.

Dubash, N.K. (2002). *Tubewell Capitalism: groundwater development and agrarian change in Gujarat.* Oxford University Press. New Delhi, India.

Fujita, K. and Hossain, F. (1995). Role of groundwater market in agricultural development and income distribution: a case study in a Northwest Bangladesh village. *The Developing Economies*, 33(4): 442–463.

GOB (Government of Bangladesh) (various years). *National Minor Irrigation Census of Bangladesh*, 1995–1996 and 1999–2000. Dhaka, Bangladesh.

GOI (Government of India) (1993). *Report on Census of Minor Irrigation Schemes, 1993*, Volume 1. Ministry of Water Resources. Minor Irrigation Division. New Delhi, India.

GOI (Government of India) (several years). *Agricultural Census.* Ministry of Agriculture. New Delhi, India.

Hernández-Mora, N.; Martínez Cortina, L. and Fornés, J. (2003). Intensive groundwater use in Spain. In: M.R. Llamas and E. Custodio (eds.), *Intensive Use of Groundwater: Challenges and Opportunities.* Balkema Publishers. The Netherlands. Pp. 387–414.

Janakarajan, S. (1994). Trading in groundwater: A source of power and accumulation. In: M. Moench (ed.), *Selling water: conceptual and policy debates over groundwater markets in India.* VIKSAT and others. Ahmedabad, India.

Kanhert, F. and Levine, G. (1989). Key findings, recommendations and summary. *Groundwater irrigation and rural poor: options for development in the Gangetic basin.* World Bank. Washington DC, USA.

Meinzen-Dick, R. (1996). *Groundwater markets in Pakistan: participation and productivity.* International Food Policy Research Institute. Washington DC, USA.

Palmer Jones, R. (1994). Groundwater markets in South Asia: a discussion of theory and evidence. In: M. Marcus (ed.), *Selling water: conceptual and policy debates over groundwater markets in India.* VIKSAT and others. Ahmedabad, India.

Palmer Jones, R. (1999). Slowdown in agricultural growth in Bangladesh. In: B. Rogaly, B. Harriss-White and S. Bose (eds.), *Sonar Bangla? Agricultural Growth and*

Agrarian Change in West Bengal and Bangladesh. Sage Publications. New Delhi, India.

Palmer Jones, R. (2001). Irrigation service markets in Bangladesh: private provision of local public goods and community regulation. Paper presented at *Symposium on Managing Common Resources: what is the solution?* Lund University, Sweden, 10–11 September 2001. (http://www.sasnet.lu.se/palmer_jones.pdf).

PPSGDP (Punjab Private Sector Groundwater Development Project) (2001). *Legal and Regulatory Framework for Punjab Province*. Technical Report 45. Lahore, Pakistan.

Shah, T. (1993). *Groundwater Markets and Irrigation Development: Political Economy and Practical Policy*. Oxford University Press. Bombay, India.

Shah, T. (1997). *Pump irrigation and equity: machine reform and agrarian transformation in water abundant Eastern India*. Policy School, Working Paper 6. The Policy School Project. Anand, India.

Shah, T. and Raju, K.V. (1988). Groundwater markets and small farmer development. *Economic and Political Weekly*, 19: A23–A28.

Shah, T.; Deb Roy, A.; Qureshi, A.S. and Wang, J. (2001). *Sustaining Asia's Groundwater Boom: an overview of issues and evidence*. IWMI's contribution to International Conference on Freshwater in Bonn. 3–7 December 2001. German Development Institute, Bonn, Germany; and International Water Management Institute, Colombo, Sri Lanka.

Shah, T.; Scott, C.; Kishore, A. and Sharma, A. (2002). *Energy-Irrigation Nexus in South Asia: Approaches to Agrarian Prosperity with Viable Power Industry*. Draft Paper. IWMI Tata Water Policy Research Program. India.

Strosser, P. and Meinzen-Dick, R. (1994). Groundwater Markets in Pakistan: an analysis of selected issues. In: M. Moench (ed.), *Selling water: Conceptual and Policy Debates over Groundwater Markets in India*. VIKSAT and others. Ahmedabad, India.

WAPDA (Water and Power Development Authority) (1989). *PC-II performance for monitoring Salinity Control and Reclamation Projects (SCARPs) for the period 1989–93*. SMO Publication 100. Lahore, Pakistan.

Legal issues of intensive use of groundwater

Antonio Embid
Faculty of Law, University of Zaragoza, Spain
aembid@unizar.es

ABSTRACT

It is very difficult to give a general and abstract description of groundwaters, since such a description is closely linked to the physical characteristics and historical evolution of each country. However, there are common trends in modern water law, such as the prevalence of public ownership of groundwaters and the scant attention given to private ownership; other common trends are the extensive intervention by the public authorities into public and private water, as well as legislation that is heavily orientated towards environmental protection. The general characteristics might also include the value of hydrological planning in the regulation of water management and the problems in adapting this management to the limits of a particular basin, although this normally applies to surface waters. Neither can it be said that law has exploited all the possible consequences of the principle of the hydrological cycle which would then call for a substantially similar legal framework for surface and ground waters, and the promotion of joint management for all of them. In the international context, one can find many more conventional norms devoted to surface than to ground waters, where these regulations are an exception. At any event, the growing use of groundwaters means that we can certainly expect an improvement in the legal regulations over the next few years, at a domestic as well as international level.

Keywords: *groundwater ownership, intensive use, overexploited aquifers, users communities, groundwater protection.*

1 INTRODUCTION

It would be a good idea to begin this paper by establishing some of the characteristics of the object of this study that have specific consequences for the consideration of their legal framework. Thus, from the legal viewpoint, which is what we are concerned with here, a methodical and ordered discussion on groundwaters is a noticeably complex problem, much more so than with surface waters. This might, in principle, be termed an apparent paradox, since, within the context of modern regulations for continental waters, almost all of them based on the principle of awareness of the hydrological cycle, there is no reason why there should be substantial differences between the legal framework for surface waters and that of groundwaters. However, this is hardly ever the case, which leads one to think that, at least in the context of law, not all conclusions have as yet been drawn from this evident circulation and transferability of water that is involved in the hydrological cycle, and which forms the basis of many contemporary legal systems, both expressly (such as specifically laid down in Spanish Law in article 1 of the revised Water Act, approved by Executive Order 1/2001, of 20th July) and implicitly.

What is probably the case is that in the context of Law, the myths and ignorance concerning groundwaters, which have been around for a long time, and still survive to a certain extent, are being prolonged, albeit artificially of course. In spite of the fact that scientific and technical advances have meant that the understanding and management of groundwaters is virtually comparable to that of surface waters, this is not wholly reflected in the legal system, which of course provides complications when putting it into practice (Caponera, 1992; Burchi, 1999). Thus, there is much more scope, as far as user activity and water availability is concerned, in the field of groundwater than of surface water, while at the same time, there are greater complications for the public authorities in monitoring possible abuse in the field of groundwater.

As we shall see later on, these discrepancies and difficulties are noticeable in all sectors of the regulation system under consideration: these include problems of ownership (private or public ownership), the powers of the Public Authorities (regulatory, policing powers) closely linked to environmental monitoring, how the users organise themselves (compulsory or voluntary, and under what scheme), and the economic/financial system, if a payment to the Authorities for water use is involved. As we shall see throughout this paper, there are specifics in the regulatory framework which may be difficult to explain in the light of technical criteria, but easily understandable from other viewpoints, especially if we bear in mind the historical evolution of regulatory frameworks in which there have always been special measures affecting groundwater.

This also involves another type of consequence that directly affects an article like this. This consequence is the difficulty in finding homogeneous regulatory systems in most of the content with regard to institutions and their effective regulatory schemes. Law in general (although what I am about to say is applicable to water law, it is applicable to any other type of law) has a substantially national basis, at least in the current phase of our political and cultural evolution. Water law, with its wide-ranging viewpoints and nuances, can only be explained in terms of physical characteristics (the regulatory system of a country with abundant rainfall is not the same as that of a country with regular drought problems) and, above all, in terms of how each country has evolved historically, an aspect that is often unrepeatable, impossible to transpose to other countries, and also very hard to place within a rational framework.

This means that legal science has a characteristic that is not normally found in other sciences. Thus, in the field that concerns us here, knowledge of hydrology, abstraction techniques, the ways of defining aquifers, the principles of coordinated and integrated management between surface and groundwaters, the chemical composition of waters, the struggle against pollution, etc. are universal; experts in the various fields, regardless of their countries of origin or places referred to in their research, immediately have common tools (starting with the terminology and the same sources of knowledge) which helps communication between them enormously, and gives rise to the seamless interchange of solutions and, especially, means to tackle problems.

But in law, you do not get this terminological or conceptual homogeneousness, and of course, the sources of knowledge are completely different in their origin and presentation. This means that when jurists want to find or construct transfrontier elements of communication – as is the case today in most fields – we are forced to give very general descriptions, quoting only the most pertinent national legal regulations, and they are pertinent for our purposes because that is where groundwaters receive the highest profile because of their generalised use.

However, there are certain causes or problems that are moving towards a process of standardisation between the various national legal systems. These are mainly aspects referring to the environment. Prevention principles applied to the quality of groundwaters and the ecosystems linked to them, or sustainable development principles in general, have given rise to a certain type of social acceptance (by users of, or those generally involved in, groundwaters or simply continental waters) and acceptance by the public authorities which, in the end, are incorporated into law, as may easily be seen in various legal reforms that have taken place in recent years and which usually have this common characteristic. This is where we shall gradually find – in a process that will become more intensive in the future – the necessary line to follow in the standardisation of groundwater law. Of course, this is something that will take time to achieve, but it is, nevertheless, worth pursuing.

At any event, it should always be borne in mind that within the specific field of environmental protection, there are clear trends in legal regulations to apply many more precautions with regard to the use of groundwaters than with surface waters. Outflows are a clear example of this, in which the regulations for groundwaters are much stricter than those for surface waters. In European countries, this is closely linked with the Community Regulations, and the recent Directive 2000/60/EC, which I shall examine in more detail later, makes a distinction in the environmental objectives laid down in article 4 – in its handling of groundwaters as against surface waters, something which should be borne in mind at all times.

Finally, what is true is that, as the most reliable sources show, the use of groundwaters is increasing everywhere (*cf.* Shyklomanov, 1997). There are ever more irrigation areas using water that was originally groundwater (and studies show, furthermore, that these waters are economically and socially the most profitable, because they call for more economy and rationalisation from users) and good many urban supply services use water originating from groundwater because it is usually of better quality and fresher than surface water, and thus is easier to treat. All this adds to the need to examine the legal reality of waters that are increasingly to be found in the economic life of widely diverse countries, now that science and technology have, in the second half of the 20th century, made it easier to abstract this resource, thereby affording more profitable usage.

2 REGULATORY FRAMEWORK

I shall now touch on matters that are always to be found in any text about law: the defining of the sources of knowledge, which I shall put simply here under this heading of the *regulatory framework*. In some countries, groundwater is mentioned in the Constitution, although this does not happen very often. For example, groundwaters are mentioned in the Constitutions of Brazil (1988), and Mexico (1917), with modifications in this respect in 1943. It is more common for Constitutions to refer to water resources in general, including, of course, references to surface as well as groundwater. This is the case with the Constitution of the Netherlands (1956). Elsewhere, especially in countries whose government structure is based on political decentralisation, water may be mentioned in the sections on power-sharing between the State (Federation) and the decentralised States. This is how the Spanish Constitution (1978) should be read, with art. 149, 1.22, which I shall come back to later. A completely different, and more complex, matter would be to examine mentions of groundwaters that might be found in the Constitutions or Statutes of Autonomy (or similar designations) of Federated States (*Ländern*, *Regioni*, *Comunidades Autónomas*, etc.), a subject that cannot be gone into here for reasons of space.

[Of course, despite the absence of any express mention of water, we should never ignore any references to natural resources (of which water is one) or the environment in general which each Constitution might contain, especially the more modern Constitutions in which the environment is fully incorporated into their content. A good example of what I mean is to be found in art. 45 of the Spanish Constitution and its reference to the "rational use" of natural resources, where the Constitution is in line with the European Communities Treaty, in which there is also a reference to the "rational use" of natural resources, *cf.* art. 174].

It is hard to draw any conclusions, beyond what has already been expressed, on the presence or absence of references to water in the highest legal instrument of different countries. At any event, the basic legal framework for groundwater, with regard to its configuration and direct application, may be found in the ordinary legislation which should, logically, be guided by what the Constitution says for reasons of normative hierarchy. However, what sometimes happens is that the texts of the Constitutions are by no means models of legal precision; they are open to various interpretations and may even be contradictory when it comes to the various references to water that they contain. This, as one may imagine, causes problems when trying to decide what the content of ordinary legislation should be. Without doubt, this vagueness is caused by the fact that most Constitutions are by way of an accord or political pact, in which this type of effect sought for more than that of a wholly coherent legal norm that may be seamlessly interpreted.

A prime example of what I have just said is the Constitution of Mexico, whose unusually extensive art. 27 is open to various interpretations with regard to an issue as basic as the ownership of groundwaters, and which needs to be interpreted in terms of the wishes of the drafters of the Constitution and, especially, what ordinary legislation subsequently said (see the 1992 National Water Act, and its doctrine in the work by Farías, 1993) in favour of groundwaters being considered as national (public) water, although there were references in this Constitution to the possibility of private acquisition.

Thus, the predominant role is played by ordinary legislation, if we are required to define – as mentioned in the heading – the general characteristics of the *regulatory framework*, and in most cases we come across texts that regulate jointly and uniformly all continental waters, surface as well as groundwaters. In *old* regulations (by *old*, I mean water laws that appeared at the end of the 19th century and the first seven decades of the 20th century, which still account for many countries' water legislation), there are substantial differences in regulation between surface and groundwaters, not least their ownership (Moreu, 1996; Pérez Pérez, 1998), while more modern legislations are guided by similarities in their legal framework. It is noticeable that many countries modified their water legislation at the end of the 20th century (Spain did so in 1985, France in 1992, Mexico in 1992, Italy in 1994, Brazil in 1997, etc., to mention just some of the more representative examples), which also enabled them to incorporate environmental considerations, today an essential component of the legal framework governing water or any other natural resource.

At any event, when talking of *ordinary* legislation, we should not just be thinking of what is usually termed in each country as *The Water Act*. Today, there are also regulations applicable to groundwaters in specific environmental impact assessment legislation, or in regulations for territorial planning or natural protected spaces (inasmuch as they affect marshlands that result from the springing of groundwaters). Even in cases where a price is established for water usage, there are norms applicable to what we are concerned with here in financial legislation.

3 THE OWNERSHIP OF GROUNDWATERS

In connection with what was discussed in the previous section, it is noticeable that old legislation used to recognise individuals' rights not only to be owners of private water resources (a situation which we might classify as being static) but also to continue acquiring the private ownership of groundwaters (a dynamic situation). In most cases, it was the owners of land who were permitted to look for groundwaters by digging wells or channels, allowing the waters to spring, and by exploiting them, incorporate them into their estate as private property (because this property extended *usque ad inferos*, in the words of the classic Roman expression within the evolution of a system so masterfully dealt with by Nieto, 1968). And so it is the ownership of the land which results in implicit ownership of the groundwater to be found beneath this land by the owner or a *prospector* other than the owner (with the owner's permission, of course). What I am describing here are concepts of ownership originating in Roman Law, translated to the Civil Code under the auspices of the Napoleonic Code, and thus twenty centuries of evolution have brought it up to very recent times (Moreu, 1996; Del Saz *et al.*, 2002).

[An exception to this can be found in some Arab countries. Based on precepts in the Koran, under common law ownership of channels, wells or springs also brings with it the ownership of a certain area of adjoining land – known as *harim* – upon which, so as not to harm the quality of water in an existing well, or reduce its quantity, it is forbidden to sink a new well. See Abderrahman (2001), with regard to the existing situation in Saudi Arabia; and Hajji (1995), with regard to the existing situation in Morocco].

However, this is a situation that has been deemed unsatisfactory because of the harmful effects that have been caused with the increasing use of groundwater, since unrestricted prospecting, appropriation, and consequently, use of groundwaters by the owner of the land has unavoidably led in many cases to the drying up of aquifers and ill-effects on the surface water courses connected to them (simply as a result of the hydrological cycle). Furthermore, a natural consequence of the water cycle is that all waters, in whatever shape or form, need to be treated the same under law. This has meant that most of the expert doctrine has for many years called for the legal framework to recognise the extension of the principle of public ownership (normal with regard to surface waters) to take in groundwaters as well (Martín Retortillo, 1987). In some cases, such as the USA (see Getches, 1997, pp. 247 *et seq.*; and Sax *et al.*, 1991, pp. 374 *et seq.*), the situation is even more complex, with the coexistence of various legal frameworks in the same country, with very different notions, even though the idea of *reasonable use* of water resources (basically theoretical albeit later incorporated into diverse legislations) is introduced, thereby imposing harsh restrictions on a mere system of private appropriation, where this concept exists, and addressing, at any event, the general interests that, one way or another, would be implied or contained in the expression *reasonable use*.

This also holds for Spain, where it was the 1866 Water Act (and subsequently the 1879 Act) that allowed the appropriation of groundwaters by the owner of the land where they were to be found or where they sprang. This legislation has strongly influenced the various legislations in Latin America, which initially, and for obvious reasons, were guided by the moral authority of Spanish legislation. For this reason, under the auspices of a Constitution such as the 1978 Spanish Constitution, which largely redefined the institution of the public domain (*cf.* art. 132 with reference especially to the maritime-terrestrial public domain), Act 29/1985 of 2nd August, concerning water resources, was passed. This Act extended the status of public domain, which surface waters had held since the 1866 and 1879 Acts, to renewable groundwaters, regardless of their renovation time (*cf.* Martín-Retortillo, 1990; 1997, pp. 107 *et seq.*). (This Act does not expressly state how the law contemplates non-renewable groundwaters although, in fact, from the viewpoint of hydrogeology, it would be true to say that, theoretically, what we have here is a specific type of groundwater, as these waters will always be renewable, regardless of the fact that this could take a good many years). Thus, in Spain it has no longer been possible to acquire private ownership of groundwaters since the 1985 Act came into force (as of 1st January 1986), although in order to preclude any claims of unconstitutionality, existing owners of private water resources were permitted to exercise, for three years, an option with the Water Authorities which consisted of continuing under the scheme of private ownership (with certain restrictions) or else switching to a scheme of concessions, with certain privileges going with the status of concession-holder. Under Judgement 227/1998 of 29th November, the Spanish Constitutional Court declared that this was perfectly in line with the Constitution; this Judgement is very important in Spanish legal history and is given priority treatment by Del Saz (1990) and De la Cuétara (1989).

The Spanish legislation is not as perfect as it may seem at first sight, and in practice leaves a lot to be desired, in that the Water Authorities (basically the traditional River Boards) have not responded rapidly to the exercising of options by title holders, nor have they been too strict in controlling new wells sunk illegally after the 1985 Act came into force, mainly because of a shortage of economic and human resources to police this. In some places – a very good example is aquifer 23 in the eastern part of Castilla-La Mancha, in the Tablas de Daimiel National Park. – this situation is contributing to the deterioration of aquifers and wide scale harming of the ecosystems linked to them; this gives rise to the paradoxical result that the source of the River Guadiana – the famous *Ojos* (Eyes) of the Guadiana – which was in fact an *overflow channel* from aquifer 23, has *shifted* to over a hundred kilometres away as a result of the fall in the aquifer's piezometric levels. Only a recent transfer of surface water from the Tagus basin, and certain restrictions in private usage, sweetened by public subsidies (known as the *Plan de Compensación de Rentas*) have managed to remedy the situation temporarily. However, this situation has given cause for grave concern, and is probably the most serious environmental problem regarding water resources that Spain currently faces.

Similar issues might be applied to France (see Varnerot, 2002) and Italy. This country's recent legislation has also tackled the public ownership of groundwaters (*cf.* Lugaresi, 1995, pp. 97 *et seq.*). This author, following the line taken by the text of the legislation, states how the transfer to public ownership takes place, because the public facet is, in fact, the interests surrounding groundwaters (Lugaresi, 1995, p. 410), and these are opinions that, as we shall see, are applicable to Italy, but also very similar to aspects to be found in other legal systems.

Elsewhere, such as Argentina, there has been progress from private to public ownership of groundwaters which, following discussions on doctrinal aspects, has become official through a reform of the 1968 Civil Code, although there are still doubtful areas and discussions to be held on doctrinal issues (*cf.* Chede, 2002).

In all systems, the transformation of the legal framework of an institution gives rise to problems involving transitory rights, not just in Spain, and these tend to spark thorny debates that can hold up the running of institutions and social progress for a good many years (Morell, 2001). When these transformations affect private ownership, it seems that the acquiring of a transitory right is an essential requirement, but at the same time, a source of conflict if the legal formulae are not adequately framed, or else the Public Authorities are simply incapable of administering this right.

4 ORGANISATION OF USERS

Modern water legislation systems tend to promote participation by users and the public in general in water management; they link this principle to the setting up of organisational formulae based on the territorial location of the river basins. There can be many forms of organisation, but the principle of participation, or even some form of self-administration is today one of the firm bases of water legislation in a good many countries. It would seem fairly clear that a community of users' interests, in the form of a certain organisation, results in better administration because of the interrelationship of powers and restrictions that accompany this situation. Full exercise of a right is only possible if the rights of others are considered, and of course it is much easier for other users to control any temptation of abuse that someone might have when exercising his own right.

With groundwaters, this self-organisation is much more necessary than with surface waters, and at the same time, harder to achieve; there is always more of a private enterprise element (sinking and exploiting a well, land transformation, the subsequent construction of irrigation infrastructures, etc.) than with surface waters. The Public Authorities tend to link the exploitation of surface waters to certain hydro works which will have been built by them (especially if these works were in the slightest measure important or costly), and this entitles them to impose a certain type of organisation for the use of water resources that have been regulated or transported by these infrastructures. With groundwaters, however, the situation is different; most of the investment usually comes from the landowners themselves, and they consider them as a natural extension of their property. As more and more groundwater is used – as a result of technical breakthroughs at the end of the 20th century – the need for organisations is seen as the only guarantee that all private rights are respected, and at the same time, that aquifers are environmentally protected. However, there is still a long way to go in this respect. What we do have are the Groundwater Technical Committees that have been set up since 1995 in the State of Guanajuato (this State in Mexico represents an advance in modern water management techniques; for an analysis, see Guerrero Reinoso, 2000; Sandoval, 2002), and in a more general form, the traditional Users' Communities in Spain, set up as Public Law Bodies, and therefore membership is compulsory, not voluntary, although we are still a long way from a situation in which these formulae, essential as they are for proper sustainable management of the resource, are a general reality.

At any event, it should be pointed out that the public or private nature of waters may determine particular characteristics of the Users' Communities established therein. Especially in Spain, it seems that the legal framework of Users' Communities which entails their consideration as Public Law Corporations – is designed for public waters. When there is still private water ownership, the Public Law Corporation formula does not seem to be appropriate. The case of Communities regarding aquifers that have been declared to be over-exploited may be a different matter, since then administrative intervention is much stronger.

5 MANAGEMENT OF GROUNDWATERS

One way or another, this analysis is leading to an appreciation of the key importance of regulating groundwater management within the context of the regulations affecting this. The idea that is being emphasised here is the wide-scale intervention by the public authorities in groundwater management, whether in systems based on private ownership (now in the minority) or public ownership (increasingly common, as say Burchi and Nanni, in press). In view of the above, we need to thoroughly examine this important issue by analysing the many variables in management.

5.1 *Generalisation of a permit system*

Thus, it is obvious in the modern water law, private ownership is no obstacle to detain the Public Authorities' powers to intervene when other aspects might be involved, such as the environmental factor, or simply when exercising a right affects the rights of other individuals. When water resources are publicly owned, their *private* use (and here I mean, use by an individual exclusive of other users) by individuals necessarily calls for an instrument (usually administrative, although in the case of certain USA States – such as Colorado – this may be judicial) issued at discretion; this is commonly known as a concession, although some legal systems use the term authorisation. But simply, what I am talking about here is a system of *permits*. Leaving aside terminological differences between one legal system and another, the use of water is only legitimate when under the auspices of a permit (regardless of whether it is in the guise of a concession or authorisation) issued by the relevant public authority.

5.2 *Hydrological planning*

It is also perfectly feasible that the use of water resources is subject to a system of administrative planning. Today, hydrological planning (*cf.* Embid, 1991) is incorporated into the most common water management tools, and great advances have been made in the field of water usage planning which have extended to areas much broader than the traditional area of hydro schemes, which was the first to use such a tool in the early 20th century. There is no doubt that Spain was the pioneer in hydro planning techniques, and they have now spread everywhere – as we shall see later – having been imposed in the European Union through Community Directives (EEC Directive 2000/60, mentioned previously, which requires all members to draw up basin hydrological plans, which are also required to include groundwaters). Despite the fact that there might be some conceptual problems with regard to applying it to private ownership waters, it is clear that this planning technique eminently suited to the realm of publicly-owned waters.

5.3 *Price for water use*

The use of groundwaters may also be linked to the payment of a certain price (designated as a tariff or by some other name) to the Public Authority. This is not a general characteristic; it only exists in some legal systems (France or Mexico, for example), whereas elsewhere the use of waters is free (a typical example is Spain), with payments being made to the Public Authorities only to cover the costs incurred previously when constructing the hydro works that have made the use of the water possible (water regulation, transport, treatment, etc.). All this could be applied to the use of groundwaters and later, when dealing with artificial aquifer recharge, I shall refer to how Spain applies what is known as the *canon de regulación* (regulation levy) to this technique, with the legal status of a tax.

Currently, payment for water – usage including groundwaters – is implemented in several countries (Brazil, for example), and this is something that is difficult to administer properly with regard to certain users. There tends to be resistance from agricultural users, while resistance from industrial or urban users is of a different hue, probably because they can afford it (upside inelasticity, in economists' and others' jargon) in most circumstances.

5.4 *The technique of aquifer over-exploitation*

Regardless of the legal status of one type of groundwater or another (in other words, whether they are public or private), the common element is that the Public Authorities have the power to intervene when there is a case of *over-exploitation* of an aquifer. This is a concept that, as everybody knows, may be vague or obscure in the field of Hydrology, but which water law uses profusely, defining it and applying a particular legal framework to it when over-exploitation occurs.

Over-exploitation is usually defined as the consuming of groundwaters in a quantity that exceeds the natural capacity for renewal in aquifers, measured over certain periods of course. Thus, when this over-exploitation is reached, the survival of the aquifer is obviously at risk, since its piezometric levels decrease, something which could affect any marshes that may be found on the surface or are natural spillways for the aquifer; it could also affect the economic uses (irrigation, supply, industrial use, etc.) that the waters had been put to. Similarly, such abstraction can lead to lower quality and thus a lower concentration of substances and nutrients in the water. All this has meant that the regulations cover a procedure whereby the public authorities can declare an aquifer to be over-exploited (Spanish law distinguishes between a provisional and a definitive over-exploitation) resulting, among other things, in the restriction of any rights that under public or private ownership individuals may have for the use of these waters. Thus, restrictions on the use of water are imposed until the over-exploitation situation is over. Similarly, when an aquifer is declared over-exploited, it is usual for the users of the aquifer to be forced to become part of an association, and an individual may not then use the water unless he is integrated into the Association, Committee, Community, Corporation (or however it is designated) of users of that aquifer.

5.5 *The territorial scope of groundwater management*

On a different note, it seems clear that intervention by the Public Authorities should be on the geographical basis of the hydro basin. This principle of water management by natural frontiers or territories, and not artificial or political frontiers or territories, dates back to the early 20th century in Spain (basically with the setting up of the *Confederaciones Hidrográficas* [River Boards] in 1926), and slowly spread to other countries. With the Constitution currently in force in Spain (1978), it is art 149, 1.22 that deals with State powers over hydro basins that extend beyond the territory of an Autonomous Community (what Spanish law calls inter-community basins) while it is the Autonomous Communities that manage the hydro basins within their territory (intra-community basins), and it is understood that, at any event, aquifers are subject to the overall regime of the basin, since they form a necessary part of the basin. This is the interpretation that was stated by the 1985 Water Act and ratified by Constitutional Court Judgement 227/1985, based on a constitutional text that was not particularly clear, it has to be said.

It may be said that the principle I am speaking of is doctrinally a starting out point that is accepted by everybody but which is often difficult to put into practice. This difficulty is increased when talking about groundwaters, because of the ignorance that has always existed over the extension of aquifers, today much more subtly defined, as we have seen. At any event, national legal systems find it easier to base their surface water legislation around hydro basins than their groundwater legislation, even though the hydrological cycle is theoretically the same in both cases, regardless of the type of water. This becomes even more complex with Federal States, when the *natural* principles of water management have to be combined with the constitutional powers of the various States. For example, in the USA, the State authorities have predominance over *natural* water management organised by hydro basins, and they must then inevitably enter into agreements (*Compacts*) with each other when these waters extend outside State limits; this may give rise to conflicts between States over the usage of different sections of the same river basin (the stock example of this is the Colorado River with its long-standing compact that in a very simple way, it must be said, divides the waters of the river among the various States it passes through until it flows, with a very low volume of water, into the Gulf of California). See Getches (1999, pp. 19 *et seq.*).

In Brazil, these complications are even more pronounced if we take into account the fact that its Constitution (1988) declares all water courses that pass through more than one State as being property of the Union (art. 20), but makes only the States responsible for groundwater management (art. 26), which thwarts any possibility of uniform management of an aquifer and, of course, of coordinated management of surface and groundwaters.

5.6 *Perimeters of protection*

A common aspect of water management is the intervention of the Hydro Authority to protect groundwater abstractions by establishing perimeters of protection around them or at certain locations along the aquifer in which abstraction might be affected. In these areas, certain activities are prohibited or restricted with the aim of avoiding any problems of water pollution (through filtration, for example) that may, in the end, affect the uses to which these waters are put. This management measure, exercised by the Water Authorities, often gives rise to not only a restriction on the rights of the landowners on whose territory the perimeter of protection is established, but also a restriction on the rights of the land planning authorities, which makes this measure especially important.

5.7 *Artificial recharge of aquifers*

Another important factor in water management is the artificial recharge of aquifers, subject to public intervention when carried out by individuals under a system of permits even though it is usually the public authorities themselves that administer this.

Its purpose is well known: to use aquifers as groundwater reservoirs by introducing, through wells or filtration, any water that has been plentiful on the surface at certain times of the year. Some of the practices and legislations regarding this are very advanced, such as that pertaining in Arizona (the 1980 Act with subsequent modifications; and for various aspects of this legislation – including the regulation of the groundwater market – see FAO, 1999, pp. 165 *et seq.*). With this technique, the Public Authority addresses, as well as *quantitative* considerations, certain precautionary principles, ensuring that the water introduced into the aquifer does not in any way degrade the environmental conditions of the aquifer. Thus, in certain circumstances, an environmental impact assessment may be necessary (see Section 6 of this article for more on this).

It should be pointed out that in Spanish law, in which there is no provision for charging for water use, just the obligation to *compensate* the State for expenses it has incurred in the construction and exploitation of hydro schemes for regulation and transport which provide water for use, there is a specific reference to *groundwater regulation* works (an expression which would take in artificial aquifer recharges), which means that the users of these *recharged* groundwaters are to pay the regulation tariff (with the legal status of a tax) to compensate the public authorities for expenses incurred specifically in this area (*cf.* art. 114 of the Spanish Water Act).

6 ENVIRONMENTAL CONSIDERATIONS. ENVIRONMENTAL IMPACT ASSESSMENT LEGISLATION

As I have already pointed out several times, the phenomenon that probably unifies most contemporary legal systems as regards groundwater and which enables them to be interpreted in a minimally standardised way, is the phenomenon of environmental protection of water and the ecosystems that are linked to it. This environmental aspect can come in many guises. Thus, the law refers to outflows that may occur with groundwater but also to abstractions and other activities that may occur in certain *sensitive* areas of aquifers, an issue that I examined in the previous section when referring to perimeters of protection for aquifers or abstractions. Similarly, there are specific references in law as to which processes affecting groundwaters must be submitted to an environmental impact assessment. We shall now look at this in more detail.

6.1 *Outflows into groundwaters*

An examination of the law and specialised studies of the subject (*cf.* Sanz Rubiales, 1997) will tell us that the regulation of outflows into groundwaters tends to be much stricter than that of outflows into surface waters. What may, in certain circumstances, be permitted with surface waters is expressly forbidden in the case of groundwaters, because of the capacity for almost limitless reproduction of a pollutant in groundwaters. As I say, this tends to be common to the various legislations and obviously plays a specific role in European Community regulations, which in turn affect the various national regulations of the fifteen member States. The Spanish Water Act is a good example of this, as it contains an article referring to outflows in aquifers and groundwaters (art. 102) which explains the need for a hydrogeological analysis prior to the authorisation of any outflow, and only when the analysis proves that there will be no ill effects will the outflow be authorised. Thus, the regulations (the Hydraulic Public Domain Regulations, first drawn up in 1986 and later modified on various occasions) address the consequences of all this and adapt it to European Law, a complex process with regard to these issues.

6.2 *Environmental impact assessment of certain measures that directly or indirectly affect groundwaters*

Ever since its origins in North American law, the environmental impact assessment has been an indispensable tool for the authorisation – or rejection or conditioned authorisation – of actions that might affect natural resources. With the environmental aspects that I am constantly emphasising, it is clear that this is a field that lends itself to these environmental impact assessments, and the law, as we shall now see, has taken note of this. In addition, it clear for all to see that certain groundwater abstraction processes have caused regrettable effects (for example, the Po delta subsidence, something that is common to all deltas) that need to be avoided in future. This is probably one of the reasons why the Spanish Act 10/2001 of 5th July, concerning the National Hydrological Plan, stipulates the drawing up of the Ebro Delta Integrated Plan to study the question of subsidence, among other things (Supplementary Provision 10), even though it seems that in this case, groundwater abstraction is not the direct cause of the problem.

At a European level, it is Directives 85/337/EC and 97/11/EC that have regulated the environmental impact assessment and ordered this procedure to be applied to certain actions. This was incorporated into Spanish law by Executive Order 1302/1986 of 28th June, concerning the environmental impact assessment, modified by Act 6/2001 of 8th May. It is this Act that stipulates when such an assessment must take place or the circumstances under which an assessment is to be made if so required by the environmental authorities in accordance with the criteria laid down by the environmental impact legislation. In all the circumstances, there are cases which may affect groundwaters directly or indirectly. These circumstances are as follows:

a) *Cases where an environmental impact assessment is compulsory:*
Under Spanish legislation, this administrative procedure must take place in the following actions:
  - Hydro resource management projects for agriculture, including irrigation or drainage projects, when they affect an area of more than 100 ha. Irrigation maintenance and improvement projects are not included.
  - Exploitations below the phreatic level, the reference level being the highest between annual variations or variations which involve a decrease in surface or deep aquifer recharge.
  - Exploitations situated in hydraulic public domain areas or buffer zones which are worked in especially sensitive areas, as set out in Directives 79/409/EEC and

92/43/EEC, or in wetlands included in the Ramsar Agreement list.
- Groundwater abstraction or artificial aquifer recharge projects, if the annual volume of water abstracted or added is equal to or higher than 10 Mm^3.
- Deep boring for water supply when the volume of water abstracted is higher than 10 Mm^3.

b) *Cases where there may be an environmental impact assessment if required by the environmental authority:*
- Hydro resource management projects for agriculture, including irrigation or drainage projects, when they affect an area of more than 10 ha or irrigation maintenance and improvement projects of more than 100 ha.
- Exploitations located in hydraulic public domain areas for abstractions higher than 20,000 m^3/yr or in buffer zones with an area of more than 5 ha.
- Groundwater abstraction or artificial aquifer recharge projects, if the annual volume of water abstracted or added is higher than 1 Mm^3.
- Water desalination facilities with a new or additional volume of more than 3,000 m^3/d.

7 GROUNDWATERS AND THE EUROPEAN UNION INTERNATIONAL GROUNDWATERS

I would like now to underline the clear role and importance of groundwaters in the context of European Union regulations, and at the same time, the problems that still exist in concluding international agreements or treaties concerning groundwaters, unlike the situation with surface waters, where there are long-standing on-going agreements in force.

In this respect, the role of European Parliament and Council Directive 200/60/EC of 23rd October is paramount. This Directive establishes a community framework for action in the field of water policies (published in the Official Journal of the European Communities of 22nd December 2000). The document is all-embracing and ambitious, aiming to almost entirely overhaul the wordy and profuse community law regarding water in force up to that time, to become practically the only source of European regulations in the future. The member States have three years as from its coming into force (i.e., 23rd December 2000) to implement this Directive which aims to attain good ecological status for surface and groundwaters in the countries of the European Union, with indications as to what the criteria for achieving this status are. With regard to groundwaters, there are various specific provisions in art. 4 of Directive 2000/60/EC which establish deadlines for the member States to meet the following objectives:

- To prevent or limit the input of pollutants into groundwater and to prevent the deterioration of the status of all bodies of groundwater.
- Protect, enhance and restore all bodies of groundwater, ensure a balance between abstraction and recharge of groundwater, with the aim of achieving good groundwater status at the latest 15 years after the date of entry into force of the Directive.
- To reverse any significant and sustained upward trend in the concentration of any pollutant resulting from the impact of human activity in order progressively to reduce pollution of groundwater.

In the field of international law concerning groundwaters, the issue is almost non-existent (*cf.* Teclaff and Utton, 1981), and there is room for a great deal of progress in the future. This is in marked contrast to the situation regarding the regulation of international watercourses, with a long tradition of bilateral (rather than multi-lateral) agreements (see FAO, 1998), based originally on regulating navigability (the early international Treaties concerning the Rhine, Danube, Meuse, etc.), now embracing many other fields, including environmental protection.

With regard to the absence of mentions of groundwaters in international treaties, it is interesting to see how a South American expert with the prestige and significance of Marienhoff (1996, p. 653) can only recall one Treaty in 1906 between the German States of Prussia, Bavaria,

Baden and Hessen, whereby the parties to the treaty undertook to compensate individuals, town councils, corporations, etc. for any damages incurred for raising the phreatic level. But that is all, despite the fact that technical breakthroughs and their potential will call for international regulations in the future.

A good example of what I mean is the use of the vast *Guaraní* aquifer (so called since 1994) which extends through Brazil, Paraguay, Uruguay and Argentina. It is clear that all these countries can find a solution for their water problems in this enormous aquifer, with a body of water of excellent quality (as opposed to the situation with its surface waters), and with the potential to bring about the development of a vast region extending over a million km^2. To cope with this situation it is clear that what is needed is an international treaty between all the countries, which is something they need to address specifically over the next few years.

Yet they will not find many guidelines in international law, as I have already said. The only guidelines they have are the 1986 *Seoul Rules*, approved by the International Law Association, whose four articles do little to regulate groundwaters. The first article gives the status of international waters to waters that go beyond the boundaries of just one State, with these groundwaters linked to those of the hydro basin within the meaning of the Helsinki Rules drawn up by the International Law Association (1972).

The second article regulates hydraulic interdependence with surface waters of the States, thus forming an international basin in terms of the Helsinki Rules, so that all States must take into account any interconnection between aquifers and surface waters for the purpose of water management.

A specific article (article 3), is devoted to the environmental protection of groundwater, whereby States are obliged to protect waters from pollution and to exchange information and cooperate with other States in the same basin.

Finally, article 4 devotes just two short lines to the need for States to consider integrated management of surface and ground waters. It should be added that there are *complementary* rules, also approved in Seoul in 1986, that do little to develop the principles described above.

8 GROUNDWATERS IN THE NATIONAL HYDROLOGICAL PLAN. THE ALTO GUADIANA CASE

I feel it is necessary to conclude this article by referring to the way groundwaters are handled in an Act that is very important for Spain, Act 10/2001, of 5th July, concerning the National Hydrological Plan. In this Act we can find certain general references to groundwaters and a specific consideration of a particular area, the Alto Guadiana. Let us examine these two aspects.

8.1 *General references to groundwaters*

The basic traditional principle of Spanish water management law based on geographical basins determines that the Act has a definition of aquifers that share various basins (art. 7) and a legal framework for their management is envisaged (art. 8). Thus, shared aquifers are those that are situated in territories stretching over two or more Basin Hydro Plans and are specifically included in an annex to the Act. The same annex also shows the allotment of resources expressed in Mm^3/yr – for each basin area.

What is more interesting is to examine the management legal framework in which an attempt is made to be consistent with the previous allotment of resources. It is stipulated that each Basin Body will manage the aquifer in its own area, but with the obligation to notify the other Basin Bodies of any decisions it adopts with regard to the resources therein. Nevertheless, and this makes good pragmatic sense, management of the aquifer may be entrusted to one of the Bodies involved, and finally a transfer between different basin areas is considered to have taken place only when it is carried out by man-made means. In this case, the legal framework regarding resource transfers regulated in the rest of the Act is applied.

Besides this, art. 29 provides for the Ministry of the Environment to implement an Action Plan for Groundwaters "to enable the sustainable exploitation of these resources", and to include programmes for improving hydrogeological expertise and the protection and planning of aquifers and groundwaters. The same article emphasises the traditional public policies of

Spanish Law for promoting the setting up of User Communities and technical assistance for drawing up aquifer exploitation plans.

8.2 *The special Alto Guadiana Plan*

Supplementary Provision 4 of Act 10/2001 of 5th July, concerning the National Hydrological Plan, orders the formation of a Special Plan for the Alto Guadiana. The purpose of this legal Provision is to maintain sustainable use of aquifers in the Upper Guadiana Basin. In this respect, the Provision envisages a legal framework for a set of measures consisting of:

a) "The re-regulation of water usage rights, to bring about the environmental recovery of aquifers.
b) The authorisation of modifications to the existing well exploitation regulatory scheme.
c) The concession of groundwaters in drought situations.
d) Other measures to bring about permanent hydro and environmental balance in this basin."

The basis of this Provision is the serious shortcomings of the above-mentioned aquifers with a deterioration that is causing problems today, problems that call into question not only the environmental status of the aquifers, but also the continuance of economic activities carried out with the extraction of these waters, mainly agriculture, and the survival of a particular natural protected area (the Tablas de Daimiel National Park) which is closely linked to these aquifers.

The language used in the Provision is very broad, but the very fact that it is mentioned gives an idea of the ambition overriding the construction of Supplementary Provision 4, which one hopes will ultimately be followed up by the appearance of a Plan which will actually respond to the expectations aroused.

Mentions of "the re-regulation of water usage rights" and of the authorisation of "modifications to the existing well exploitation regulatory scheme" may only be understood in terms of the great legal complexity surrounding groundwaters in Spain, since Temporary Provision 3 of Act 29/1985 of 2nd August, concerning Water Resources, which enabled compatibility between public and private ownership of water (depending on the holder's option on the date the Act came into force) has not been followed up with proper diligence by the Authorities to regularise situations and especially to speed up the thousands of administrative proceedings initiated in line with this Provision. Similarly, there has been a certain indiscipline in the use of groundwaters in the Alto Guadiana Basin, caused by economic exploitation by private users which has not been stopped by the Water Authorities who are always short of personal and material resources and therefore unable to respond to the strict requirements of the law in force.

At any event, the above Provision, if it is ever made specific, will need to be carefully studied because of the difficulty of establishing the area in which it is to be applied. It is not hard to predict that such a study will be made not just from a Spanish perspective, since the critical situation in which the Alto Guadiana aquifers find themselves goes beyond Spain's border, and therefore any solutions that are attempted will also be monitored outside this country.

REFERENCES

Please note: in view of the understandable profusion of bibliographical references that could be offered with regard to widely varying international legislation, the works presented here are subdivided into the following categories: a) General; b) Spain; c) Other countries. Spain is afforded a separate category because of its clear influence in South America and contains general works on water law, not only regarding groundwater law, because therein can be found many references to the theme that concerns us here.

Some of the following references are not quoted in the text, but they represent relevant reading.

a) General bibliography

Burchi, S. (1999). National regulations for Groundwater: Options, issues and best practices. In: A. Salman (ed.), Groundwater. Legal and Policy Perspectives. Proceedings of a World Bank Seminar. *World Bank Technical Paper*, 456, Chapter 3.

Burchi, S. and Nanni, M. (in press). *How groundwater ownership and rights influence groundwater intensive use management.*

Caponera, D. (1992). *Principles of Water Law and Administration, National and International*. Balkema, Rotterdam, the Netherlands.

FAO (1998). *Sources of International Water Law*. FAO Legal Office, Rome, Italy.

FAO (1999). *Issues in Water Law reform*. FAO Legal Office, Rome, Italy.

Teclaff, L.A. and Utton, A.E. (1981). *International Groundwater Law*. Oceana Publications, London, UK.

b) Spanish bibliography

De la Cuétara, J.M. (1989). *El nuevo régimen de las aguas subterráneas en España*. Tecnos, Madrid, Spain.

Del Saz, S. (1990). *Aguas subterráneas, aguas públicas. El nuevo Derecho de aguas*. Marcial Pons, Madrid, Spain.

Del Saz, S.; Fornés, J.Mª. and Llamas, M.R. (eds.) (2002). *Régimen jurídico de las aguas subterráneas*. Fundación Marcelino Botín and Ediciones Mundi-Prensa. Madrid, Spain. 331 pp.

Embid, A. (1991). *La planificación hidrológica. Régimen jurídico*. Tecnos, Madrid, Spain.

Martín Retortillo, L. (1987). Aguas subterráneas y aguas que discurren íntegramente dentro del territorio. *Revista de Administración Pública, 113*. Centro de Estudios Políticos y Constitucionales, Madrid, Spain.

Martín Retortillo, L. (1990). Las aguas subterráneas como bienes de dominio público. In *Libro Homenaje a Villar Palasí*. Civitas. Madrid, Spain. Pp. 677 *et seq*.

Martín Retortillo, S. (1997). *Derecho de Aguas*. Civitas. Madrid, Spain.

Morell, L. (2001). Las titularidades sobre aguas privadas. *Revista de Administración Pública, 154*. Centro de Estudios Políticos y Constitucionales, Madrid, Spain. Pp. 7 *et seq*.

Moreu, J.L. (1996). *Aguas públicas y aguas privadas*. Bosch. Barcelona, Spain.

Nieto, A. (1968). Aguas subterráneas: subsuelo árido y subsuelo hídrico. *Revista de Administración Pública, 56*. Centro de Estudios Políticos y Constitucionales, Madrid, Spain.

Pérez Pérez, E. (1998). *La propiedad del agua. Sistema estatal y sistema canario*. Bosch. Barcelona, Spain.

Sanz Rubiales, I. (1997). *Los vertidos en aguas subterráneas. Su régimen jurídico*. Marcial Pons. Madrid, Spain.

c) Other countries

Abderrahman, W.A. (2001). Water demand management in Saudi Arabia. In: N.I. Faruqui, A.K. Biswas and M.J. Bino (eds.), *Water management in Islam*. United Nations University Press. Tokyo, New York, Paris.

Chede, A.P. (2002). Régimen jurídico de las aguas subterráneas en la República Argentina. In: A. Embid (dir.), *El Derecho de aguas en Iberoamérica y España: cambio y modernización en el inicio del tercer milenio*. Civitas. Madrid, Spain. Vol. II, Pp. 59 *et seq*.

Farias, U. (1993). *Derecho mexicano de aguas nacionales*. Pub. Porrúa. Mexico.

Gazzaniga, J.L. and Ourliac, J.P. (1987). *Le droit de l'eau*. Iltec. Paris, France.

Getches, D. (1997). *Water Law*. West Publishing Co.

Getches, D. (1999). Resolución jurídica de los conflictos sobre aguas transfronterizas en los Estados Unidos. In: A. Embid (dir.), *Planificación hidrológica y política hidráulica. (El Libro Blanco del Agua)*. Civitas. Madrid, Spain. Pp. 19 *et seq*. (This paper has been translated to Spanish by A. Embid).

Guerrero Reinoso, V. (2000). *Decentralizacion of Water Management in Mexico by means of basin councils*. Paper presented in a Workshop in Salvador de Bahia, Brazil.

Hajji, M. (1995). *Le regime juridique des zones de protection des ressources en eau potable au Maroc*. Paper presented in a Hispano-Moroccan seminar on water law in Málaga, Spain (dir.: A. Embid).

Lugaresi, N. (1995). *Le acque pubbliche. Profili dominicali, di tutela e di gestione*. Giuffré editore. Milan, Italy.

Marienhoff, M.S. (1996). *Tratado de Derecho Administrativo, vol. VI*. Buenos Aires, Argentina.

Sandoval, R. (2002). Experiencia del Estado de Guanajuato, México. Participación social y descentralización para la gestión integrada del agua por cuencas hidrológicas. In: A. Embid (dir.), *El Derecho de aguas en Iberoamérica y España: cambio y modernización en el inicio del tercer milenio*. Civitas. Madrid, Spain. Pp. 433 *et seq*.

Sax, J.L.; Abrams, R.H. and Thompson, B.H. (1991). *Legal control of water resources. Cases and materials*. West Publishing Co.

Shiklomanov, I. (1997). *Comprehensive assessment of the freshwater of the World*. World Meteorological Organization.

Varnerot, V. (2002). L'étrange pérennité du droit de propriété sur les eaux souterraines. A propos de la décision du TGI d'Angers en date du 12 juillet 2001. *Revue Juridique de l'Environnement, 2*.

Vergara, A. (1999). *Derecho de aguas*. Editorial Jurídica de Chile. Santiago, Chile. 2 vol.

Intensive use of groundwater in the European Union Water Framework Directive

Javier Samper
Escuela de Ingenieros de Caminos, University of Coruña, Spain
jsc@iccp.udc.es

ABSTRACT

In October 2000 the European Council together with the European Parliament approved the *Water Framework Directive* (WFD), which sets a framework for the protection of inland surface waters, transitional waters, coastal waters and groundwaters with the aim of achieving a good status of water bodies by the year 2015 (European Commission, 2000). This paper addresses the issue of intensive use of groundwater in the WFD. The main objectives and the overall approach of the WFD for groundwater planning, management, protection and control are described. Implications of the WFD for European countries are discussed, with special emphasis on semiarid areas and countries such as Spain, where groundwaters are intensively used for water supply and especially for irrigation.

Keywords: *Groundwater, intensive use, European Union, Water Framework Directive, semiarid regions, sustainable water use.*

1 INTRODUCTION

As pointed out by Llamas and Custodio (2003) in the Proceedings of the *Workshop on Intensively Exploited Aquifers*, held in Madrid in December 2001, the protection and wise use of aquifers can contribute significantly to the improvement of human life and aquatic ecosystems. In the summary chapter, Custodio and Llamas (2003) indicate clearly the main facts and consequences of intensive use of groundwater and discuss the main economic, social, institutional issues. Many of their thoughts apply to the countries of the European Community.

While some voices claim that overexploitation of groundwater resources is emerging as a pan-European problem due to the increasing water demand for agriculture, industry and public supply (Balabanis, 1999), for others intensive use of groundwater in many areas of Southern Europe has brought significant social and economic benefits (Hernández-Mora *et al.*, 2003). As pointed out by Custodio (2000), the ecological cost of groundwater development should be compared with the socio-economic benefits produced by groundwater use. Groundwater development should not be rejected or seriously constrained if it is well planned and controlled. Groundwater withdrawal has provided affordable potable and irrigation water, thus improving public health and significantly contributing to alleviate malnutrition and famine. Often the adverse effects of *overexploitation* are misunderstood or exaggerated. Very often long records of declining water levels are taken as indications of groundwater overexploitation and are associated to situations of abstractions being larger than average renewable resources (see Llamas, this volume).

2 GROUNDWATERS IN THE EUROPEAN UNION

European water resources, both in terms of availability and water demand, are extremely diverse and variable in space and time. The management of water resources in the coastal zone of the Mediterranean basin is particularly problematic. Over the past 30 years the amount of irrigation from groundwater has increased in the Southern European countries. This is due, on the one hand, to the strong disequilibrium between water supply and water demand and, on the other, to the lack of an integrated approach of water resources management. As a result, water resources have been used intensively, leading in some cases to a drying of wetland habitats and terrestrial ecosystems, saline intrusion, and the degradation of groundwater quality (Balabanis, 1999; Hernández-Mora *et al.*, 2003). In Northern Europe, on the other hand, overexploitation occurs principally because groundwater resources have historically provided a low-cost source of high quality water for public water supply. However, the historical growth in the exploitation of groundwater resources is, in many cases, unsustainable. And in some catchments, this has already brought on economic, environmental and social consequences, which, in turn, have already given rise to economic, social and political pressures to resolve the conflicting demands for groundwater. Water pollution is also a major environmental problem across Europe, which calls into question the long-term reliability of many drinking water sources. Pesticide pollution constitutes a threat to Europe's groundwater pollution. Disposal sites for industrial and municipal waste are potential sources of contamination. The pollution trends and impacts of other hazardous pollutants, such as heavy metals, organic micropollutants and pathogens on water resources are largely unknown, and so are their impacts on health (Balabanis, 1999).

3 GROUNDWATERS IN THE WFD

Early European water legislation began with standards for the rivers and lakes used for drinking water abstraction in 1975, and culminated in 1980 in setting binding quality targets for drinking water. This legislation also included quality objectives for fish waters, shellfish waters, bathing waters and groundwaters. Over this period, progress was made in understanding the key issues that need to be addressed and, as a result, a fundamental rethinking of Community water policy took place in mid-1995 (Balabanis, 1999). Thus, the *Water Framework Directive* was proposed to be the operational tool by which Community members would set the objectives for water protection well into the next century. The European Council and Parliament approved on 23 October 2000 the *Water Framework Directive* (WFD) which establishes a framework for the protection of inland surface waters, transitional waters, coastal waters and groundwater with the following objectives (European Commission, 2000):

a) Preventing further deterioration and protecting and enhancing the status of aquatic ecosystems;
b) Promoting a sustainable water use based on a long-term protection of available water resources;
c) Enhancing protection and improvement of the aquatic environment through specific measures for the reduction of emissions of toxic substances;
d) Ensuring the progressive reduction of pollution of groundwater and preventing its further pollution; and
e) Mitigating the effects of floods and droughts.

The WFD pursues the following main objectives for groundwaters (European Commission, 2000):

a) Implement measures to prevent or limit the input of pollutants into groundwater and to prevent the deterioration of the status of all bodies of groundwater.
b) Protect, enhance and restore all bodies of groundwater, ensure a balance between abstraction and recharge of groundwater, with the aim of achieving good groundwater status at the latest 15 years after the date of entry into force of the WFD.
c) Implement measures to reverse any significant and sustained upward trend in the

concentration of any pollutant resulting from the impact of human activity in order to reduce groundwater pollution.

The time horizon of 15 years to reach a good quantitative status may be a too short time span for all groundwaters because some aquifers have diffusivities large enough to require decades to achieve steady conditions.

The WFD considers surface water and groundwater bodies, none of which embodies the unsaturated zone, a major component of the hydrologic cycle, especially in arid and semiarid regions. This is a major weakness of the WFD, that need to be overcome in a future revision of the WFD or corrected by State members, especially by those in the Mediterranean region.

3.1 *Definitions*

The WFD introduces some key definitions for groundwater which are of relevance for the analysis of intensive use of groundwater in the EU countries. *Groundwater* is defined as "all water which is below the surface of the ground in the saturation zone and in direct contact with the ground or subsoil" while a *body of groundwater* is defined as a "distinct volume of groundwater within an aquifer or aquifers". The term *body of groundwater* (BoG) is a new concept in many countries that is creating some confusion. Its main purpose is operational because water status (to be defined later in this paper) is evaluated at the scale of BoG. According to the WFD, bodies of groundwater must be connected to a surface water body and it should be possible to make a water budget within its boundaries in order to evaluate its quantitative status. Their size and areal and vertical extent should allow the proper management, control and protection of groundwater. A too large BoG could be split into several BoG or reversely, several small BoG could be lumped together for monitoring and control purposes.

A key definition for groundwaters in the WFD is the concept of *available groundwater resources* which is defined as "the long-term annual average rate of overall recharge of the body of groundwater less the long-term annual rate of flow required to achieve the ecological quality objectives for associated surface waters, to avoid any significant diminution in the ecological status of such waters and to avoid any significant damage to associated terrestrial ecosystems". Although plausible, this definition has the potential to lead to conflicts, because available groundwater resources in a BoG may be hardly quantifiable. Therefore, there are many doubts about its operational capacity.

3.2 *Quantitative status of groundwater*

The status of a groundwater body is good if the level of groundwater in the groundwater body is such that the "available groundwater resource is not exceeded by the long-term annual average rate of abstraction". Accordingly, the level of groundwater should not be subject to anthropogenic alterations such as would result in: 1) a failure to achieve the environmental objectives for associated surface waters; 2) any significant decrease in the status of such waters; and 3) any significant damage to terrestrial ecosystems which depend directly on the groundwater body. Alterations to flow direction resulting from level changes may occur temporarily, or continuously in a spatially limited area, but such reversals should not cause saltwater or other intrusion, and should not indicate a sustained and clearly identified anthropogenically induced trend in flow direction likely to result in such intrusions. Member States must prepare maps of the groundwater quantitative status, colour-coded in accordance with the following regime: good = green; poor = red (European Commission, 2000). It should be noticed that the quantitative status is defined in terms of groundwater levels and discharge fluxes. Its definition is somehow confusing and arbitrary because it will be extremely difficult to quantify. After all, *the status is good or poor*, is a poor measure of the status of groundwater. It will hardly be operational or useful in semiarid regions where it is extremely difficult to quantify the amount of groundwater discharge required to achieve the environmental objectives of surface waters. If applied with an extreme view, all the natural groundwater discharge could be considered as needed to achieve the environmental objectives of surface waters. In such case, there would be no available groundwater resources.

3.3 *Groundwater chemical status*

The parameters for determining the chemical status are conductivity and concentrations of pollutants. In the WFD there is no mention to the concentration of major and minor natural constituents of groundwater. It refers to future European regulations for the definition of chemical status which are currently under discussion and will eventually lead to a future directive for establishing strategies to protect groundwater against pollution (European Commission, 2003). This is the so-called *Groundwater Daughter Directive* which aims at: 1) Setting up criteria for assessing good groundwater chemical status; 2) Setting up criteria for the identification and reversal of significant and sustained upwards trends; and 3) Establishing measures to prevent and limit groundwater pollution with the aim of achieving a good groundwater chemical status.

3.4 *Full-cost recovery*

One of the novel aspects of the WFD is the consideration of the principle of recovery of costs of water services. Member states must take account of the principle of recovery of the costs of water services, including environmental and resource costs, having regard to the economic analysis and in accordance with the *polluter pays* principle. They must ensure by 2010 that water-pricing policies provide adequate incentives for users to use water resources efficiently, and thereby contribute to the environmental objectives of the WFD. Member States may have regard to the social, environmental and economic effects of the recovery as well as geographic and climatic conditions.

Accounting for total cost will have a tremendous effect on water used for irrigation, because irrigation water makes up around 80% of the total water use in Mediterranean countries such as Italy and Spain. The analysis of water pricing policies in the irrigation sector is an essential element for judging resource and agricultural policies, identifying policy gaps and inconsistencies and suggesting guidelines to design legal and planning proposals. However, little progress can be made in this area unless proper irrigation cost valuations are carried out. A recent report of the Organization for the Economic Cooperation and Development (OECD, 2002), shows that irrigators face private and communal costs, but can also impose significant externalities to society in the forms of water pollution leading to ecosystems degradation and excessive use of scarce water resources. Irrigated agriculture also supports economic development in rural areas, providing jobs and supporting agro-food industries in areas which otherwise would become depopulated. In addition, irrigated agriculture helps maintain rural landscapes. Although significant progress has been made in the last decade to identify and define water subsidies, many OECD countries still promote growth in their irrigated areas, and do so acknowledging that at least 50% of the development costs will never be recovered. In some OECD countries, there is a remarkable growth in the use of market incentives that perform with moderate success in water allocation services.

Water pricing can perform two functions. One is to ensure that the service costs are completely borne by the users, allowing public water agencies to focus on actions and policies that reduce hydrological risks and restore water bodies' environmental quality. The other is to promote water use efficiency, narrowing the gap between private and social benefits and ensuring that farmers make appropriate use of their water resources. However, there are few examples across OECD countries of water pricing policies put in place to deliver benefits related to both functions. Subsidies and adverse incentives are still present in many OECD countries. While it is still difficult to measure them, most countries recognise that large social benefits would accrue if incentives were realigned to ensure that irrigators are made more responsible for their actions and infrastructure. For instance, agricultural policies in the European Union significantly distort both irrigators' selection of crops and incentives to invest in irrigation equipment. Similar disincentives come from pay-per-hectare schemes, which are overwhelmingly used in Spain although it is not always possible or cost effective to shift to volumetric rates.

3.5 *Overall remarks about the WFD*

The WFD focuses mainly on water quality as stated in preliminary statement 19: "This Directive aims at maintaining and improving the aquatic environment in the Community. This purpose is primarily concerned with the quality of the waters concerned. Control of quantity is an ancillary element in securing good water quality and therefore measures on quantity, serving the objective of ensuring good quality, should also be established". In fact, the WFD barely touches groundwater quantity aspects and quantity requirements are mostly related preserving the ecological status of water bodies. The WFD aims at achieving *sustainable use of water* (see point 14 of the WFD in European Commission, 2000).

The WFD recognises that water policy must take into account the interaction between surface waters and groundwaters within the entirety of the respective river basins. More particularly, with regards to the surface waters, the Directive aims to protect the aquatic ecology, the drinking water resources and bathing water, providing mechanisms for renewing all the quality standards established for chemical substances at the European level. With regards to groundwater, the Directive aims to limit over-abstraction, prohibit direct discharges to groundwater; and set a requirement to monitor groundwater bodies so as to detect changes in chemical composition and to reverse any anthropogenically induced upward pollution trend. The Directive emphasizes strengthening the role of citizens and the involvement of any interested parties on water policies by establishing a network for the exchange of information and experience among water professionals throughout the Community.

The WFD defines measures, programmes and provisions which are different for surface and groundwaters. From a management point of view, they must be treated differently. In this respect, the overall approach of the WFD is adequate. It states clearly that the task of ensuring a good status of groundwater requires early action and stable long-term planning of protective measures, owing to the natural time lag in its formation and renewal. It adopts a pragmatic approach because in cases where a body of water is so affected by human activity or its natural condition is such that it may be unfeasible or unreasonably expensive to achieve good status, less stringent environmental objectives may be set on the basis of appropriate, evident and transparent criteria. In any case, all practicable steps should be taken to prevent any further deterioration of the status of waters.

The WFD emphasizes the coordination of surface water and groundwaters. The objective of achieving good water status should be pursued for each river basin, so that measures in respect of surface water and groundwaters belonging to the same ecological, hydrological and hydrogeological system are coordinated.

The final version of WFD introduced exemptions from the requirement to prevent further deterioration or to achieve good status under specific conditions, if the failure is the result of unforeseen or exceptional circumstances, in particular floods and droughts, or, for reasons of overriding public interest, of new modifications to the physical characteristics of a surface water body or alterations to the level of bodies of groundwater, provided that all practicable steps are taken to mitigate the adverse impact on the status of the body of water.

Further details on water policy in the European Union can be obtained from the official web page of the European Commission (http://www.europa.eu.int). The European Environment Agency (http://www.eea.eu.int) is in charge of providing the European Commission with the information needed for making sound and effective policies to protect the environment and support sustainable development.

The 15 Member States, Norway and the European Commission agreed on 2 May 2001 in Fiskebackskill (Sweden), to create a Forum for a *Common Implementation Strategy* in order to improve the sharing of information and the views on the implementation of the *Water Framework Directive* throughout Europe (European Commission, 2001). CIRCA is an interest group on *Implementing the Water Framework Directive* which is accessible through the Internet (http://europa.eu.int/comm/environment/water/waterframework/information.html). Guidelines and technical recommendations for water monitoring have been also prepared by the European Environment Agency (Bogestrand *et al.*, 1998).

4 SOME FACTS ABOUT GROUNDWATER IN SPAIN

Spain has a population of 40 million people. It covers an area of around 500,000 km^2. Groundwater amounts to about one third of the total average runoff of 110 km^3/yr. Until 1986, groundwaters were privately owned. After the 1985 Water Act, groundwaters are managed by Water Authorities which have proved to be unable to deal with the huge amount of existing wells which probably exceed a few million. Official recording of water wells has not yet been completed. Therefore, statistics on groundwater users and amounts for different uses are highly incomplete.

Water supply of one fifth of the population relies on groundwater. This is one of the lowest ratios throughout Europe (Hernández-Mora *et al.*, 2003). Groundwater is used for irrigation of about 1 million ha, that is around one third of the total irrigated surface. About 80% of the total water use is devoted to irrigation. Some of the best National Parks such as Doñana and Daimiel as well as many other wetlands rely on groundwater. River Basin Water Plans which were approved in 1999, consider aquifers only from a quantitative point of view.

Little attention has been paid to groundwater quality because water quantity is still considered to be the first priority. Water quality objectives, standards and management are defined for all types of water with no distinction for the slow motion of groundwaters. As a result, protection and control of groundwater pollution is not perceived a relevant issue. Therefore, no protection measures or actions have been taken. To the best of our knowledge no protection zones have been defined yet for water supply wells in Spain.

Some of the main aquifers in Spain show significant nitrate pollution. The *European Union Nitrates Directive* 91/676 on groundwater pollution by nitrates was transposed to the Spanish laws with great delay (1996), and its implementation was performed with many difficulties and *no major enthusiasm*. Aquifers showing signs of nitrate pollution have almost *a license to be polluted*. Water managers perceive that there is no point in their protection. The implementation of the *Nitrates Directive* has not been particularly successful in many European Union States which have been tardy in enacting measures to implement the Directive and often failed to implement the Directive correctly (Grossman, 2001). Farmers often have objected to new manure and fertilizer management responsibilities required to control nitrates in groundwater and surface waters. The discussion of voluntary nitrate programs and implementation of the *Nitrates Directive* in many countries has shown how agricultural, political, and regulatory complications have accompanied nitrates regulation. In addition to the reluctance of some Member States to implement the *Nitrates Directive*, structural problems connected with intensive livestock production make correct implementation of the Directive difficult in some Member States. Efforts in various Member States to implement the Directive have already resulted in reduced nitrate concentrations.

The *Spanish National Water Plan* was approved as a Law on June 2001 (Government of Spain, 2001). This Law was supposed to provide solutions to unsolved questions in River Basin Plans, establish coordination measures among River Basin Plans and account for the principles of the WFD. However, it ended up being a law approving a large and highly-controversial water transfer project (Samper, 2001). The *National Plan* does not follow or attach to the principles of the WFD. Although the *Water Plan* barely touches groundwaters, there are some reasons for hope, because the final version of the *National Water Plan Act* includes provisions for: 1) Initiating a *National Action Program for Groundwaters*; 2) Fostering the formation of Groundwater Users Associations; 3) Performing education campaigns on water resources and their sustained use; and 4) Monitoring and research of wetlands to promote their recovery. Nevertheless, activities and funding on groundwater aspects by Spanish Agencies are still very scarce.

5 IMPLICATIONS OF THE WFD FOR SPANISH AQUIFERS

Enforcement of the WFD in Spanish aquifers will require major efforts from water authorities and groundwater stakeholders. In addition to common European initiatives for the implementation of the WFD such as the *Forum for the Common Implementation Strategy*, many activities are being carried out at the national and local levels. The *Spanish Chapter of the International Association of Hydrogeologists* organized on November 2002 a Conference on *Present and Future of Groundwaters in Spain in the light of the Water Framework Directive*, which took place in Zaragoza (AIH-GE, 2003). Technical sessions addressed the following key issues: 1) Characterization of groundwaters; 2) Groundwater and the environment; 3) Groundwater protection; 4) Groundwater monitoring and control; 5) Correction measures and programmes; 6) Participation of stakeholders in groundwater management; and 7) Wastewater reutilization.

Characterization of groundwaters in Spain will require: 1) Delimitation of bodies of groundwater; 2) Evaluation of the quantitative status of such bodies; 3) Evaluation of the chemical status of groundwater bodies; and 4) Classification of groundwater bodies into two classes, those with a good status and those with a risk of not achieving a good status by year 2015. While the former must be characterized every 6 years, the latter need an additional and more detailed characterization, with special emphasis on human impacts on groundwaters. Programmes for monitoring water status should be undertaken in order to establish a coherent and comprehensive overview of water status within each river basin district. For groundwaters such programmes must cover monitoring of chemical and quantitative status. These programmes must be operational by 2006.

The WFD stresses the need to integrate surface and groundwaters and evaluate their relationships and connections. The interactions of streams, lakes, and wetlands with groundwaters are governed by the positions of the water bodies with respect to groundwaters flow systems, geologic characteristics of their beds, and their climatic settings (Sophocleous, 2002). Consequently, for a thorough understanding of the hydrology of water bodies, all these factors should be taken into account. In arid and semiarid regions of Spain, stream-aquifer interactions are extremely complex and poorly known. Their analyses require detailed and costly monitoring along vertical sections of the aquifer and in the unsaturated zone. Water fluxes along rivers and wetlands are poorly known except for a few exceptional cases such as Doñana and Tablas de Daimiel National Parks. Understanding and quantifying the interactions of surface waters and groundwater will be a major challenge for the application of the WFD in coastal and detritic aquifers in Spain. Intensive groundwater use has ecological implications on streams and wetlands. Because of the interdependence of surface water and groundwater, and water-reliant ecosystems, changes in any part of the system have consequences for the other parts (Sophocleous, 2003). Therefore, groundwaters cannot be managed in isolation from the rest of the environment. For example, what may be established as an acceptable rate of groundwater withdrawal, with respect to changes in groundwater levels, may reduce the availability of surface water to an unacceptable level. In addition, many of the effects of groundwater development manifest themselves slowly over time, so that pumping decisions today may affect surface water availability many years in the future. Consequently, a comprehensive, long-term, and integrated approach to the management of groundwater resources is required if the water quality and supply are to be sustained in a long term and other ecosystems dependent on water are to be protected.

For the most part, loosing streams are in many cases a major source of groundwater pollution. Unsaturated zone monitoring is currently very scarce. Large investments will be needed to understand and quantify the interrelationships of aquifers and surface waters in these regions.

The application and implementation of the full-cost recovery principle will be beneficial in Europe and will counteract current perverse subsidies. It will also favor groundwaters in Spain which for the most part have been developed by private owners who actually paid for the full cost of drilling and pumping. This is not the case of areas irrigated with surface waters which were funded by large public investments. Farmers rarely paid back for the full cost (Hernández-Mora and Llamas, 2001). A recent study of water used for irrigation in the region of Andalucia in Spain revealed that (Llamas *et al.*, 2001): 1) Areas using groundwaters are more efficient in water use than areas relying on surface waters. While 4,000 m^3/ha are required in groundwater, 7,600 m^3/ha are used in surface water; and 2) For the same amount of water used, areas irrigated with groundwater produce 5 times more and generate 3 times more jobs than areas irrigated with surface water.

Overall, the WFD will be beneficial to groundwaters in Europe. In particular, in Spain the WFD will promote the implementation of aquifer protection measures and the participation of stakeholders in water planning and management and contribute to *unveil* the advantages of groundwater if the full-cost recovery principle is fully implemented.

In addition, the WFD should be beneficial to hydrogeologists in Spain because it will promote the development of national action programs. Only some programs contained in the *White Paper on Groundwater* (1994) have been developed and at a very slow pace. The number of hydrogeologists in River Basin Authorities which currently is insufficient, will increase because additional opportunities will arise to achieve the needs of WFD. The implementation of the WFD will require the active participation of groundwater stakeholders. Recent experiences in the Upper Guadiana Basin in Spain (Hernández-Mora *et al.*, 2003) clearly illustrate that the complex problems related to intensive use of groundwater can be solved with participatory management schemes and coordination between water authorities and groundwater stakeholders.

6 CONCLUSIONS

The WFD approved in 2000 defines a framework for the protection of waters within the European Union. It aims at achieving good quantitative and chemical status of groundwater bodies by the year 2015. Its objectives and overall approach are fine. It contains measures and goals such as the implementation of the full-cost recovery principle, the protection of groundwaters, and the active participation of stakeholders that have the potential to be beneficial for groundwaters in Europe. However, its overall approach to quantitative aspects of groundwater is weak. Its definition of *available groundwater resources* may lead to conflicts because available groundwater resources may be hardly quantifiable. The concept of *quantitative status of bodies of groundwater* is confusing, extremely difficult to quantify and therefore hardly useful in semiarid regions where it is difficult to quantify the amount of groundwater discharge required to achieve the environmental objectives of surface waters. With an extreme view, this definition could lead to situations in which all natural groundwater discharge could be considered as needed to achieve the environmental objectives of surface waters and therefore there would be no available groundwater resources.

ACKNOWLEDGEMENTS

Support for this work was provided by a research project funded by *Xunta de Galicia* (PGIDT00PX 111802). The final version of this paper includes suggestions, comments and remarks derived from discussions maintained during the SINEX meeting in Valencia. The author wants to thank all the people who participated in the discussions. Thanks are also given to Andrés Sahuquillo for his remarks.

REFERENCES

AIH-GE (2003). *Presente y Futuro de las Aguas Subterráneas en España y la Directiva Marco Europea.* Proceedings of the Conference organized by the Spanish Chapter of the International Association of Hydrogeologists. Ed: Instituto Geológico y Minero de España. Madrid, Spain. 521 pp.

Balabanis, P. (1999). Water in Europe. Research achievements and future perspectives within the framework of European research activities in the field of environment. *Revista CIDOB d'Afers Internacionals*, 45–46, April 1999.

Bogestrand, J.; Grath, J.; Lack, T.J. and Nixon, S. (1998). *EuroWaterNet: The European Environment Agency's Monitoring and Informational Network for Inland Water Resources*. Technical guidelines for implementation. Technical Report No. 7. European Environment Agency. Copenhagen, Denmark.

Custodio, E. (2000). Groundwater-dependent wetlands. *Acta Geologica Hungarica*, 43(2): 173–202. Budapest, Hungary.

Custodio, E. and Llamas, M.R. (2003). Main common concepts, relevant issues and some suggestions.

In: M.R. Llamas and E. Custodio (eds.), *Intensive Use of Groundwater. Challenges and opportunities*. Balkema Publishers. Lisse, the Netherlands: 457–462.

European Commission (2000). Directive 2000/60/EC of the European Parliament and the Council of 23 October 2000 establishing a framework for community action in the field of water policy. *Official Journal of the European Communities*, L327: 1–72.

European Commission (2001). *Common Strategy on the Implementation of the Water Framework Directive*. Strategic document.

European Commission (2003). *Proposal for a Directive of the European Parliament and of the Council establishing strategies to protect groundwater against pollution*. Draft. June 2003.

Government of Spain (2001). *Ley 10/2001 de 5 de julio del Plan Hidrológico Nacional (National Water Plan)*. BOE No. 161, 6 July: 24228–24250.

Grossman, M.R. (2001). *Nitrates from Agriculture in Europe: the EC Nitrates Directive and its implementation in England*. Available at the following web address: (http://www.bc.edu/bc_org/avp/law/lwsch/journals/bcealr/27_4/01_TXT.htm).

Hernández-Mora, N. and Llamas, M.R. (eds.) (2001). *La economía del agua subterránea y su gestión colectiva*. Fundación Marcelino Botín and Ediciones Mundi-Prensa. Madrid, Spain. 550 pp.

Hernández-Mora, N.; Martínez Cortina, L. and Fornés, J. (2003). Intensive Groundwater Use in Spain. In: M.R. Llamas and E. Custodio (eds.), *Intensive Use of Groundwater. Challenges and opportunities*. Balkema Publishers. Lisse, the Netherlands: 387–414.

Llamas, M.R. and Custodio, E. (eds.) (2003). *Intensive Use of Groundwater. Challenges and opportunities*. Balkema Publishers. Lisse, the Netherlands. 478 pp.

Llamas M.R.; Fornés, J.; Hernández-Mora, N. and Martínez Cortina, L. (2001). *Aguas subterráneas: retos y oportunidades*. Fundación Marcelino Botín and Ediciones Mundi-Prensa. Madrid, Spain. 529 pp.

OECD (Organization for the Economic Cooperation and Development) (2002). *Transition to full-cost pricing of irrigation water for agriculture in OECD countries*. Environmental Directorate and Directorate for Food, Agriculture and Fisheries. 45 pp.

Samper, J. (2001). ¿Por qué es reprobable el Plan Hidrológico Nacional? In: *El Plan Hidrológico Nacional a Debate*. Fundación Nueva Cultura del Agua. Ed. Bakeaz: 165–168.

Sophocleous, M.A. (2002). Groundwater-surface water interactions: the state of the science. *Hydrogeology Journal*, 10(1): 52–67.

Sophocleous, M.A. (2003). Environmental implications of intensive groundwater use with special regard to streams and wetlands. In: M.R. Llamas and E. Custodio (eds.), Intensive Use of Groundwater. Challenges and opportunities. Balkema Publishers. Lisse, the Netherlands: 93–112.

CASE HISTORIES

Making intensive use of groundwater more sustainable – key lessons from field experience

S. Foster[(1)], H. Garduño[(2)], K. Kemper[(3)], A. Tuinhof[(4)] and M. Nanni[(5)]
World Bank/Global Water Partnership. Groundwater Management Advisory Team – GW-MATE
[(1)] *gwmatefoster@aol.com*
[(2)] *hgarduno@mexis.com*
[(3)] *kkemper@worldbank.org*
[(4)] *a.tuinhof@acaciainstitute.nl*
[(5)] *marcellananni@libero.it*

ABSTRACT

The intensive development of groundwater has led to major benefits in many developing nations through improving social welfare and economic productivity. However, significant concerns have emerged over counterproductive competition between users and the sustainability of the resource base itself, pointing to an urgent need for increased investment in resource management. Groundwater management is among the more important and highly complex of natural resource challenges facing society. To be fully effective it requires integration of appropriate levels of policy framework, regulatory provision, technical diagnosis, institutional arrangements, social participation and economic measures. But it is also necessary to recognize that *perfect may become the enemy of good* and that in practice progress will only be made if an incremental and adaptive strategy is adopted.

The paper, which is based on field experience from on-going projects in parts of China, Argentina and the Yemen, does not attempt to provide a simple blueprint for groundwater resource management. To do so would be unrealistic given the prevailing wide hydrogeologic and socioeconomic diversity. The approach adopted is to review the factors preventing effective management, the underlying technical concepts and the necessary legal and institutional framework. The importance of strengthening the capacity of local regulatory agencies to implement regulations and providing groundwater users with sufficient incentives and powers to participate in a more sustainable era of groundwater management is highlighted.

Keywords: *aquifer safe-yield, groundwater management, aquifer management organisation, real groundwater savings, groundwater over-development.*

1 BACKGROUND TO PAPER

1.1 *Historical perspective on groundwater development*

Many nations, especially those located in regions with more arid climate, have witnessed large increases in the use of groundwater resources from their major aquifers during the past 20–40 years. This has been both for agricultural irrigation and for urban water-supply, and has yielded major social and economic benefits. However, in certain cases negative impacts of the aggregated effect of intensive groundwater development have become evident. Some developing countries are now suffering regional aquifer depletion, with a significant proportion of their irrigation of agricultural crops being dependent upon the abstraction of non-renewable groundwater reserves.

Groundwater exploitation is not new, but abstraction on the large-scale is. The potential for groundwater exploitation changed radically with the spread of advances in rotary drilling technology, rural electrification, the turbine pump and geological knowledge, most notably during the 1960s and 1970s. In some countries governments encouraged groundwater development (in the absence of any controls and through provision of a variety of subsidies), unaware of, or disregarding, the availability of renewable resources. This situation has created an added obstacle to the subsequent implementation of resource use constraints.

1.2 *Scope of groundwater resource management*

Groundwater resource management has to deal with balancing the availability of a complex resource with increasing demands of water and land users, who can pose a threat to resource availability and sustainability. Calls for groundwater management do not usually arise until a decline in well yields and/or quality affects one of the stakeholder groups. Groundwater management is *as much about managing people as it is about managing water*, or perhaps more precisely stated *about enabling the people to have rights and assume duties in the management of water*.

There is a recurring assertion by some that aquifers cannot be managed in isolation, since groundwater and surface water are hydrologically-connected. This argument is relatively weak, however, since in most instances where surface water resources are being managed, no significant account is taken of connected groundwater resources. And today many aquifers are under such stress that pragmatism dictates to tackle them directly, whilst not neglecting the basic principles of integrated water management, especially if the hydraulic interconnection is relatively direct.

1.3 *Characteristics of field project areas*

Groundwater resource management initiatives from on-going projects in parts of Argentina, China and Yemen are presented. A summary of the project field areas in terms of their groundwater resource status and water resource administration system is given in Tables 1A and 1B. The main groundwater resource issues and management action-plan are then discussed in more detail area-by-area within Text Boxes A, B and C, and generic lessons drawn from them (and other field experience) in the main text.

It should be noted that in the case of the North China Plain (Text Box B) and that of the Sana'a Basin-Yemen (Text Box C) physically-unsustainable depletion (*mining*) of aquifer reserves has arisen on an unplanned basins, and current groundwater resource management efforts are directed at bringing order to the situation. But at the same time it has to be recognised that the existing framework of local groundwater resource administration is much more developed in China than in the Yemen, and thus initial priorities of the new aquifer management initiative will be distinct.

In contrast to the other two field project area, the Mendoza Northern Oasis of Argentina (Text Box A) is not experiencing widespread and continuous aquifer depletion, but in part is highly sensitive to groundwater quality deterioration through various salinisation processes. This necessitates urgent groundwater resource management actions to control the spread of groundwater salinisation in an aquifer system which is a vital source of irrigation water for the cultivation of high-value, but salt-sensitive, crops, which are the cornerstone of the local economy.

2 INFLUENCE OF HYDROGEOLOGIC AND SOCIOECONOMIC DIVERSITY

The hydrogeological factors controlling groundwater availability and the socioeconomic factors affecting resource use vary widely between locations and also evolve with time. The interaction of these sets of factors, together with the local institutional context, determines the applicability of resource management options. It is thus necessary to develop management approaches that can be tailored closely to specific situations, and that can adapt effectively as the larger socio-economic context changes.

Amongst the key factors which determine the initial level of groundwater management response appropriate at a given time in a specific location are:

- need and capacity for major socioeconomic adaptation to water resource scarcity;
- degree of evolution of institutional framework and scope for its rapid consolidation;
- level of water-user and other stakeholder participation;
- availability of groundwater storage and level of excessive groundwater abstraction;
- susceptibility of the groundwater resource system to rapid degradation by salinisation;
- social and environmental significance of groundwater level-related services.

Whilst no simple blueprint for the promotion of groundwater resource management can be provided, it is considered useful to draw a series of ***generic lessons*** from on-going field experience, which will be relevant (now or in the future) in most areas of intensive groundwater resource utilisation where national and/or local water resource agencies are striving to improve the sustainability and productivity of groundwater use. These lessons are ordered (somewhat arbitrarily) under the following headings:

- the diagnosis of groundwater resource status;
- approaches to groundwater resource management;
- groundwater management institutions and tools.

Although in reality they are all closely interlinked.

GENERIC LESSON 1	***unrealistic to provide a* simple blueprint *for groundwater resource management because of wide hydrogeologic, socioeconomic and institutional diversity***

3 DIAGNOSIS OF GROUNDWATER RESOURCE STATUS

3.1 *Variability of aquifer recharge rates*

A fundamental consideration in groundwater management is the sustainability of resource development. In this regard estimation of contemporary aquifer recharge rates is of central importance. Furthermore, understanding aquifer recharge mechanisms and their linkages with land use is an essential ingredient of integrated water management.

The average rate of aquifer replenishment can exhibit marked long-term variation if certain changes occur in the recharge area. These changes include riverflow control or diversion, modifications to surface water irrigation infrastructure and technique, changes in natural vegetation or crop type, reduction in urban mains water leakage and in-situ wastewater percolation, lowering of water-table due to groundwater abstraction, etc., and such considerations must always feature in resource assessments.

GENERIC LESSON 2	***the common paradigm of* constant average rates of contemporary recharge *is false and can lead to serious* double resource accounting *in more arid regions***

In practice the quantification of natural recharge is subject to significant methodological difficulties, data deficiencies and resultant uncertainties (Text Box C) due to:

- wide spatial and temporal variability of rainfall and runoff events;
- widespread lateral variation in soil profiles and hydrogeological conditions.

Nevertheless, preliminary estimates based on available data are useful for initial management decisions, and can be refined subsequently through monitoring and analysis of aquifer response to medium-term abstraction. It is vital that sufficient effort goes into monitoring and short-term economies in this respect are likely to prove counterproductive in the long run.

A

COUNTRY	ARGENTINA	CHINA		YEMEN
Location	Mendoza Northern Oasis	North China Plain Piedmont	North China Plain Heilongang	Sana'a Basin
Aquifer Type	multi-layered alluvial outwash complex	thick unconfined sands and gravels	thin unconfined sand lenses with deep confined sands	Cretaceous Sandstone overlain locally by alluvial sands and gravels
Aquifer Storage	moderately large but towards margins increasing salinity	extremely large, but at distance from upstream limits an increasing proportion is of brackish quality		probably moderate, but depends on average storativity of sandstone
Average Rainfall	180–220 mm/yr	550–600 mm/yr	500–550 mm/yr	200–250 mm/yr
Source of Primary Recharge	Major riverbed infiltration from large rivers, unlined canals and excess surface water irrigation	excess rainfall, riverbed and canal infiltration, and excess surface water irrigation, but latter two have greatly decreased due to upstream flow diversion		periodic infiltration of runoff in wadi beds and during spate irrigation
Pumping Wells (Study Area)	250–300 (230 km^2)	2,000 (220 km^2)	300–350 (300 km^2)	400–500 (320 km^2)
Aquifer Depletion	periodic, but not long-term	averaging 0.5 m/yr	averaging 2 m/yr	few data, probably 3–5 m/yr
Salinisation Risk	very high in upper aquifer horizon and threatening deeper layers	for most part low	very high in both shallow/deep aquifers, but mechanisms differ	apparently low
Land Subsidence	fairly low hazard	generally low hazard	major hazard	low hazard

Table 1A & B. Summary of groundwater resource status and administration arrangements in field project areas.

B

COUNTRY	ARGENTINA	CHINA	YEMEN
Location	**Mendoza Northern Oasis**	**North China Plain**	**Sana'a Basin**
Groundwater Uses	continuous/ supplementary irrigation of commercial grapevines, fruit orchards, small town/minor industry supply	irrigation of cereal crops/ fruit orchards/market gardening, urban and industrial supply	only source for capital city together with small-scale irrigation of qat/grapes
Legal Provisions	long-established water law and more recent specific provisions for groundwater	national water law revised; related county regulations on abstraction permits/policy/planning need up-dating	water law just approved and regulations being drafted
Regulatory Agency	autonomous provincial-level planning and regulatory agency	county water resource bureaus, supported by provincial and national agencies	embryonic basin-level management agency
Abstraction Rights	licenses exist for most part, need for streamlining to make reallocation feasible	abstraction licenses introduced in 1993 but action still needed to make them a true management tool	only through customary tribal *wadi irrigation rights*
Stakeholder Participation	well-established WUAs, but no aquifer-oriented institution as yet	some village-level WUAs but no aquifer-oriented institutions as yet	
Groundwater Fee	fee in range US$ 15–40/yr	nominal charge of US$ 0.1/m^3	no abstraction charges
Economic Subsidies	some via rural electricity tariffs, but much larger subsidy for surface water-supply	indirectly on cereal grain production	indirectly via limited subsidies on diesel fuel for some well pumps
Institutional Assets	strong water regulatory agency with good user database and research support	decentralized *bottom-up* institutional approach well developed	research support under strong development
Management Impediments	lack of powers/resources to reallocate resources; WUAs surface water focused	insufficient *top-down* coordination between neighbouring county resource regulatory agencies	no adequate legal and institutional arrangements yet in place

TEXT BOX A: MENDOZA NORTHERN OASIS OF ARGENTINA

Diagnosis of Groundwater Salinity Problem

The Mendoza Northern Oasis is an extensive irrigation area in a very arid region, underlain by Quaternary alluvial deposit. This aquifer system is mainly recharged by the Mendoza and Tunuyan rivers, directly on highly-permeable alluvial fans as they emerge from the adjacent hills and indirectly via the irrigation canals and irrigated fields over a larger area. With increasing distance from its upstream margin the aquifer system exhibits marked layering, with three aquifer units separated by interbedded aquitards.

The state water resources authority has taken a planned approach to allocation for irrigated agriculture and urban water-supply, and is attempting to integrate groundwater more consistently into a hydraulic infrastructure of long history by:

- encouraging irrigation well drilling (where feasible) in areas outside the command of existing surface water canals.
- permitting irrigation well drilling in areas where existing canals cannot provide a reliable supply during periods of maximum plant demand.

But this has encountered problems in a few areas where aquifer behaviour has been different to that initially expected.

The current focus is in the Departamento San Martin where a *groundwater use restriction zone* of 23,180 ha (with a current abstraction of 60–85 Mm^3/yr) was declared in 1995 as a result of an emerging problem of aquifer salinisation (Inset A1). Increasing groundwater salinity was initially detected in the water-table aquifer beneath well drained soils at 5–15 m depth, and appeared to be the result of soil concentration in irrigation return waters (not phreatic evapotranspiration). More recently farmers have constructed deeper wells to tap semi-confined aquifers which exhibited excellent quality. But heavy pumping resulted in salinity being drawndown into the second aquifer, threatening its use for irrigation of high-value, salinity-sensitive crops, such as export-quality viniculture and fruit trees.

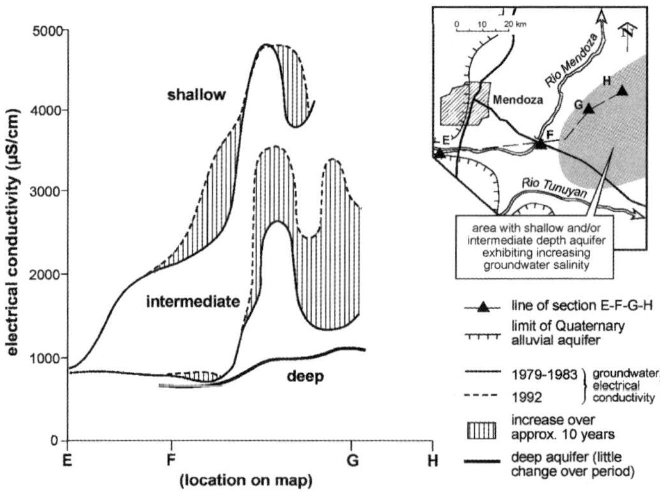

Inset A1. Evolution of groundwater salinity in the Quaternary multi-aquifer system of the Mendoza Northern Oasis.

The induced downward migration of saline groundwater appears to be related to:

- the low natural horizontal hydraulic gradient and thus small groundwater through flow of the semi-confined aquifers;
- the fact that downward vertical leakage is more readily induced by pumping than increased horizontal groundwater flow.

It is tentatively concluded that 70% of the supply pumped from the second aquifer is derived from above. The existence of poorly-constructed and/or highly-corroded wells (many abandoned) is further aggravating the groundwater salinisation problem.

Approach to Groundwater Management

A number of management measures are being taken to reduce the rate of salinisation of the semi-confined aquifers including:

- substantially reducing pumping of groundwater from deeper aquifers to reduce the rate of downward leakage by transferring *additional surface water* into the critical area, with irrigators relinquishing groundwater abstraction rights.
- improving the efficiency of surface-water irrigation and making real water savings by lining secondary/tertiary canals and introducing pressurised irrigation systems.
- promoting a fund for real water-savings in groundwater-based irrigation (which will be partially recovered through reductions in the use of the electrical energy subsidy).

The introduction of the *groundwater use restriction zone* is viewed as successful in terms of taking control of abstraction, while allowing a degree of flexibility on the replacement of wells. Some special (relatively unusual) local social characteristics:

- groundwater users include a high percentage of recently-established major international wine and fruit producers, using modern high-efficiency irrigation.
- surface water users are dominantly long-standing smallholders using traditional low-efficiency irrigation techniques.

Make water resource reallocation from latter to former in the interests of aquifer management encounter strong opposition. The *sale* of excess surface water allocations (but not of rights) is permitted, but with irrigation modernisation costs of around US$ 1,000/ha this is not a strong enough incentive for surface water irrigators to invest in water-saving measures. And the water resources regulator lacks the legal powers and financial resources to transmit excess surface water to areas without rights, to subsidise demand management and to reduce rights in riparian areas of inefficient water use.

Local irrigation-water management in a substantial part of the *groundwater use restriction zone* is the responsibility of a water-user association (Inset A2). This association is based on the primary irrigation canal infrastructure, and for some years now has made an important contribution to the allocation and operation of surface water resources and underlying groundwater. But there are significant impediments to its involvement in groundwater resource management. This is the result of incompatible boundaries and the fact that neighbouring *groundwater only users* are reluctant to contribute to running costs. The formation of an aquifer management organisation is considered the best way forward, in which the existing water-user association is one representative.

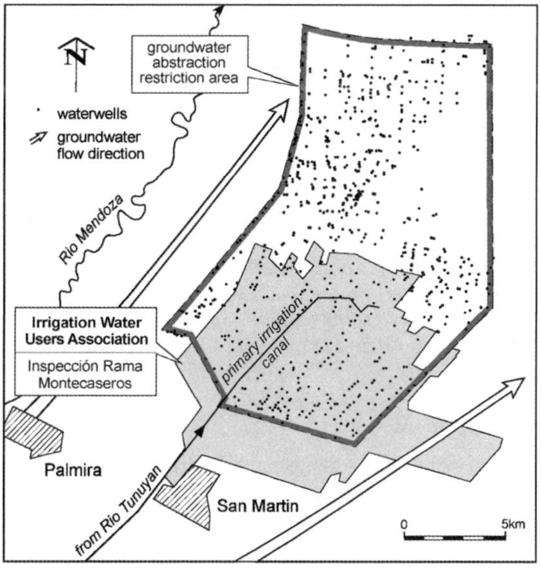

Inset A2. Map of the Departamento de San Martin-Mendoza showing the water-user association area and neighbouring groundwater use restriction zone.

TEXT BOX B: NORTH CHINA PLAIN

General Status of Groundwater Resources

A very extensive Quaternary aquifer system occupies very large tracts of the densely-populated North China Plain the national centre of wheat and maize production, and an extremely important industrial region.

There are two distinct hydrogeological settings on the North China Plain (Inset B1):

- gently-sloping piedmont plain with major alluvial fans below a mountain escarpment.
- main alluvial plain, the so-called Heilongang, with many abandoned river channels.

Below the Heilongang the sequence includes a brackish-water body, locally overlain by thin lenses of phreatic fresher groundwater (associated with surface water channels and major irrigation canals), and everywhere underlain by a deep confined freshwater aquifer whose groundwater is believed to have been emplaced in a colder wetter epoch more than 10,000 years BP.

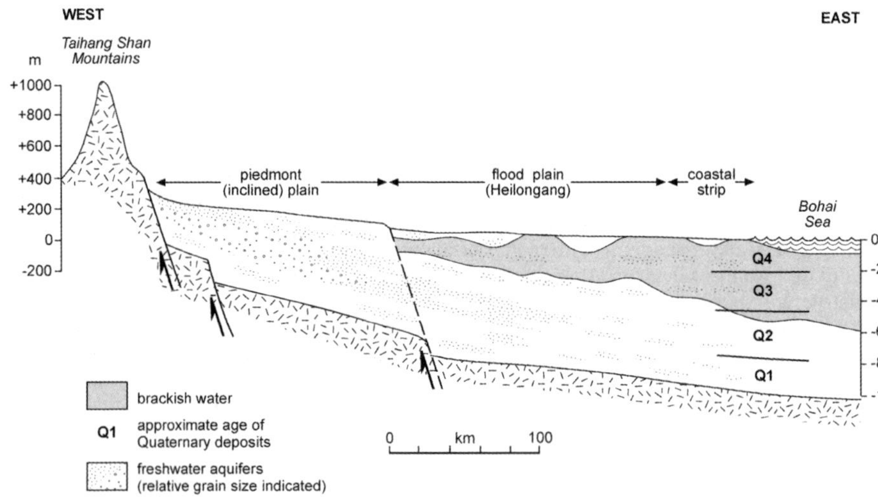

Inset B1. Cross-section of the North China Plain showing the general hydrogeological structure.

Intensive exploitation of groundwater over the past 20–40 years has reaped large benefits, in terms of farming employment, poverty alleviation, grain production, and potable and industrial water-supply, but has also encountered increasing difficulty due to:

- on the piedmont plain continuous long-term water-table decline of around 0.5 m/yr (suggesting an average recharge deficit of 40–90 mm/yr), with much greater rates in some urban centres (Inset B2).
- more recently the deep confined aquifer has been rapidly developed for urban and industrial water-supply, and where the shallow aquifer is thin or absent for agricultural irrigation, resulting in average value of groundwater level decline of more than 2 m/yr (caused by relatively modest rates of abstraction of essentially fossil groundwater) but resulting in lowering of the overlying brackish water interface into the deep freshwater aquifer at rates of 0.5–1.5 m/yr and serious land subsidence in some urban centres.

Approach to Resource Management

Two pilot groundwater management areas – Fangshan – District and Guantao County – representative of the two hydrogeological settings above, are the focus of current efforts to develop effective approaches to stabilising aquifer water-levels and to achieving a more sustainable regime of groundwater use.

Improved irrigation and cultivation measures for cereal crops can substantially reduce non-beneficial evapotranspiration and appear capable of effecting *real water savings* of 50–80 mm/yr. If adopted widely, they could thus reduce the rate of decline of the deep confined aquifer and stabilise the shallower water-table aquifer. But since they depend heavily on water-user participation they are likely to take

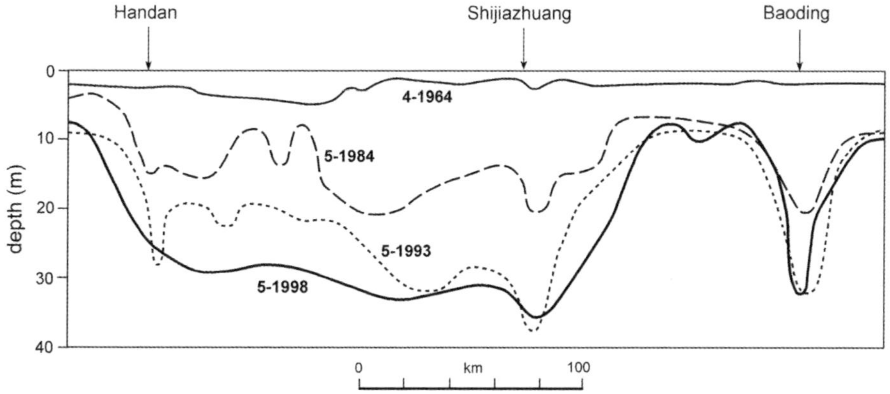

Inset B2. Historical evolution of water-table decline in the Quaternary North China Plain Aquifer on a long-section of the piedmont plain.

many years to implement fully. A more radical option would be to ban the cultivation of irrigated cereal crops:

- in groundwater-critical areas around cities.
- using the deep confined freshwater aquifer.

thus reducing the overall proportion of land under irrigation, and enlarging the area with dryland cropping. There is also potential to introduce higher-value, lower water-demand, crops (e.g. drip irrigation of fruit trees and horticultural greenhouse cultivation) but there remain significant market, transport and storage limitations on these options.

Complementary supply-side actions are also being evaluated and where feasible implemented, such as:

- artificial aquifer recharge with excess surface run-off (in summer months of wetter years) using pre-prepared sections of ephemeral surface watercourses and canals.
- using wastewater for non-sensitive irrigation purposes (such as commercial woodland and forage crops) with concomitant reduction in groundwater abstraction.

Institutional Needs for Resource Management

In China day-to-day management is carried out at the local level by the County (or District) Water Resources Bureau (CWRB). This highly-decentralised (and in effect *bottom-up*) approach is potentially a major asset, presenting opportunity for close interaction with groundwater users.

The success of agricultural water-saving measures in reducing aquifer water-level decline depends upon reductions in irrigation lamina being directly translated into permanent reductions in well abstraction, and not simply being used to expand the overall area under irrigation or to increase water-use in other sectors. For this purpose there are various institutional issues that must be addressed:

- farmers need to grasp the *real water saving* concept.
- provision of economic incentives for installation of more efficient irrigation technology.
- a consistent relation must be established between groundwater resource estimates and permitted abstraction rates for village-level water-user associations.
- a realistic *groundwater resource fee* needs to be imposed, generating finance for aquifer monitoring needs and serving as an incentive for reducing abstraction.
- more uniform enforcement of existing groundwater rights through local water abstraction measurement, aquifer monitoring, and information dissemination.
- broader participation in groundwater resource management, including representation of urban and industrial users, as well as irrigation water-user associations.

TEXT BOX C: SANA'A BASIN OF YEMEN

Status of Groundwater Resources

The Sana'a Basin is an arid highland area of 3,200 km^2, which under present hydrological conditions is an almost-closed basin. The population numbers 1.8 million, and irrigated cultivation has grown rapidly since 1985 from some 2,000 ha to 23,400 ha. All past attempts to calculate the groundwater resources balance have been fraught with large errors due to lack of adequate data, but on the simplified basis estimates of consumptive groundwater use can be compared with new aquifer replenishment (Inset C1), avoiding the need to evaluate various water recycling components in detail.

Estimates of consumptive groundwater use by the dominant irrigated crops, qat and grapes (at present 50% and 35% respectively), remain rather uncertain. Both are irrigated only 4–5 times per year but receive total lamina of 750–850 mm/yr. Recent work puts gross groundwater abstraction at 208 Mm^3/yr, with some 20% as irrigation return, suggesting a net consumptive use of 166 Mm^3/yr. An earlier study, however, put abstraction at 149 Mm^3/yr (for 21,600 ha irrigated area) with 30% irrigation return, and by adjustment would imply a present consumptive use of 119 Mm^3/yr. The level of non-beneficial evapotranspiration associated with current irrigation and cultivation practices is also rather uncertain. The gross abstraction for the Sana'a conurbation (population 1.3 million) is put at 21–24 Mm^3/yr, and, since most is returned to the ground, the net urban consumption is only 2–3 Mm^3/yr, or 6–7 Mm^3/yr including sewage effluent use.

Aquifer recharge, as always, is a difficult parameter to quantify. Moreover, confusion may occur when comparing estimates, due to lack of clarity over area and aquifer under consideration, and whether recycling is included. Considering new replenishment only, previous estimates fall into two distinct groups: a *lower set* in the range 25–38 Mm^3/yr (Inset C1), derived from analysis of hydrometeorological data for the Quaternary Alluvium and Cretaceous Sandstone aquifer outcrop (Inset C2), and a *higher set* which include an allowance for deep recharge through thick Tertiary Volcanics.

PARAMETER (2002 ESTIMATE)	RANGE OF ESTIMATES (mm/yr) MINM	MAXM
Active Aquifer Replenishment	+ 25**	+ 38**
Consumptive Use of Groundwater	– 173	– 119
Mining of Aquifer Storage*	– 148**	–81**

**net value, since in some parts of urban area groundwater levels are rising*

*** if groundwater inflow from the southwest and southeast were proven then replenishment would increase by 20–25 Mm^3/yr and rate of mining would decrease*

Inset C1. Range of recent estimates for the groundwater resource balance of Sana'a Basin – Yemen.

A number of conclusions on the resource imbalance are reached:

- there is currently substantial mining of Cretaceous Sandstone groundwater storage (Inset C2), principally by irrigated agriculture.
- the scale of the imbalance is still very uncertain and could be anywhere in the range 56–148 Mm^3/yr, having grown from zero at some time during 1978-86.
- even with widespread real water-saving measures and wadi artificial recharge measures, it is unlikely that the replenishment-consumption imbalance can be closed without significant reduction in irrigated area.

There is little reliable monitoring of groundwater levels, but regularly-measured deep boreholes in the vicinity of the Sana'a Wellfields (north of the city) give average rates of aquifer depletion of 3.7 and 4.2 m/yr respectively during 1981-93, and hearsay from the major Wadi Al-Sirra irrigation area suggests water-levels are now falling at 3–6 m/yr. No evidence exists to bear out any serious risk of salinisation, with most groundwaters having EC in the range 300–700 μS/cm (except in the immediate Sana'a urban area), and it appears likely that the entire basin was *well flushed* during Pleistocene episodes of much wetter climate.

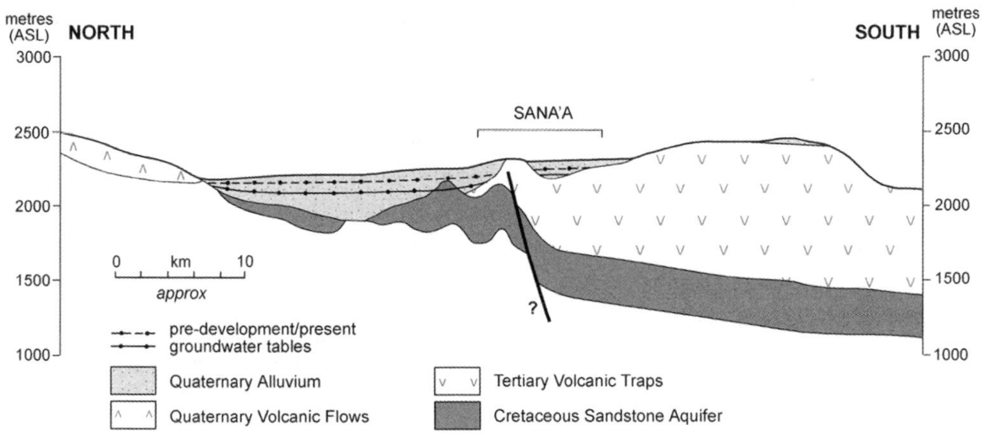

Inset C2. Hydrogeological cross-section of Sana'a Basin illustrating the occurrence of the Cretaceous Sandstone and Quaternary alluvial aquifers.

Groundwater Resources Management Strategy

The main objective has to be to increase the usable life of groundwater storage, and hence postpone the need for long-distance import of water and allow time for a shift to a less water-based economy. In effect the water resources strategy hinges on the mining of aquifer storage to *buy time* for the necessary socio-economic adaptation required, but the difficulties in achieving such adaptation must be recognized and dealt with early enough.

Reliable answers to the questions *how long will groundwater stocks last* and *what measures can best reduce the rate of aquifer depletion* are critical to defining a realistic water resources strategy for the basin, and require:

- quantification of the exploitable groundwater storage of the main Cretaceous Sandstone aquifer, especially those strata that have recently, and will soon be, drained, to improve prediction of the availability of non-renewable resources.
- pilot studies on agricultural water-demand reduction, through improved irrigation technology to make real-water savings, refocused agricultural subsidies and constraints on irrigation expansion.
- supply-side enhancement, through construction of appropriate structures to enhance wadi-bed recharge small *check dams* being likely to be most effective and a useful vehicle to engage local communities in practical water resources conservation – but there remains significant uncertainty about the availability of surface run-off for recharge and the overall net basin gain in groundwater recharge that would be achieved.
- urban wastewater management to improve the treatment process and agricultural reuse of the growing volume of Sana'a sewerage effluent.

New legislation has been approved and regulations are currently being prepared that provide the institutional framework for successful implementation of these measures and the following related actions:

- establishment of a basin water agency to be proactive in resource management and to nucleate an aquifer management organization involving representatives of all users.
- campaigns of public awareness on the status of groundwater resources and the need for their proactive management.
- formalisation of stakeholder participation in groundwater management through the establishment of rural water-users association with groundwater use rights.
- anticipation of the new socioeconomic situation that eventually and inevitably will arise through reduced water-supply availability.

3.2 *Aquifer safe yield*

All too often the so-called *average aquifer recharge rate* is taken as its *safe yield*. This rather persistent fantasy fails to consider:

- the variability of aquifer recharge rates described above;
- the potential need to maintain aquifer discharges or water-levels in the interest of other water users, of sustaining aquatic and/or terrestrial ecosystems and of preventing coastal saline intrusion.

So-called *safe yield* is clearly bounded by the contemporary long-term average rate of aquifer recharge but should also involve:

- value judgements about the importance of maintaining a proportion of natural discharges from the aquifer system;
- consideration of consumptive use (Text Box C) and catchment export (as opposed to uses generating local effluent and/or return flows), since groundwater recirculation via urban water-supply and irrigated agriculture commonly causes confusion in many aquifer water balances.

GENERIC LESSON 3	***the balance between* active new replenishment *and* consumptive plus export use *is the most meaningful basis for evaluation of the groundwater resource balance***

Such factors are rarely considered adequately – thus *statements of aquifer safe yield* are usually too simplistic. Nevertheless, target rates of tolerable groundwater consumptive use need to be defined, and resource evaluation must distinguish between:

- discharge to freshwater systems, since these may be required to sustain downstream water-supply or river ecosystems;
- discharge via natural vegetation, including that sustaining ecologically and/or economically valuable freshwater wetlands and brackish lagoons;
- discharge to saline areas, including coastal waters, salt lakes and *playas*.

The related term *aquifer overexploitation* is not capable of rigorous scientific definition. Some regard an aquifer as overexploited when its groundwater levels show evidence of *continuous long-term* decline. But all groundwater development involves drawdown and (in large aquifers of low transmissivity/storativity ratio) this process can have a time-lag of various years before a new equilibrium is established, which in some cases might be mistaken for continuously-declining aquifer water-levels. Others take it to mean that long-term average groundwater recharge is less than abstraction. But even this definition may not be workable because of difficulties over:

GENERIC LESSON 4	***although not capable of rigorous definition, the term* aquifer overexploitation *should not be abandoned altogether because of its clear register at public and political level.***

- specifying the period and area for which the balance should be evaluated;
- major temporal variation in aquifer recharge components.

But, in practice, the main concern is with the consequences of intensive groundwater exploitation, rather than with its absolute volumetric level.

GENERIC LESSON 5	***major concern must be that* the overall cost of negative impacts of groundwater abstraction should not exceed the net benefits of groundwater use**

3.3 *Aquifer susceptibility to degradation*

The overdraft of aquifer storage (on whatever time-base) can have a series of consequences (Table 2). Most widespread is competition for available groundwater due to continuously declining aquifer water-levels. In some instances there is evidence of increasing social inequity where deeper (larger-capacity) bore-holes lower the water-table and increase the cost of (or eliminate the access to) water for users of shallower wells.

In addition to reversible interference with wells and springs, more severe side-effects may

Table 2. Consequences of excessive groundwater exploitation.

CONSEQUENCES OF EXCESSIVE ABSTRACTION		FACTORS AFFECTING SUSCEPTIBILITY
REVERSIBLE INTERFERENCE	• pumping lifts/costs increase • borehole yield reduction • springflow/baseflow reduction	- *aquifer response characteristic* - *drawdown to productive horizon* - *aquifer storage characteristic*
	• phreatophytic vegetation stress • aquifer compaction • transmissivity reduction	- *depth to groundwater table* - *aquifer compressibility*
IRREVERSIBLE DETERIORATION	• saline water intrusion • ingress of polluted water • land subsidence and related impacts	- *proximity of saline/polluted water* - *vertical compressibility of overlying/interbedded aquitards*

also occur (Table 2). Most notable are those involving the encroachment of saline water, commonly through lateral intrusion from the sea (if coastal hydraulic gradients are reversed) and from above (in layered aquifers when upward hydraulic gradients are inverted) even at moderate levels of resource exploitation (Text Box A). These effects are usually quasi-irreversible and the ingress of saline water terminal for most uses. In major urban areas where groundwater abstraction often exceeds the average rates of local recharge, associated changes in hydraulic head distribution within aquifers can induce downward leakage of polluted water.

GENERIC LESSON 6	***groundwater management must critically assess the susceptibility of aquifer systems to adverse side-effects when subjected to temporary and long-term overdraft***

3.4 *Potential of aquifer storage*

The vast storage of many groundwater systems is, in some senses, their most valuable asset, which needs to be exploited in a strategic fashion. On the one hand important components of groundwater value (such as pumping costs, individual accessibility for the poor, sustaining some freshwater wetlands and dry weather streamflow) depend on depth to water-table and not on volume in storage. On the other hand, in many situations aquifer storage is the only major source of freshwater in drought, and ways need to be found to exploit this resource whilst mitigating the impacts on groundwater-level related services.

Moreover, there is no fundamental reason why exploitation of non-renewable aquifer storage should not be considered. But for this to be socially-sustainable reliable evaluation of the following factors is essential:

- the rate of groundwater mining that can be achieved for the period in question;
- the potential scale of any internal aquifer effects and external environmental impacts;
- the level of interference with all existing and potential groundwater users;
- the economic benefits of the proposed use compared to potential future uses.

In addition, *what comes after* question, needs to be addressed in terms of socioeconomic transition that will be required.

GENERIC LESSON 7	***if exploitation of non-renewable groundwater reserves is carefully-planned it can form part of a logical social development strategy***

All too often, however, this is not the case and a sequence of progressive overdraft of aquifer storage is embarked upon in an anarchical or unplanned fashion, with negative long-term consequences for all groundwater users.

4 APPROACH TO GROUNDWATER MANAGEMENT

4.1 *Influence of National policy factors*

While groundwater management always has to be conducted at local aquifer level, there are a number of national policy issues which can have a pronounced impact on groundwater abstraction (Table 3), but are seldom considered as instruments of groundwater management. Among these, subsidies on rural electricity, well pumps and grain or milk prices are probably the most significant as far as groundwater resources are concerned. In general terms it is always recommendable that these subsidies be reviewed, and consideration be given to re-targeting the revenue involved into water-saving technology and/or assisting only the most-needy members of the community.

The case of Mendoza-Argentina (Text Box A) illustrates well the frequent complexity of the economics of rural water-supply provision. Despite a typical rural electricity subsidy to all groundwater users of around US$ 240/yr, at present most irrigators pay around 7 times more for groundwater than surface water (Table 4). This is because:

- historically all surface water irrigation infrastructure has either been wholly or partially subsidized;
- groundwater users have to finance well drilling and pump purchase.

GENERIC LESSON 8	***national energy and food policy pressures and management can exert an overriding influence on groundwater development strains***

Additionally, it is important at national level that groundwater resources are properly considered in strategic planning. Addressing the policy question of *what services are most required from groundwater* is necessary to provide targets for local management action, but it is one which is frequently ignored or not even recognised.

Table 3. Range of measures capable of influencing groundwater use and management.

TERRITORIAL LEVEL OF ACTION	TYPE OF INTERVENTION	
	DEMAND-SIDE MANAGEMENT MEASURES	SUPPLY-SIDE ENGINEERING MEASURES
Local (micro)	• incremental reallocation/reduction in some groundwater rights • sanctions for non-compliance with abstraction controls • agricultural real-water savings • urban real water-saving measures	• general water harvesting techniques • appropriate artificial recharge structures • urban wastewater recycling/ reuse
National (macro)	• national water-pricing policy • re-targeting agricultural crop subsidies • banning cultivation of certain highly water-consuming, low-value crops • reducing food import restrictions • re-targeting electricity/diesel subsidies • eliminating well construction subsidies • authorising realistic groundwater resource charges	• major surface water transfer schemes

4.2 *Demand-versus-Supply side intervention*

It is always essential to address the issue of constraining the demand for groundwater abstraction (so-called demand-side measures) since these are often volumetrically more significant to achieving the groundwater balance and in the more arid and densely-populated areas will always be required in the longer run. Real groundwater resource savings in irrigated agriculture can be obtained if abstraction regulations (set-up in consultation with users and enforced by the regulatory agency) are closely associated with economic incentives for the introduction of water-saving technology. Similar considerations apply in urban areas, where in certain circumstances water mains leakage and wastewater seepage actually comprise useful aquifer recharge, whilst in others real groundwater savings can be made by reducing them since they are lost to brackish water bodies or create drainage problems.

GENERIC LESSON 9	**demand-side interventions** ***will usually make a more significant long-term contribution to stabilising aquifer water balances than*** **supply-side interventions**

But complementary local supply-side measures, such as rainwater harvesting and artificial aquifer recharge with excess surface run-off, should always be encouraged, especially where conditions are favourable. They are often important in terms of building better relationships with groundwater users and can provide an initial focus for their participation in aquifer management (Text Box B & C). There may be institutional impediments to overcome, especially in relation to the financial, administrative and legal basis for using surface runoff for the benefit of local groundwater users as opposed to downstream interests.

4.3 *Real water-saving in irrigated agriculture*

This subject must assume great importance given that, in most regions, agriculture is the predominant consumer of groundwater resources. Moreover, excess irrigation often forms an important component of aquifer recharge, and one in turn normally available to other groundwater users or as baseflow in downstream rivers. It follows that while increasing irrigation efficiency represents *an energy saving* (since less pumping will be required), it does not necessarily represent *a water resource saving* (because the water would anyway have returned to the aquifer). In many instances improvements in irrigation water-use efficiency and the cultivation of higher-value crops (whilst generating important improvements in water-use productivity and farmer incomes) lead to a deterioration in the groundwater resources balance as a result of:

- substituting increased field-level evaporation/evapotranspiration (as occurs with

Table 4. Summary of typical costs of groundwater and surface water to agricultural irrigation users in the Mendoza Oasis.

COST COMPONENT	GROUNDWATER WELLS	SURFACE WATER CANALS
Permission to Construct Well	US$ 20–40/yr/well	not applicable
Water Resource Fee		US$ 10–15/yr/ha irrigated
Operation & Maintenance Costs	in part partly covered by DGI/IC fee, but also allowed for below	some local costs included in DGI/IC fee, but others met by local government
Capital Depreciation Allowance	allowed for in calculation of below	hydraulic infrastructure provided by local government without cost recovery
Typical Equivalent Cost to Irrigator	US$ 0.015/m^3**	US$ 0.002/m^3

* *expressed as US$ but based on Argentina pesos in June 2002 (exchange rate about 3.0)*

** *after deducting subsidy of about US$ 240/yr*

spray irrigation) for major groundwater irrigation-return flows (occurring in flood irrigation);

- making feasible the expansion of the irrigation command and area actually under cultivation (due to the capacity of pressurized water delivery);
- the introduction of higher-value crops making it viable for farmers to deepen wells and to pump groundwater against greater hydraulic heads.

GENERIC LESSON 10	***a common fallacy is to assume that increasing* irrigation water-use efficiency *invariably leads to* groundwater resource conservation, *in practice the reverse is currently often the case***

Thus only those modifications to irrigation and cropping practices that reduce non-beneficial evaporation, or water loss to saline water bodies, actually represent *real water savings*. The former include evaporation from the irrigation distribution system, soil evaporation from between crop rows, evapotranspiration by the crop itself which is not effective in producing yield, increased direct phreatic evapotranspiration by unwanted vegetation, evaporation during spray irrigation, etc. These should be the primary targets for demand management aimed at conserving groundwater resources (Text Box B).

There is generally considerable scope for agricultural water-savings by:

- *engineering measures*: such as irrigation water-distribution through low-pressure pipes (not open earth canals) and application by drip and micro-sprinkler technology.
- *management measures*: to improve irrigation forecasting, water scheduling and soil moisture management.
- *agronomic measures*: such as deep ploughing, straw and plastic mulching, and the use of improved strains/seeds and drought-resistant agents.

If larger water-savings are needed, then consideration should also be given to changes in crop type and land use. There is significant potential to introduce higher-value, lower water-demand, crops through direct substitution and by greenhouse cultivation (Text Box B). An even more radical option would be to place a ban on the cultivation of certain types of irrigated crop in the most critical groundwater areas.

The success of agricultural water-saving measures in reducing decline in aquifer water-levels depends directly on these savings being translated into permanent reductions in well abstraction rights and actual pumping. It is essential that water-savings be not used to expand the irrigated area or to increase water use in other sectors. This will require a flexible system of abstraction rights and clear incentives for users to act in the collective interest of resource conservation.

At the urban-rural interface resource reallocation to more productive commercial and industrial uses can be best promoted through schemes whereby the corresponding municipality finances improvements in agricultural irrigation (generating real water savings) in return for abstraction rights over a proportion of the groundwater saved.

5 MANAGEMENT INSTITUTIONS AND TOOLS

5.1 *Appropriate institutional framework*

Groundwater resource management is a many-facetted process best carried out through the collaborative effort of some form of groundwater regulatory agency (GWRA) and aquifer management organisation (AMOR), involving representatives of local water-users associations (WUAs) and other stakeholder groups. While national water laws generally exist and in many cases articulate such basic principles clearly, it is important that national ministries take action to enable, to empower and to invest in such a framework.

The territorial levels at which groundwater management responsibilities and actions are focused is of considerable significance. It should be that necessary to ensure effective interaction with groundwater users and potential polluters. However, other critical functions (such as the identification of groundwater bodies or aquifer management areas (AMAs) based on natural groundwater recharge and flow boundaries) may need to be performed at a higher level, especially where very extensive aquifer systems are present. For example, the scale of the North China Plain is such that for effective resource administration AMAs need to be delineated (Figure 1) and AMORs set-up by the provincial water resource administration, whilst day-to-day groundwater management functions are mainly handled by the county water resources bureaus.

GENERIC LESSON 11	**top-down** ***and*** **bottom-up** ***approaches must be reconciled to achieve effective groundwater resource management***

The leadership of a GWRA is absolutely essential. Many such agencies exist but are focused more on resource development issues. They need political support, financial investment and capacity building to assume their groundwater management and protection duties, including such functions as monitoring aquifer behaviour, maintaining management databases, promoting stakeholder participation and public awareness, and enforcing legal compliance.

GENERIC LESSON 12	***a government agency having the legal mandate and political backing to act as*** **groundwater guardian** ***is critical to successful groundwater resource management and quality protection***

5.2 *Knowledge of groundwater use and users*

Another key management function required of the GWRA is the detailed databasing of groundwater use and users. This is never a trivial issue and often a formidable task, given the very large number of dispersed users generally involved. For irrigation water use it can be greatly aided by remote sensing of land use, correlation of waterwells with electricity meters, systematic well census and its facilitation by WUAs.

GENERIC LESSON 13	***to mobilize effectively on groundwater management it is essential to have a systematic database of water-users or*** **user groups,** ***their use patterns and socioeconomic characteristics***

A key administrative issue is the volumetric measurement of groundwater abstraction. The use of water meters amongst irrigation users is

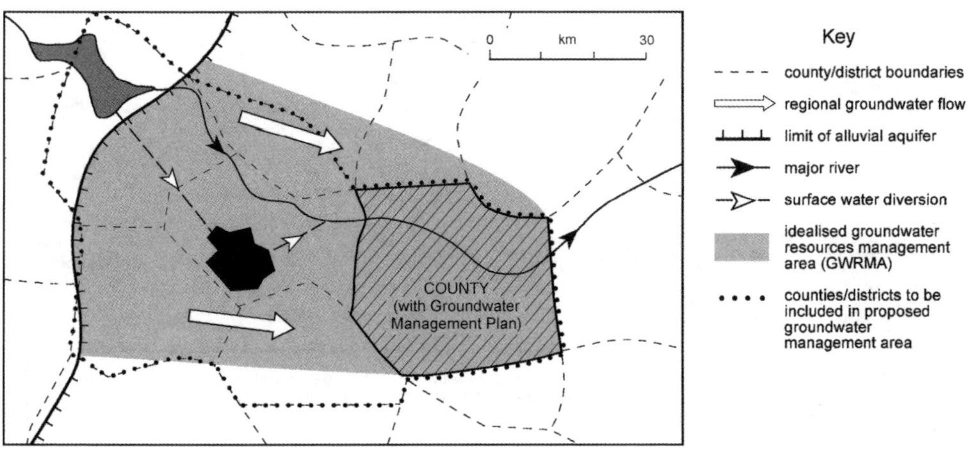

Figure 1. Theoretical example of consolidation of county-level groundwater resource administration units into logical aquifer management areas.

rarely successful unless requested and maintained by users, and the use of satellite imagery to delineate irrigated areas and crop types is an alternative (Figure 2). But the best prospect for effective groundwater resource fee collection is the correlation of rural electricity consumption with groundwater pumping and charging through electricity accounts. Good pilot experience of this approach also exists in Mendoza-Argentina (Text Box A), but a major effort is needed to harmonise and ratify the respective inventories of electricity meters and well owners/operators, and to periodically recalibrate electrical energy consumption and groundwater abstraction for representative electro-mechanical pumping plant.

Experience also suggests that regulating the construction of waterwells (and thus reducing the drilling of illegal wells) is the most effective way of controlling the growth of groundwater abstraction. This is best approached by building good relations with waterwell drilling contractors, through their formal licensing and the provision of well construction guidelines and specific hydrogeological advice in return for data returns.

GENERIC LESSON 14	***regulating well construction – their numbers, depths and diameters – is a key component in the control of groundwater abstraction***

5.3 *Mobilising stakeholder participation*

Resource regulation is central to groundwater management, but is rarely sufficient alone because of the large number of geographically-dispersed abstraction points involved. Unless broad social support exists, enforcement is often politically problematic. The introduction of demand-side measures implies changing activities that take place daily and affect livelihoods and lifestyles. Thus users play a critical role and regulators need to work collaboratively with them in formulating strategy and setting objectives.

Figure 2. Detailed survey of irrigation water-use in part of the Mendoza Northern Oasis.

Whilst WUAs have often proved effective in the equitable sub-allocation of water-use rights, they usually lack the perspective and influence to address broader aquifer management needs. This function is better performed by some form of intermediate institution and is leading to the formation of AMORs (Figure 3), which have representation of various WUAs, municipal water companies and industrial groundwater abstractors, together with the GWRA and/or other regulatory institution. Both AMORs and WUAs require significant investment, but return on this will be partners in groundwater resource management and monitoring.

Problems may arise, and need to be addressed, where the logical hydrogeological demarcation of an AMOR departs markedly from local political boundaries and those of long-standing WUAs built-up around surface water irrigation systems (Text Box A). Public relations are extremely important – AMORs, WUAs and the general public need to be kept informed on groundwater resource availability and vulnerability, the status of groundwater resource exploitation and the benefits of sound resource administration.

GENERIC LESSON 15	***delineating aquifer management areas (AMAs) and promoting aquifer management organizations (AMORs) with balanced stakeholder participation are key management steps***

5.4 *Consolidation of groundwater abstraction rights*

A groundwater abstraction rights system is needed to give users the degree of security required to invest in water-use efficiency and thus create a favourable social environment for resource management, as well as to ensure the availability of groundwater for poverty alleviation and sustaining the environment. Furthermore it provides the legal basis for resource reallocation to higher-value uses, with the possibility of compensation on reallocation or of rights trading. It should also make provision for the possibility of systematic

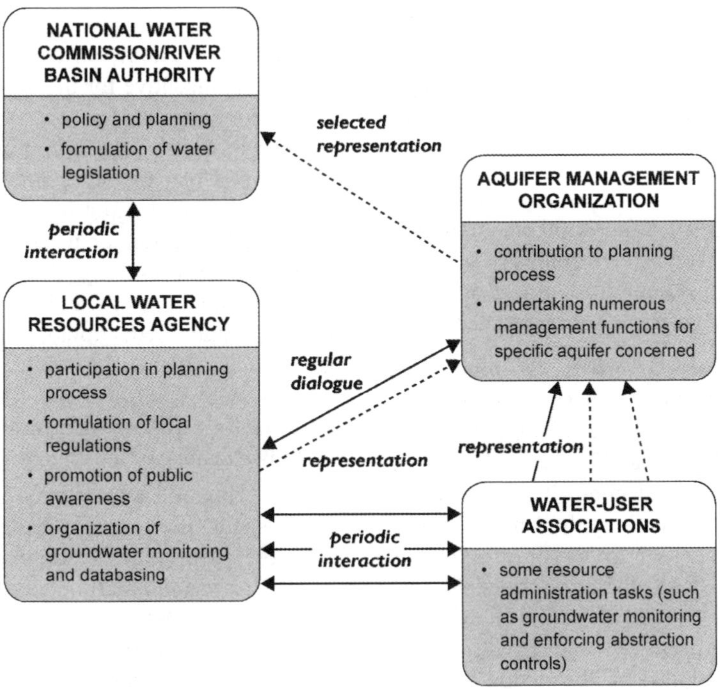

Figure 3. General scheme of institutional interaction in participatory groundwater resource management.

and gradual reduction of rights allocations by the GWRA in critical resource situations and/or the creation of aquifer conservation zones in which no new abstraction rights will be granted.

GENERIC LESSON 16	***establishment of groundwater abstraction rights is essential for mobilising user participation in resource management and eventually for resource reallocation***

Not withstanding the fact that a fully-fledged abstraction and use rights system is central to groundwater resource management, again it will be necessary in practice to be pragmatic (particularly when very large numbers of users are involved) by:

- issuing aggregated abstraction permits to rural WUAs (or village associations) representing reasonably homogenous groups of rural groundwater users;
- exempting very small domestic users from the need to hold abstraction permits;
- developing user databases as a useful first and then gradually promoting the rights system in close collaboration with WUAs once some enforcement capacity is in place.

Other common impediments to the introduction or consolidation of groundwater abstraction and use rights are:

- the need to deal with customary rights;
- the existence of large numbers of illegal wells, which has to be confronted by bringing them into the system or closing them down depending on antecedent circumstances.

In some urban conurbations, municipal groundwater abstraction has been constrained by regulatory agencies, because of concern about aquifer saline intrusion and/or land subsidence. But there is little point controlling municipal abstraction if private self-supply is not similarly controlled. If there is no overall enforcement of groundwater abstraction regulations, water-supply cost increases and/or shortages often lead to an explosion of private well drilling with aggregate levels of groundwater abstraction increasing. In effect, what occurs is the replacement of a moderate number of municipal supplies (capable of being systematically controlled, monitored, protected and treated), by a very large number of shallower, largely uncontrolled, unmonitored and untreated sources.

5.5 *Economic management instruments*

The economic characteristics of groundwater have played a major role in the emergence of management problems. Groundwater is generally undervalued, especially where exploitation is uncontrolled. In this situation the exploiter of the resource receives all the benefits of groundwater development, but (at most) pays only part of the costs (Figure 4) – usually the recurrent costs of pumping and the capital cost of well construction (although these are sometimes subsidized), but never the costs of associated externalities (such as baseflow reductions, saline intrusion, wetland impacts, land subsidence, and loss of strategic or drought groundwater storage).

Undervaluation leads to inefficient patterns of groundwater use. In many instances groundwater is allocated to low-value uses (such as the production of grain or fodder crops in arid regions), while higher value uses are only partially met. In addition, because in-situ values associated with groundwater are rarely reflected, undervaluation reduces the incentive for investment in resource conservation.

There is an array of economic instruments for groundwater management, among which water rights fees (or abstraction charges) are best known. Abstraction charges have the major advantage of providing direct incentives to undertake water savings, and can be structured in ways to promote reductions in consumptive use. Nevertheless, it may be politically difficult to raise them to levels which represent a strong incentive for efficient water-use. Approaches to groundwater abstraction charging include:

- metering well abstractions, but in rural areas this will only generally be operationally feasible when meter installation is requested by WUAs to aid their function;
- use of indirect measurements, such as rural electricity metering or satellite imagery of cropping patterns, but these may be challenged legally on precision grounds;

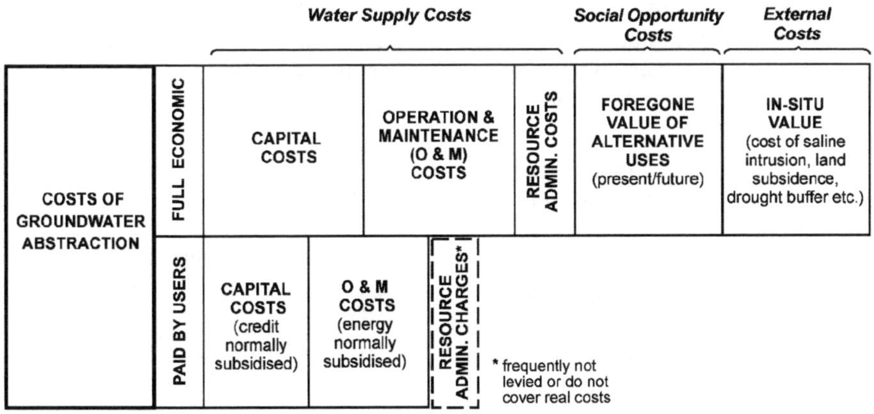

Figure 4. Measuring the costs of groundwater use.

- levying on basis of installed pump capacity, but this has the disadvantage of discouraging sensible provision for drought as opposed to normal use.

GENERIC LESSON 17	***abstraction charging is an important demand management tool but a transparent and acceptable basis for determining use is an essential basis for effective implementation***

Informal water markets are fairly widespread in developing nations. These markets involve local transactions between well owners and other users adjacent to each well, and are different from the trading of legally-defined volumetric water rights. The latter recognises differences in water value between sectors (such as industrial water-supply versus agricultural irrigation), but can only be introduced after:

- an effective groundwater rights system is in place;
- groundwater resource availability is reasonably well quantified.

Moreover, trading prices do not reflect the insitu value of the resource nor the third-party cost resulting from its exploitation, thus while adding value trading provides no incentive for reducing abstraction. GWRAs may also experience difficulty constraining the market to allow for resource uncertainty and future social need.

GENERIC LESSON 18	***groundwater markets are useful to facilitate resource reallocation but are not an instrument of resource conservation***

6 CONCLUDING REMARKS

An improved approach to groundwater is on the horizon, with a clear shift from a laissez-faire attitude towards more proactive regulatory agencies setting clear incentives to stakeholders to work together for better management of the resource. This is leading to an *awareness shift*, and instead of being the *invisible resource* groundwater is now becoming of more prominent concern.

The new management style involves:

- bringing groundwater users into the decision-making process;
- improved communication of groundwater value and management needs, based on sound sociological assessment and targeted specifically at groundwater users, the political and bureaucratic level, and the general public;
- recognition that, in the longer run, a finely-tuned balance of participatory, regulatory and economic tools will be needed.

GENERIC LESSON 19	***use of a single* management tool *will seldom be sufficient to achieve effective long-term groundwater resource management***

Moreover, despite the complexity of groundwater systems and the frequent institutional impediments, progress on groundwater resource management can always be made by recognising that:

- current stakeholder capacity to respond is a critical factor, and in some situations the overriding priority will be to understand, inform and mobilise the groundwater users;
- implementation of management tools will be severely hindered if they are selected with little regard to existing GWRA administrative capacity and current stakeholder ability to comply;
- a *parallel track approach* with realistic goals needs to be adopted, in which management tools are tested against existing institutional capability whilst capacity building programmes are introduced;
- perfect is the enemy of good.

GENERIC LESSON 20	***it is always feasible to make incremental improvements in groundwater resource management***

Given the rate of degradation of some aquifers, there may not be sufficient time for groundwater users to go naturally through the slow process of collective organisation. Instead, it would be advisable for existing GWRAs to disseminate information on resource status and to provide an institutional framework for users to collaborate in the management task, since this will be the most realistic route in terms of lead-time and transaction-cost in many countries. Management of a highly-stressed aquifer is not the same as that of a little-used one. The costs of developing an effective management approach for the former will generally be less than the socioeconomic and environmental cost of continued excessive abstraction.

ACKNOWLEDGEMENTS

This paper is a by-product of the operations of the World Bank-Groundwater Management Advisory Team (GW-MATE), which was set-up in response to the call of the 2nd World Water Forum (The Hague/March 2000) for a "shift in emphasis from groundwater resource development to management". The GW-MATE is financed by trust funds from the Dutch and British governments, and is a Global Water Partnership Associate Programme.

The paper attempts to draw generic lessons from the promotion and/or implementation of groundwater resource management through World Bank-funded projects in widely-spread parts of the developing world. In these projects the GW-MATE provided advice and assistance to the staff of national counterpart organizations. Their work and the enthusiasm of the corresponding World Bank project task managers is gratefully acknowledged especially – Doug Olson, Liping Jiang, Lin Wang, Javier Zuleta, Armando Llop, Peter Koenig and Satoru Ueda. The opinions expressed in the paper are, however, those of the authors and not necessarily of the World Bank or its client governments.

REFERENCES

Since most of the relevant national reference material used develop this paper is unpublished, conventional bibliographic referencing is not used, but the following sources represent relevant further reading:

Burke, J. and Moench, M. (2000). *Groundwater and society: resources, tensions and opportunities*. UN-DESA Publication.

Foster, S. and Kemper, K. (eds.) (2002-03). GW-MATE Briefing Note Series – Sustainable Groundwater Management: Concepts and Tools.
Accessed via: *www.worldbank.org/gwmate*.

Foster, S. and Kemper, K. (eds.) (2002-03). GW-MATE Case Profile Collection.
Accessed via: *www.worldbank.org/gwmate*.

Foster, S.; Lawrence, A.R. and Morris, B.L. (1997). Groundwater in urban development: assessing management needs and formulating policy strategies. *World Bank Technical Paper, 39*.

Foster, S.; Chilton, P.J.; Moench, M.; Cardy, W.F. and Schiffler, M. (2000). Groundwater in rural development: facing the challenges of supply and resource sustainability. *World Bank Technical Paper, 463*.

Garduño, H. (2001). Water rights administration – experience, issues and guidelines. *UN-FAO Legislative Study, 70*.

Kemper, K. (in press). *Groundwater: from laissez-faire exploitation to management*. World Bank.

Moench, M.; Dixit, A.; Janakarajan, S.; Rathore, M.S. and Mudrakartha, S. (2003). *The fluid mosaic – water governance in the context of variability, uncertainty and change*. ISET, Kathmandu, Nepal.

Aquifer overexploitation in Libya – the Gefara plain case

Omar M. Salem
General Water Authority (GWA), Tripoli, Libya
gwalibya@hotmail.com

ABSTRACT

Libya is a large country occupying a great part of the Sahara desert. It depends almost completely on its groundwater resources. In recent years, excessive use of water has lead to a continuous decline in water levels of up to few meters per year in the coastal aquifers, leading to a serious deterioration of water quality as a result of seawater intrusion and the inflow from the adjacent salty aquifers.

The *Gefara* plain covers an area of about 15,000 km^2 of NW Libya and stretches over 290 km along the Mediterranean coast. It hosts more than 40% of the Libyan population and provides about 60% of the total agricultural output. The plain is threatened by desertification. Several groundwater aquifers varying in age from Triassic to Quaternary are in existence. Historically, the Quaternary aquifer is of particular importance due to its shallow depth and good quality.

A piezometric network consisting of 51 wells (formerly over 260) is periodically monitored. In few piezometers, water quality is also checked and changes analyzed. In almost every ten years, an overall assessment of the water situation is conducted and a close estimate of the total water withdrawal is concluded on the basis of field surveys.

The annual natural recharge was estimated at 200 Mm^3, or about 10% of the total annual precipitation. Two major studies of seawater intrusion were conducted in the last two decades and the rate of advance of the seawater front around Tripoli is currently ranging from 50 to 550 m/yr. Future scenarios for seawater intrusion are modeled.

Short and long-term remedies are proposed, few of which have already been implemented.

Keywords: *aquifer, hydrogeology, groundwater, seawater intrusion.*

1 INTRODUCTION

Libya is a North-African country with a surface area exceeding 1.5 million km^2 and a Mediterranean coast of nearly 2,000 km. It falls within an arid to hyper-arid zone dominated almost entirely by a desert climate except for a narrow strip along the Mediterranean. Precipitation is extremely limited over most of the country and areas receiving more than 100 mm/yr do not exceed 7% of the total area, as shown in Figure 1.

High air temperatures and consequently high evaporation rates characterize most regions. Table 1 summarizes the main climatic features in different parts of the country.

The Libyan population has grown from 1.51 million in 1964 to 5.12 million in the year 2000, which reflects high growth rates of over 3%. Growth rates are currently stabilizing and are expected to undergo a steady decrease in the coming decades. Tables 2 and 3 show the population growth evolution and its future trends.

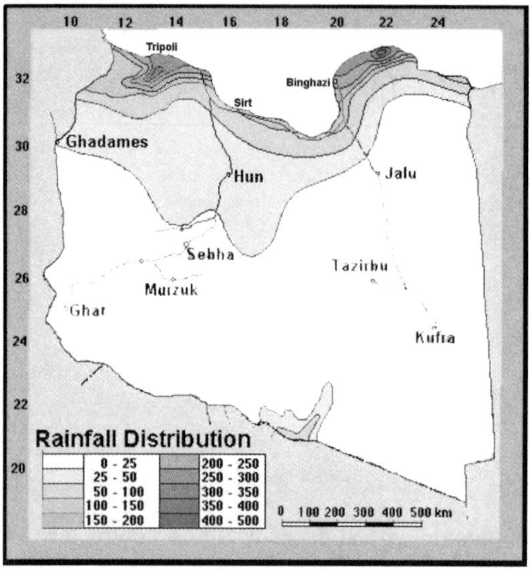

Figure 1. Rainfall distribution.

Table 2. Population growth*.

Year	1964	1973	1986	1995	2000
Population (Million)	1.51	2.05	3.42	4.39	5.12

* Not including foreigners.

Table 3. Population projections*.

Year	2005	2010	2015	2020	2025
Population (Million)	6.7	7.8	9.0	10.3	11.7

* Including foreigners.

Table 1. Climatic data (Salem, 1988).

Station	Temp. (°C)					Relative Humidity (%)	Evaporation (Piche) (mm/yr)	Rainfall (mm)
	Daily av.	Average max.	Average min.	Abs. max.	Abs. min.			
Tripoli	19.9	24.7	15.0	46.0	-0.6	63	1,499	338
Gadames	21.6	29.5	13.6	50.6	-8.0	33	5,183	33.8
Sebha	22.6	30.1	15.1	46.5	-4.6	34	5,658	8.8
Hun	20.7	29.1	12.2	47.2	-6.9	48	3,921	30.3
Bengazi	20.0	25.4	14.6	45.6	0.6	65	2,414	273.5
Kufra	23.0	30.8	15.2	46.2	-3.3	30	6,119	2.3

2 HYDROGEOLOGY

Libya depends almost entirely on groundwater resources from five main basins as shown in Figure 2.

Only the northern three basins receive considerable annual recharge from local storms and wadi runoff. The remaining two major sedimentary basins in the south, namely the *Murzuq* and the *Kufra-Sarir* depend to a great extent on the huge quantities of paleo waters stored within their thick and porous sandstone formations (Salem, 1992).

2.1 *The Gefara plain*

The *Gefara* plain occupies the NW part of Libya and covers an area of more than 15,000 km^2, which is only 1% of the total area. It has a triangular shape with its base along the Libyan-Tunisian border and stretches along the Mediterranean coast for nearly 290 km. Southward, the plain is bordered by the *Jabal Nafusa* mountains with elevation ranging from 600 to 800 m.a.s.l. At the foot of the Jabal, the elevation is in the order of 200 m.a.s.l. sloping gently northward towards the coast.

The plain shares more than 40% of the total population and hosts the largest Libyan city, Tripoli with over 1 million inhabitants.

In the year 2000, the total net irrigated area in Libya was estimated at 313,000 ha of which the *Gefara* plain accounts for 142,000 ha or 45%.

For the same year, the estimated total groundwater abstraction for all sectors was in the order of 4,441 Mm^3, 23% of which comes from the *Gefara* alone.

The Gefara plain is receiving an annual recharge estimated at 200 Mm^3/yr (Pallas, 1980). It consists of a number of aquifers, the most important of which belong to the Triassic (the *Azizia* limestone and the *Abu Shayba* sandstone aquifers), the Miocene (Lower and Middle Miocene), and the Quaternary as shown in Figure 3 (Kruseman and Floegel, 1980).

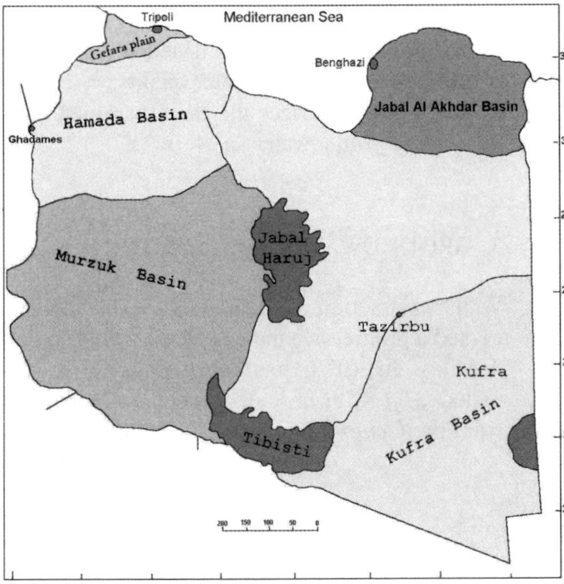

Figure 2. Groundwater basins.

2.1.1 *The Quaternary aquifer*

The first encountered aquifer is the Quaternary, which includes the Upper Miocene, the Pliocene and the Quaternary formations of the central and Northern Gefara. They are hydraulically connected and behave as one unconfined unit. In Eastern and Western *Gefara*, the Quaternary is of limited extension and thickness. The aquifer is recharged mainly by direct rainfall and *wadi* runoff and to a lesser degree by lateral inflow from the south and upward leakage from the Lower and Middle Miocene.

2.1.2 *The Miocene aquifers*

The Miocene aquifer includes both the Lower and Middle Miocene water bearing

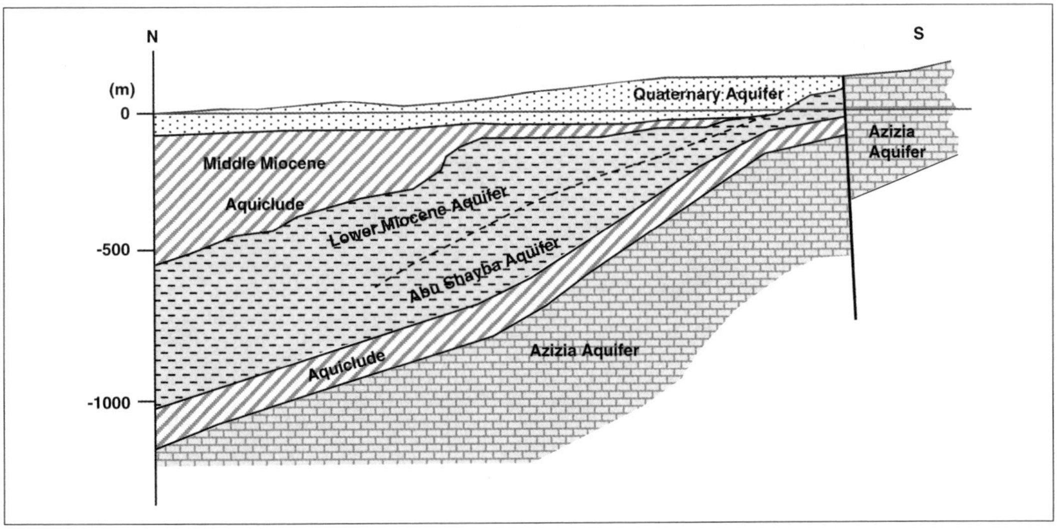

Figure 3. A North-south cross-section showing the main aquifers of the Gefara plain.

formations covering the central and Northern *Gefara*.

The Lower Miocene aquifer consists of sandy to dolomitic limestone in West-central *Gefara* and occurs at a depth of 250 to 390 m, dipping westward to reach a depth of more than 485 m with thickness reduced to about 80 m. The relatively high salinity of this aquifer limits its use especially in Western *Gefara* despite its free flowing condition. In this zone, the aquifer receives lateral inflow from the Triassic limestone (*Azizia*) in the south. In East-central *Gefara*, the Lower Miocene maintains a thickness of 150 to 200 m and is in hydraulic contact with the underlying *Abu Shayba* aquifer of the Upper Triassic.

On the other hand, The Middle Miocene aquifer is confined between two layers of clay with depth to its top ranging from 10 to 120 m and a thickness of 125 to 200 m.

2.1.3 *The Triassic aquifers*

Two aquifers belong to this group, namely the *Abu Shayba* sandstone aquifer of the Upper Triassic and the *Azizia* limestone aquifer of the Middle-Upper Triassic.

The *Abu Shayba* aquifer underlies the Lower Miocene in the East-central *Gefara* where it reaches a maximum thickness of 350 m. It becomes unconfined in the south near the *Jabal Nafusa* foothill and is totally missing in Northwestern *Gefara*.

The *Azizia* aquifer, made of dolomitic limestone, is unconfined in the south-central part of the *Gefara* and is highly exploited as a result of its relatively high yield and better properties. Elsewhere, the aquifer is confined and occurs at great depths with lower water quality.

Table 4 summarizes the main parameters of the Gefara groundwater aquifers.

3 GROUNDWATER ABSTRACTION

Water abstraction at the national level has witnessed a progressive increase during the last few decades. In 1978, total abstraction was estimated at 1,471 Mm^3, of which, 532 Mm^3 are from the *Gefara* plain alone. In the year 1998, a

Table 4. Aquifer parameters (Kruseman and Floegel, 1980).

Aquifer	Thickness (m)	Water level (m.b.g.s.)	Transmissivity (m^2/d)	Storage Coefficient	TDS (mg/L)
Mio-Plio-Quaternary	10 – 90	10 – 80	170 – 8,640	8×10^{-2}	1,000 – 2,000
Middle Miocene	100 – 200	35 – 90	45	7×10^{-3}	3,000 – 4,000
Lower Miocene	50 – 100	34 – 97	170 – 1,700	10^{-4}	2,500 – 4,500
Abu Shayba	150 – 400	10 – 145	2 – 1,700	10^{-4}	1,000 – 2,000
Azizia	200 – 350	165 – 176	600 – 17,000	5×10^{-2}	1,500 – 2,000

Table 5. Water abstraction evolution (Pallas and Salem, 1999).

	1959/62	1972	1973	1975	1978	1992	1998
AUTHOR	USGS	GEFLI	FAO	GEFLI	FAO	MacDon.	Pallas
Agriculture use	195	313	336	475	435	802	900
Municipal use	15	65	?	92	97	200	141*
Total	210	378	?	567	532	1,002	1,041

* Groundwater abstraction only without the contribution of GMRP.

nationwide survey has resulted in total withdrawals of 4,441 Mm^3 with the *Gefara* plain representing 23% or 1,041 Mm^3. Table 5 shows the water abstraction evolution in the *Gefara* plain.

The agricultural sector is the main user of water and accounts for more than 80% of the total consumption. In the *Gefara* plain, hundreds of wells are being drilled every year to meet the growing demand. Irrigated areas have increased from 100,000 ha in 1978 to over 140,000 ha in the year 1998, despite the regulations that restrict the drilling of additional wells in highly affected areas. A high percentage of the drilled wells are completed in the Quaternary and Miocene aquifers, particularly in the northern and central parts of the plain where agricultural activities are more intense.

4 GROUNDWATER MONITORING

A systematic monitoring of groundwater levels in selected wells has started at early stages of the last century. A well-established network of piezometers was initiated in the early seventies, immediately after the creation of the General Water Authority in 1972. The initial network consisted of 262 observation wells tapping the major aquifers in the *Gefara*. At present, only 51 wells are operational as many of the shallow ones have either been damaged or become dry. Some wells are equipped with automatic recorders while others are measured manually using electric tapes. In most cases, water level measurements are collected four times a year and groundwater decline maps along with individual well hydrographs are prepared as shown in Figures 4 and 5.

The annual recharge from local rainfall and *wadi* runoff was estimated from previous studies to fall between 135 and 360 Mm^3 in central and Eastern *Gefara*. Other sources of recharge originate at the southern borders of the basin in the form of lateral inflow from adjacent aquifers, namely the Upper and the Lower Cretaceous of the *Hamada* basin. Such a recharge is not well quantified at present but is estimated from regional mathematical models to be in the order of 170 to 270 Mm^3/yr. Inflow from the sea was also estimated by a 1993 model (Mott MacDonald, 1994) at 166 Mm^3/yr. Table 6 shows a simulated flow balance components for the pre-development period of 1949 and the year 1992/93.

Future projections reveal that water demand will far exceed the available recharge as shown in Figure 6.

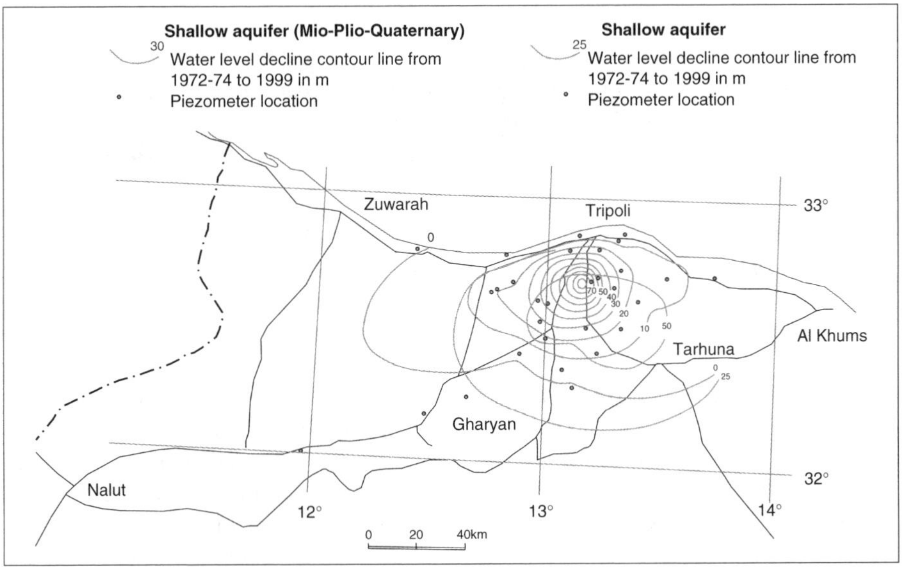

Figure 4. Water level decline in the deep aquifer in the Gefara plain from 1972–74 to 1999.

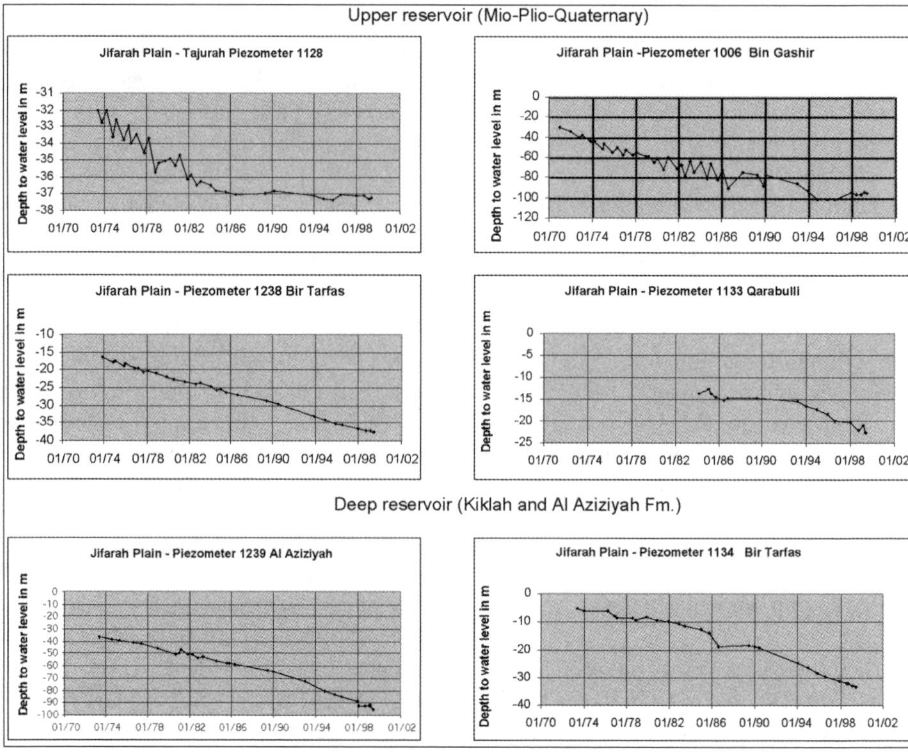

Figure 5. Well hydrographs for selected piezometers in the Gefara plain (Pallas and Salem, 1999).

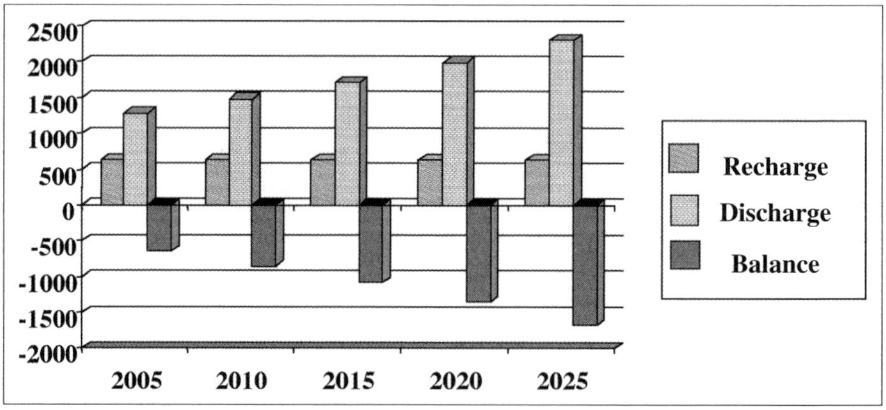

Figure 6. Water balance in the Gefara plain.

5 WATER QUALITY DETERIORATION

Apart from the continuous decline in water levels and its enormous economic impact, water quality deterioration has also been a major problem in recent years. A steady increase in salinity due to seawater intrusion necessitated the shutdown of the major municipal water supply wellfields located South of Tripoli. Other wells drilled for industrial supplies were either abandoned or equipped with desalination units. Groundwater desalination has become a common practice in several coastal cities and private or public installations. Figure 7 reflects the

Table 6. Flow balance components (Mott MacDonald, 1994).

Flow balance component (Mm^3/yr)	Pre-development Conditions			1992 / 93		
	Upper Aquifer	Lower Aquifer	Total	Upper Aquifer	Lower Aquifer	Total
Inflows to the aquifer system						
Recharge: Rainfall	147.1	9.9	157.0	74.3	9.6	83.9
Recharge: *Wadi* runoff	38.0	0.0	38.0	9.7	0.0	9.7
Irrigation/municipal losses	4.0	0.4	4.4	264.8	25.8	290.6
Total recharge	189.1	10.3	199.4	348.8	35.4	384.2
Boundary inflows: Southern	0.0	267.6	267.6	0.0	217.3	217.3
Boundary inflows: Coast	0.0	0.0	0.0	166.1	0.0	166.1
Total boundary inflow	0.0	267.6	267.6	166.1	217.3	383.4
Release from storage	0.0	0.0	0.0	356.7	227.3	584.0
Leakage	114.6	0.0	114.6	22.9	0.0	22.9
TOTAL INFLOW	303.7	277.9	581.6	894.5	480.0	1,374.5
Outflows from the aquifer system						
Abstractions: Agriculture	-8.8	-1.0	-9.8	-569.4	-232.6	-802.0
Abstractions: Municipal	-5.6	-0.3	-5.9	-136.0	-64.0	-200.0
Total abstractions	-14.4	-1.3	-15.7	-705.4	-296.6	-1,002.0
Boundary outflows: Coast	-251.5	-123.3	-374.8	-161.8	-131.5	-293.3
Boundary outflows: Tunisia	-3.8	-38.5	-42.3	-2.2	-28.8	-31.0
Total boundary outflow	-255.3	-161.8	-417.1	-164.0	-160.3	-324.3
Loss from *sebkhas*	-34.5	0.0	-34.5	-25.5	0.0	-25.5
Gain in storage	0.0	0.0	0.0	0.0	0.0	0.0
Leakage	0.0	-114.6	-114.6	0.0	-22.9	-22.9
TOTAL OUTFLOW	-304.2	-277.7	-581.9	-8,594.9	-479.8	-1,374.7
Model imbalance	-0.5	0.2	-0.3	-0.4	0.2	-0.2

effect of salinization in the *Swani* municipal wellfield in the period from 1976 to 1993 before its closure.

Increase in salinity has also resulted from deep drilling, as the shallow aquifer can no longer sustain high demands for water. This situation has forced many farmers to shift to salt tolerant crops.

5.1 *History of seawater intrusion*

Seawater intrusion was subject to several studies, two of which were aimed at defining the intrusion front and simulating its future behavior in conjunction with the assumed rates of water abstractions, using appropriate mathematical models.

Seawater intrusion was first observed in the 1930s in the northern part *of Suq El Juma à* area (East of Tripoli) when a seawater channel was extended to provide the salt making ponds with seawater. Since then, growing demand for irrigation water in *Suq El Juma à* and *Tajura* lead to the lowering of the water table and therefore the advance of the seawater front for 2 km to cover an area of 15 km^2 by 1957 (Cederstrom and Bastaiola, 1960).

By the late 1960s and early 1970s, further increase of groundwater abstraction has forced the seawater front another 1 km inland in *Tajura, Suq El Juma à* and Tripoli (GEFLI, 1972).

Previous studies indicated that until 1975, groundwater flow was from south to north. Since then and as a result of the drilling of hundreds of wells equipped with modern submersible pumps, a reverse hydraulic gradient was created pushing the seawater intrusion front further south in the area around Tripoli.

In 1976, the *Swani* wellfield, consisting of 52 wells tapping the Quaternary aquifer south of

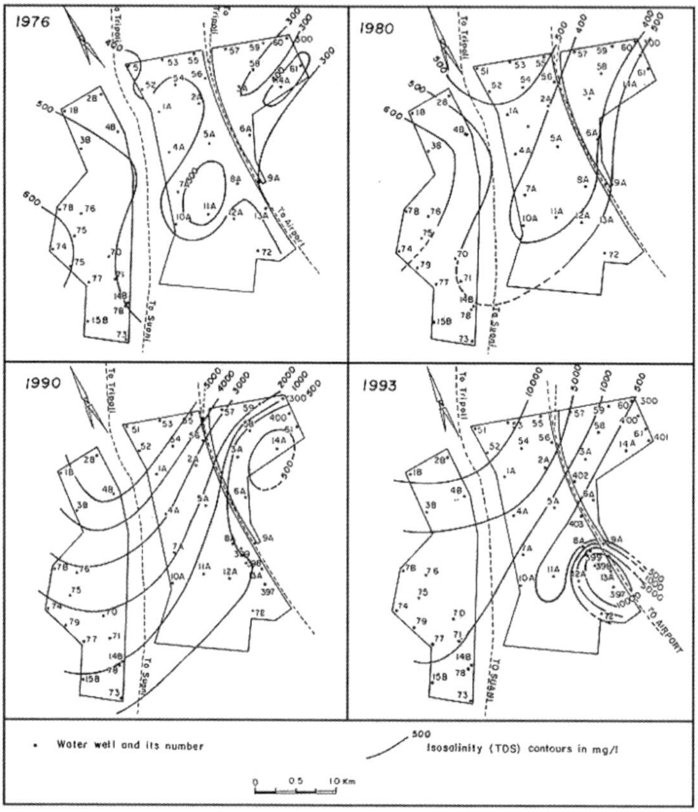

Figure 7. Changes in salinity in the Swani wellfield (El Baruni, 1995).

Tripoli was inaugurated for the supply of the city with domestic water. The continuous pumping from the wellfield strongly affected the water table decline in the area and resulted in the advance of the seawater intrusion front by more than 3 km southward by the year 1979 (Floegel, 1979). Seawater intrusion had also extended towards *Gargaresh*, *Janzur*, and El-Maya, West of Tripoli.

From 1979 to 1995, high rates of population growth and fast urbanization, forced irrigated agriculture to move further south and necessitated the development of an additional wellfield consisting of 87 wells southeast of Tripoli. Considerable decline of Quaternary aquifer water table was therefore witnessed resulting in the advance of the seawater intrusion front by 6 km in Tajura, 9 km in *Gargaresh*, 2 km in *Janzur* and 5 km in *El Maya* and the total area affected by seawater intrusion reached 250 km^2 (ERCO, 2002) as shown in Figure 8.

The highest advance of the seawater intrusion front in the Tripoli area took place from 1979 to 1995. During this period, the advance rate was estimated at 200 to 400 m/yr especially in the central zone south of Tripoli, while the least advance was in the area of Janzur due partly to the high elevation and the existence of a forest belt to the south.

6 RECOMMENDED SOLUTIONS

Libya has already adopted the principle of using its non-renewable water resources in order to narrow the gap between supply and demand, in a way that ensures aquifer safety and guarantees its continuity. This approach would allow for drastic changes in the desalination technology to become acceptable for all purposes. This policy requires continuous monitoring of aquifer behavior and application of suitable mathematical

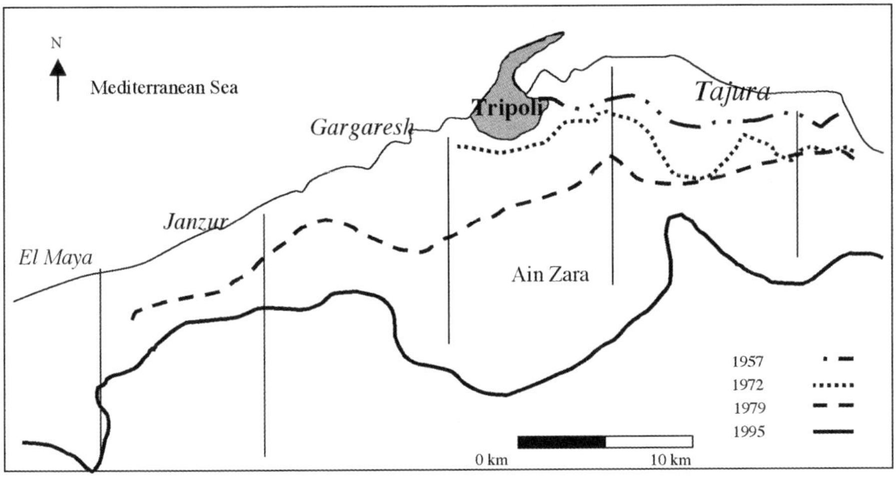

Figure 8. Seawater intrusion situation (ERCO, 2002).

models in order to secure a minimum drawdown with the least changes in water quality. Previous studies and master plans have recommended certain measures to reduce water consumption. Most of these measures are translated into laws and regulations and adopted as long-term policies. Among such remedies are the ban of high water consuming crops, the classification of several areas as restricted zones for additional expansion in irrigation, and the shut-down of wells located at or near the seawater intrusion front. Other measures include the encouragement of the erection of additional desalination plants for domestic and industrial water supplies, and most important of all the call for inter-basin water transfer. The latter is currently under implementation and is expected to provide slightly over 1,000 Mm^3/yr to the *Gefara*, mainly from the Murzuq basin with little contribution from the *Hamada*. New desalination plants of high capacity will be constructed in the coming years.

7 CONCLUSION

Like all other groundwater basins in Libya, the *Gefara* plain is undergoing a severe case of water quantity depletion accompanied by deterioration in water quality. This situation is caused by the excessive pumping of groundwater to cope with the high population growth and the escalating rates of demand. Although difficult to reverse, the deficit in the water budget can be narrowed through several remedial measures, many of which Libya has already started to implement.

ACKNOWLEDGEMENTS

The author would like to thank Mr. S. El-Baruni and Mr. R. Al-Futaisi for providing some useful data. Mr. A. Doma and Mr. A. Ayoubi helped in the preparation of the manuscript on computer.

REFERENCES

Cederstrom, D.J. and Bastaiola, H. (1960). *Groundwater Resources of the Tripoli area, Libya.* USGS Joint Services. Open File Report.

El-Baruni, S. (1995). Deterioration of Quality of Groundwater from Swani Wellfield, Tripoli, Libya 1976–1993. *Hydrogeology Journal*, 3(2).

ERCO (Engineering Research and Consultancy Office), Faculty of Engineering, Al Fateh University (2002). *Study of Seawater Intrusion in NW Libya.* General Water Authority (GWA), Tripoli, Libya.

Floegel, H. (1979). *Seawater Intrusion Study, Field Report.* SARLD and FAO, Tripoli, Libya.

GEFLI (1972). *Soil and Water Resources Survey for Hydro-Agricultural Development, Northern Zone.* General Water Authority (GWA), Tripoli, Libya.

Kruseman, G.P. and Floegel, H. (1980). Hydrogeology of the Jifarah, NW Libya. *Geology of Libya, Vol. 2*. Al Fateh University, Tripoli, Libya.

Mott MacDonald (1994). *General Plan for the Utilization of the Great Man-Made River Waters, Phase II. Final Water Management Plan*. GMRWUA, Tripoli, Libya.

Pallas, P. (1980). Water Resources of the Socialist People's Libyan Arab Jamahiriya. *Geology of Libya, Vol. 2*. Al Fateh University, Tripoli, Libya.

Pallas, P. and Salem, O. (1999). Water Resources Utilization and Management of the Socialist People's Libyan Arab Jamahiriya. *Proceedings, International Conference on Regional Aquifer Systems in Arid Zones – Managing Non-Renewable Resources*. UNESCO.

Salem, O. (1988). Groundwater in the Socialist People's Libyan Arab Jamahiriya. In: Groundwater in North and West Africa. *Natural Resources / Water Series, 18*. UN. New York, USA.

Salem, O. (1992). *Hydrogeology of the Major Groundwater Basins of Libya*. OSS. Cairo, Egypt.

Lessons from intensively used transboundary river basin agreements for transboundary aquifers

Shammy Puri
IAH Commission on Transboundary Aquifer Resource Management (TARM)
shammypuri@aol.com

L. Gaines [(1)], A. Wolf [(2)] and T. Jarvis [(3)]
Oregon State University, USA
[(1)] *gainesl@science.oregonstate.edu*
[(2)] *wolfa@geo.orst.edu*
[(3)] *jarvisw@onid.orst.edu*

ABSTRACT

Research on transboundary waters, transboundary water law, and mitigation of transboundary water conflict, has traditionally focused almost exclusively on surface water supplies. The hydrologic link between groundwater and surface water is recognized but understood at a reconnaissance level even in the most studied basins in the world.

Transboundary aquifers are poorly understood by policy makers due to the uncertainty associated with the *invisible* resource. Uncertainties in recharge mechanics and the flow regime within different types of aquifers provide technical challenges to policy development and spatial considerations in groundwater management. Conflicts over water quantity and quality are certain to escalate with increased reliance on groundwater to meet demands for drinking water, agricultural and industrial uses, highland forests and wetlands.

Because groundwater management in the international arena is in its infancy, this paper builds upon the analysis of transboundary river basin agreements. Transboundary aquifer agreements should be developed when resources are threatened because of their actual or perceived intensive use. What are the lessons that nations can learn for the sound management of their transboundary aquifers? History has shown that institutional capacity has a tendency to make international borders become areas of cooperation as opposed to areas of conflict.

Keywords: *Transboundary aquifers, international shared aquifers, environmental conflict resolution, groundwater resource management, aquifer protection, water security.*

1 INTRODUCTION

With groundwater becoming an increasingly important source of water for the agricultural, industrial and domestic sectors over the past 60 years, it now accounts for a substantial volume of all water consumption in many regions of the world. As demand continues to grow groundwater occurring across international borders will increase in significance. In arid regions as some countries struggle to develop sufficient quantities of water for long term sustainable livelihood,

[1] Megacities dependant on at least 25% of their demand on groundwater include Mexico City, Calcutta, Tehran, Shanghai, Buenos Aires (Morris *et al.*, 2003).

there are suggestions that extracting and moving groundwater over vast distances may be possible for export and economic gain (NGWA, 2002). Over half the world's 23 megacities,[1] are substantially dependant on groundwater and with increasing demands, extraction of groundwater from distant sources could well result in a *race to the pump*, with little attention to consequences.

Intranational and international conflicts over the right to use water traditionally focus on surface water and have been predicted to increase over the next 15 years as many countries press against the limits of their available water (Postel and Wolf, 2001). According to Puri *et al.* (2001), the hidden nature of groundwater and the lack of clarity in international law governing shared aquifers invite misunderstandings leading to potential conflict. With few countries regulating the use of groundwater and the uncertainties associated with managing a hidden resource, conflicts over surface water will appear like a warm summer shower compared to the hurricane of disputes over groundwater. Likewise, examples of unintentional degradation of groundwater by agricultural and industrial activity in nearly every country are manifold. The environmental damage to shared aquifers from past military activities is just starting to be recognized. The threat of damage to the hidden resource as a result of future aggressive activities has not been evaluated. Given the hydrogeological uncertainties[2] associated with the *invisible* resource, it comes as no surprise that sound and equitable groundwater management in the international arena is in its infancy (Puri *et al.*, 2001; Matsumoto, 2002).

As the competition for once plentiful resources grows, threats to shared water resources could be perceived, presumably due to their more intensive utilisation. This paper seeks to draw on lessons from the analysis of past and current transboundary river basin agreements. Arguably such agreements come about when resources may be at threat, because of their actual or perceived intensive use. Formulation of agreements will reduce or prevent future disputes. What are the lessons that nations can learn for the sound management of their transboundary aquifers?

2 BACKGROUND

The dimensions of internationally-shared aquifers range from several hundreds of km^2 to several tens of thousands of km^2. As depicted on Figure 1, the principal components of transboundary aquifers include subsurface flow which is intersected by an international boundary. Water transfers from one side of the boundary to the other may occur naturally or due to capture by wells located on one side of the boundary. In many cases the aquifer might receive the majority of its recharge on one side, and the majority of its discharge would occur on the other side. The subsurface flow system at the international boundary itself can be visualized to include regional, as well as the local, movement of water (Puri *et al.*, 2001; Matsumoto, 2002).

3 TRANSBOUNDARY WATER RESOURCES AND AGREEMENTS

While a substantial body of research has been undertaken on transboundary waters, transboundary water law, and mitigation of transboundary water conflict, this research has focused almost exclusively on surface supplies. 145 countries are riparian to the world's 261 international river basins (Wolf, 1999). These basins cover nearly one-half of the earth's surface area, account for an estimated 60% of global freshwater flow, and are home to approximately 40% of the world's population (Wolf *et al.*, 1999). By analogy a large percentage of the world population also resides in lands underlain by transboundary aquifers because almost 90% of all accessible freshwater is found in aquifers (Shiklomanov, 1999). Unlike transboundary surface water and river basins, transboundary aquifers are poorly understood by policy makers who develop legislative frameworks and establish property rights. In many nations and cultures, groundwater is much less amenable to management through legislation, as traditionally property rights tend to include groundwater as a private, rather than a public good (Burke and Moench, 2000).

[2] Uncertainty derives from insufficient data to adequately define hydrogeological parameters.

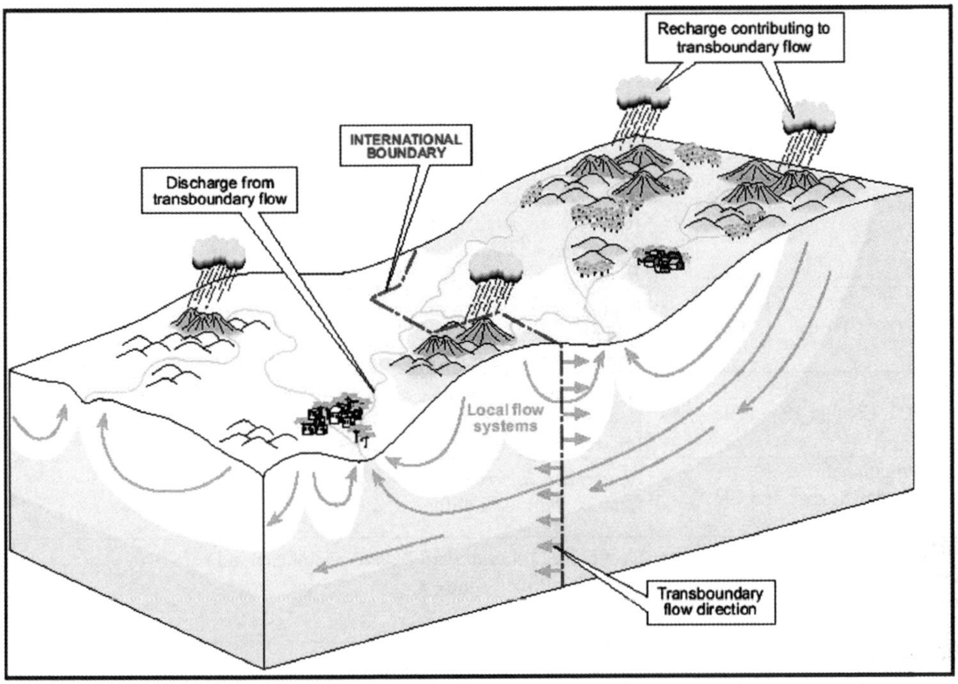

Figure 1. Schematic illustration of a transboundary aquifer (Puri *et al.*, 2001).

More than 3,600 water-related treaties have been signed by riparian countries in international river basins between the years 805 and 1984 (Wolf, 1998). On the whole, international surface water resources are characterized by a history of tremendous, and impressively resilient, cooperation. The only recorded incident of an overt war over water was 4,500 years ago between two Mesopotamian city-states, Lagash and Umma, currently known as Southern Iraq (Wolf, 1998).

While the impact of groundwater withdrawal may be contained within administrative boundaries, few state or provincial water laws address transboundary groundwater management, due either to its *invisible* nature, or to the hydrogeotechnical challenges in exactly predicting the spatial and temporal changes. Part of the problem is associated with recognizing the different types of aquifers; sand and gravel transmit and store groundwater differently than groundwater stored in fractured rocks or in karst. Both Puri *et al.* (2001) and Matsumoto (2002) underscore that current international law does not adequately define groundwater and aquifers, much less the dynamics of spatial flow.

The largest empirical study of water cooperation and conflict, the Transboundary Freshwater Dispute Database (TFDD), was completed by Oregon State University in 2001 (Wolf *et al.*, 2003). The TFDD documents 1,831 events of water conflict/cooperation over the last 50 years – 1,228 cooperative, 507 conflictive, and 96 neutral or non-significant. A review of international water law specifically addressing groundwater listed in Table 1 reveals that many of the agreements were developed in the past 50 years, and have only recently adopted a definition of an aquifer. Matsumoto (2002) inventoried treaties listed in the TFDD and summarized the following listing of treaties recognizing groundwater:

- 35 treaties developed between European countries;
- 13 treaties in Africa;
- 10 treaties in the Middle East and Asia;
- 4 treaties in North America;
- No treaties developed in South America.

The challenges posed to policy makers is that groundwater flow dynamics are typically known only at regional levels, sometimes even in the most studied aquifers in the world. Locally

Table 1. Summary of International Groundwater Law (Modified after Matsumoto, 2002).

International Law	Date	References to Groundwater
Helsinki Rules	1966	• Defines underground water as part of international drainage basin. • Ignores confined aquifers.
Seoul Rules	1986	• Defines international drainage basin as "an aquifer intersected by the boundary between two or more states that does not contribute water to, or receive water from, surface waters of an international drainage basin constitutes an international drainage basin for the purposes of the Helsinki Rules".
Bellagio Draft Treaty	1989	• Recognizes the hydrologic connection between surface water and groundwater. • Transboundary aquifer is considered part of an international basin.
Agenda 21, Chapter 18	1992	• Groundwater has parallel status as surface water as *freshwater bodies*. • Recommends holistic freshwater management. • Ignores transboundary freshwater management.
Non-Navigational Uses of International Watercourses (Draft)	1994	• Recognizes *International Watercourse*. • Ignores confined aquifers.
Non-Navigational Uses of International Watercourses (Convention)	1994	• Recognizes *International Watercourse* listed in Draft. • Ignores confined aquifers.
Non-Navigational Uses of International Watercourses (Resolution)	1994	• Recognizes confined aquifers. • Water management rules outlined in Draft may be applicable to transboundary confined aquifers.
Convention on the Protection and Use of Transboundary Watercourses and International Lakes	1999	• Recommends integrated water resources management, inclusive of groundwater. • Recommends extending water resources management to transboundary issues.

groundwater flow dynamics can cross geopolitical boundaries in unpredictable ways, due to hitherto unmapped barriers imposed by faulting, by conduits imposed by unpredicted fractures, or seasonally due to basin switching, typically in aquifers drained by karst. In large regional aquifers the limited estimation of annual recharge and the uncertainties in the subsurface flow regime from natural heterogeneity reveals some of the challenges associated with not only policy development, but also with the spatial considerations in groundwater management. Livingstone *et al.* (1996) have shown that groundwater *captured* by wells may be drawn from recharge occurring across rivers, if they are underlain by aquitards; many rivers in the world comprise international boundaries.

Protection of groundwater quality has only been addressed in the past few years. Of the many international water treaties, few have monitoring provisions, and hardly any have enforcement mechanism (Chalecki *et al.*, 2002).

While groundwater resources would not so far appear to have been the main focus of transboundary conflict, in regions of shortage, competition for water coupled with lack of hydrogeological data may sometimes lead to political tensions. Insufficient data in complex groundwater systems compounds the difficulty for the decision maker to conceptualise aquifers flows and to fully appreciate the issues. Given the uncertainty in defining exact groundwater flow, together with the doubts of the hydraulic continuity between groundwater and surface water resources, conflicts over water quantity and quality can escalate; threats increase to reliance on groundwater that meets demands for drinking water, agricultural and industrial uses, and maintains natures *green* reservoirs – forests and wetlands. Likewise, there are many examples of transboundary aquifers where recharge is received on one side of an international boundary while natural discharges and sometimes better yields are on the other side of the boundary.

4 AGREEMENTS FOR INTENSIVELY USED AQUIFERS

As regards aquifers in which intensive use has already occurred, there is only the model *Bellagio Treaty* that has been developed. While its general format provides a good basis for application to all aquifers, it was developed for application in the specific case of the Mexico-USA conditions, though it has never been implemented. Consequently, there remains no experience to gauge the success of such treaties, nor of their actual shortcomings.

An important contribution to the sound management of transboundary aquifers was made through the *UN ECE Convention on Protection and Use of Transboundary Water Courses and Lakes* (also known as the *Helsinki Convention*), which by 1999 was ratified by 25 Member countries, was driven by the need to address the quality of surface waters in Europe. However its scope includes aquifers. The European Union (EU) Water Framework Directive, though only relevant to the EU, also has some lessons. Both provide several lessons for potential application elsewhere.

As noted above, successful institutions for management of intensively used transboundary resources are those that have moved from the *allocation of water* to *sharing the benefits* of water. These institutions have the ability to develop positive-sum agreements, rather than the limited and often intractable division of water – a dynamic natural resource.

5 INSTITUTIONAL NEEDS AND COOPERATIVE MANAGEMENT

How can we learn from previous experiences in view of the continuing demand for water in general, and for aquifers, many of them transboundary, in particular? Despite very large storage in some aquifers and slow response times, intensive use of transboundary aquifers can be expected to be inevitable. The time is ripe for the water resources community to address this important issue. Existing efforts in international transboundary water management is replete with case studies and lessons learned for cooperation. These lessons learned suggest that the international community might consider focusing more attention on the specific institutional needs of individual basin communities by assisting riparian countries in the development of cooperative management networks that take into account key factors as suggested by Giordano and Wolf (2001), and Llamas and Custodio (2002):

1) *Adaptable management structure.* Effective institutional management structures incorporate a certain level of flexibility, allowing for public input, changing basin priorities, and new information and monitoring technologies. The adaptability of management structures must also extend to non-signatory riparians, by incorporating provisions addressing their needs, rights, and potential accession.
2) *Clear and flexible criteria for water allocations and water quality management.* Allocations, which are at the heart of most water disputes, are a function of water quantity and quality, as well as political fiat. Thus, effective institutions must identify clear allocation schedules and water quality standards that simultaneously provide for extreme hydrological events; new understanding of basin dynamics, including groundwater reserves; and changing societal values. Additionally, riparian states may consider prioritizing uses throughout the basin. Establishing catchment-wide water precedents may not only help to avert inter-riparian conflicts over water use, but also protect the environmental health of the basin as a whole.
3) *Equitable distribution of benefits.* This concept, subtly yet powerfully different from equitable use or allocation, is at the root of some of the world's most successful institutions. The idea concerns the distribution of benefits from water use – whether from hydropower, agriculture, economic

development, aesthetics, or the preservation of healthy aquatic ecosystems – *not* the benefits from water itself. Distributing water use benefits allows for positive-sum agreements, whereas dividing the water itself only allows for winners and losers.

4) *Concrete mechanisms to enforce treaty provisions.* Once a treaty is signed, successful implementation is dependent not only on the actual terms of the agreement but also on an ability to enforce those terms. Appointing oversight bodies with decision-making and enforcement authority is one important step towards maintaining cooperative management institutions.
5) *Detailed conflict resolution mechanisms.* Many basins continue to experience disputes even after a treaty is negotiated and signed. Thus, incorporating clear mechanisms for resolving conflicts is a prerequisite for effective, long-term basin management.

6 SPECIFIC ISSUES THAT RELATE TO AQUIFERS

There are many contrasts between transboundary rivers and aquifers. Some of these are listed in Table 2 and these peculiarities need to be accounted for in the application of the above suggestions.

7 WHAT IS BEING DONE?

Transboundary aquifers are the subject of the ongoing work by the International Shared Aquifer Resources Management (ISARM), program that was initiated by the United Nations Educational, Scientific, and Cultural Organization (UNESCO), in partnership with FAO, and UN ECE in 2000. ISARM anticipates publishing a worldwide inventory of Transboundary Aquifer Systems in 2006. Transboundary aquifer systems

Table 2. Comparison between physical characteristics of transboundary surface water and groundwater and institutional issues.

Transboundary Rivers	Transboundary Aquifers	Institutional lessons for aquifers
Long linear features.	Bulk 3-dimensional systems.	Responsibility must be *basin/aquifer wide*.
Use of resources generally limited to vicinity of the river channel.	Resources may be extracted from and used extensively over outcrop and subcrop.	As above, but must address diverse users, e.g. industry as well as irrigation.
Replenishment always from upstream resources.	Replenishment may take place from any, or all of 3-dimensions.	The resource planning mandate has to be wide.
Rapid and time-constrained gain from replenishment.	Replenishment could be slow, net gain can be drawn upon over longer periods.	Planning horizon must be related to the aquifer response time.
Abstraction has an immediate downstream impact.	Abstraction impact can be much slower –can be tens of years.	As above.
Little impact on upstream riparian sites.	Could have an equal impact on both upstream and downstream riparian sites.	A mandate for multi national linkage of institutions.
Pollution transported rapidly downstream.	Slow movement of pollution.	Relate to the response time.
Pollutant transport invariably downstream, upstream source may be unaffected.	Pollutant transport controlled by local hydraulics; an operating well may induce *upstream* movement towards itself.	Both qualitative and quantitative responsibility needed.

currently being assessed by ISARM include the following:

- the Guaraní Aquifer (South America);
- the Nubian Sandstone Aquifers (Northern Africa);
- the Karoo Aquifers (Southern Africa);
- the Vechte Aquifer (Western Europe);
- the Slovak Karst-Aggtelek Aquifer (Central Europe);
- the Praded Aquifer (Central Europe).

The Madrid *Workshop on Intensively Exploited Aquifers* held in 2001 provided ideas and suggestions to improve water management where there is an intensive use of groundwater. The

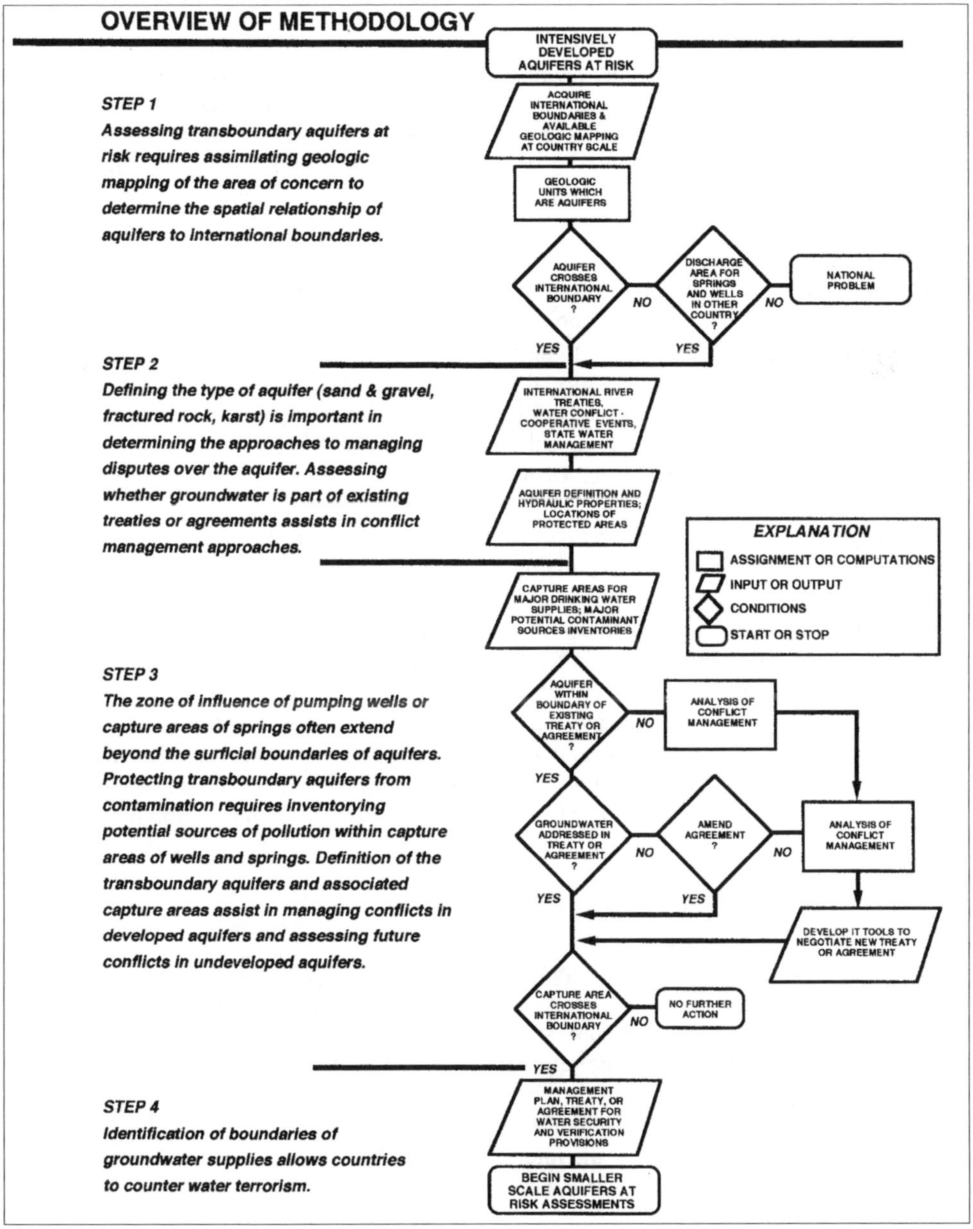

Figure 2. Preliminary assessment for intensively exploited transboundary aquifers.

Valencia *International Symposium on Intensive Use of Groundwater* held in 2002 built upon these issues (Llamas and Custodio, 2002).

8 SUGGESTIONS FOR FUTURE WORK

By building upon the ongoing work, coupled with recognition of the identified institutional issues, suggestions for future work recognize that potential conflicts over transboundary aquifers will require a holistic approach to address ecosystem-wide, multidisciplinary, and multimedia issues. Groundwater science is at the core of transboundary groundwater issues because international interests and options are not easily defined without the assistance of specialists who can interpret causal chains (Renevier and Henderson, 2002). Geographic Information Systems (GIS) provided some of the tools to identify common policies and potential conflict for the TFDD covering surface water, especially in situations involving large amounts of information. Information Technology (IT) enables international parties to prepare themselves for negotiation, providing a forum for value creation, allows more direct observation and participation by non-negotiators, affords negotiators currently unavailable opportunities for discussions between negotiations, and enhances post negotiation implementation (Freid and Wesseloh, 2002). Real world visualizations can assist in the process of negotiating and managing international water resource systems (Chalecki *et al.*, 2002). Position-oriented negotiation can become difficult without the aid of some computer-based information management tools (Lotov *et al.*, 1998).

The suggested path for future work seeks to address this and other significant knowledge gaps and provides guidance for the development of transboundary groundwater policy through global scale research which may be applied to the specific international basins. The global analysis should involve an inventory of the geographic delineation of the world's transboundary aquifers which has been completed by the International Association of Hydrogeologists (IAH). This world map of aquifers could be linked to the database of existing international water treaties and institutions developed by the TFDD. Through an analysis of the treaties, an assessment will be made to determine which transboundary aquifers currently have strong institutional capacity for their shared management, which have weak capacity, and which have no current capacity. Drawing from this analysis and existing knowledge on intranational groundwater and international surface water management, lessons could be derived for bolstering institutional capacity for transboundary aquifer governance and incorporating groundwater management within international water agreements. The results will also be used to assess potential conflicts, begin planning for conflict management, and possibly develop templates for water security management plans.

An overview of the activities and methodologies for implementing the suggestions for future study are summarized in Figure 2. The stepwise process is multidisciplinary, integrating expertise in hydrogeology, geographic information science, computer science, and conflict resolution.

9 CONCLUSIONS AND RECOMMENDATIONS

In conclusion transboundary aquifers are not yet intensively developed – but the time for this is coming soon. We need to develop approaches that will be robust and can be used in the new global reality. The lessons learned from transboundary agreements for surface waters provide a good basis for commencing this task.

REFERENCES

Burke, J.J. and Moench, M. (2000). *Groundwater and Society: Resources, Tensions and Opportunities*. UN DESA, New York, USA.

Chalecki, E.L.; Gleick, P.H.; Larson, K.L.; Pregenzer, A.L. and Wolf, A.T. (2002). *Fire and Water: An examination of the Technologies, Institutions, and Social Issues in Arms Control and Transboundary Water-Resources Agreements*. Pacific Institute for Studies in Development, Environment, and Security (www.pacinst.org).

Freid, T.L. and Wesseloh, I. (2002). Integrating Information Technology into Environmental Treaty Making. In: L. Susskind, W. Moomaw and K. Gallagher (eds.),

Transboundary Environmental Negotiation – New Approaches to Global Cooperation. Jossey-Bass, A Wiley Company. San Francisco, California, USA.

Giordano, M. and Wolf, A. (2001). *Sharing waters: post-Rio international transboundary water management*. Revised from an original report prepared as a contribution to the United Nations World Water Assessment Programme. UNESCO.

Hammond, A.G. (2003). How do you write "Yes": A study on the effectiveness of online Dispute Resolution. *Conflict Resolution Quarterly*, 20: 261–286.

Livingstone, S.; Franz, T. and Guiger, N. (1996). Managing Groundwater Resources Using Wellhead Protection Programs. *Geoscience Canada*, 22: 121–128.

Llamas, M.R. and Custodio, E. (2002). Intensively exploited aquifers: main concepts, relevant facts, and some suggestions. UNESCO, IHP-VI. Series on Groundwater No. 4.

Lotov, A.V.; Bushenkov, V.A.; Kamenev, G.K.; Loucks, D.P. and Camara, A.S. (1998). Water Resource Conflict Resolution Based on Interactive Tradeoffs Display. In: D.P. Loucks (ed.), *Restoration of Degraded Rivers: Challenges, Issues and Experience*. Kluwer Academic Publishers. The Netherlands.

Matsumoto, K. (2002). *Transboundary Groundwater and International Law: Past Practices and Current Implications*. Unpublished Research Paper. Oregon State University, USA. 67 pp.

Morris, B.L.; Lawrence, A.R.L.; Chilton, P.J.C.; Adams, B.; Calow, R.C. and Klinck, B.A. (2003). *Groundwater and its susceptibility to degradation: a global assessment of the problem and options for management*. Early Warning and Assessment Report Series, RS. 03–3. United Nations Environment Programme. Nairobi, Kenya. 126 pp.

NGWA (National Groundwater Association) (2002). Transboundary Aquifers Issues. (www.ngwa.org/sig/transummary.htm).

Postel, S.L. and Wolf, A.T. (2001). Dehydrating Conflict. *Foreign Policy*. September-October issue.

Puri, S.; Appelgren, B.; Arnold, G.; Aureli, A.; Burchi, S.; Burke, J.; Margat, J.; Pallas, P. and von Igel, W. (2001). *Internationally shared (transboundary) aquifer resources management, their significance and sustainable management: a framework document*. International Hydrological Programme. UNESCO. Non Serial Publications in Hydrology. 71 pp.

Renevier, L. and Henderson, M. (2002). Science and Scientists in International Environmental Negotiations. In: L. Susskind, W. Moomaw and K. Gallagher (eds), *Transboundary Environmental Negotiation – New Approaches to Global Cooperation*. Jossey-Bass, A Wiley Company. San Francisco, California, USA.

Shiklomanov, I.A. (1999). *World Water Resources and their use*. UNESCO and State Institute of Hydrology. St. Petersburg, Russia.

Wolf, A.T. (1998). Conflict and cooperation along international waterways. *Water Policy*, 1: 251–265.

Wolf, A.T. (1999). The Transboundary Freshwater Dispute Database Project. *Water International*, 24: 160–163.

Wolf, A.T.; Natharius, J.A.; Danielson, J.J.; Ward, B.S. and Pender, J.K. (1999). International river basins of the world. *Water Resources Development*, 15: 387–427.

Wolf, A.T.; Yoffe, S. and Giordano, M. (2003). International Waters: identifying basins at risk. *Water Policy*, 5: 29–60.

Opportunities and challenges of intensive use of groundwater in Sub-Saharan Africa

Christine Colvin[(1)] and Bettinah Chipimpi[(2)]
Water Programme, CSIR, Stellenbosch, South Africa
[(1)] *ccolvin@csir.co.za*
[(2)] *bchipimpi@csir.co.za*

ABSTRACT

The key imperative for water use in Sub-Saharan Africa (SSA) is to alleviate poverty and enable development in the region. Intensive use of groundwater in SSA offers significant opportunities to enable rural and urban sustainable development. The potential role of aquifers in providing a high assurance of good quality water supply during times of drought and dry seasons should be explored and tested within the region. This role realises the high strategic value of aquifers in the region.

The key challenge in successfully implementing intensive groundwater use is developing capacity to manage and monitor intensive use. Appropriate technology and sustainable financing are also critical management constraints on the implementation of intensive use. Resource related challenges relate to the high degree of spatial and temporal variability in groundwater recharge and availability in secondary aquifers in the arid and semi-arid areas of SSA. Predicting sustainable yields and the short term impacts resulting from intensive use will be more difficult if climate change occurs. However, the very challenges that are inherent in intensively using and managing groundwater resources in SSA provide the motivation for use of drought-buffered aquifers in this way.

The SSA region can be summarised as experiencing increasing poverty, reliance on agriculture, rapid urbanisation, lack of basic water and sanitation provision and high rates of illness linked to water quality.

Keywords: *groundwater, Sub-Saharan Africa, rural water management, drought preparedness.*

1 INTRODUCTION

This paper represents a personal view of the opportunities and challenges of intensive groundwater use in Sub-Saharan Africa (SSA). Section 1 sets the context for intensive groundwater use and management within the region with respect to the biophysical and socio-economic environment. Opportunities for optimal application of intensive groundwater use are discussed in Section 2. Key issues affecting the sustainability and challenges of intensive groundwater use are outlined in Section 3.

1.1 *Biophysical context of the region*

1.1.1 *Climate*

The SSA is defined by spatial rainfall variability, with the Sahara desert, receiving less than 100 mm/yr of precipitation, forming its northern boundary. Mean annual rainfall reaches 2,500 mm/yr in some areas of West and Central Africa. The spatial rainfall gradient is steepest in Ethiopia, Southern Angola and the Sahel regions (including Mali, Niger and Chad) where rainfall varies by more than 1,000 mm/yr over a distance of 750 km (UNEP, 2002). These high gradient areas are most vulnerable to relatively small changes in the position of the Inter Tropical Convergence Zone (ITCZ) (UNEP, 2002) and variations related to climate change.

Coefficients of rainfall variability exceed 200% in the desert areas, 40% in the semi-arid regions and between 5 and 20% in the wettest areas (UNEP, 2002). Many areas are known to have a strong response to El Niño Southern Oscillation (ENSO) (Schulze, 1999), which contributes to the high inter-annual variability in rainfall. Periodic drought is therefore a regular climatic feature in SSA and is expected to become more frequent and severe as climate change is felt. Africa as a whole contributes less than 4% of global CO_2 emissions and 17% of total greenhouse gas emissions (Watson, 2001). Its vast forests act as a critical sink for CO_2, however, Africa is expected to suffer most from predicted climate change scenarios (Watson, 2001).

The rate of evaporation in SSA is high and in excess of 2,000 mm/yr in large areas, resulting in high aridity indices for many areas (for example in South Africa, PE/MAP is >20 in arid areas and 3 even in moist regions) (Schulze, 1999). This also results in a lower than average conversion of rainfall to run-off (world 35%, Africa 20%, South Africa 9%) and high evaporative losses from dams.

1.1.2 *Hydrogeology*

Almost two thirds of SADC's (Southern Africa Development Community) aquifers are secondary aquifers, with around 27% of the surface area comprised of porous aquifers; 3% of karst; 26% fractured sediments, basalt and dykes; and 44% crystalline basement and metamorphic units (excluding Mauritius and the Seychelles) (from SADC WSCU, Web Page http://www.sadcwscu.org.ls). Very few data are available which characterise volumetric consumption of groundwater by aquifer type or per consumptive uses. Even less is known about ecological dependency on groundwater in the region.

The draft hydrogeological map of the world shows that much of SSA is underlain by low and moderate yielding aquifers. This is typical of basement areas where secondary aquifers predominate. Despite expected low to moderate aquifer yields, the potential and importance of groundwater supplies here is significant. In a typical basement area of Zambia, an analysis of 700 mainly low yielding groundwater points indicated that: 32% were low yielding but satisfactorily met domestic requirements; 27% were low yielding as a result of inappropriate borehole design and could be improved; 21% were low yielding as a result of poor siting; 20% were potentially high yielding as they existed but had inadequate pumping capacity (Lovell *et al.*, 1999).

The strategic value of groundwater in SSA is primarily related to the quality and reliability of supply over widely distributed areas during dry periods, rather than volumes available. This strategic importance is often lost in volumetric, catchment water balances. Groundwater may represent only between 3 and 5% of renewable water resources, but represent over 90% of the stored water of the catchment. Groundwater therefore represents a resource buffered from the inevitable effects of rainfall seasonality and drought, whose value is related to its assurance of supply during crisis periods.

1.1.3 *Trans-boundary water resources*

Africa has more international rivers than any other continent. This is largely due to a historical legacy of international boundaries drawn by the former colonial powers with little regard for established natural or ethnic boundaries (Grey and Sadoff, 2002). Some 85% of Africa's surface water resources are comprised of large river basins shared by several countries (Ashton, 2002). It has been estimated that the percentage of the African population experiencing water scarcity or deficit will increase from 39% currently, to 76% by 2025 (Ashton, 2002). Trans-boundary resources in a climate of increasing scarcity are seen to pose a significant risk of conflict over water in the region.

Trans-boundary aquifers identified in SSA include (UNESCO–ISARM, 2002):

- The Iullemeden system shared by Mali, Niger and Nigeria.
- Different Karoo-Kalahari systems shared by Namibia, Botswana and South Africa and by Angola-Zambia respectively.

– The Chad Aquifer systems shared between Chad, Cameroon, Central African Republic, Libya, Niger, Nigeria and Algeria.
– The Benin-Togo Coastal Aquifer.
– The Djibouti-Ethiopia shared basalt aquifer.
– The Kenya-Tanzania basalt aquifer.
– The Merti aquifer (Kenya-Somalia).

The majority of these are in semi-arid and arid areas of SSA which are currently affected by drought and are predicted to be negatively impacted by climate change. Regional level cooperation will be necessary for future shared use of these important resources.

1.2 *Socio-economic context of the region*

The 51 states of SSA are characterised by great socio-economic diversity, in addition to their biophysical diversity. Nigeria is the most populous with an estimated population of around 120 million, while 10 other countries in the region have 1 million or less (Sparks, 2001). Life expectancy is 52 years for the region, but falling in many countries, such as Zimbabwe, as a result of the AIDS pandemic (Sparks, 2001). In the late 1990s, 90% of the world's HIV positive children lived in SSA and an estimated 14,000 people are infected each day (FAO, Web Page http://www.fao.org/Focus/E/aids/aids6-e.htm2003). Infant mortality in SSA is the highest in the world at 91‰ (before 5 years of age) (Sparks, 2001).

1.2.1 *Economy*

Almost 40% of Africans live below the poverty line and Africa is the only region in the world where poverty is predicted to rise during this century if adequate measures are not urgently taken (UNEP, 2002). SSA accounts for less than 2% of global GDP (Gross Domestic Product) and has exhibited an average annual GDP growth rate of around 3% since 1961 (Sparks, 2001).

The United Nations Secretary General's Millennium Report states that "nowhere is a global commitment to poverty reduction needed more that in Africa south of the Sahara, because no region of the world endures greater human suffering".

Currently 75% of Africans live in rural areas, however rapid urbanisation is forecast to increase the proportion of city dwellers to 50% by 2025 (Sparks, 2001). Agriculture contributes to 40% of the region's GDP, uses 88% of water consumed and employs more than 60% of the labour force (UNEP, 2002). Typical of the global average, between 40 and 60% of African irrigation water is lost to seepage and evaporation. This results in wastage of precious resources, salinization of water and soil as well as water logging.

Groundwater is seen as being particularly important for irrigation in Botswana, South Africa, Gambia, Kenya, Nigeria, Togo, Uganda, Zimbabwe and Zambia.

1.2.2 *Water resources management*

Around 100,000 Mm^3 of water are consumed annually in Africa (UNEP, 2002). That represents an estimated 3% of the volume available annually from groundwater and surface water supplies (UNEP, 2002). Available water resources are distributed unevenly both geographically and seasonally. As a consequence, less than half of Africa's population has access to clean water supply or sanitation. In SSA 51% of the population has access to safe drinking water and 47% to sanitation (UNEP, 2002). In general, better services are provided in cities. It is estimated that around 80% of illness in Africa's least developed countries are associated with this failing (Sparks, 2001).

The combination of poverty and hydrologic variability has historically compromised the ability to adaptively manage water resources in Southern Africa. Figure 1 simplifies the links between poverty, lack of human, financial and infrastructural capital, and lack of coping capacity. This has necessitated external assistance to manage crises, which further undermines the development of sustainable IWRM as the experience of management is lost to the region.

Efforts to address the negative *status quo* are underway at regional and national levels within SSA. Africa wide, the New Partnership for African Development (NEPAD) aims to eradicate poverty and to place African countries, both individually and collectively, on a path for sustainable

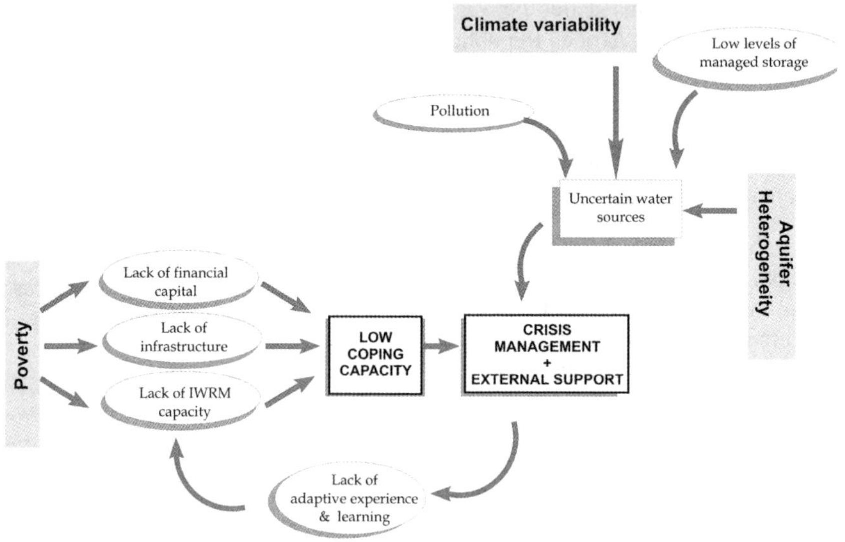

Figure 1. Water management in Sub-Saharan Africa undermined by economic and hydrologic drivers.

growth and development, and at the same time to participate actively in the world economy and body politic.

Within this framework, the African Ministerial Conference on Water (AMCOW) identified the following key issues at the *Bonn 2002 International Conference on Freshwater*: governance in the water sector; inter-governmental dialogue for water security; financing; management capacity building; technology transfer; urban water demand; basic human needs; gender representation in water management; linking catchment environments to the coast. In 2002, AMCOW decided on the following key elements in an African Regional Programme of Action:

- Develop a regional support programme to strengthen water sector reforms.
- Seek support for the establishment of a Regional Water Technology Collaborating Centre for Africa.
- Develop a regional strategy and plan for technical/advisory services.
- Promote actions which will realise the African Water Vision.
- Designate the African Water Task Force to play the role of *Regional Water Watch*.

The Accra declaration, which informed the AMCOW contribution to the World Summit on Sustainable Development, focussed on paving the way for clear policies and commitment to eradicate poverty, reduce water-related diseases and achieve sustainable development in Africa (Africa Water Task Force and Local Organizing Committee, 2002). AMCOW aims to reduce the number of people without basic water and sanitation services by 75% by 2015 and ensure that almost all have their basic human needs met by 2025. South Africa has progressed reasonably well since the democratic elections in 1994, spending US$ 5.2 million on providing water for 7.2 million people (Kasrils, 2002). This rate of provision indicates that the AMCOW targets can be met, but the sustainability of the existing provision is yet to be proven.

Within the Southern African Development Community (SADC) a Water Sector Coordination Unit (WSCU) has been set up to facilitate regional cooperation in sustainable water management. In West Africa, UNEP, UNESCO and ECA (Economic Commission for Africa), are working together to implement a project on groundwater assessment and management. This focuses particularly on the mitigation of the adverse impacts of urban pollution of groundwater used as a potable source of water for African cities.

Regional institutions are seen to be fairly strong and should improve as the implementation of various programmes cements the policy intentions. National and catchment levels of water management competency are highly variable in terms of supporting legislation, funding and capacity. Local water management at a village or farm level, particularly in the rural areas, often operates effectively despite difficult economic and physical conditions.

2 OPPORTUNITIES FOR INTENSIVE USE OF GROUNDWATER

It should first be mentioned, that while there are critical opportunities for intensive use of groundwater in SSA, the resource in many areas is currently under-used and opportunities for even basic sustainable use of aquifers are unrecognised at a regional and national level. Many discussions about the future and current use of African's water resources do not explicitly mention groundwater and reference to river basins and surface catchments indicates that aquifers have not been sufficiently considered in terms of their critical role in integrated water resources management. At thc local farmer and village level, there is often a high awareness of groundwater potential in alluvial and other superficial aquifers, and a high level of dependence during the dry season. Seasonal intensive use of groundwater in many of the semi-arid areas of the region is already common practice, guided by the indigenous knowledge base in the absence of scientific hydrogeological information.

The unique characteristics of groundwater resources need to be related to the most pressing needs of the SSA to achieve maximum positive impacts of intensive use. These characteristics which distinguish groundwater resources from surface water or rain water harvesting are:

- Protection from pollution at surface, especially from disease causing microbes.
- Significant stored resources, offering greater assurance of supply and availability during dry and drought periods.
- Wide distribution, allowing local control of abstraction and management of the resource.

In considering opportunities for intensive use of groundwater, the acceptability of intensive use needs to be established. Intensive use often implies a higher than normal risk of consumptive use close to or beyond sustainable limits. However, the risk is justified by attempting to realise optimal benefits for the majority of resource stakeholders. The key imperative for water use in SSA in the future is to alleviate poverty and enable development in the region. SSA, as described above in Section 1, can be summarised as experiencing increasing poverty, lack of economic growth, reliance on agriculture, lack of basic water and sanitation provision, high rates of illness linked to water quality, rapid urbanisation and high spatial and seasonal variability in rainfall. Figure 1 showed how these poverty issues can be related to water management, the legacy of water resources and policies which attempt to minimise risk.

Priority applications for the intensive use of groundwater which link aquifer characteristics to the needs of SSA are:

- *Drought preparedness* (and relief) in rural agricultural environments.
- Urban and peri-urban intensive use for *basic human needs*.
- Supporting the development of rural and urban sustainable livelihoods.

Access to groundwater is perhaps the most critical factor enabling many rural populations to maintain sustainable livelihoods (Burke and Moench, 2000). Assured water supplies greatly reduce the risk poor farmers face when investing in agricultural inputs and secure water supplies enable them to increase yields, income levels, savings and capital formation substantially (Burke and Moench, 2000).

The International Hydrological Programme (IHP) and SADC are exploring the possibility of establishing a regional centre for Integrated Drought Management. The facility would aim to build on existing initiatives and serve as an integrating institution, with a strong focus on policy coordination and information dissemination (UNESCO, 2001). This forms part of the SADC WSCU's long term programme to increase drought preparedness in the region through improved monitoring, collaboration and effective management.

Intensive groundwater use presents an opportunity for devolution of water resources management and local empowerment as the stakeholders develop the ability to manage risks posed to their livelihoods. Proven benefits of community controlled water supply include: time savings for productive members of the community; health benefits; additional income opportunities related to household level activities; local skills development; community coherence and confidence; increased general economic momentum (WWAP, 2001).

Opportunities for intensive urban use may present higher risks and greater challenges, but the potential benefits to the growing urban population should justify these. Contamination risks and impacts are significantly higher in urban areas due to the concentration of sanitation and other impacting activities. One example of potential urban use is in Cape Town, South Africa, where much of the urban development is underlain by the shallow, primary Cape Flats aquifer. It is generally accepted that this aquifer is currently under-used and higher levels of use could meet urban needs, including basic human needs and garden irrigation. A fully integrated development of water and land-planning is required with the testing of Water Sensitive Urban Design principles in a developing city.

Conjunctive use of groundwater often represents the most efficient use of water resources in semi-arid areas. Very low levels of dam storage are used in Africa (Grey and Sadoff, 2002). For example, South Africa has approximately 10% of the per capita dam storage of the USA. However, given the extremely high rates of evaporation experienced in much of the region and high aridity indices (Schulze, 1999), surface storage of precious water resources is often extremely ineffective. Opportunities for artificial recharge to different primary and secondary SSA aquifers have been tested with good results and it is hoped that this practice will be more widely employed in the region (Murray and Tredoux, 2002). Unused aquifer storage capacity can, for the most part, be developed at a significantly lower cost than surface storage facilities, and without the environmental problems frequently associated with surface storage. The overall costs of artificial recharge operations are often less than half the capital cost of conventional water supply alternatives, therefore in the developing economies of SSA, artificial recharge offers a cost effective as well as a water effective option.

Artificial recharge requires good aquifer characterisation and a fairly high level of water management capacity to be used effectively, and this will be a limiting factor in many areas.

3 CHALLENGES OF INTENSIVE GROUNDWATER USE WITHIN SUB-SAHARAN AFRICA

The main challenges in enabling the sustainable intensive use of groundwater resources in SSA can be summarised as relating to first order, or resource based challenges, and second order, or management related challenges (Mathieu and Trother, 2001).

3.1 *Understanding groundwater resources*

The key resource based challenge is to understand the long term sustainable yield for groundwater and impacts associated with intensive use during critical periods. These are made particularly difficult by resource characteristics such as:

- The predominance of secondary aquifers with significant heterogeneity and associated difficulties in predicting aquifer behaviour.
- The predominance of arid and semi-arid conditions with high coefficients of inter-annual variation for precipitation and aquifer recharge.
- Increasing unpredictability of groundwater recharge as a result of expected climate change impacts.
- Variations in natural water quality.

Recharge as a proportion of annual precipitation declines rapidly below 500 mm/yr of rainfall (Beekman *et al.*, 1996). This makes sustainable yield prediction in semi-arid environments and the management of the long term impacts of intensive groundwater use extremely difficult. Recharge becomes more dependent on individual intense events under these situations, and is more difficult to predict than annual averages.

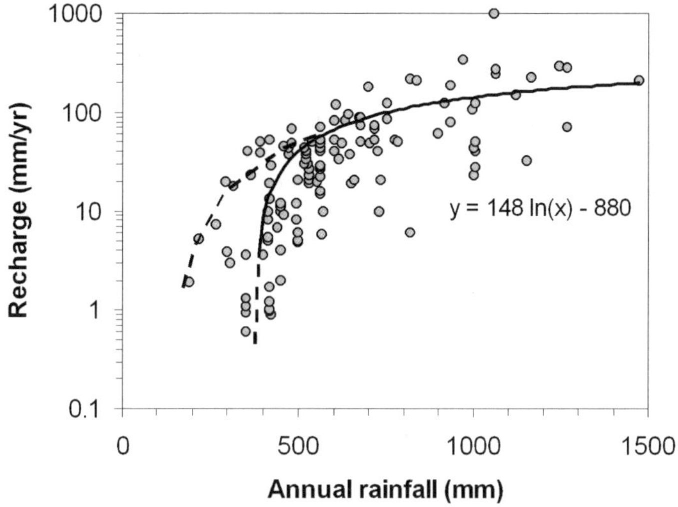

Figure 2. Recharge rates in Southern Africa (after Beekman *et al.*, 1996).

A long term study of the Stampriet basin in Namibia and Botswana showed that long term declines in confined water levels were suddenly reversed as a result of a single month's rainfall which exceeded 20 mm (Tredoux *et al.*, 2002).

A 20% decrease in mean rainfall volumes could translate to an 80% decline in recharge for areas that currently receive 500 mm/yr of rainfall or less (Figure 2) (Cavé *et al.*, 2003). This has serious implications for areas expected to receive lower rainfall as a result of climate change and presents a serious hazard for those dependent on intensive groundwater use in these areas.

In addition to understanding long term aquifer yield characteristics, the short term impacts of intensive use should also be understood and monitored. A current challenge in this regard is understanding the nature of ecological dependency on groundwater, particularly during periods of drought when both terrestrial and aquatic ecosystems may be most sensitive to falling groundwater levels and reduced rates of discharge.

3.2 *Managing the resource*

Potentially the most significant obstacle to intensive groundwater use in SSA is the lack of hydrogeological capacity in the region. Intensive use often requires high risk levels of abstraction during which adaptive management responses may be necessary to maintain the integrity of the resource. This requires planning and direction from hydrogeologists and a reasonable level of hydrogeological awareness amongst water managers, supply engineers, farmers and local villagers who may be responsible for managing abstraction and monitoring. In most areas inadequate capacity exists at all of these levels to enable sustainable intensive use.

Similarly, part of the management cycle should incorporate monitoring data, learning and feedback to improve the understanding of the resource under extreme conditions and to assist in future drought preparation. In most of SSA, current monitoring infrastructure and databases are insufficiently maintained. In addition, continuity in the hydrogeological knowledge base is seen as a significant problem in SSA (Chilton, 2000) under pinned by capacity limitations and the need to bring in temporary foreign expertise. Discontinuous data and learning experience on water supply and assessment projects has resulted in a lost opportunity for incremental learning (Chilton, 2000).

SADC has recently developed guidelines for minimum standards to be adopted in the development of groundwater resources in the regions (SADC, 2000). These include guidelines on data capture and reporting, as well as borehole siting, design and testing. The challenge is now to implement these guidelines and ensure that national and catchment databases are sufficiently accommodating groundwater data.

An enabling policy and legal framework is required at a national as well as regional level to enable intensive use. The approach for protection and allocation of use of water resources within South Africa, includes a Classification of water resources based on current and future acceptable impact on the resource within sustainable limits (Colvin *et al.*, 2002). The preliminary Classification method for groundwater recommends an assessment of the vulnerability of the resource to over-exploitation. Over-exploitation is taken to mean use, typically abstraction, of the aquifer beyond its sustainable limits in the long term. Protection against long term over-exploitation should not preclude short term intensive use of groundwater, where significant benefits are expected.

Conflict prevention over water resources has focussed primarily on cooperative management of internationally shared resources. However, the regional cooperative framework seems strong enough to handle present aspects of this threat (Ohlsson, 1995). The most likely scenario for future conflict relates to the failure to meet internal challenges facing governments in the region (Ohlsson, 1995). This should be addressed through both improved monitoring of resources, so that debates can focus on data and established levels of confidence in the translation of monitoring data to system understanding, and stakeholder decision-making in resource management and allocation.

The sustainability of borehole and well infrastructure is a critical issue in many developing countries. Significant attention needs to be given to investing in appropriate technology that will enable intensive groundwater use during periods of drought. Again, local capacity building to ensure adequate maintenance is essential. In the severe drought of 1991–1992 in Malawi, the effects of drought were exacerbated by inadequate installation and maintenance of boreholes and wells. Relatively few boreholes and wells actually dried up, and those that did were in the mountainous areas tapping aquifers of limited thickness and extent (Calow *et al.*, 1997). However, in many areas groundwater was not accessible for use during the drought due to infrastructural failure.

Financing is a serious control on the provision of water services generally, including the intensive use of groundwater resources. It has been estimated that around US$ 47,000 million will be required during 1997–2007 to provide 40% of Africans with high level services and 60% with basic services (Janssens *et al.*, 1997). The amount and structure of financing required is a significant current constraint, with a shift towards public-private partnerships and a reduction in government control being encouraged. In countries where governments have controlled all levels from policy development to supply operations, e.g. Zimbabwe, significant inefficiencies have resulted (Mtetwa, 1999).

4 CONCLUSIONS

Intensive use of groundwater in SSA offers significant opportunities to enable rural and urban sustainable development. The potential role of aquifers in providing a high assurance of good quality water supply during times of drought and dry seasons should be explored and tested within the region. This role realises the high strategic value of aquifers in the region.

The key challenge in successfully implementing intensive groundwater use is developing capacity to manage and monitor intensive use. Appropriate technology and sustainable financing are also critical management constraints on the implementation of intensive use. Resource related challenges relate to the high degree of spatial and temporal variability in groundwater recharge and availability in secondary aquifers in the arid and semi-arid areas of SSA. Predicting sustainable yields and the short term impacts resulting from intensive use, will be more difficult if climate change occurs. However, the very challenges that are inherent in intensively using and managing groundwater resources in SSA provide the motivation for use of drought-buffered aquifers in this way.

ACKNOWLEDGEMENTS

We would like to acknowledge the support of the organisers of the SINEX conference who have enabled our participation in this important event; colleagues in the Groundwater Group of the CSIR, South Africa, particularly Lisa Cavé, Hans Beekman, Gideon Tredoux, Freeternity Rusinga, Simon Hughes and Pannie Engelbrecht; colleagues in the wider CSIR who have supported the production of this paper and broad thinking on African issues, particularly Johan de Beer and Pete Ashton; the broader community of hydrogeologists, particularly within the IAH, who continue in many ways to support the interests of African hydrogeologists and the needs of the developing world.

REFERENCES

Africa Water Task Force and Local Organizing Committee (2002). *Water and Sustainable Development in Africa.* Regional Stakeholders' Conference for Priority Setting. Conference Executive Summary and Accra Declaration. Accra, Ghana, 15–17 April 2002.

Ashton, P.J. (2002). Avoiding conflicts over Africa's water resources. *Ambio,* 31(3): 236–242.

Beekman, H.E.; Gieske, A.S.M. and Selaolo, E.T. (1996). GRES: Groundwater recharge studies in Botswana 1987–1996. *Botswana Journal of Earth Sciences*, Vol. III, 1–17.

Burke, J.J. and Moench, M.H. (2000). *Groundwater and Society: Resources, Tensions and Opportunities.* Themes in groundwater management for the twenty-first century. DESA, UN Dept. of Economic and Social Affairs with the support of the Economic Commission for Latin America and the Caribbean (ECLAC). ISET (Institute for Social and Environmental Transition). 170 pp.

Calow, R.C.; Robins, N.S.; Macdonald, A.M.; Macdonald, D.M.J.; Gibbs, B.R.; Orpen, W.R.G.; Mtembezeka, P.; Andrews, A.J. and Appiah, S.O. (1997). Groundwater management in drought-prone areas of Africa. *Water Resources Development*, 13(2): 241–261.

Cavé, L.; Beekman, H.E. and Weaver, J.M.C. (2003). Impact of climate change on groundwater resources. In: Xu and H.E. Beekman (eds.), *Groundwater Recharge Estimation in Southern Africa.* UNESCO-IHP. ISBN: 92-9220-000-3.

Chilton, P.J. (2000). *Groundwater Research Priorities for the SADC Region.* Report of a Workshop at the IAH 30th Congress, Cape Town, November 2000. Technical Report IR/01/27, Hydrogeology Series, British Geological Survey, Natural Environment Research Council: 1–8.

Colvin, C.A.; Hughes, S.; Clarke, S. and Silinga, A. (2002). *Classification of Groundwater under the RDM: A Case Study for the Kammanassie-Oudtshoorn Area.* WRC draft report, South Africa.

Grey, D. and Sadoff, C. (2002). *Water Resources and Poverty in Africa: Essential Economic and Political Responses.* A paper for discussion. Presented by the World Bank to the African Regional Ministerial Conferences on Water (ARMCOW): 3–13.

Janssens, J.; Kriss, P.; Sorel, C. and Watson, P. (1997). *Public-Private-partnerships in Urban Water Supply and Sanitation.* HABITAT/UNDP Consultations on Partnership in the Water Sector for Cities in Africa. Cape Town, South Africa, 8–10 December 1997: 1–9.

Kasrils, R. (2002). Budget speech for Ministry of Water Affairs and Forestry.

Lovell, C.J.; Kremer, A.; Moriarty, P.B.; Dube, T.; Macdonald, D.M.J. and Lombe, F. (1999). Integrating Productive Water Points into Rural Water Supply as a means of coping with drought. *Proceedings of International Conference on Integrated Drought Management.* UNESCO.

Mathieu, P. and Trother, J. (2001). *Water Scarcity, Vulnerability, Livelihoods and Security: the Role of Institutions.* Printed from IHDP Newsletter Update, 2001/2, Article 6.

Mtetwa, S. (1999). Experiences in the Water Resources Development and Management for Sustainable use in Zimbabwe. *Natural Resources Forum*, 23: 31–42.

Murray, E.C. and Tredoux, G. (2002). *Pilot artificial recharge schemes: testing sustainable water resource development in secondary aquifers.* Water Research Commission, Pretoria.

Ohlsson, L. (1995). *Water and Security in Southern Africa.* Publication on Water Resources, No. 1. Department for Natural Resources and the Environment, University of Gothenburg, Sweden.

SADC (Southern Africa Development Community) (2000). *Development of Minimum Common Standards for Groundwater Development in the SADC Region.* Draft Report No. 2 for the SADC Water Sector Coordination Unit, Lesotho.

Schulze, R.E. (1999). Freshwater Resources in Africa. In: J.H.C. Gash, E.O. Odada, L. Oyebande and R.E. Schulze (eds.), *Proceedings of a Workshop.* Nairobi, Kenya, October 1999: 1–146.

Sparks, D. (2001). *Economic trends in Africa south of the Sahara 2000.* 30th Edition, 2000 Europa Publications: 11–19.

Tredoux, G.; Kirchner, J.; Miller, R.M.; Yamasaki, Y.; Christelis, G.M. and Wierenga, A. (2002). Redefining the recharge behaviour of the Stampriet Artesian Basin, Namibia. *Proceedings of IAH Conference "Balancing the Groundwater Budget".* Darwin, Australia, May 2002.

UNEP (United Nations Environment Program) (2002). *Action Plan for the Implementation of the Environment Initiative of the New Partnership for Africa Development, Part I.* Working Draft 1, 7 June 2002: 1–10.

UNESCO (2001). *Integrated Drought Management – Lessons for Sub-Saharan Africa.* International Hydrological Programme, UNESCO Division of Water Sciences. Brochure: 1–15.

UNESCO–ISARM (2002). *Workshop Report and Recommendations*. International Workshop on "Managing Shared Aquifer Resources in Africa". Tripoli, Libya, 2–4 June 2002.

Watson, R.T. (2001). IPCC Special Report on the Regional Impacts of Climate Change – An Assessment of Vulnerability. Chapter 2.2, Regional Climate. Intergovernmental Panel on Climate Change (UNEP, WMO) (Web Page http://www.grida.no/climate/ipcc/r egional/010.htm).

WWAP (World Water Assessment Programme) (2001). *Water Security: A Preliminary Assessment of Policy Progress since Rio.* Contribution to the International Conference on Freshwater (Bonn, Germany, December 2001) and the World Water Development Report. 25 pp.

Overview of the groundwater resources of Canada

Alfonso Rivera [(1)] and Miroslav Nastev [(2)]
Geological Survey of Canada, Quebec, Canada
[(1)] *arivera@nrcan.gc.ca*
[(2)] *miroslav.nastev@nrcan-rncan.gc.ca*

ABSTRACT

The availability of groundwater and surface water varies considerably in Canada. Canada holds about 20% of the world's freshwater; three major rivers of Canada rank among the top 10 rivers of the word, and the country has over 32,000 lakes with areas greater than 10 km^2. Close to 1 million km^2 of space is covered by Canada's freshwater lakes, ponds and rivers (excluding the Great Lakes). These numbers, however, do not tell the whole story. Approximately 60% of Canada's freshwater drains north, whereas 90% of the population live in the South in a small fringe along the Canadian-USA border. In many cases, and in particular in rural areas, groundwater is the main source for domestic and agricultural uses. This poses some problems to manage the abundant high-quality water demands, having the second highest consumption in the world with 350 L/d, per capita, after the USA.

There is ample information available on surface water, but there is only scattered information available regarding Canada's groundwater resources. The groundwater resources of Canada have not received the public and political attention required for their effective and modern management on a long-term sustainable basis. Regardless of the vast amounts of surface water, some 30% of the Canadian population depends on groundwater and this percentage is constantly growing. With the increase of groundwater exploitation, other looming issues are quickly emerging in Canada: competing demands and users; groundwater pollution; important aquifers extend over political boundaries, including international boundaries; fragmented and competitive management of groundwater; non-existent inventory of groundwater resources; growing water demand in the USA; and climate changes. In addition, recent incidents of *E.Coli* and *TCE* contamination have brought to question the *abundant mentality* relative to water. Increased public awareness of groundwater issues has spurred efforts to collect basic groundwater data and conduct geological mapping to assist management of Canada's groundwater resources at a national scale.

Three groundwater examples are presented in this paper: the issue of transboundary water, water exports, and a case study of regional groundwater assessment of sustainable yield (in pre-development conditions), in an area north of Montreal.

Keywords: *Groundwater sustainability, transboundary waters, water exports, strategic resource, groundwater pollution.*

1 SUSTAINABILITY OF GROUNDWATER RESOURCES IN CANADA

The term *safe yield* is commonly used to quantify the amount of groundwater that can be withdrawn on sustainable basis. However, its definition as "the maximum amount of water that can be withdrawn from a groundwater basin without producing undesired effects" is less useful as a measure of sustainability today than when it was first coined in the early twentieth century. The reason for this is that by defining a single target volume, "a maximum amount of water", the perception is fostered that groundwater is to be used solely as a commodity. The fact is that the amount of groundwater which can be withdrawn "without producing an undesired result" will vary spatially and temporally depending on a number of interrelated factors. These include hydrogeologic setting, climate and climate change, land use and land-use change, groundwater quality and groundwater quality change, and the like. It also will depend on the current and future availability of surface water resources. Very importantly, the amount of groundwater that can be withdrawn for use as a commodity will also depend on the amount that must be allocated to protect the *common good* that is, the environment and ecosystem function. Sophocleus (2000), correctly states that many uses and environmental values (of groundwater) depend on the depth of water – not the volumetric amount (that is) theoretically available.

It is also important to recognize that social and economic factors often play a controlling role in the decision about the *best use* of a groundwater resource. While some countries have put forward strong social and economic reasons for the use of non-renewable groundwater resource, e.g. semi-arid countries, Canada is only beginning to learn about its vast groundwater resources. This has come only after taking into consideration the importance of groundwater as a *common good* and the social and economic realities facing a small community, nation or broad international region that we can arrive at a true measure of the amount of groundwater that is available for use.

This paper provides an overview of the groundwater resources in Canada, and provides examples of the issues facing this country as it strives to meet growing demands for a safe and sufficient supply of fresh water.

2 IMPORTANCE OF GROUNDWATER IN CANADA

Canadians are fortunate to possess an abundant supply of surface water and perhaps even greater quantities of high quality groundwater.

Table 1. Distribution of fresh surface and groundwater use in Canada in 1991 (WRI, 1998).

	Industry	Agriculture	Domestic	Total
Surface water	71%	11%	17%	44.1 km^3
Groundwater	14%	43%	43%	1.0 km^3
Total	70%	12%	18%	45.1 km^3

Many aquifers in Canada are found in coarse fluvio-glacial sediments deposited at the end of the last ice age. Shallow granular aquifers provide most of the water supply for the Kitchener-Waterloo region in Ontario, and the Fredericton area in New Brunswick. In Manitoba, the Carberry aquifer (a long-buried delta of the ice-age Lake Agassiz) is a prime source for irrigation water. A major sand and gravel aquifer located in British Columbia's Fraser Valley is widely used for municipal, domestic and industrial water supply.

On the Prince Edward Island, groundwater found in the thick fractured sandstone unit provides the entire water supply of the province. In the cities of Winnipeg, Manitoba, and Montreal, Quebec, substantial fractured rock aquifers are used for industrial water supply.

In 1991, Canada used about 45 km^3 of freshwater, with 44 km^3 coming from surface waters and only 1 km^3 from groundwater (WRI, 1998). These data, however, can be misleading with regard to the overall importance of groundwater resources as currently (2002) 30% of the population (10 million people) rely on groundwater for their water supply. Small domestic wells located in rural areas account for most of the groundwater withdrawals in Canada. The rate at which groundwater is being withdrawn is constantly increasing. Households and agriculture are the main users of groundwater in Canada (Table 1).

The geographical distribution of groundwater use in Canada ranges from 0.1% in the Northern territories to 100% on the Prince Edward Island (Figure 1).

An *abundant mentality* has developed in Canada regarding surface water. Thus, while ample information is available about its surface waters, there is only scattered information available regarding Canada's groundwater resources. The sustainable yield of major regional aquifers

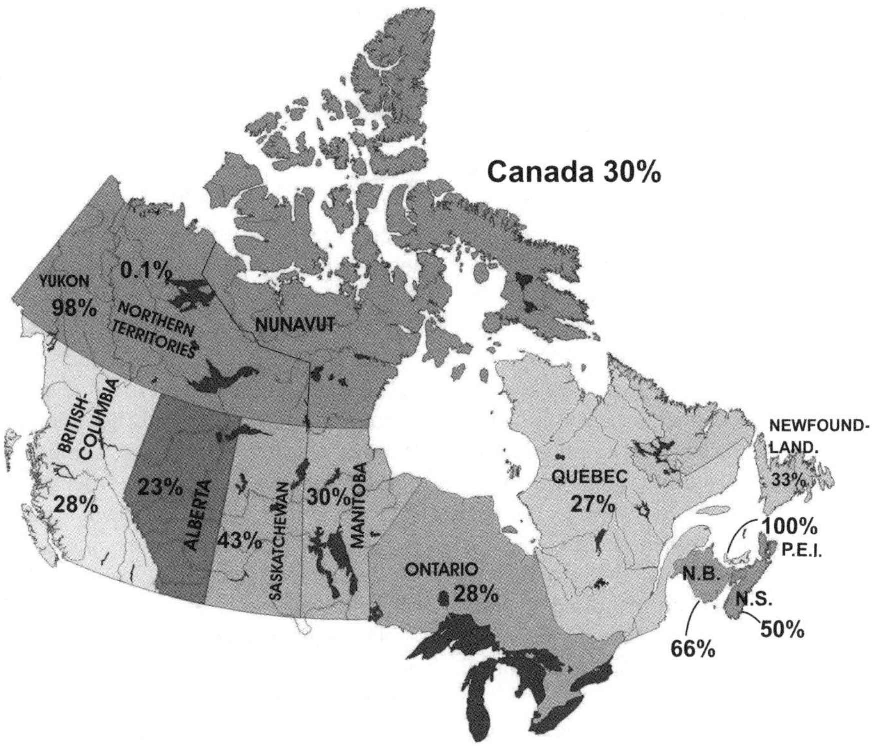

Figure 1. Groundwater use in Canada.

systems in Canada is not yet known and there is no unified, consistent approach in mapping major aquifers and quantifying the groundwater resources. Information regarding the underground diffusion rates, surface water/groundwater interaction, recharge and discharge rate, and storage capacity is needed for the development and the adoption of effective and sustainable extraction practices and for the protection against contamination.

Public awareness in groundwater dramatically increased since the *E. Coli* accident that killed 7 people in Walkerton, Ontario in 2000. The *TCE* contamination in Shannon in Quebec has led to a change in mind and strategy regarding groundwater. In many regions of Canada, there is a mounting concern about groundwater depletion, and instances of aquifer contamination. Recognition of the provincial and territorial responsibility for groundwater has resulted in more emphasis on groundwater monitoring, management, and regulations within provincial governments.

Several looming issues are likely to further awareness of Canada's groundwater resources:

- Increase in water demands. Groundwater use increased from 10% in 1970 to 30% in 1998;
- Groundwater depletion, and instances of aquifer contamination;
- Recognition of large knowledge gaps in the country's groundwater resources;
- Bulk water exports. Exports of water to the USA under the North American Free Trade Agreement (NAFTA) and to other countries;
- Climate change impact and adaptations.

Groundwater currently poses administrative problems in Canada because groundwater resources belong to, and are managed by, the provincial governments. Thus, jurisdictional issues prevent Canada from having a unified, consistent knowledge of its overall groundwater resources. The legal and jurisdiction framework for groundwater management is fragmented, inconsistent, and incomplete. Groundwater management practices vary from jurisdiction to jurisdiction, and in some cases, do not exist at all. This problem has been acknowledged for long but only recently a framework of collaboration for groundwater studies at the national scale has been designed (Rivera *et al.*, 2001).

Presently, the Geological Survey of Canada (a federal agency) in conjunction with its provincial partners is developing plans to map and conduct groundwater research in several major aquifer systems across the country. This is the first major inventory of Canada's groundwater resources since 1967. The plan will focus on providing basic geological and groundwater data essential to manage Canada's groundwater resources at the national scale. Its culmination is proposed as a National Groundwater Management Strategy (Rivera *et al.*, 2001). In this framework, jurisdictions and researchers agree on a long-term commitment to study groundwater resources of Canada with a unified vision. The next 10 years will see the development of the first consistent inventory of the groundwater resources in Canada.

The International Joint Commission (IJC) has also recognized groundwater as an issue to be fully addressed within the context of Canada-USA shared waters in the 21st century. In the 2000 report (IJC, 2000), the IJC made a call to all governments (federal, provincial) to enhance groundwater research in order to better understand the role of groundwater in the Great Lakes Basin.

3 AN EXAMPLE OF TRANSBOUNDARY WATER AND WATER EXPORTS

Contrary to its two North-American neighbors, the USA and Mexico, Canada does not have obvious problems as a consequence of the intensive use of groundwater. Canada's main attention is concentrated in keeping the quality of its surface and ground waters at the highest standard possible, and in overcoming the knowledge gaps of its groundwater resources. In the process of assessing water quality and quantity, it has become obvious that both surface and groundwater resources are in most cases hydraulically interconnected and the need for evaluating surface water/groundwater interactions is becoming urgent. In addition, Canada is concerned about transboundary water issues, both between provinces and internationally with the USA, and more recently about water exports.

3.1 *Transboundary water*

There is no competition in Canada for groundwater resources between provinces or internationally. The most important cases of transboundary aquifers with potential competition are located in the Prairie provinces of Alberta, Manitoba and Saskatchewan. There are nineteen aquifers spanning inter-provincial boundaries in the Prairies (Plaster and Grove, 2000). When an aquifer extends across the border of two jurisdictions, conflict may arise when one jurisdiction depletes groundwater resources that affect the quantity and quality of water available to the other jurisdiction.

The equitable and *reasonable* use of shared waters is the most essential principle considered when negotiating a groundwater apportionment method for the inter-provincial aquifer of the Prairie Provinces. Other factors considered are: the priority use; the sustainable yield of the aquifer; the joint apportionment of surface water and groundwater (though an appropriate method for incorporating surface water/groundwater interaction is yet to be developed); the specification of pumping locations and amounts; the existing Prairies agreement (changes in surface water levels are included in water balances for aquifer interacting with inter-provincial lakes or streams); and the provincial allocation methods.

The current international practices on transboundary aquifers in North America are managed by the USA-Canada International Joint Commission, and the USA-Mexico International Boundary and Water Commission (IBWC).

The IJC was established under the 1909 Boundary Waters Treaty. The Treaty provides the principles and mechanisms to help prevent and resolve disputes, primarily those concerned with surface water quantity and quality along the international boundary between Canada and the United States. The 1909 Treaty did not mention groundwater; it was only in 1977 when transboundary aquifers were first considered by the IJC.

There are three major transboundary aquifers between Canada and the USA: the Abbostford aquifer located between the Lower Fraser River valley in British Columbia and the Nookack River valley in Washington; the Poplar River aquifer of which a third is located in southern Saskatchewan and two thirds in Montana; and the Franklin aquifer located along the Château-gauy valley between Quebec and New-York.

Although the exploitation of those international transboundary aquifers is important and has consequences for both countries (e.g. decline in water levels and water quality), there have not been major disputes or competition. Local task forces or sub-commissions have jointly developed long-term strategies for the effective management of those highly sensitive aquifers.

In recent years, the attention has been shifted to groundwater in the Great Lakes region shared by Canada and the USA. The IJC has emphasized the need for additional work in the region required to better understand the implications of consumption, diversions and removal of surface water, and sharing groundwater from other basins along the boundary (IJC, 1999).

The IJC report states the importance of groundwater contribution to stream flow and to surface water levels in the Great Lakes. Groundwater recharge comes mainly from percolation and precipitation in the Great Lakes basin. Groundwater withdrawal greater than the recharge rate, causes the decline of water levels in aquifers and thus reduce the amount of water discharging to the Great Lakes. When the amount of withdrawal is sufficient, water is drawn from streams or lakes into the groundwater system. This is indicative of the inextricable link between ground and surface waters.

Although there is uncertainty and a lack of adequate information about withdrawals of groundwater, it is estimated that about 5% of overall water use in the basin is accounted by groundwater. Groundwater use does not currently appear to be a major factor relative to Great Lakes levels. It is nevertheless a matter of considerable concern and importance to the more than 20% of the basin's population who rely on groundwater (IJC, 2000).

Finally, it has been estimated that groundwater recharge into the Great Lakes, south of the border (USA), is done indirectly through streams and rivers flowing into the Great Lakes. Data for groundwater input into the Great Lakes, north of the border (Canada), are scarcer. The average groundwater component of streamflow ranges from 48% for Lake Erie to 79% for Lake Michigan (Grannemann *et al.*, 2000). Lake Michigan is the one receiving the most of groundwater flow. Although small in comparison to the amount of water stored in the Great Lakes, groundwater directly and indirectly contributes about 80% of the water flowing from the watershed into Lake Michigan. Groundwater is also very important to the Great Lakes ecosystem. On the basis of these data, it is evident that groundwater is an important component of the hydrologic cycle for the Great Lakes Region.

3.2 *Water exports*

Estimates of Canada's supply of fresh water vary from 5.6%, to 9%, to 20% of the world's supply, depending on how one defines *fresh water* – whether it means *available*, *usable*, or merely *existing*. One study says Canada has 20% of the world's fresh water – ranking it at the top – but only nine per cent of *renewable* fresh water (Environment Canada, 2001).

It has been said that water will be "the oil of the 21st century," or "liquid gold," and that it will be the reason for future wars. Whatever happens with regard to global water, and the environmental, economic and political fallout, Canada, no doubt, will be a major player. Talks have intensified during the past few years on whether Canada should take advantage of its bountiful fresh water supply by selling it – like gas, oil and timber.

The House of Commons held televised hearings starting in September 2001 on *freshwater security*, to examine the pros and cons of exporting Canada's water to other countries. Canada exports bottled water to other countries, but shipments of bulk water are not allowed. There is also the issue of whether, under the terms of the General Agreement on Tariffs and Trade (GATT) and the North American Free Trade Agreement, water is a *vital resource* like the air we breath, or a *commodity* which can be sold and traded. There is a sharp divide on what to do about Canada's water.

In Canada, the water resources belong to the provinces and the federal government has no jurisdiction on that matter. When it comes to water exports, however, the issue has to be dealt with internationally, thus bringing federal government into play. Nevertheless, some provinces are defying Ottawa and the rest of Canada with plans for bulk fresh water exports.

The province of Newfoundland, eastern Canada, has made plans in early 2001 to sell water from the Gisborne Lake near the south coast of Newfoundland. About 500,000 m^3 would be skimmed from the lake each week and ship in bulk to overseas customers. It is argued that "draining 500,000 m^3 of water would lower the lake level for only 2.5 cm which would be replenished naturally within 10 hours" (CBC, 2001). The province government is very enthusiastic about the plans and would go for it alone, regardless of the federal government's opinion.

Environmentalists in Canada argue that allowing Gisborne Lake water to be sold in bulk would make Canadian water a *commodity* which falls under the terms and conditions of GATT and NAFTA.

A similar situation happened two years earlier when the province of Ontario issued a permit to collect Great Lakes water and ship it in bulk to Asia. The permit was issued to a private company, allowing it to export up to 600 million liters of Lake Superior water by 2002. There was such a public outcry on both sides of the border, that the permit was withdrawn.

Other examples exist across Canada, and no doubt, they will continue to defy Canadian's position on water exports. Nevertheless, some critics regard the federal hearings as an indication that Canada is about to change its policy on prohibiting bulk water sales. Some Canadians even talk about diversion (e.g. diverting rivers flow to the south).

Other critics argue that debate over exporting Canada's water is a useless exercise. They say there is no international market for Canadian water. Even if there were, the cost of collecting and shipping Canadian water to distant markets would be prohibitive, far more expensive than drinkable water recovered by new-generation desalination plants.

Whatever the outcome, the provincial and federal governments are preparing themselves for future eventualities by trying to estimate the value of water (e.g. water prize), and by inventorying their other hidden water resource: aquifers.

4 A CASE STUDY OF A REGIONAL SUSTAINABLE YIELD ASSESSMENT, NORTH OF MONTREAL

A comprehensive case study was recently undertaken to estimate the sustainability and vulnerability of the groundwater resources in the St-Lawrence Lowlands of south-western Quebec (Nastev *et al.*, 2002). This case study is part of a series of regional aquifer assessments within a national groundwater inventory program currently lead by the Geological Survey of Canada. The study area, depicted in Figure 2, is located north of Montreal and encompasses approximately 1,500 km^2. The main project objectives were the delineation and characterization of regional aquifers, quantification and protection of the groundwater resource, and delineation of zones suitable for future groundwater development of this fast-growing region in the neighborhood of the second largest city in Canada.

The applied methodology is schematically presented in Figure 3. A multidisciplinary approach was applied including: stratigraphic, lithological, structural and geophysical mapping of fractured bedrock and overlying Quaternary deposits; measurement of groundwater levels in wells and piezometers intercepting regional aquifers; hydraulic testing; geochemical surveys; recharge and discharge estimations; inventory of the current groundwater use and projections; building a conceptual model for the regional groundwater flow; and development of a three-dimensional numerical model. The numerical model was calibrated under steady-state conditions against the piezometric measurements, stream baseflow estimations and direct recharge measurements. The areal recharge and the hydraulic conductivity of the aquifer units were considered as calibration parameters.

The generation of a comprehensive database allowed the first estimate of the aquifer natural

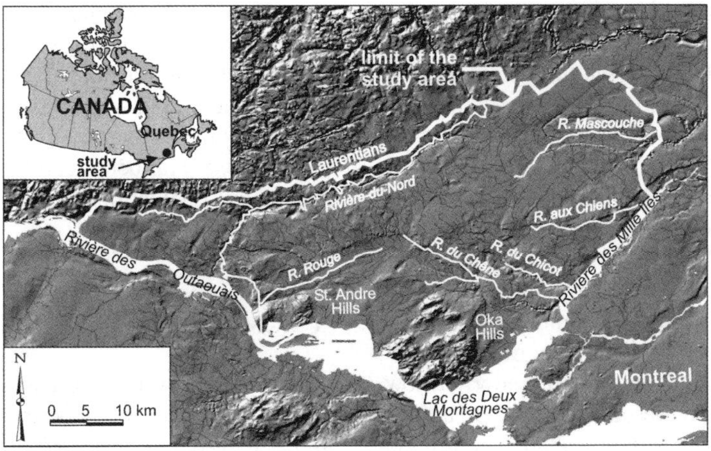

Figure 2. Location of the study area of the St-Lawrence Lowlands hydrogeological mapping project with the digital elevation model as background.

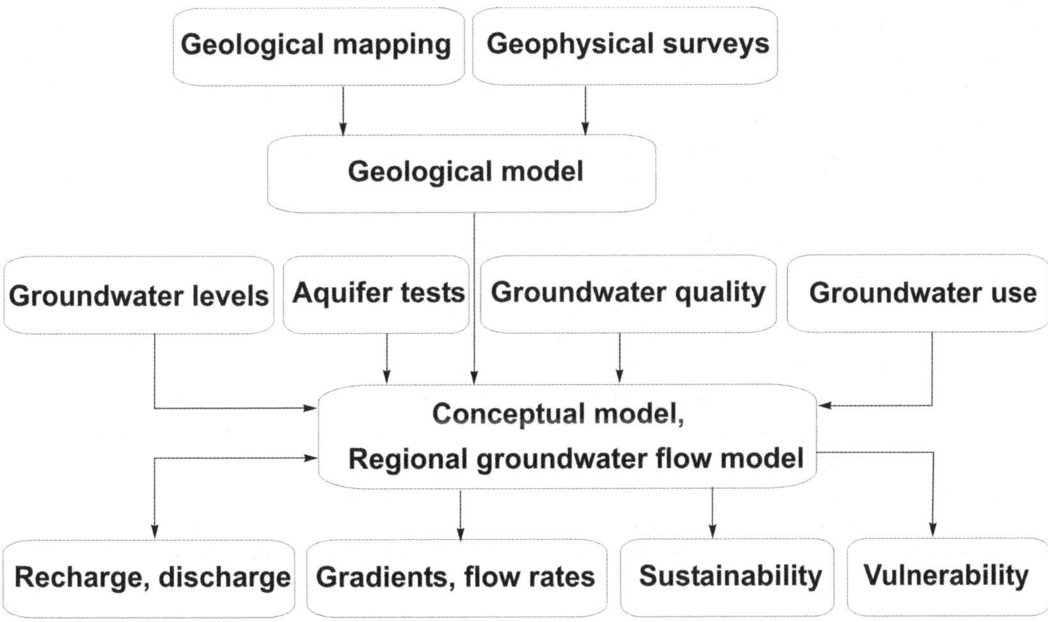

Figure 3. Applied methodology.

basin yield. It turned out that the aquifer system, composed of fractured sandstone, dolostone and limestone, combined with strata of coarse glacial and fluvio-glacial deposits (sand and gravels), is still in early-development conditions. That is, the aquifer system is in hydrodynamic equilibrium; the current groundwater exploitation, essentially, municipal and private wells, rock quarries, and bottlers, do not disturb significantly the annual groundwater flow conditions. The annual withdrawal rate of all groundwater uses combined is in the order of 18 Mm^3, accounting for approximately 20% of the annual recharge to the regional aquifer. The results show that the current pumping conditions provoke a median regional drawdown of 0.7 m compared to the pre-development conditions (no pumping at all).

The results show that the total areal recharge is 94 Mm^3/yr or 77 mm/yr, approximately 8% of the total precipitation in the region. The total inflow to the region, including the induced recharge and underflow rate from the Laurentians, amounts to 111 Mm^3/yr. The current withdrawal of 18 Mm^3/yr (14.8 mm/yr) represents 16.3% of the total regional groundwater flow. A rough estimate of the groundwater volume stored in the region is in the order of 3,000 Mm^3; thus, the total annual renewal groundwater flowing in the region, represents a very small amount of the groundwater stored (~4%). The groundwater discharge to rivers and streams accounts for about 70% of the recharge, representing a very important source of baseflow in the area. Figure 4 shows the groundwater balance of the region in a schematic cross section.

The 3D calibrated numerical model was used to simulate various scenarios of the groundwater extraction in order to estimate the sustainability of the regional aquifers. Uniform withdrawal rates were applied. The simulated drawdowns increased faster with higher withdrawal rates. Doubling of the withdrawal rate leads to a three-fold increase of the median drawdown to 2.2 m. The zones least sensitive to the imposed withdrawal are the discharge zones situated along the major streams. The most advantageous zones for groundwater development in the region were delineated combining the zones less sensitive to groundwater extraction with areas having superior groundwater quality.

It was estimated that the current groundwater use in the area is sustainable, as the simulated drawdowns are lower than the average piezometric fluctuations in the region. As a first approximation, it was calculated that increasing the groundwater extraction by a factor of 2 to 3 would not have adverse effects in the water balance, those numbers represent approximately the natural fluctuations of the annual groundwater levels.

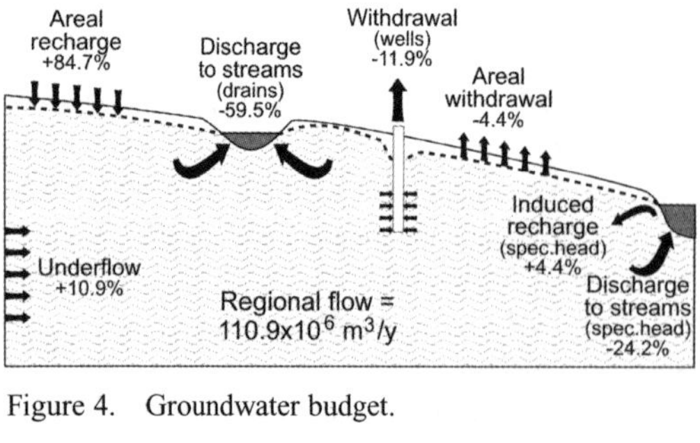

Figure 4. Groundwater budget.

5 SUMMARY AND CONCLUSIONS

Use of groundwater in Canada will continue to increase in the coming years. Sustaining such usage will require that surface and groundwater be managed conjunctively in order to meet demands during droughts or periods of exceptionally high usage. Faced with increasing demands for water resources, and with the uncertainty caused by the effects of regional and global climate change, better predictive models are needed to select appropriate water management options.

At present, Canada, arguably the most water-rich country in the world, does not have obvious problems from intensive use of groundwater. However, recent incidents of *E. Coli* and TCE contamination have brought into question the *abundant mentality* relative to water. Increased public awareness of groundwater issues has spurred efforts to collect basic groundwater data and conduct geological mapping needed to manage Canada's groundwater resources at a national scale.

Groundwater is, probably, the largest source of fresh water in Canada, but is still poorly understood. Knowledge about its occurrence, distribution and quality is needed in order to make informed decisions about its availability for use. Education is needed so that citizens will understand the consequences of the casual disposal of wastes or the inappropriate placement and use of wells. It is recognized that social and economic realities may force Canada to exploit the commodity valuation of its groundwater. It is hoped, however, that lessons learned in other countries will bring alternate solutions to ensure the sustainability of groundwater resources in harmony with the natural environment in Canada.

Given the uneven distribution of surface water in Canada, groundwater resources may become a strategic resource for Canadians. Groundwater in Canada is in general abundant and of good quality; aquifers are generally in hydrodynamic pre-development conditions. Most groundwater exploitation, however, is done in shallow aquifers that are very vulnerable to pollution. On the other hand, aquifers in Canada generally provide the base flow of rivers and represent main sources of water for wetlands and riparians zones. This equilibrium should be assessed prior to any future groundwater development.

REFERENCES

CBC News (2001). The future of Canada's water. CBC News, June 2001.

Environment Canada (2001). Water. available on-line at http://www.ec.gc.ca/water/en/e_quickfacts.htm.

Grannemann, N.G.; R.J. Hunt; J.R. Nicholas; T.E. Rilley and T.C. Winter (2000). The importance of Groundwater in the Great Lakes Region. USGS Water-Resources Investigation Report 00-4008.

IJC (International Joint Commission) (1999). Letter of reference to the International Joint Commission from the governments of Canada and the United States on the consumption, diversion and removal of water. February 10, 1999.

IJC (International Joint Commission) (2000). Protection of the waters of the Great Lakes. Final Report to the Governments of Canada and the United States. February 22.

Nastev, M.; A. Rivera; R. Lefebvre and M. Savard (2002). Hydrogeological context and numerical simulation of regional groundwater flow in the St-Lawrence Lowlands of South-Western Quebec. Submitted to the Hydrogeology Journal.

Plaster, K. and G. Grove (2000). A review of transboundary groundwater apportionment. PPWB Report no. 155; National Water Research Institute, Environment Canada.

Rivera *et al.* (2001). Canadian Framework for Collaboration on Groundwater. Multi-agency document to be submitted to provincial and federal governments. National Ad-hoc Committee on Groundwater; http://cgq-qgc.ca/cgsi/.

Sophocleous, M. (2000). From safe yield to sustainable development of water resources – the Kansas experience. Journal of Hydrology, 235:27-43.

WRI (1998). World Resources, 1998-99. Reported in the World Commission on Water for the 21st century. Regional consultations North America.

Intensive use of groundwater in Latin America

Emilia M. Bocanegra
CIC – Centro de Geología de Costas y del Cuaternario. UNMDP, Mar del Plata, Argentina
ebocaneg@mdp.edu.ar

ABSTRACT

In Latin America, the availability of fresh water responds to a marked geographic heterogeneity, ranging from regions of high pluviometry with an important use of surface water to regions in which groundwater is not only the main, but the sole source of supply. Intensive use of groundwater has allowed free access to drinking water in urban as well as in rural areas, and has promoted the development of productive-economic activities which have improved the quality of living for the entire population. Groundwater exploitation in large cities such as Buenos Aires, Sao Paulo, Bogota, Lima and Mexico share certain common features associated with the variations in the piezometric levels, in some cases due to management strategies and climatic influence, as well as the degradation of the quality of groundwater caused by a variety of anthropic activities.

Arid and semiarid regions in western Argentina and the Northeast of Brazil make intensive use of groundwater resources, mainly for agricultural activities, which has caused a significant depression in the piezometric levels.

In the areas of most significant tourist activities and/or population density in the coastal regions of the Atlantic and Pacific, there exists extensive marine intrusion due to intensive exploitation. Mar del Plata (Argentina), Recife (Brazil) and Costa de Hermosillo (Mexico) are typical cases of sea water intrusion.

Rational management of groundwater resources in the region, makes imperative the implementation of adequate regulations and the development of research projects.

Keywords: *intensive use, groundwater, Latin America, large metropolis, arid regions.*

1 INTRODUCTION

Latin America and the Caribbean have a surface of 20,438,000 km^2 and a population of 447.4 million inhabitants, with a political distribution of one country in North America, seven countries in Central America, 22 countries in the Caribbean and 13 in South America. The region cannot be considered homogeneous in so far as the countries show significant differences in size, population density, economic development, culture, governmental system, climate, geology and surface and subsurface hydrology.

In most countries, the main economic activity is agriculture, with a significant increase in industrial activity during the last decades. In Mexico and South America, the urban population varies between 75-80%, due to a higher availability of health care services, education and jobs in larger cities. Whereas in Central America, only 20% of the population live in large cities, urban population increase is fast, so that by the year 2030 only 28% of the citizens will live in rural areas.

The high foreign debt exerts an important impact on the economic development of the region, and thus, budget negotiations for the settlement of payments generate an overwhelming pressure on productive activities and a reduction in health services and education, causing – among other things – greater levels of poverty and an increase in the population who suffer from unfulfilled basic needs, unemployment and impoverishment 7% of the people who live in extreme poverty in the world live in Latin America.

Adequate quantity and quality of water availability comprise the main environmental factor for the socio-economic development of the region. Precipitation distribution varies significantly, ranging from wide arid and semiarid regions with few millimetres of rainfall per year, to tropical areas with 7,700 mm/yr. Water supply is provided by means of surface water, groundwater and, to a lesser extent, direct precipitation.

Water usage varies in the different regions. In Mexico, agriculture demands 86% of water, industry 8% and domestic consumption 6%. 10% of the urban population and 40% of the rural population do not have ready access to safe water (NAC, 1999). In Central America, 83% of the urban population and 42% of the rural population have access to water supply. Nevertheless, taking into account that in rural areas less than 5% of the water supply is quality controlled, about 20 million Central Americans drink water of uncertain quality. In the Caribbean, high population density, changes in the use of land, inadequate disposal of sewages and climatic hazards (hurricanes, floods) have caused a serious degradation of hydric resources (CATHALAC, 1999). In South America, 70% of water demand is for irrigation, 17% for domestic consumption, and 13% for the industrial activity (mainly in Brazil and Argentina). In spite of the fact that the average consumption is of 200 L/d per habitant, in some countries this figure is much higher, whereas in others there is a real water scarcity, with values truly inferior to average consumption (SAMTAC, 1999).

The larger reserves of groundwater are located in sites where surface resources are abundant; nevertheless, in arid or semiarid regions, groundwater represents the economic and productive support of development. In South America groundwater occurs in porous, fissured and karst media; with a predominance of the first. In Central America and the Caribbean karstic aquifers rank in second place. Groundwater availability in South America, estimated as the subterranean discharge according to the hydric balance performed by Hernández (1990), is of about 3,740 km^3 and constitutes 14.6% of groundwater volumes annually available in the planet.

2 BENEFITS AND CONFLICTS OF CURRENT EXPLOITATION OF GROUNDWATER IN THE REGION

The use of groundwater in the countries of the region has different features depending on hydric condition and the availability of surface and groundwater resources, as well as on the socio-economic and cultural situation.

Historically speaking, the first exploitations were designed to meet human needs, later incorporating supplies for industry, and, last of all, agriculture. Nowadays, agricultural use predominates over any other, and Latin America shows a clear trend to reach the planet average use of water.

The benefits of water use include a wide range of improvements, among them the following are worth mentioning:

- Water availability for multiple uses in arid and semiarid areas in the region.
- Access to safe drinking water in urban, suburban and rural areas, with a positive impact on people's health.
- Increase of irrigation areas, greater food production, increase in job opportunities, and decrease in poverty.
- Increase in industrial activity and decrease in levels of unemployment.

In some cases, groundwater exploitation, associated with planning errors in land use and in the management of hydric resources, leads to certain conflicts and negative impacts, among which we can mention:

- Water quality deterioration due to salinization or contaminated water migration.
- Groundwater quality deterioration due to sanitation *in situ* in areas without water supply and sewage systems.

- Conflict of uses, mainly between agricultural and industrial activities and domestic use, increasing piezometric drawdown and greater extraction costs, as well as deterioration of groundwater quality caused by the incorporation of agrochemical and industrial effluents.
- Piezometric level decline producing land subsidence and land collapse affecting building structure stability.
- Recovery of piezometric levels affecting urban facilities as a result of the abandonment of contaminated or salinizated wells.

These benefits and conflicts created by the intensive use of groundwater are partially or totally produced in the countries of the region. Our discussion encompasses different cases associated with the intensive use in different scenarios: large metropoles, arid and semiarid regions, and coastal areas.

3 INTENSIVE USE IN LARGE METROPOLIS

3.1 *Buenos Aires and the Greater Buenos Aires area (Argentina)*

The Argentine's capital, Buenos Aires, and its conurbation have a population of 11.5 million inhabitants in a total area of 3,800 km^2. Intensive exploitation of groundwater is performed mainly for industrial and human consumption by means of domestic wells and public water supply services. Exploitable groundwater comes from the Pampeano and Puelches aquifers, reaching maximum depths of 70 m. It is a multilayer aquifer system, hydraulically connected to the water table. This makes the contamination of the superficial aquifer – caused by the lack of sewage systems and of the elimination of industrial untreated effluents – reach the deepest aquifers.

The hydrodynamic evolution for both main aquifers was clearly differentiated in the 1970s, with a W–E regional flow direction for the Pampeano and areas with depression cones in the Puelches aquifer, according to the piezometry elaborated by Consejo Federal de Inversiones-EASNE (1972). After the 1970s, the overexploitation of the aquifer led to the depression of the upper aquifers, the Pampeano aquifer and the phreatic levels due to the hydraulic connection of the system, an effect which has been examined in the piezometry carried out by Hernández (1978). This led to an encroachment of the saline water/fresh water interface towards the continent and an inversion of the regime of flow in the area and the risk of pollutant migration into the aquifer (Santa Cruz *et al.*, 1996).

From the 1990s, there has been a slow recovery of the piezometric levels, and currently groundwater is a few centimetres above ground level in many sectors of the Buenos Aires conurbation. According to Santa Cruz and Silva (2002), the substitution of pumping wells for water extraction from the Río de la Plata (River Plate), the infiltration through cesspools, the decrease in groundwater industrial use due to the economic recession and to the increase of rainfalls in the area, all together represent the most significant causes for the rise in the piezometric level in the metropolitan area of Buenos Aires. The efficient intervention of the public sector is imperative in order to introduce an integrated and sustainable management of the hydric resources, reassigning groundwater as a function in the environmental equilibrium typical of humid regions.

3.2 *Metropolitan area of Sao Paulo (Brazil)*

The metropolitan area of Sao Paulo contains a 17.5 million population in a territory of 5,680 km^2. Groundwater resources play a pivotal role in the complementary supply of the private sector and constitute 11% of total consumption. The hydrographic basin of Alto Tiête, on which the metropolitan region rests, encompasses two systems. The Crystalline Aquifer System made up of granitic as well as metamorphic rocks with a variable permeability depending on the alterations of the rock, and the Sedimentary Aquifer System, ranging from unconfined to semiconfined in character, which – in spite of occupying only 25% of the basin – it is the most exploited of the two (Hirata *et al.*, 2002).

It is estimated that a large number of industries, residential areas and enterprises feed off the aquifers through 7,000 operating wells. There seems to be an increase at a rate of 450 private perforations per year since, due to the high charge public enterprises levy on water service, a private 200 m well can be recouped in six months. Currently, there exists interference between wells and a significant decrease in the piezometric levels leading to a 50% reduction in the aquifer thickness. This has led to a rise in water costs, due to a greater hydroelectric energy consumption, the deepening in perforation and changes in the pumping systems. If this trend persists, the exploitation of this aquifers will exceed recharge in the next 15 years. This loss of the resources will lead to a substitution of wells by the services offered by the water supplier firm. This may pose several problems to the public water supply service, which is currently working to its limit (Hirata *et al.*, 2002).

3.3 *Lima (Peru)*

The city of Lima, with its 6.4 million citizens, lies on the Rimac and Chillón basins, and takes its water from an aquifer made up of quaternary alluvial deposits with an extension of nearly 390 km^2. Maximum exploitation was reached in 1997 with 12 m^3/s by means of both 378 wells and a filtrating gallery performed by the Public Service Supplier, as well as approximately 700 industrial and private wells. Intensive exploitation together with the progressive reduction of recharge sources due to changes in land use – from agricultural to urban – have produced a decrease in the phreatic levels, which in the most critical areas ranges from 10–50 m, at a rate of 1–5 m/yr (Quintana and Tovar, 2002).

By numeric simulation, has been determined a maximum exploitable discharge for the city of Lima is of 8 m^3/s being the safe yield 6 m^3/s. The actions taken in order to achieve equilibrium and the recovery of the exploitable reserves were:

a) Rational use of water: the establishment of about 700,000 domestic meters which led to a monthly consumption reduction from 100–150 m^3 to 36 m^3.
b) Conjunctive use of underground and superficial water.
c) Artificial recharge generated from the river bed to the aquifer.

As a result of these actions taken, extractions fell 25% in four years, reaching 9 m^3/s in August 2001 (Quintana and Tovar, 2002).

3.4 *Metropolitan area of the City of Mexico and Federal District*

The city of Mexico and its conurbation contain a densely concentrated population of above 18 million inhabitants in an area of 400 km^2 located in the southern region of the Large Mexico Basin. Water supply for multiple activities is provided by perforations on granular and fissured aquifers and, to a lesser extent, from the aquitard. The latter constitutes the shallowest area and provides a confined and semiconfined character to the underlying aquifers, protecting them temporarily from superficial water contamination (Huizar Alvarez and Carrillo Rivera, 1998). It is thought that there exist more than 10,000 wells in this area. Among the most important problems related to groundwater use in the metropolitan area of the city of Mexico and mainly in the Federal District are: the high rate of consumption and wastage, the great leakages, the growing decline of piezometric levels, the continuous increase in extraction depths, up to 7.5 m soil subsidence due to clay compactation of the aquitards and the consequent damage to urban facilities, the need of importing water from other basins, not to mention the contamination associated with solid and liquid wastes.

One of the main external and most proximal source is the basin of the Lerma River, located at about 100 km from the city of Mexico. Its aquifer supplies more than 6 m^3/s, 9% of the total demand. Intense extraction has produced a drawdown in the piezometric levels and a progressive reduction of groundwater recharge stemming from the lakes of the area. Policy definition and enforcement of regulations of groundwater exploitation are essential to protect the basin of the Lerma River (Garfias *et al.*, 2002).

4 INTENSIVE USE IN ARID AND SEMIARID REGIONS

Groundwater reserves in arid and semiarid areas are, in many cases, the only or main sources of supply. They ensure a pluriannual regulation of the resources and help to overcome droughts by compensating for the lack of superficial hydric resources. Intensive exploitation cases in arid regions of western Argentina and in semiarid regions of the Northeast of Brazil are herein analysed. The socio-economic development in both regions is based on groundwater availability.

4.1 *Province of Mendoza (Argentina)*

The Province of Mendoza has an arid climate, and its hydric resources come mainly from snowfalls in the high mountains, from frozen water in the pervious profile of rocky formations, and from water stored in glaciers, which generate several rivers and permanent streamflows, reaching about 320 m^3/s.

Thanks to water development, irrigation oases have developed along the plains occupying 2% of the total surface of the province, where agricultural, urban and industrial centres have been established and 98.5% of the population lives. Both groundwater and surface water provide for water needs.

Hydrogeologic basins, formed by quaternary sediments, constitute groundwater reservoirs which are mostly recharged by means of permanent-flow rivers. These reservoirs are exploited through 22,000 wells, most of which are located in cultivated areas. The highest rate of groundwater consumption goes for agricultural activities, 85% of which supplies vineyards and fruit trees, as well as vegetable farms, pasture land and forestry. Notwithstanding, because of the desertic climate of this province, the use of groundwater for human consumption has gained relevance. The most exploited area is the Mendoza Norte Basin, with nearly 17,000 perforations and a mean density of 6 wells/km^2 in the cultivated area, and in some sectors the density is even higher. The depths of these perforations varies between 40 m to more than 30 m, although most of the ones that exploit the phreatic aquifer have been abandoned for their high salinity and nitrate content – due to agricultural, municipal and industrial contaminants. Gradually, users have started to exploit deeper aquifers, which has lead to economic loss generated by the abandoned wells not yet amortised, supplementary investments in wells and pumps necessary for deeper aquifers, costs from supplementary pumping, and the complete loss of small farms due to the water salinization.

Atomised and anarchic exploitation of groundwater resources has given rise to several cases of overexploitation, aquifer salinization and pollution. This undesirable effects constitute an imminent threat to the provincial hydric patrimony and holding assets acquired at its expense, as well as to the environmental sustainability of the irrigation oases. The joint management of surface water and groundwater and the protection of the aquifers is a necessary measure to improve this situation and finally attain sustainability (Alvarez, 1997).

4.2 *The Brazilian Northeast*

The Northeast of Brazil has an area of 1.6 million km^2 corresponding to 18.3% of the national territory and a population of 44.8 million inhabitants, equivalent to 28.5% of the population in the country. In this region, the climate is semiarid in 70% of the area. Regarding groundwater resources, (Costa, 2001) most of the area is dominated by a fissured aquifer of crystalline rocks of very low permeability. Well yields are low, with mean depths of 50 m, producing discharges of about 1.5 m^3/h. Another serious problem in this aquifer is salinization reaching values of 5,000 mg/L due to intense evaporation and insufficient precipitations. The total number of wells perforated in this aquifer is of nearly 50,000, though 30% of them are deactivated due to low production and poor water quality.

Sedimentary basins occupy 45% of the Northeast; along with coastal and interior basins, there is the Parnaiba basin, with an area of 500,000 km^2 and a maximum thickness of 5,000 m being the largest reservoir of fresh water in the Northeast.

Conjunctive water use meets irrigation demands (59%), human consumption (22%), industrial and agro-industrial consumption (13%), and animal consumption (6%). Groundwater participation in this demand is complementary in some areas, although in others, mainly in interior cities, it is almost exclusive. Priority uses in order of importance are: human supply (urban and rural), animal, industrial and irrigation. Irrigation is, in volumetric terms of groundwater, the last user in this ranking is striking, because it is privately exploited since the government sponsor mostly large irrigation projects of surface water. The North of Minas Gerais is the exception to this rule, as 89% of the irrigation water comes from underground resources.

In many areas of the Northeast, intensive use of groundwater runs the risk of resource depletion, salinization and soil subsidence. This is mainly seen in exploited aquifers with rather low storage volumes such as the calcareous aquifers in the central region of Bahia and the northern region (Costa, 2001).

5 INTENSIVE USE IN COASTAL AREAS

Coastal areas are sites where populations usually settle and develop their productive activities on large scale due to the possibility of maritime communication, the development of natural resources and the tourist attractions coastal landscapes possess. The developmental potential of coastal areas speeds up city development and the establishment of small towns. This gives rise to an increasing demand for basic needs, among which the supply of drinking water stands out, and groundwater use becomes utterly essential.

The use of coastal aquifers is common practice throughout Latin America, ranging from areas of dry climates such as those in Peruvian coastal area (with mean annual precipitations of 4 mm to 40 mm) to the coast of Brazil, continental areas and the Caribbean islands of humid tropical climates.

Three cases in which the intensive use of coastal aquifers has favoured socio-economic development of the population are herein presented; however, the impact on this resource shows that new management strategies must be pursued.

5.1 *Mar del Plata (Argentina)*

Mar del Plata is the main tourist city of Argentina, located on the Atlantic coast at 400 km South of the city of Buenos Aires. Its population is 600,000 inhabitants, which triples during the summer. For urban, agricultural and industrial water supply, the city feeds exclusively on groundwater.

The aquifer has a top unconfined and multilayer behaviour. It consists of Pleistocen-Holocen loess sediments of silty-sandy and very fine-grained sandy textures. Its thickness varies from 70 m to 100 m, and it rests on miocene marine sediments in some areas. In other areas bedrock is made of eopaleozoic orthoquartzites which show an alternation of tectonic grabens and horsts.

The groundwater exploitation criteria in use since the beginning of the twentieth century are based on the need for drinking water in a city with the high demographic growth rate of 100,000 inhabitants per decade, which, in the last five decades, has led to the development of several management strategies.

Since 1945, as a consequence of an unrestricted exploitation, an accelerated saline intrusion has been registered, reaching a velocity of 150 m/yr and in some wells an increase in the chloride content of 2,000 mg/L/yr. Extraction continued on a larger scale, reaching 137 wells under exploitation and 29 abandoned wells in 1970.

The intensive use in the urban radius led to the development of aquifer salinization in that sector. Due to this, a new geometric design for its exploitation was planned, starting gradually in 1969–1970 with a battery of wells along the Provincial Route 2, which joins the city of Buenos Aires to Mar del Plata and its transversal branches. There was a significant drawdown of water levels in this period, in the wells near the exploitation axis located in the rural area, which jeopardized the water use in this region as pumping costs increased.

Water quality in the suburban and rural area underwent a noteworthy degradation caused by nitrate concentration exceeding 90 mg/L, as well as by bacteriological contamination in private wells located in suburban neighbourhoods without sewage system. Furthermore, there was a 10 m recovery of the piezometric levels in the urban area as a consequence of the abandonment of wells with high chloride content. This led to basement infiltration in some buildings of the city, creating the need for continuous drainage, and some premises became unutilisable. Apart from this, certain problems in the structure stability of buildings arose, as well as wall and floor damage in the downtown area.

As a consequence, since 1992 a new area of exploitation in the South of the city was proposed, and, in early 1993, a new management strategy was designed, which had as its main objective to achieve an integral but rational development of hydric resources. This new production strategy was designed based on two fundamental concepts: flow balance for the hydrologic basin and pumping suitability fit for the critical discharge. This exploitation methodology, while maintaining the equilibrium *supply-extraction*, avoided the accelerated drawdown of the piezometric levels and thus prevented the quick process of *deepening-expansion* of the depression cone. The distance between wells, which historically was 400 m, reached 600–700 m and, therefore, reduced the interference which produced less drawdowns and greater efficiency in pumping systems. As a control measure, chemical and bacteriological quality controls are performed two or three times a year in the urban, periurban and rural areas with no drinking water supply or sanitation services. This data are also used in order to expand the network, giving top priority to more densely populated neighbourhoods of poor water quality. (Bocanegra *et al.*, 1997; Bocanegra *et al.*, 2001).

5.2 *Recife (Brazil)*

The city of Recife, capital of the state of Pernambuco in the Northeast of Brazil, occupies an area of 112 km^2, and is home to 1.6 million people. Groundwater meets the needs of 40% of the current domestic supply through 4,000 wells. In the last 25 years, there has been a decrease in the permanent resources available from the confined aquifer, whose top is at 60 m below sea level, producing a decline of the piezometric levels of more than 100 m in the last decade.

The overexploitation of this aquifer has led to the groundwater salinization in a region occupying about 10% of the Coastal area of Boa Viagem and in the downtown neighbourhoods of the city of Recife. As a consequence of water level drawdowns, the compacting of a 150 m thick sedimentary package could take place, along with an imminent risk of sea water intrusion, as the city is located 1–5 m above sea level. This could cause damage to civil constructions, such as breakage of water pipes and cracks in the sewage system, among other catastrophic consequences.

The current deficit in water input and output in the system is of 1.2 m^3/s, or 37.8 Mm^3/yr in the plains of Recife. If the current exploitation regime continues, the aquifer will become depleted in this area within the next 10 years unless measures are taken to avoid this. The only solution seems to be the artificial recharge through the injection of duly treated surface water, extracted from the Capibaribe River (Costa, 2002).

5.3 *Costa de Hermosillo (Mexico)*

The aquifer of the Costa de Hermosillo, Sonora, is located on the Mexican North Pacific and belongs to an exorreic basin located on the western slope of Mexico whose superficial water drain towards the Gulf of California. The climate in the region is dry, with an average annual precipitation of 200 mm.

The hydrogeologic bedrock consists of crystalline rocks alternated with tectonic graben and horsts reaching depths above 1,500 m. Filling materials formed a multilayer aquifer in which there are semiconfined layers, but, as a whole, it behaves as an unconfined aquifer.

The aquifer was first exploited in 1945 with 17 wells, in 1965 it reached its highest extraction volume with about 1,100 Mm^3/yr; and since then extractions have been reduced until reaching the current value of 550 Mm^3/yr.

The loss of the aquifer hydraulic potential caused a depression cone and inverted the flow from the coast to the continent. Consequently, restrictions on groundwater exploitation were adopted: 105 affected wells were grouped according to salinity, and their extraction was steadily reduced until attaining a volume similar to the estimated recharge. This action could not stop the drawdowns of the aquifer dynamic levels nor the saline intrusion that has affected a great part of the aquifer, up to 30 km in the Northeast and more than 20 km in the South. It has been concluded that marine intrusion acts as a recharge coming from the sea, and it is being directed towards the center of the aquifer, where pumping actions were historically located. This suggests that if current conditions continue, marine intrusion would completely mix with the fresh water of the aquifer. (Medina *et al.*, 2002).

6 FINAL COMMENTS

Aquifers are an essential part of the patrimony for the socio-economic development of Latin America and the well-being of its population. The information and evaluation of groundwater resources as well as legislation regarding this subject matter vary from one country to another. In general there is a dispersion and/or lack of data, poor organization of information, and legislation is often inadequate or absent when it comes to solving problems. Regardless of this, the perception that it is imperative to generate projects and programs on aquifer rational management and protection exists. This management must cover three fundamental transversal axes: environmental, economic and social. Furthermore, the access to safe water is a right contemplated by international laws and by the majority of national constitutions.

However, the tendency to privatise public water service suppliers in Latin America is leading to prioritize company economic benefits over those of social well-being. Privilege is granted to big consumers or sectors with higher economic incomes, who perform investments in the private companies or water supply concessionaires.

The State should respect, assert and protect the right to safe water for its population. Provide water services is their duty, not a charity. Water is a resource subjected to the law of supply and demand law, but the State should ensure that each citizen has access to safe drinking water and thus exercise the administrative, legal and political power it has to attain this goal.

This is the long road Latin America has ahead.

REFERENCES

Alvarez, A. (1997). Recursos hídricos de la cuenca Mendoza Norte – Información general en base al conocimiento actual. Instituto Nacional del Agua, Mendoza, IT-182.

Bocanegra, E.M.; Massone, H.E. and Cionchi J.L. (1997). Sustentabilidad y gestión de recursos hídricos subterráneos. Mar del Plata como caso de estudio. I Congreso Nacional de Hidrogeología. Bahía Blanca, Argentina. 433–444.

Bocanegra, E.M.; Cardoso, G.; Custodio, E.; Massone, H.E.; Martinez, D.E. and Raposa De Almeida, R. (2001). Los acuíferos costeros del litoral atlántico sudamericano: la explotación en el Estado de Rio de Janeiro (Brasil) y la Provincia de Buenos Aires (Argentina). XI Congreso Latinoamericano de Geología. Montevideo. Uruguay. 18–22.

CATHALAC. (Water Center for the Humid Tropics of Latin America and the Caribbean) (1999). Vision on Water, Life and the Environment for the 21st Century: Regional Visions: Central America and the Caribbean. In: World Water Vision, Making Water Everybody´s Business. Bozena Blix and Subhrendu Gangopadhyay Webmasters.

Consejo Federal de Inversiones-EASNE. (1972). Contribución al estudio geohidrológico de la región NE en la Prov. de Buenos Aires. Serie Técnica 24. Buenos Aires.

Costa, W.D. (2001). Agua subterranea no Brasil: um enfoque na regiao semi-arida. – Taller Aguas Subterráneas y Gestión Integrada de Recursos Hídricos. Global Water Partnership – INA. Mendoza. Argentina. 26–27 July 2001. 31 pp.

Costa, W.D. (2002). A sobre-explotação dos aqüíferos costeiros em Recife-PE. Groundwater And Human Development. Bocanegra, E – Martínez, D – Massone, H (eds.). 792–804.

Garfias, J.; Franco, R. and Llanos, H. (2002). Análisis de la vulnerabilidad intrínseca y su adecuación mediante un modelo de flujo con trazado de partículas para evaluar la vulnerabilidad del acuífero del Curso Alto del río Lerma, Estado de México. Revista Latino-Americana de Hidrogeología, No 2:115–126.

Hernández, M. (1978). Reconocimiento hidrodinámico e hidroquímico de la interfase agua dulce-agua salada en las aguas subterráneas del estuario del Plata (Partidos de Quilmes y Berazategui, Buenos Aires). VII Congreso Geológico Argentino, Neuquén. Actas II: 273–285.

Hernández, M. (1990). La crisis hídrica y las expectativas de los países sudamericanos. Nuevos recursos hídricos. Latinoamérica, Medio Ambiente y Desarrollo. IEMA. Colec. Encuentros 8: 133–136. Buenos Aires.

Hirata, R.; Ferrari, L.C.; Ferrari, L.M.R. and Pede, M. (2002). La explotación de las aguas subterráneas en la cuenca hidrográgica del Alto Tiete (Sao Paulo, Brasil): crónica de una crisis anunciada. Boletín Geológico y Minero. 113 (3): 273–282.

Huizar Álvarez, R. and Carrillo Rivera, J. (1998). Panorama general del agua subterránea en la Ciudad de México. Memoria del Simposio Internacional de Aguas Subterráneas. México. 61–68.

Medina, M.R.; Saavedra, R.M.; Montaño M.M. and Gurrola, J.C. (2002). Vulnerabilidad a la intrusión marina de acuíferos costeros en el Pacífico Norte Mexicano; un caso, el acuífero Costa de Hermosillo, Sonora, México. Revista Latino-Americana de Hidrogeología. No 2: 31–51.

NAC. (North American Consultations) (1999). Vision on Water, Life and the Environment For the 21st Century. In: World Water Vision, Making Water Everybody's Business. Bozena Blix and Subhrendu Gangopadhyay Webmasters.

Quintana, J. and Tovar, J. (2002). Evaluación del acuífero de Lima (Perú) y medidas correctoras para contrarrestar la sobreexplotación. Boletín Geológico y Minero. 113 (3): 303–312.

SAMTAC. South American Technical Advisory Committee (2000). South America Regional Vision and Framework for Action. In: World Water Vision, Making Water Everybody's Business. Bozena Blix and Subhrendu Gangopadhyay Webmasters.

Santa Cruz, J. and Silva Busso, A. (2002). Evolución hidrodinámica del agua subterránea en el conurbano de Buenos Aires, Argentina. Boletín Geológico y Minero. 113 (3): 259–272.

Santa Cruz, Silva Busso, A.; Amato, S.; Guarino, M.; Villegas, D. and Cernads, M. (1996). Explotación y deterioro del acuífero Puelches en la región metropolitana de la República Argentina. Agua em revista, CPRM, Brasil. Año 2, Vol. 1: 48–57.

Intensive use of aquifers by mining activity in Northern Chile

Orlando Acosta
Hydrogeology Group. Technical University of Catalonia, Barcelona, Spain
hidrogeologos@terra.es

ABSTRACT

In the Andean highland of the Tarapacá Desert (Iquique Province) most of the Chilean mining richness can be found, especially copper. In this zone, despite semi-arid conditions, there are abundant groundwater resources in several and vast aquifers of sedimentary and volcanic type. However these aquifers are located in closed basins containing a wetland area that acts like a final sink in the groundwater circulation system. Many of these wetlands support ecosystems that house conspicuous endemic species; and also, they are crucial sites for the resting and nesting of some migratory birds. This condition is recognized and protected by the Chilean environmental laws. On the other hand the current legal context in force, considers water rights grants from the government to private ones, as a domain character and not as a concession. This and the previous situation, shape circumstances in which the environmental government agencies must act in a holistic way in the technical and legal evaluation of mayor mining projects that would like to make an intensive use of groundwater. In this context, the Environmental Impact Evaluation System is the instance that regulates the cooperation between the public and the private sector. Sometimes, it is difficult to get off from these evaluation processes the political pressure and the normal social-economic expectations caused by projects that generate employment and economic dynamism in the area.

Keywords: *aquifers, exploitation, mining, salares, Chile.*

1 ANTECEDENTS

Mining is one of the richest sectors of the Chilean economy. 43% of export embarkation from Chile correspond to mining products, of which nearly 80% is copper. The great copper deposits of the country represent 35% of the worldwide reserves, allowing Chile to occupy the leading position of extraction in the world with 30% of world production (data 1998, source: Mining Ministry of Chile). The Great Mining of copper is developed principally in the north of the country, in the vast terrains of Tarapacá and Atacama deserts. In this area apart from red metal there are also other important metallic mining resources, like molybdenum, nitrate, iron, lithium and sulphur exploited in mining of large and average scales.

Since the colonial age, the extreme climatic and orographic conditions, characteristic of the mountainous areas in Los Andes, have conditioned the mining activity developed there. The designated Great North, which includes the Tarapacá desert and the Atacama Desert, corresponds to an uninterrupted 1,500 km long arid zone, bordered in the west with the Pacific Ocean and in the east with the massif of Los Andes mountain range (Figure 1). This region corresponds to one of the driest deserts on the earth, with average precipitation that range from 0.6 mm/yr in the coastal area (hyper arid zone), to 250 mm/yr at an altitude of 5,000 m in the middle of Los Andes (semi arid zone).

In this zone, the superficial waters are scarce and are essentially developed as an internal

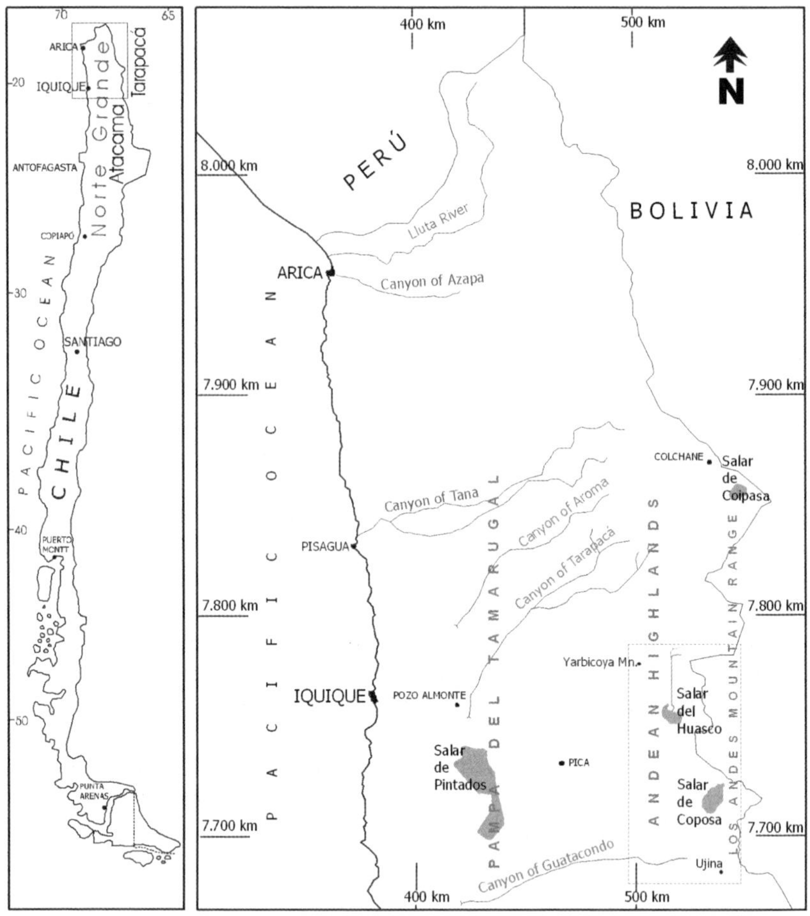

Comparative land extensions: Chile - 760,000 km^2; Tarapacá - 59,000 km^2; Catalonia - 32,000 km^2

Figure 1. General location map of the mining activity zone in the Andean highland of Tarapaca Desert.

drainage in closed basins where finally infiltrate or recharge a wetland. Nevertheless, numerous and extensive hydrogeological systems exist, which store abundant groundwater resources in sedimentary aquifers and fractured volcanic rocks (Galli and Dingman, 1962; Hargis and Montgomery, 1981; Díaz, 1999; Acosta, 2003).

Chilean mining has historically been a strong consumer of water resources, preferably for concentration processes or mineral refinement (Díaz, 1999), with an even more sensitive demand in the north of the country, due to a lack of renewable hydrological resources. Recently it has become more and more clear that the technical and economical feasibility analysis of the large mining projects, should contemplate hydrological studies and the evaluation of environmental impact, which verifies not only the existence of sufficient availability of water during the lifespan of the project, but furthermore the sustainability of the extractions in environmental terms, including a meticulous evaluation of the potential impacts and a budget of eventual mitigation measures. Many times the antecedents that these studies contributed have been determined in the materialization and in the design or engineering of the projects.

During the 1990s in the Andean highland of the Tarapaca Desert (Iquique Province), important private copper mining projects, which make intensive use of groundwater resources in various highland basins, were initiated. In some of these projects the demand for water resources is higher than the usual because the water is use also as a vehicle of mineral transport. Mineral combined with water and aids form a fluid (pulp), which descends through a 150 km long mining duct from the labour area, at an altitude of 5,000 m.a.s.l., to the embarkation port in the coastal area.

2 PECULIARITIES OF THE GROUNDWATER EXPLOITATION IN THE ANDEAN HIGHLAND CLOSED BASINS OF IQUIQUE BY PRIVATE COPPER MINING

The copper mining operations that unfold in the Iquique province are situated in the south part of the Andean highland. This zone has an average altitude of 4,000 m.a.s.l. and a minimum temperature of −20°C, factors that for many years have limited human activity and presence in the area. Currently with exception to the mining settlements, this zone is practically unpopulated. Only a few families are dedicated to cattle ranching (llamas and alpacas) with an extensive shepherds system of extremely low ecological impact.

In spite of the peripheral and wildness of the area, the central administration early understood the importance of the water resources stored in the highland, as shown the Brüggen (1918), Galli and Dingman (1962) and Sayes (1978) studies, some of a few, and the rich statistics of hydrometeorological data produced by means of a control network, which in 1969 already controlled the principle points that are currently measured (DGA, 1999).

In the Andean highland these water resources are generally situated in topographically closed basins, which have an internal drainage network, with a wetland as a sink in the circulation system of groundwater. In these wet areas permanent and temporary lagoons are generated, in which an accumulation of salts by evaporation occurs. When this salt is exposed on the terrain surface form *salares*, the name given to these wet areas. Many of these salares sustain ecosystems that harbour conspicuous endemic species, like flamencos, the Andean goose or *trepadores*. These are also crucial locations for the resting and nesting habits of some migratory birds of the northern hemisphere.

Towards the end of the 1970s and in the beginning of the 1980s the realization of firsts hydrogeological studies began, these were specifically orientated to evaluate the availability of water resources for particular mining projects. As a result of this during the next two decades various mining companies presented water use rights petitions for the exploitation of groundwater, for pumping rates that in some cases corresponded to the estimated value of recharge (maximum flow rate susceptible to grant as the exploitation right according to the standards of this period).

This incident confronted the public and private agents of water resource to various new situations: To the administration, it imposed the necessity to consider the availability of resources, not only at the level of water work but also within the limits of aquifer unit and the urgency to define a strategy and a sustainable exploitation rate of groundwater, attending to the delicate ecosystems which they support; to the users, the requirement is to evaluate with more precision the real potentiality of the aquifers and the possible environmental impacts of an intensive use of them, and moreover, to consider in expenses of the project, the eventual measures of mitigation; and to both of them, the duty to establish good relations for the management of resources in the single user basins that were established.

Alongside this it started to harbour a feeling of concern in society, which saw how a significant amount of the groundwater resources of the province were involved to private mining companies. The positive vision that they have from theses projects as generators of job and economic dynamism, is opposed in the social conscience with the normal anxiety of a community that is aware of the restricted availability of water in the region, being less than 1,000 m^3/hab/yr, a threshold considered internationally as restrictive for the economic development of the countries (DGA, 1999).

The author estimates that the use of water, for consume purposes, in the Iquique Province corresponds to: 65% in mining, 25% in domestic use and 10% in agriculture. This situation demonstrates the relative importance of the mining sector as a consumer of water resources, especially if we compare this with the national panorama: according to the Chilean Water Agency (DGA, 1999), the mining and industrial uses represent only 18% of the demand in the country, more than the drinking supply of water at 4.4%, however much more lower agricultural use which is 84%.

The mining companies of the province, recognizing the scarcity and high costs of the water resources extraction have made important efforts to minimize water loss. Currently private copper mining requires between 0.4 m^3 and 0.8 m^3 of fresh water for each dry ton of material in the processing plant. This index is extremely good and will hopefully improve being even more adjusted when operations are expanded (Grilli, 2002).

3 WATER LEGISLATION IN CHILE AND PUBLIC WATER MANAGEMENT

If the first legal dispositions referring to the use of water in Independent Chile do date back to the year 1819, it was not until the first of January 1857 when to come into effect the Civil Code, that important orders in water subjects were established. It could be emphasized that in this early code the waters were consecrated as national goods of public use, over which the appropriate authority could grant water use concessions.

The obvious necessity to compile all the legislations relevant to water into a single legal body was finally materialized in the enactment of the first water code in the year 1951, and for the first time the term *water use rights* was implicated. In 1967 were introduced important modifications that subsequently emerged in the 1969 code. This second regulatory body reinforced the concept of water as a public good; declaring of public use all previously grants as private property and thus, susceptible of expropriation; and established the water use right only as a real administrative right.

From 1979, a process that changed the legal regime of water was initiated in Chile. The political constitution of 1980 and some acts by virtue of anterior laws, arranged that private users of the water use rights would be officially given the corresponding property.

This legal transformation was finally materialized on the 29th of October 1981, with the publication in the Official Journal of the Order Law N01122 that contains the current water code. This new normative body established that water is a national good of public use and consequently is not susceptible to private domain; however, individuals were granted with use rights of their own with no extinct clause and no expiry date.

In conformance with this code, the Chilean Water Agency (DGA) should constitute water use rights when it was legally proper and resource availability exists. In the groundwater field, leaving aside some exceptions, the administration has circumscribed this availability only to a renewable resource (the recharge) reserving stored water as a strategic good.

This legal framework configures demanding circumstances in which the Administration should operate using a holistic vision in the technical and legal evaluation of mayor mining projects that assume to make intense use of aquifers which nourish wetlands of ecological importance. Previous situation is especially true taking into account the fact that some of these exploitations would be carrying out exercising the entitlement and ownership of the legally constituted water use rights.

4 ENVIRONMENTAL LEGISLATION

The biological importance and the delicate hydrological balance of these Andean ecosystems started being recognized and protected during the middle of the 1980s, due to an incipient environmental legislation, then dispersed in various legal bodies.

In 1984, the Chilean Government ratified the Ramsar Convention about wetlands of international importance being the habitat of aquatic birds, and in 1996 the Salar de Surire, in the Parinacota Province and the Salar del Huasco, in the Iquique Province, were declared Ramsar Sites.

In 1992, the water code (1981) was perfected by means of establishing the prohibition of new groundwater exploitation in areas that nourish *vegas* (water meadows) and those named *bofedales* (high Andes freatofites mix) in the regions of Tarapacá and Antofagasta, recognizing in this way the importance of the mentioned ecosystems.

Later on the DGA defined the limits of these zones exactly and specified by resolution (N0186/96) that this prohibition also extended to the vegas and bofedales situated on private property. Many of these protected vegas and bofedales areas can be found surrounding the salares and others are linked to springs or natural permanent streams.

In 1994 it was enacted the Environmental General Bases Law, that within its dispositions established the creation of the Environmental Impact Evaluation System (SEIA), an instance that regulates all the susceptible activities, which cause environmental damage. Following this, in 1997 it was published the respective order that highlights and deepens in nature of projects or activities susceptible to causing an environmental impact, which should be submitted to the SEIA. In the text are mentioned activities of drainage or desiccation of vegas and bofedales, of all given dimensions, situated in the regions of Tarapacá and Antofagasta. Moreover, it also considers drainage or desiccation activities of whatever natural water bodies like those of lakes, lagoons, reservoirs, marshes, muskeg, water meadows, wetlands or bofedales, whose affected surface is equal or superior to 10 ha, addressing the regions mentioned earlier.

5 CASES OF INTENSIVE USE OF GROUNDWATER IN SALARES

5.1 *Salar de Coposa basin*

Towards the end of 1998, the private Collahuasi Project commenced in the Iquique Province, this was designed to produce close to 400,000 tons per year of fine copper, of which about 50,000 are contained in cathodes and the rest in concentrate. The project presents three principle areas of influence: The mountain range, the port area and the mining duct area. In the mountain range are the three copper deposits of the project and the installations for its development.

The water required for the project is extracted from 8 wells situated in the Salar de Coposa, after approving the corresponding study of environmental impact, sanctioned in the SEIA framework, which does not contradict the stipulated in the DGA resolutions which granted the water use rights. This basin has a surface of 1,116 km^2 of which 85 km^2 correspond to the salar surface (Figure 2). The salar is made up of a chloride centre surrounded by salt and sandy silt. Its average altitude is 3,730 m.a.s.l. The principal lagoon has a very changeable extension, which oscillates together other small lagoons, around 7 km^2. The average precipitation over the basin it estimated close to 150 mm/yr with an average temperature of 5°C (Risacher *et al.*, 1999).

The total average consumption of water from the project has been evaluated in 650 L/s, and even though a maximum pumping capacity of 1,400 L/s was installed in the Salar de Coposa basin, a flow rate of 1,050 L/s was contemplated as a maximum extraction import for the temporary starting periods and resetting of operations (CMDIC SCM and GP Consultores, 1998).

In the environmental evaluation phase of extraction impact concerning the salar, the

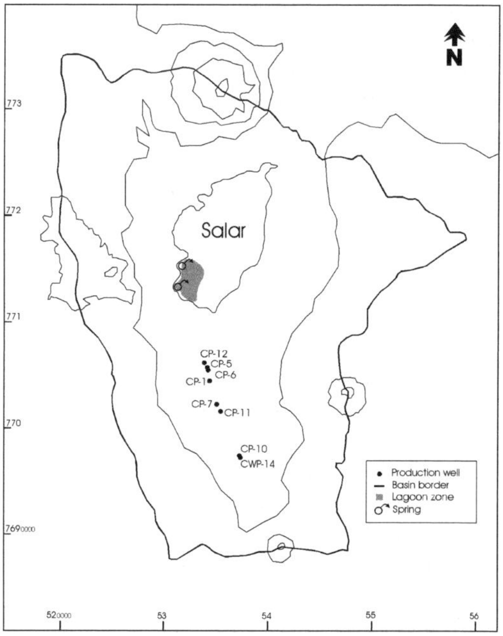

Figure 2. Location map of the Salar de Coposa basin and groundwater extraction wells.

numerical models of simulation employees delivered estimated decreases of between 0.4 m and 0.6 m in the piezometric surface of the superior aquifer in the wetland area, after 25 years of continual exploitation with a pumping rate of 867 L/s (CMDC SCM and GP Consultores, 1998). The latter is the maximum extractive rate authorized by virtue of the water use rights those of which the company owns. However in the setting up act of such rights, the administration established that its practice should give credit to the Environmental General Bases Law and they would be effectuated in a way of maintaining the natural flow rate of the principal spring, the water levels of the major lagoon and the phreatic levels of the sector.

At present, the Environmental Regional Commission and the mining company itself are carrying out the phase of impacts monitoring and control of the aquifer exploitation with respect to the declines in the piezometric levels and the chemical quality of water. The extended transient period that follows the changes in balance terms (whose duration depends on the size of the aquifer and its permeability and coefficient storage), is a crucial phase for a detailed evaluation of the effects of exploitation, and in adopting eventual means of correction, since in general the hydrogeological information that it generates, did not exist beforehand.

5.2 *Salar del Huasco basin*

The Salar del Huasco basin is situated 90 km to the northwest of the Salar de Coposa and it has a surface of 1,500 km^2, within which there are 6,000 ha that correspond to the Salar del Huasco wetland, declared a Ramsar Site (Figure 3). The salar has a 51 km^2 surface consists of salty silt and salt crust. Its altitude is close to 3,780 m.a.s.l. The total extent of the principal lagoon (permanent lagoon) and others temporary lagoons, oscillates around 5 km^2. It is estimated that the average precipitation on the basin is close to 160 mm/yr, whilst the average annual temperature in the longitudinal valley zone is 4.5°C (Acosta, 2003).

Currently, the Salar del Huasco basin is being studied with a view to the exploitation of its groundwater resources, whose water use rights have recently been solicited to the DGA. The planned expansion of the Collahuasi Project requires additional water resources to those available in the Salar de Coposa. It is intended to extract a flow rate close to 900 L/s from 14 production wells placed in various sectors of the basin (Figure 3). The Salar del Huasco aquifer is a complex hydrodynamic system formed by two

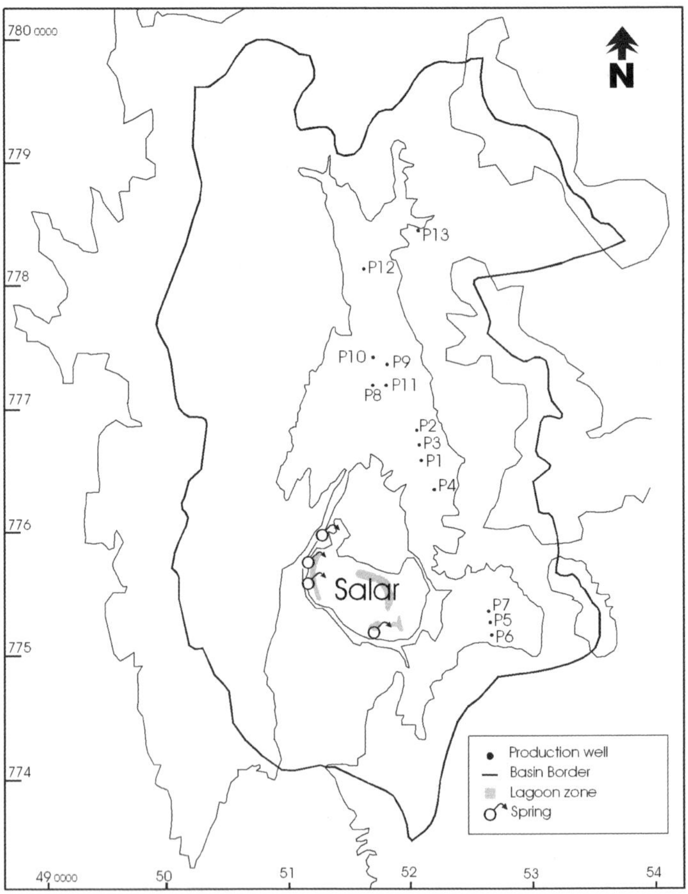

Figure 3. Location map of the Salar del Huasco basin and groundwater extraction wells.

sedimentary hydrogeological units (Quaternaries), which interact hydraulic, chemistry and thermally with a third one, in this case, volcanic (Tertiary-Plio Quaternary), which defined a fractured medium of acceptable transmissive characteristics (Acosta, 2003). In Figure 4 is shown a schematic figure that represent the functioning of the Salar del Huasco aquifer system, which also results quite paradigmatic in respect to other basins in the sector.

The analysis available in some studies of regional character (DGA, 1987; JICA-DGA, 1995), allow the evaluation of a total recharge to the system that varies between 700 L/s and 900 L/s. Nevertheless, recent studies undertaken by the author about the infiltration properties of the superficial soils of the Salar del Huasco, demonstrate that the climatic and edaphologic conditions of the High Andean Steppe of northern Chile, plays an important modulator role about the direct recharge of rain water, creating a water balance scene in the edaphic soil favorable to the infiltration and distinct to the usual one in the majority of the arid zones studied in the world. The evaluation has permitted to estimate a value of effective global recharge between 1,200 L/s and 1,900 L/s (Acosta, 2003).

On the other hand, a regional flow model has given a prime evaluation of the impacts that a 25 years long extraction of 900 L/s will produce. It predicts declines in the piezometric surface of 0.2 m in the principal lagoon zone, 0.3 m in the centre of the salar and also 1.6 m on the east border of the salar. With respect to the discharge of springs it hopes for approximately a 5% reduction in the flow rate. Moreover, it will produce a 10% decrease in the evaporation from the salar and an induced recharge from the Collacagua River to the superior aquifer, due to the greater hydraulic gradients. At present a flow model updating is being developed including new and valuable hydrogeological information (Acosta, 2003).

6 FINAL REMARKS

In the Andean highlands of northern Chile numerous and extensive hydrogeological systems exist, which store abundant resources of groundwater in sedimentary aquifers and volcanic fractured mediums. These aquifers are the unique source of water that allows the exploitation of the vast mining deposits that exist there, in a region that does not present many other comparative advantages for the development of economic activities. Nevertheless, the exploitation of these groundwater resources has to be compatible to the preservation of the salares,

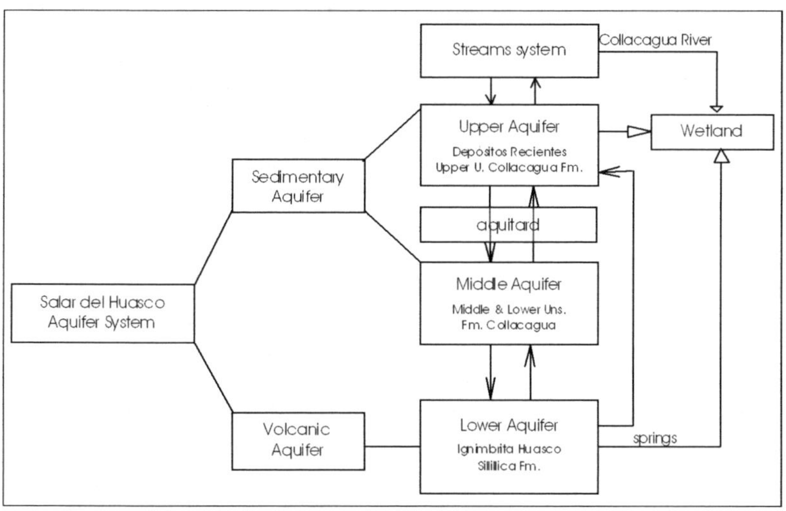

Figure 4. Schematic figure showing how the Salar del Huasco aquifer system works (Acosta, 2003).

acknowledging them as important virgin ecosystems which are redoubts of biodiversity that must be preserved.

The fulfillment of both objectives is possible with a reduced ecological impact and great private and social benefits, as long as the subject is tackled with the seriousness that it should be. Frequently certain irresponsible postures adopted by environmentalists or some politicians, who tend to demonize any activity that can affect the environment, slowing up the environmental evaluation process and attaching mistaken concepts in the conscience of citizens, who are becoming more and more sensitive (but not necessarily more informed) in these areas and whom many times public agencies fail in communicate with them.

The overexploitation of aquifers that are frequently associated to something ethically bad, do not necessarily have to be this way for a certain time, but a period in the evolution towards sustainable development (Custodio, 2002).

To decide what grade of aquifer overexploitation and environmental impact is permissible, a detailed and updated consideration of the effects of the extractions and the correction means that will eventually be adopted is required. To take this decision the rules or general standards and the consideration of a few indirect observations are not sufficient alone.

Observations of control are necessary, a good understanding of the aquifer and behaviour calculations or modeling and all of this in a framework of several objectives and politics established by a management institution, with the implication of those that have an interest in groundwater taking into account the environmental and social conditions (Custodio, 2002).

ACKNOWLEDGEMENTS

Many of the ideas expressed in this paper are the result of the author's collaboration and interaction with several colleagues at the Chilean Water Agency, at the INVEREX Company, at the Geology Group of the Doña Ines de Collahuasi Mining Company (CMDIC), and also at the Hydrogeology Group of the Technical University of Catalonia, Barcelona. Some data and ideas have been taken from the Study of Hydrogeologic Characterization of the Salar del Huasco Basin that INVEREX and the author developed for the CMDIC. Nevertheless, the opinions expressed in this paper are the exclusive responsibility of the author. Useful comments have been received from Dra. Laura Vitoria (University of Barcelona), whose critical revision helped to improve the final written version, as well as reviews by Nikki Wilkinson.

REFERENCES

Acosta, O. (2003). Impactos de las extracciones de agua subterránea en el Salar del Huasco (norte de Chile): caracterización hidrogeológica, evaluación de la recarga y aplicación de un modelo numérico. *Tesis de Máster. Universitat Politècnica de Catalunya*, Barcelona, Spain (in progress).

Brüggen, J. (1918). Informe sobre el agua subterránea en la rejión de Pica. *Sociedad Nacional de Minería. Santiago.*

CMDIC SCM and GP Consultores (1998). Modelación Tridimensional del Flujo de Aguas Subterráneas, Cuenca del Salar de Coposa, I Región, Chile.

Custodio, E. (2002). Aquifer overexploitation: what does it mean?. *Hydrogeology Journal*, vol. 10 (2), pp 254–277.

DGA (1987). Balance Hídrico de Chile. *Dirección General de Aguas* - Ministerio de Obras Públicas, República de Chile.

DGA (1999). Dirección General de Aguas, *1969–1999, 30 Años, Memoria.* Ministerio de Obras Públicas, República de Chile. Santiago de Chile.

Díaz, G. (1999). Vida Bajo el Desierto. *Revista Vertiente*, Órgano Oficial de Difusión del Capítulo Chileno de la Asociación Latinoamericana de Hidrología Subterránea para el Desarrollo. Año 4, N04, july 1999. pp 66–71.

Galli, O.C. and Dingman, J.R. (1962). Cuadrángulos de Pica, Alca, Matilla y Chacarilla – con un estudio sobre los recursos de agua subterránea. Provincia de Tarapacá, Scale 1:50,000, *Instituto de Investigaciones Geológicas de Chile.*

Grilli, A. (2002). Disponibilidad de recursos hídricos y explotación de recursos mineros. GP Consultores Ltda., Recursos Hídricos y Medio Ambiente. *www.gpconsultores.cl.*

Hargis and Montgomery Inc. and Geomar Ltda. (1981). Hydrogeological Investigation of the Salar del Huasco Basin. Interim Report, August 4, 1981.

JICA-DGA (1995). The Study on the Development of Water Resources in Northern Chile. Supporting *Report B: Geology and Groundwater.* Pacific Consultants International – Tokyo. Convenio de cooperación, Japan

International Cooperation Agency. Dirección General de Aguas.

Risacher, F.; Alonso, H. and Salazar, C. (1999). Geoquímica de Aguas en Cuencas Cerradas: I, II y III Regiones – Chile, Vol II. Convenio de Cooperación *Dirección General de Aguas*, Universidad Católica del Norte and Institut de Recherche pour le Développment. S.I.T. N051.

Sayes, J. (1978). Cuadrángulos Collacagua y Laguna del Huasco, Provincia de Iquique, I Región: scale 1:100,000. *Instituto de Investigaciones Geológicas de Chile*, 43 pp.

IMPACTS

The impact of intense urban and mine abstraction of water on groundwater resources of the regional Triassic aquifer systems (Southern Poland)

Andrzej Kowalczyk
Faculty of Earth Sciences, University of Silesia, Sosnowiec, Poland
kowalcz@ultra.cto.us.edu.pl

ABSTRACT

The Triassic carbonate formation covering the area of about 4,000 km^2 is the most important and valuable source of potable water for the Upper Silesia urban-industrial region (Southern Poland). It is made up by five major aquifer systems. Since the end of the 19th century it has been intensively drained by Zn-Pb ore mines, numerous well fields and, indirectly, by coal mines. At the end of the 1990s total abstraction of groundwater ranged from about 9 to 10.6 m^3/s (773,000–893,000 m^3/d). The flow regimes of these aquifer systems have changed markedly due to water-level declines of about 50–70 m in well fields and 100–260 m in mining areas. A significant increase in recharge that occurred in these aquifers was caused by: 1) induced leakage from overlying aquifers in confined parts of the aquifer systems; 2) capture of aquifer discharge to streams; 3) induced percolation of surface water from streams; 4) induced lateral inflow from neighbouring aquifer systems; 5) leakage from water mains and sewers. The paper summarizes the effects of intensive drainage of groundwater on flow regimes and groundwater resources of these 5 aquifers systems.

Keywords: *Triassic aquifer, mine abstraction, sources of recharge, changes of water balance, groundwater resources, Poland.*

1 INTRODUCTION

The Upper Silesia urban industrial region (Southern Poland) constitutes one of the most industrialised areas in Europe. It results from a huge concentration of mineral deposits, including hard coal, zinc and lead ore and other raw materials. The population is about 3.9 million inhabitants within the area of *ca.* 6,600 km^2.

Triassic carbonate formation covering the area of about 4,000 km^2 (Figure 1) is the most important and valuable source of potable water for Upper Silesia. It is made up by five aquifer systems. Within the Triassic formation there are four regions with Zn–Pb ore mining activity in the Middle Triassic beds, and within two of them there is the system of two-level exploitation of Zn–Pb ores and hard coal deposits in Carboniferous.

Groundwater is intensively exploited by numerous wells for public supply and for industrial purposes, and by mining drainage as well. The flow regimes of the Triassic aquifer systems have changed markedly due to water-level declines and changes of hydraulic gradients caused by this pumping. Large increases in recharge have occurred in some of the aquifer systems as a result of induced leakage from overlying aquifers or capturing aquifer discharge to streams and to neighbouring aquifers. Induced lateral inflow from neighbouring aquifers and percolation of surface water from streams as well as leakage from water mains and sewers are additional sources of water supplying this pumping.

Computer-based models to simulate the flow in four of the five aquifer systems have been developed (Kowalczyk, 2003). It was simulated the steady-state groundwater flow throughout the full extend of the aquifers. These simulations provided groundwater budgets for both predevelopment and development conditions. The results of modelling allowed identifying the main sources and the changes in recharge and discharge rates and changes in groundwater resources caused by pumping. The models were designed using the MODFLOW computer code (McDonald and Harbaugh, 1988). The paper summarizes the effects of intensive groundwater drainage of flow regimes and groundwater resources in the examined aquifer systems.

2 HYDROGEOLOGICAL SETTING

The investigated Triassic carbonate formation belongs to the Silesian-Cracow Monocline, which discordantly overlies the folded and faulted Paleozoic basement. The Muschelkalk and the Roethian dolomite-limestone complex is the regional aquifer of the porous-fissure-cavernous type (Różkowski, 1990; Motyka, 1998), which is the main groundwater reservoir for the Upper Silesia region. The thickness of the water-bearing formation varies from 20 to 200 m. Its nearly impervious substratum comprises marl, clay, mudstone, siltstone of Early Triassic, Carboniferous and Permian age. The carbonate complex is covered,

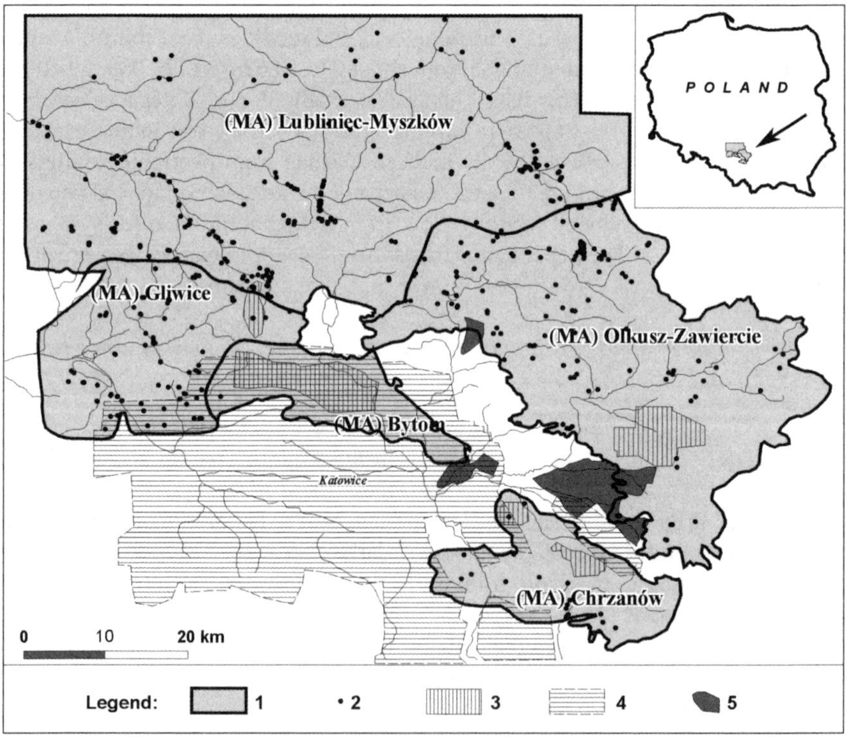

Figure 1. Location of wells and mining areas within the Triassic major aquifers in the Upper Silesia region (Southern Poland).

Legend:
1) Extend and boundary of the Triassic major aquifer (MA) systems; 2) Wells; 3) Zn-Pb ore mining areas; 4) Hard coal mining areas; 5) Sand open pits.

on the area of about 2,700 km^2, by Upper Triassic and locally by Tertiary clays, two formations of very low permeability. In the outcrop areas, the Triassic water-bearing complex is covered by permeable Quaternary deposits. Shallow Quaternary and Upper Jurassic aquifers are present mainly in the northern and eastern parts of the area.

3 GROUNDWATER ABSTRACTION

It is estimated that groundwater from Triassic aquifer is exploited by approximately 500 wells and by mine workings. The well yield varies from 5 to 700 m^3/h, being above 100 m^3/h for 45% of wells. Groundwater abstraction by Zn–Pb ore mines (still working or abandoned) and quarries, and indirectly by hard coal mines, was developed to a large scale at the end of the 19th century, but in some areas it lasted for more than 400 years (Różkowski and Wilk, 1980). Maximum groundwater abstraction from the Triassic formation reaching about 360 Mm^3/yr was noticed in the mid-1980s (Figure 2) and about 50% of total has been drained by mining activities.

A gradual decrease, by about 31%, of groundwater abstraction by wells was observed in the discussed period and recently it has been about 39% of the total withdrawal. It is caused by decreasing demand for water mainly for industrial purposes and for public supply as well. Over the same period groundwater drainage by mines varied from 137 Mm^3 in 1995 to about 200 Mm^3 in 2000 (Figure 2). It was caused by a very dry season in the beginning of the 1990s and by a very high precipitation in the end of the 1990s. At present, the highest water withdrawal is in the Olkusz-Zawiercie aquifer; it amounts to 176 Mm^3/yr, and the lowest abstraction is in Chrzanów aquifer – 26 Mm^3/yr.

4 PIEZOMETRY AND GROUNDWATER FLOW

The piezometric surface configuration illustrated in Figure 3 results mainly from the difference in hydraulic heads among recharge and discharge areas. The pre-development regional pattern of groundwater flow has been much disturbed by intensive groundwater abstraction in large well fields and mining areas, which became the centres of huge man-made drainage. Large scale and long-term abstraction have caused the groundwater to decline by *ca.* 30–70 m, in well fields, and to 100–260 m in mining areas (Figure 3). Extensive and very deep cones of depression were created, largest of which, caused by mining drainage in the Olkusz region, cover an area of about 500 km^2. Piezometric levels have declined to such an extent that the confined aquifer in pre-development conditions is actually unconfined in some areas.

Taking into account geological structures and actual piezometry within the Triassic formation, five major Triassic aquifers (MA) have been determined in the discussed region: Lubliniec-Myszków (MA 327), Gliwice (MA 330), Bytom (MA 329), Chrzanów (MA 452), Olkusz-Zawiercie (MA 454) (Różkowski, 1990) (Figure 3).

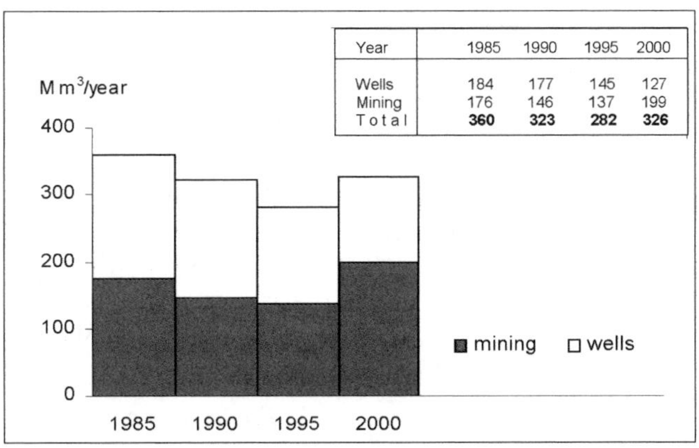

Figure 2. Abstraction of groundwater by wells and mining activities from the Triassic major aquifers in the period from 1985 to 2000.

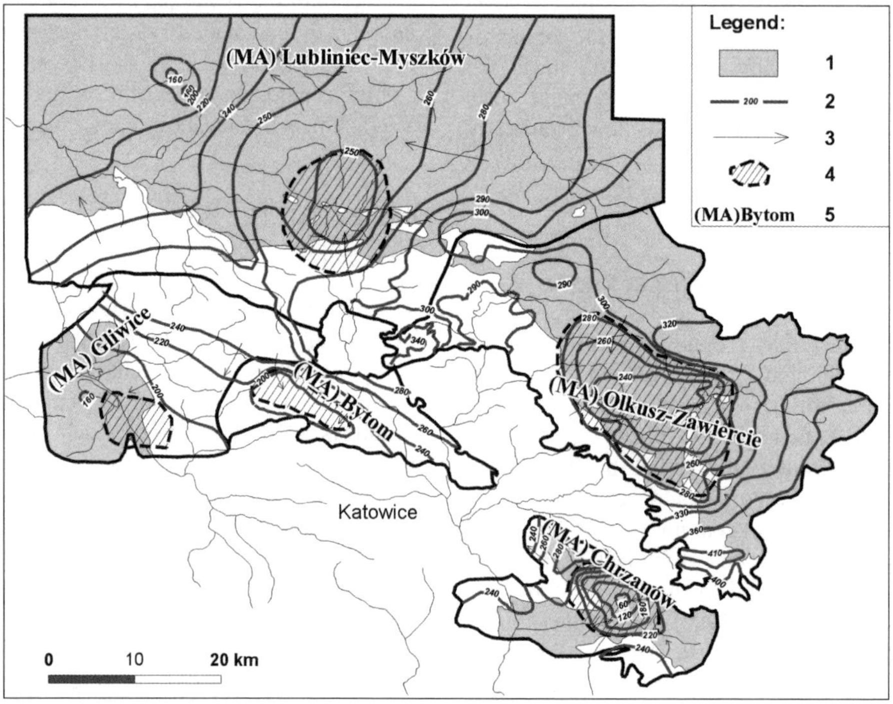

Figure 3. Hydrogeological sketch of the Triassic major aquifer systems.

Legend:

1) Extend of confining units; 2) Groundwater contour lines (in m.a.s.l.); 3) Groundwater flow directions; 4) Areas of maximum groundwater level decline; 5) Name of major aquifer (MA).

A decrease in groundwater withdrawal and simultaneous increase in precipitation in the late 1990s resulted in the rise of piezometric levels, from 3 to 5 m on average, and over 10–20 m in some well fields.

Unfortunately, the pre-development groundwater levels in the examined Triassic formation are generally unknown because of lack of observation records during long-lasting groundwater abstraction period in some areas. They were only reproduced by simulation on the numerical models performed in four out of five aquifer systems mentioned above.

5 RECHARGE–DISCHARGE REGIMES OF TRIASSIC AQUIFER SYSTEMS

Mathematical models for four major aquifer systems were developed using the Modflow computer code to simulate steady-state groundwater flow, to establish water balance and to assess groundwater resources (Kowalczyk, 2003). For the fifth major aquifer – Bytom–, the water budget was established basing on the results of the study of recharge sources which are controlled and equilibrated by mining drainage measured in mining activities (Kropka, 2002a, 2002b). The pre-development conditions were reproduced on the models only for three major aquifers: Lubliniec-Myszków, Gliwice and Chrzanów. Their water budgets are established very roughly and with high uncertainty because of lack of piezometric data for the pre-mining or pre-development conditions.

5.1 *Pre-development conditions*

Groundwater flow regimes examined for the three aquifer systems vary greatly among

Table 1. Pre-development recharge rates and nature of discharge in the major Triassic aquifers.

No	Major aquifer	Areal extend km^2		Precipi-tation	Pre-development recharge rate		Nature of discharge
		of the aquifer	of the confined part	mm/yr	m^3/s	mm/yr	
1	Lubliniec-Myszków	2,023	1,908	826	1.64	25	outflow to adjoining aquifer systems, diffuse upward leakage, to rivers and streams
2	Gliwice	356	91	787	0.94	95	outflow to neighbouring aquifer systems, to streams and springs
3	Chrzanów	273	134	779	0.97	112	outflow to neighbouring aquifer systems, to streams and springs
4	Olkusz-Zawiercie	902	–	810	–	–	to rivers, streams and springs
5	Bytom	180	–	802	–	–	to neighbouring aquifer systems, to rivers, streams and springs

their systems, depending on the topography, the nature and the extent of unconfined and confined areas, and the hydraulic properties of the aquifers and confining units. There is a wide variation between the pre-pumping rates of the recharge and discharge of the Triassic aquifers listed in Table 1. At one extreme there is the Lubliniec-Myszków aquifer, confined almost in the total extent of the area, with much differentiated transmissivity and with relatively small pre-development recharge and discharge rate (29 mm/yr). About 44% of recharge rate occurred as vertical leakage from overlying aquifers. In contrast to this, the pre-development recharge and discharge rates in the Chrzanów and Gliwice aquifers, which are partly confined and have rather high transmissivity, are relatively high (112 and 95 mm/yr respectively).

For two other Triassic aquifers listed in Table 1 the pre-development recharge–discharge rates cannot be assessed. According to Sawicki (2000), the rate of regional pre-development discharge within Olkusz-Zawiercie for multi aquifer system (Triassic and shallow aquifers) is equal to 278 mm/yr. But the share of the discharge from Triassic aquifer cannot be estimated.

5.2 *Changes in the recharge–discharge regimes due to groundwater abstraction*

Groundwater budgets for the three major aquifer systems for the pre-development and recent development periods are shown in Figure 4. The sources of water supplying pumping from five aquifers are summarized in Table 2. These budgets indicate that groundwater abstraction has caused significant changes in sources and rates of recharge and discharge and in total groundwater resources in all examined aquifers.

In order to explain the changes in the recharge rates we assume that the mentioned aquifer systems are approximately in equilibrium, so the losses of water from storage are not taken into account.

Generally, changes in recharge rates for the Triassic aquifers, shown in Figure 4 and in Table 2, are due to transfer of groundwater from shallow aquifers or outcrop areas to deeper parts of those systems. The main transfer mechanisms are the following:

- Induced vertical leakage across confining units; for example in the Lubliniec-Myszków aquifer it has increased by about 125% in relation to the pre-development period.
- Induced lateral inflow from neighbouring aquifer systems; this is a case of smaller aquifers such as Gliwice and Chrzanów.
- Water infiltration from surface water courses; it is confirmed by hydrologic measurements for various small streams and rivers as well. It is an important component of recharge in the Bytom (Kropka, 2002b) and Olkusz-Zawiercie aquifers (Motyka and Różkowski, 2001).

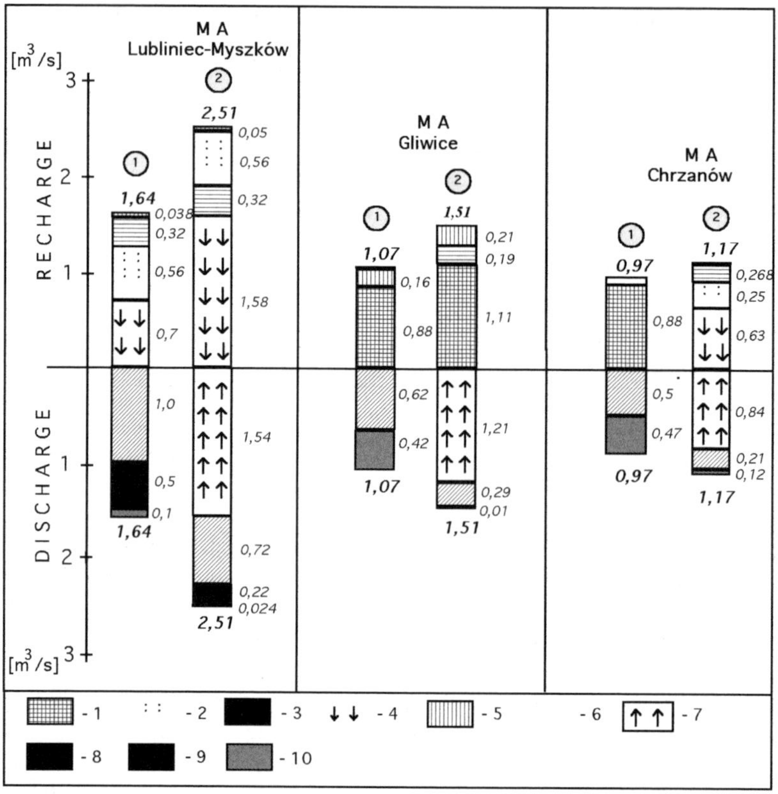

Figure 4. Groundwater budgets before development (1), and after development (2), of major Triassic aquifers.

Legend:

Recharge: 1) Downward leakage from overlying aquifers or percolation from outcrop areas; 2) Infiltration of precipitation from the surface; 3) Lateral inflow; 4) Downward leakage; 5) Losses of water from rivers and streams; 6) Non identified sources of recharge from the surface.

Discharge: 7) Pumping; 8) Lateral outflow; 9) Upward leakage; 10) Discharge to surface water courses.

Table 2. Sources of groundwater supply and total recharge-discharge rates of major Triassic aquifer systems.

No	Major aquifer system	Pumpage		Induced leakage and percolation	Induced lateral inflow	Reduction in natural discharge	Total recharge-discharge rates of major Triassic aquifers		References
		m³/s	mm/yr	m³/s	m³/s	m³/s	m³/s	mm/yr	
1	Lubliniec-Myszków	1.55	24	0.91	0	0.64	2.51	39	Kowalczyk (2003)
2	Gliwice	1.21	88	0.4	0.11	0.7	1.51	110	Kowalczyk (2003)
3	Chrzanów	0.84	97	0.21	0.2	0.63	1.17	136	Kowalczyk (2003)
4	Olkusz-Zawiercie	5.6	196	5.6			6.5	227	Kowalczyk (2003)
5	Bytom	1.41	246	1.41			1.41	248	Kropka (2002a)
Total		10.6	–	–			13.1	–	

Changes in natural discharge caused by groundwater abstraction, as shown in Figure 4, consist in decreasing lateral outflow to neighbouring aquifer systems, reduction in upward leakage from Triassic to Quaternary shallow aquifer (in the Lubliniec-Myszków aquifer system) and reduction in the base flow of streams supplied by the aquifer systems.

Results of the study indicate that the former recharge areas in the Chrzanów, Olkusz-Zawiercie and Bytom aquifers have become discharge areas. And *vice versa*, the former discharge areas have become recharge areas. It is due to the changes in hydraulic gradients resulting from heavy groundwater abstraction by mining activity at a very deep level.

6 SOURCES OF WATER SUPPLYING PUMPING AND GROUNDWATER RESOURCES OF TRIASSIC AQUIFER SYSTEMS

The sources of water supply pumping from discussed aquifer are presented in Table 2. There are two main sources:

- Induced downward leakage across confining units and percolation from outcrop areas.
- Reduction in natural discharge as explained above.

In the Lubliniec-Myszków Triassic aquifer, where the direct outcrop area comprises about 5% of the areal extension, about 60% of the pumping was supplied by the leakage from shallow aquifers. Decrease in natural discharge by diversion of flow into streams and springs is the main source of pumping from four other Triassic aquifers listed in Table 2. The pumping from the Gliwice aquifer is also supplied by increased recharge caused by an expansion in the size of the recharge area. The simulations for Gliwice and Chrzanów aquifers indicate the reduction in natural discharge by about 58 and 75% accordingly. An important source of pumping is the loss of water from surface water. Its share in pumping rates from local drainage centres ranges from 8% in Olkusz mining area (Sawicki, 2000) to 21% in Bytom mining area. (Kropka, 2002b). Additional sources of supplying pumping are anthropogenic water, i.e. the losses of water from water supply network and, to a smaller degree, sewage infiltration in rural and urbanised areas. For example, the leakage from water supply network was roughly estimated as *ca*. 30–35% in the Tarnowskie Góry (Kowalczyk *et al.*, 2000a) and 35–40% in the Bytom urbanised area (Kropka, 2002 b).

The groundwater resources are represented by recharge–discharge rates of the Triassic aquifers, and for the examined multi-aquifer systems, they are listed in Table 2. The pumping is generally close to or higher than pre-development recharge rates. Renewable groundwater resources of these 5 major Triassic aquifers increased remarkably, maximum by about 52% in the Lubliniec-Myszków aquifer. The last one is almost totally confined with a very small pre-development and present recharge rates. Consequently, its groundwater resources constitute only a small fraction (30%) of the total groundwater resources of the multi-aquifer system. Four other aquifers are only partly confined, with important outcrop areas, with higher pre-development recharge rates and they are subject to a large and very deep drainage by mining and wells. Their groundwater resources are supplied mostly by reducing natural discharge to rivers and to neighbouring aquifers and also by other sources, already mentioned.

Additional sources of renewable groundwater resources are significant in all 5 aquifer systems. These are the losses of water from rivers and streams, which vary from 7 to 10% of the total recharge rates among these aquifer systems. The losses from water supply network are less important at regional scale. The only exception is the Bytom aquifer of a small areal extent, in which the share of the water supply leakage and sewage amounted to about 0.095 m^3/s (Kropka, 2002b), i.e. *ca*. 6–7% of the total renewable groundwater resources of this aquifer.

7 FINAL REMARKS RELATED TO GROUNDWATER MANAGEMENT

Groundwater abstraction has greatly altered regional flow regimes of the examined Triassic major aquifer systems, which is due to water-level declines. Increased percolation from outcrop areas, induced leakage from shallow aquifers, capturing natural drainage to spring and streams and losses of water from surface water courses also result from groundwater abstraction. The changes of groundwater budgets and a very important increase in groundwater resources are observed. The reduction in base flow of rivers and streams and disappearance of springs is a negative environmental aspect of the groundwater exploitation.

For these reasons these aquifer systems are sometimes qualified in Poland as overexploited (Różkowski, 1990; Różkowski *et al.*, 1996; Witkowski and Kowalczyk, 2000), without considering other environmental impacts and the economic and social benefits resulting from mining exploitation of natural resources and from industrial activity of the region.

A very important aspect of the groundwater management in the area of the Triassic aquifer system, not discussed in this paper, is the deterioration of water quality. This is observed in the centres of man-made drainage such as active and abandoned but still drained Zn-Pb ore mines and well fields, also in urban areas. The negative changes of groundwater quality in mining areas and in some well fields consist in significant increase of SO_4^{2-} content, TDS (total dissolved solids) and also some heavy metals content (Motyka and Witkowski, 1999; Witkowski and Kowalczyk, 2000; Kowalczyk *et al.*, 2000b). This is caused by overlapping impact of various sources of pollution and by natural geochemical processes related to the transformation of the deepened unsaturated zone in the aquifer. The problem is connected with the commencement of the closure or gradual flooding of Zn–Pb ore mines, which led to a significant groundwater level rise in some areas. In the urban areas the decrease in demand for water supply and increase of groundwater contamination have caused the gradual closure of wells. Abandoning wells and flooding of mine workings will cause further deterioration of groundwater quality in some mining areas (Czop *et al.*, 2001) and well fields. As a consequence of the observed and further possible deterioration of quality of groundwater resources the implementation of the holistic and integrated water management in the examined region is needed.

REFERENCES

Czop, M.; Motyka, J. and Szuwarzyżski, M. (2001). Sulphates in groundwater inflowing to the Zn-Pb Trzebionka mine. (In Polish). In: T. Bocheńska and S. Stasko (eds.), *Współczesne Problemy Hydrogeologii* (X)1: 291–299.

Kowalczyk, A. (2003). Formation of groundwater resources in carbonate aquifers of the Silesia-Cracow Triassic under a human impact. (In Polish). Wydawnictwo. Uniwersytetu, Śląskiego, Katowice: 196.

Kowalczyk, A.; Rubin, H. and Rubin, K. (2000a). Hydrogeological problems of the Tarnowskie Góry urbanized area. In: A.T. Jankowski, U. Myga-Piątek and S. Ostaficzuk (eds.), Environment of the Upper Silesia region. (In Polish). Wydział Nauk o Ziemi. Uniwersytet Śląski: 28–38.

Kowalczyk, A.; Motyka, J. and Szuwarzyński, M. (2000b). Groundwater contamination as a potential result of closing down of the Trzebionka mine, Southern Poland. *Mine Water and Environment. Proceedings of the IMWA International Congress*. Katowice-Ustroń: 299–308.

Kropka, J. (2002a). Problems related to groundwater balance assessment for urban and mining areas. (In Polish). In: *Proceedings of the XIV Conference: Problemy wykorzystywania wód podziemnych w gospodarce komunalnej. Gospodarowanie zasobami wód podziemnych*. Częstochowa, 11–12 April 2002: 86–93.

Kropka, J. (2002b). Quantitative analysis of supply from anthropogenic sources to mine workings of closed zinc-lead ore mines in the Bytom Trough (Southern Poland). Uranium in the Aquatic Environment. *Proceedings of the International Conference: Uranium Mining and Hydrogeology III and the IMWA Symposium*. Freiberg: 1075–1082.

McDonald, M.C. and Harbaugh, A.W. (1988). *MODFLOW, a modular three-dimensional finite difference groundwater flow model*. U.S. Geological Survey, Open-file report 83–875, chapter A1.

Motyka, J. (1998). A conceptual model of hydraulic networks in carbonate rocks, illustrated by examples from Poland. *Hydrogeology Journal* 6: 469–482.

Motyka, J. and Różkowski, A. (2001). Impact of water losses from Biała Przemsza river on the amount of water flowing to mine workings of the Pomorzany Zn–Pb ore mine (Southern Poland). In: *Proceedings of the XI Czech National Congress of Hydrogeologists*. Ostrava: 215–218.

Motyka, J. and Witkowski, A.J. (1999). Sulphates in groundwater of the karst-fractured Triassic aquifers in areas of intensive mining drainage (the Olkusz and the Bytom regions). *Mine, Water and Environment, Vol. I. Proceedings of the IMWA International Congress*. Sevilla, Spain: 189–195.

Różkowski, A. (ed.) (1990). *Fissure-karst groundwater basins of the Cracow-Silesian monocline and problems of their protection*. Monograph CPBP 04.10. Vol. 57 (in Polish). Wydawnictwo SGGW-AR. Warszawa, Poland: 123.

Różkowski, A. and Wilk, Z. (eds.) (1980). Hydrogeological conditions of the zinc lead ore deposits of the Silesia-Cracow region. (In Polish). *Pr. IG, no 310*. Warszawa, Poland: 319.

Różkowski, A.; Kowalczyk, A.; Kropka, J.; Rubin, K. and Witkowski, A. (1996). Hydrogeology of Triassic carbonate complex of Silesia-Cracow Monocline. *Proceedings of the International Conference on Karst-Fractured Aquifers – Vulnerability and Sustainability*. Katowice-Ustroń: 205–221.

Sawicki, J. (2000). *The changes of natural infiltration of precipitation into aquifers as the result of a deep mining groundwater drainage*. (In Polish). Ofic. Wyd. Pol. Wrocław. Wrocław: 1–174.

Witkowski, A.J. and Kowalczyk, A. (2000). Changes of quantity and quality of groundwater in the Triassic carbonate formation in Upper Silesia, Poland. Groundwater: Past Achievements and Future Challenges. *Proceedings of the XXX IAH Congress*. Balkema Publishers. Rotterdam, the Netherlands: 847–852.

Priority rights for nature: a true constraint on aquifer over-exploitation?

Eran Feitelson
The Hebrew University of Jerusalem, Jerusalem, Israel
msfeitel@mscc.huji.ac.il

ABSTRACT

The over-exploitation of aquifers is an outcome of political–economic processes. This paper suggests that a coalition between environmentalists and water managers should be formed to counteract these processes. To this end red lines need to be established. These may be backed by both groups if the threats to the aquifer and to aquatic ecosystems correlate well. For such a coalition to be effective some institutional modifications will usually be needed, so as to allow the green bodies a greater say in water management. One specific option is outlined for the Israeli case. A discussion of this proposal in the Israeli case raises, however, several concerns and limitations. One, it pertains only to a limited set of aquifers, where there are clear well-known relationships between water tables and highly valued groundwater dependent ecosystems (GDEs). Two, there is a need for an alternative source of water for domestic consumption. Three, it is questionable whether this approach is suitable for shared aquifers, especially if the GDEs lie primarily on one side of the border (the side of the lower riparian).

Keywords: *ecosystems, management, politics, water quality, Israel.*

1 INTRODUCTION

The cost of pumping from aquifers is often lower than that of large scale diversions of surface water, and can be done at the local level. In such circumstances the exploitation of groundwater, and particularly shallow aquifers, is very attractive (Shah *et al.*, 2000). However, such exploitation often comes at the expense of nature. Essentially, any consumptive human use of groundwater is likely to result in a decrease in natural outflow. This is true even if a safe yield is maintained. Any change in the natural outflow is likely to have some effect on natural systems.

In many parts of the world, and particularly semi-arid areas, aquifers are often over-exploited. Under such a situation the water tables drop, and are likely to result in cessation of spring flow and desiccation of wetlands, and of the ecosystems that were dependent on these springs or wetlands, or on high water tables. This has occurred in Israel on a wide scale (Ortal, 2001).

The principles of sustainable management of aquifers are well known[1]. Yet, in many places they are not adhered to. The question that needs to be asked, therefore, is why are these principles not followed? There can be two principle answers to this question: ignorance and politics. Ignorance can take two forms. It can either stem from a lack of knowledge, or from the inability to communicate the principles to decision makers. In the case of Israel none of these would apply, as the aquifers have been well studied, and the dangers often raised. Thus, as Dery and Salomon (1997) have argued the cause for

[1] See Haddad *et al.* (2001a) for a brief review.

over-exploitation, at least in the Israeli case, is the politics associated with it.

In the next section the political economy of aquifer over-exploitation is outlined. Essentially, it is suggested that the power of users, particularly farmers, combined with the invisibility of aquifers lead to excessive allotment of property rights over groundwater, at the expense of nature, and that decision makers try to avoid any infringement on such allocations, even in droughts. The implications of this over-allotment for water quality and for nature are usually discussed as an externality imposed by over-pumping. However, such over-pumping is the logical outcome of the political game in democratic societies, and hence should be analyzed in apolitical context.

This paper suggests that the impacts on nature should be brought to the fore in the political arena, and that there is a common interest between environmental bodies and water managers to do so. This insight can serve as a basis for a coalition that may have a better chance to advance sustainable management principles. However, there are inherent tensions between the goals of water managers and of green bodies. These are explored for the Israeli case in the latter part of this paper.

The Israeli case is of general interest, as Israel has been one of the first countries to fully utilize its water resources, it is heavily dependent on groundwater, but has a relatively high capacity to address its water issues. Israel is often upheld as a paragon of wise water management. Indeed, it has implemented many advanced measures, such as active aquifer recharge and conjunctive use, facilitated by well developed monitoring systems and centralized control. The issues raised in the Israeli case may thus be illustrative for a wider set of situations, or clarify issues that may arise in the future in less stressed regions.

2 THE POLITICAL ECONOMY OF GROUNDWATER EXPLOITATION

Aquifers are classical commons. Pumpers have an incentive to pump all they can as long as the marginal value product of the water they pump is greater or equal to the marginal pumping costs. If the marginal pumping costs do not include the social costs (the costs of the externalities imposed by the pumpage) the outcome will lead to over-pumping as the shadow value of water rises. In semi-arid areas where water is scarce and the shadow value is high, an open-access regime, whereby landowners have unfettered access to groundwater, will lead to over-pumping. To avoid this very common scenario, it is necessary to regulate pumpage.

There are two ways to regulate pumpage, both of which require monitoring of pumpage. The first, often recommended by economists, is to impose a levy on the pumpage so as to internalize its externalities. This is likely to encounter stiff opposition from all pumpers, as it is widely perceived as a new form of tax. The second alternative is to issue rights. These can be in the form of either common property rights at the community level, or private property rights issued to individual pumpers.[2] When communities are very stable, and have the ability to exclude others from use of the groundwater, stable community rights can evolve in long societal processes and be maintained (Ostrom, 1990). However, when aquifers are highly stressed and demand is rising rapidly, such systems are often undermined.

Once property rights have been assigned, and demand is growing, the value of such property rights escalates. Hence, all parties have an incentive to obtain additional rights, so as to be able to benefit from the stream of rents they generate. The ensuing efforts are termed in the economic literature as *rent seeking*, and are considered wasteful from a societal perspective.

As a result the institutions established to assign and monitor groundwater use rights are likely to be under continuous and escalating pressures to assign additional rights to various existing or potential pumpers. This will be especially acute during drought periods, when the value of water rises. Thus, the pressures for additional pumpage will rise precisely at the time when, from a sustainable management perspective, the

[2] A discussion of the various forms of property rights is beyond the scope of this paper. For reviews of the definitions of different forms of property rights and their functioning see Bromley (1991) or Berkes (1996).

authorities should limit the pumpage. In democratic societies (as well as in many undemocratic ones) these pressures are likely to be manifest in the political arena.

To understand the outcomes, it is necessary to analyze the different parties to the ensuing game from a power perspective. The pumpers, usually landowners, will apply pressures for additional pumpage. Clearly, the better connected ones and more powerful ones have a greater probability to obtain such rights. Who these pumpers may be will vary according to the local circumstances. However, from a societal perspective it is clear that the outcome is likely to be regressive.

Increasing pumpage, especially in times of drought, may increase the risk for the quality of the groundwater (due to salinization) and is likely to come at the expense of nature.

However, the risk is a function of the extremity of the drought and the extent of over-pumping. Moreover, it is unclear when would the implications for nature and water quality be felt. In contrast, pumpers feel any cut in entitlements immediately and directly. From the decision makers' perspective it is thus a certain loss. Following Kahneman and Tversky's (1979) prospect theory, Bromley (1991) suggests that decision makers will prefer a probabilistic loss to a certain loss. In discussing the question of pumpage from aquifers in drought years in Israel, Dery and Salomon (1997) suggest in a similar vain that decision makers try to avoid the certain political cost of reducing pumpage. Therefore, they prefer to over-pump the aquifer, rather than to cut allocations to farmers, in contrast to the advice of water managers.

Environmental organizations and water management institutions have historically been at odds over the development of water projects (Reisner, 1986, for example). However, in the case of groundwater, there is a potential for a coalition between the two groups. Water managers often oppose excessive pumpage, as they do see the sustainable management of the resource as a primary goal, and are increasingly concerned over the water quality implications of excessive pumpage and the virtual irreversibility of resulting

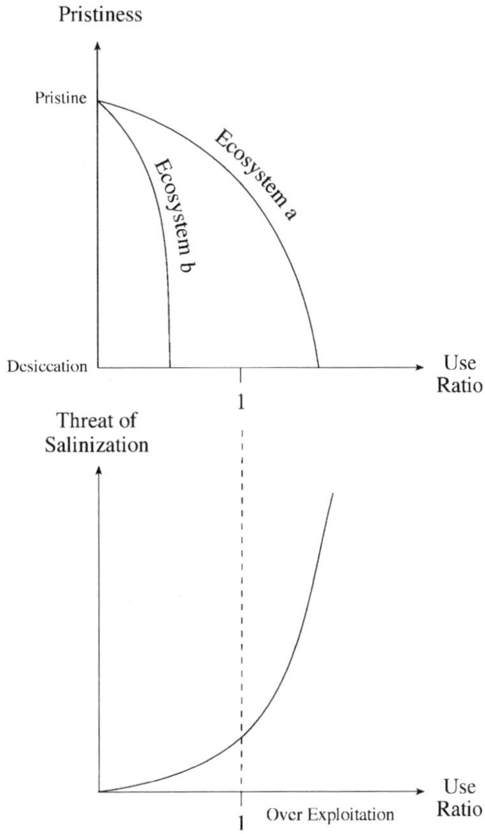

Figure 1. Threats to aquifers and GDEs.

aquifer salinization. High pumpage rates and subsequent drops in water levels threaten many natural systems (Sophocleous, 2003). Hence, environmentalists also have a cause to support actions that will preclude a decline in water levels. As the level of exploitation (or long-term use ratio) rises the interests of the two groups may merge. This is illustrated in Figure 1. When the use ratio is low water managers largely will support further pumpage, if it is seen as economically beneficial. However, as the use ratio rises the threat of salinization rises[3] and hence water managers will seek to implement controls on pumpage. In some cases (such as that of ecosystem *a* in Figure 1) the threat of salinization may be well correlated to the threat of desication (that is both rise rapidly around the same use ratio),

[3] This threat can be expected to rise rapidly once the use ratio exceeds one, as then the long term withdrawals exceed long term replenishment at that point.

and hence the two groups can form a coalition to prevent further pumping. In other cases, however (such as in the case of ecosystem *b* in Figure 1), desiccation may occur even when the threat of salinization is low. Hence, by the time water managers will seek to control further exploitation the natural systems may already be lost. In such situations, where protection of the ecosystem requires that low use ratios be maintained, environmentalists and water managers may not be able to agree on the desired level of exploitation, and thus will not be able to form such a coalition of interests. To examine this hypothesis we turn now to the Israeli case.

3 THE ISRAELI CASE: BACKGROUND

Israel is heavily dependent on groundwater. Two of the three main reservoirs are the Mountain aquifer, a karstic aquifer, and the coastal aquifer, a sandstone aquifer with an annual yield of 250 Mm^3.[4] There are several smaller aquifers, the most important of which for the purpose of this paper is the western Galilee aquifer, a karstic aquifer with an annual yield of approximately 150 Mm^3. All these aquifers are fully utilized.

The Israeli system is managed under the 1959 Water Law, which nationalized all the water. A Water Commissioner was established under this law, who allocates all water in Israel to one of several uses: residential, agricultural, industrial, public or commerce. Nature was not seen as a legitimate water use under the 1959 law. A national water company, Mekorot, is in charge of the national and regional water conveyance. While existing pumpers were grandfathered in the law, pumpage is regulated by licenses issued by the Water Commissioner.[5] All pumpage and usage is monitored. The Water Commissioner was originally placed under the responsibility of the Minister of Agriculture, who traditionally comes from the agricultural sector.

This highly centralized system could be expected to prevent any abuse of the water resources. The Water Commissioner has essentially all the authority and information needed to prevent over-pumping. Yet, since the mid-eighties the aquifers have been over-pumped. Dery and Salomon (1997) argue that this was an outcome of the reluctance of the commissioners to cut allocations to farmers in drought years. However, a closer look at water allocations reveals that during the eighties water allocations to agriculture actually increased, despite the diminishing importance of agriculture in the national economy, suggesting that it was also an outcome of rent-seeking by farmers.

This practice came under severe criticism in a special State Comptroller report issued following repeated droughts in the late eighties and 1990, which resulted in the conspicuous lowering of the water level in Lake Kinneret, as well as in all main aquifers, to the administratively determined *red lines* (State Comptroller, 1990). In this report the State Comptroller faulted the Water Commissioner for taking an excessive risk in managing the water resources and suggested that the office be separated from the Ministry of Agriculture. This report resulted in the firing of the acting water commissioner and the appointment of a new commissioner, who for the first time did not come out of the agricultural sector. Subsequently water rates were raised and water consumption reduced. However, after an exceptionally rainy winter in 1991/92 all water reservoirs were re-filled, and the practice of over-pumping eventually resumed.

The widespread development of water resources in the fifties resulted in the drying of most coastal streams and capture of most springs. As a result almost 97% of the wetlands and aquatic ecosystems in Israel were desiccated by 1980 (Ortal, 2001). Since the establishment of the Nature Reserve Authority in 1963 efforts were made to protect some of the remaining ecosystems. One of the main impediments to these efforts was the lack of water allocations for nature. As nature was not an established water use and thus did not possess any water

[4] The third main reservoir is Lake Kinneret (Sea of Galilee). The Mountain aquifer is composed of three sub-aquifers, the two most important from an Israeli perspective are the north-eastern and western sub-basins. This aquifer is shared with the Palestinians. For more about it see: Harpaz *et al.* (2001).

[5] For further details about the Israeli system and its operation, see: Arlosoroff (2001).

rights the Nature Reserve Authority reached adhoc agreements with the Water Commissioner to allocate water to several nature reserves. The legal standing of these rights is, however, inferior, as it is dependent on the administrative decision of the Water Commissioner, and is not established by law.

4 CRISIS AND OPPORTUNITY: GROWING RECOGNITION OF NATURE BY WATER MANAGERS

The winter of 1998/99 was particularly dry. This winter followed three years of over exploitation of the aquifers. The extreme drought year led to a precipitous decline in the water level of the western mountain Aquifer and Lake Kinneret. While rainfall in subsequent years was around or slightly below the multi year average, recharge was substantially below average. As a result the water levels in Lake Kinneret and all main aquifers (except the coastal aquifer) dropped to the *red lines* or below them.

During the first two years of the crisis the government proved unable to cut the allocations to farmers, despite its legal power to do so, due to arguments over the compensation that should be paid. As a result the crisis worsened. To break out of the impasse, the government decided in March 1999 to embark on a desalination program. In subsequent decisions the amounts to be desalinated were increased. The plan for the first desalination plant was approved in 2000, and the winner of the tender was announced in September 2001. The price of desalinated water at the seashore, as determined by this tender, is US$ 0,54 per m^3. This low price encouraged the government to push for additional desalination, thereby opening new options for long term water planning.

As a result of the crisis a parliamentary inquiry commission was formed. Its report, submitted in June 2002, includes many recommendations including emergency regulations for the interim years, until desalination comes on line, organizational changes, modifications of the law, ratification of a national water plan (several were prepared but shelved in the previous fifteen years), supply enhancement and capacity building. Interestingly, several of its recommendations suggested that the role of nature in the management of water be strengthened.

In partial response to the mounting pressure the Water Commissioner published concurrently, in June 2002, a new draft of a strategic ten year plan for the water sector. This plan, prepared by leading water professionals in the commissioner's office, proposes that desalination be expedited so that pumping can be drastically curtailed, in order to allow water levels in the aquifers to rise and thus re-establish a strategic storage capacity (that was lost due to the excessive pumping in the mid-nineties) (Water Commissioner, 2002). One by-product of this new policy is likely to be the rejuvenation of the desiccated GDEs. The plan also, for the first time, explicitly acknowledges the need to allocate freshwater to nature.

The Water Commissioner also initiated an examination of possible modifications of the existing Water Law. One of the ideas raised in this examination is the establishment of water rights for nature. Hence, it seems that water rights for nature has indeed become part of the sanctioned discourse (the discourse seen as politically viable) within the water community. The new emerging question is, therefore, how will these water rights be determined, and will they indeed help constrain pumpage from the aquifer.

5 CAN NATURE SERVE AS A CONSTRAINT IN ISRAEL?

One of the implications of the crisis has been widespread desiccation of GDEs. In a review of the conditions of the remaining GDEs in Israel, Ortal (2001) found that by mid-2001 most have been desiccated or damaged since the outset of the drought in 1998. This has occurred in all parts of the country, except the Golan Heights.[6]

This situation has prompted the Nature Reserves and National Parks Board (a public

[6] As a result of an exceptionally rainy winter in 2002/03, particularly in the northern part of the country, many of the streams began to flow again. However, it is too early to tell at the time these lines are written (early May 2003) to what extent this will ameliorate the damages of the drought.

body with a policy advisory role) to request the Nature and Parks Authority (NPA) to identify the minimal water needs of nature reserves and national parks, so they can serve as a basis for defining water allocations to nature. In response the NPA commissioned a study to estimate the quantities of water that would be needed for nature protection and landscape conservation in Israel (Shacham *et al.*, 2002).

There are two possible sources of water for nature in Israel. The first is reclaimed wastewater. These are intended mainly to rejuvenate the coastal streams, which have been desiccated in the fifties and subsequently polluted by untreated sewage flows from municipal, agricultural and industrial sources (Bar-Or, 2000). However, reclaimed wastewater does not have the properties of the original freshwater, and thus cannot be expected to re-generate the original ecosystems. Neither can they substitute for freshwater in the remaining GDEs. For this reason freshwater allocations would still be needed for the GDEs. Yet these GDEs have been severely damaged by the lowering of the water tables. The (Nature Reserve and National Park Authority) NRNPA has resorted to digging up ponds to keep the water bodies in nature reserves. But this cannot be seen as a long-term option, and changes the overall view of the reserves, as the water bodies remain in what become essentially man-made ponds.

Shacham *et al.* (2002) estimated the needed amount of freshwater for nature alone (excluding landscape) to be on the order of 600 Mm^3 (excluding the lower Jordan River). Most of this water can later be utilized, and thus need not be considered a loss from the national water planning perspective.[7] But, as the nature reserves receive the water first, the allocations for nature may perhaps be viewed as priority rights.

A further question is how this water should be supplied to the nature reserves. During the drought water was pumped in several places from boreholes to the nature reserves. However, this contradicts much of what nature reserves stand for. Ideally, the water should flow freely to the nature reserves. This will require the maintenance of high water tables. The question is: how high? The ability to form a coalition between environmental bodies and water managers is largely predicated on the answer to this question.

Clearly, if a GDE is to be maintained at a pristine state the possibility to utilize the aquifer on which the GDE is based will be greatly reduced, and perhaps even negated (see Figure 1). Thus, proposals to maintain pristine GDEs are likely to be rejected by those concerned with water supply. In semi-arid states it is generally infeasible to keep major sources intact. Moreover, in Israel, which has been continually habituated for over 10,000 years, and which is densely populated today, all so-called natural areas have in fact been greatly modified by human intervention over millenniums. Hence, there are no real *pristine* GDEs in such an environment. Still, it is possible to seek some balance between free flows to the nature reserves and the exploitation of the aquifers – between environmental bodies and water managers.

A review of the actual GDEs (mapped in Figure 2) shows that hardly any are dependent on the coastal aquifer. In contrast, many such GDEs are based on the Western Galilee aquifer, Lake Kinerret and the North-eastern mountain aquifer.[8] The western mountain aquifer, which is the main sub-basin on which Israel relies in that aquifer, has only two large outflows – the Yarkon and Taninim streams. The relatively saline Taninim springs continue to receive a base flow even at high exploitation levels. That is not true of the Yarkon springs. As a result Shacham *et al.* (2002) found that maintaining a perennial flow in the Yarkon springs has a benefit/cost ratio below one.

If higher water tables are maintained in the western Galilee aquifer, an additional desalination plant will most likely be needed north of Acre (Shacham *et al.*, 2002; Water Commissioner, 2002). While this clearly will impose a cost, it can be compared to the value of water in nature reserves. This valuation study has not been conducted yet. Given the values derived for less attractive open spaces in Israel lately, it is possible that the value of water in nature in these areas will indeed be greater than the cost of additional and earlier desalinization.

[7] They suggest that the net amount needed (that cannot be used later) can be as little as 30 Mm^3.

[8] As the Eastern mountain aquifer has been largely allotted to further Palestinian use it is not discussed in this paper.

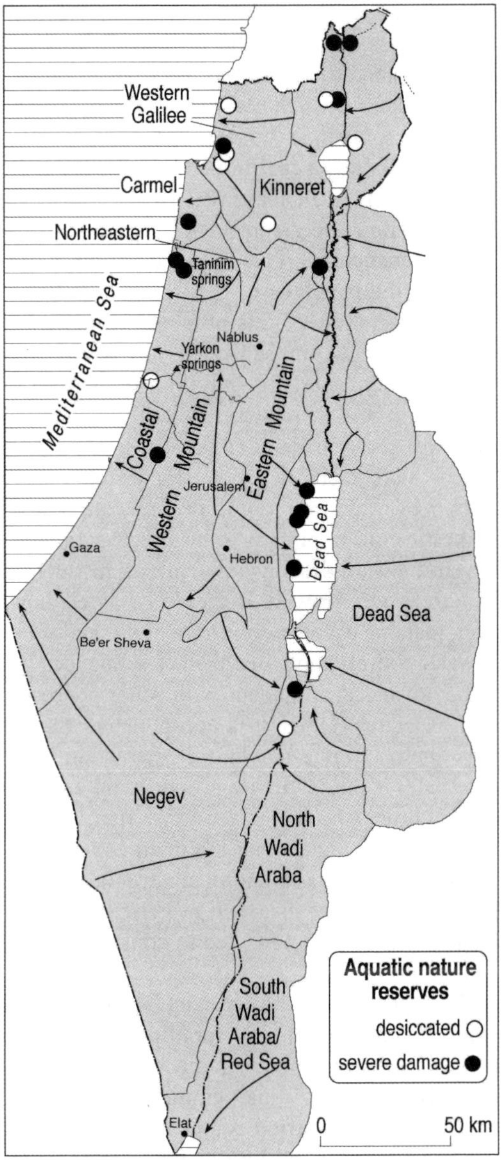

Figure 2. Aquatic nature reserves affected by the drought and groundwater basins.

6 DISCUSSION

The importance of aquatic ecosystems, the adverse implications of over-pumping for GDEs, and hence the need to allocate freshwater to maintain aquatic ecosystems, as well as water-based leisure activities in parks and nature reserves have now been recognized not only by environmentalists, but also by water managers in Israel. Moreover, first steps to advance this new perceived need have been taken. But it is still unclear how the allocations for nature will be defined, by whom and on which basis. Perhaps more worryingly, the priority these allocations will have *vis-à-vis* other users, particularly when pressures escalate during droughts, have not been discussed yet. This last issue needs to be discussed within the institutional framework of the water sector.

One of the main problems in the institutional structure in Israel is the lack of a system of checks and balances. Essentially, the power to determine *red lines* and to alter them is concentrated in the hands of the water commissioner. While the commissioner needs to consult with the Water Board, established under the 1959 Water Law, he needs not accept their advice. Moreover, this board is composed largely of farming interests, as these were seen at the time to be the main interests affected by water policies. Thus, this board only re-enforces the tendency of decision makers to defray any cuts in allotments. For this reason the parliamentary inquiry commission suggested that it be re-structured to provide checks and balances, and that representatives of green bodies be included in it.

This recommendation of the parliamentary inquiry commission can be extended even further. It can be stipulated that if the *red lines* established to assure a base flow to GDEs in nature reserves is breached the water commissioner will need to consult with the Nature Reserves and National Parks Board. This is a 28-member public board, currently chaired by the author, that is composed of representatives of several government ministries, public bodies, local jurisdictions (both urban and rural), and of academics and public people (chosen by the Minister of the Environment). As a public body it is geared toward the consideration of general interests *vis-à-vis* nature protection interests, mainly with regard to terrestrial systems. It can, conceivably, thus consider also the water supply demands *vis-à-vis* the protection of GDES, just as it considers housing or other infrastructure requirements *vis-à-vis* other nature protection considerations.

The brief description of the layout of GDEs shows that if such a checks and balance system is implemented, it will pertain only to a limited part of Israel's water system. Hence, it should not be viewed as a constraint on all water supply activities in Israel. Actually, the two largest aquifers (coastal and western mountain) are likely to be exempt from it.

Still, this suggestion raises several issues. The first is how it should pertain to shared aquifers. All sub-basins of the mountain aquifer are shared with the Palestinians. In recent work a per-capita based allocation of the water in these aquifers has been proposed (Haddad *et al.*, 2001b). If this is indeed accepted it is difficult to see how a constraint on overall use based on GDEs within one side only (Israel) will indeed be accepted as a basis for managing the aquifer.

A second issue that is likely to arise is who should pay for the additional desalinization that will be needed in order to maintain higher water tables so as to protect GDEs. One argument is likely to be that as the purpose is to maintain GDEs the *green* bodies (primarily the NPA) should pay. However, given the limited funding available to such bodies, and particularly the NPA, such a demand is likely to undermine the whole idea. The counter argument is that as water is not *supplied* to the GDEs, but has been there naturally. Hence, there should not be any payment for maintaining what has been. Actually, GDEs should be compensated for the water taken from them. This is clearly a normative question, the discussion of which is beyond the scope of this paper. However, it is important to note the existence of an ethical overtone to this argument, as it will need to be addressed.

A third concern that should be raised is the spatial limitations of this approach. It pertains only to cases where there is a clear, well-known, relationship between water table levels and outflows. In aquifers that no longer support valuable GDEs (such as the coastal aquifer in Israel), or where the relations are largely unknown, it will be either meaningless or impossible to implement this approach.

7 CONCLUSIONS

The over-exploitation of aquifers is often a logical outcome of political-economic processes. To counter-act these processes this paper suggests that a coalition between environmentalists and water managers should be formed. To this end *red lines* need to be established so as to protect GDEs, wherever such *red lines* can be backed by both environmental bodies and water professionals. For such a coalition to be effective some institutional modifications will usually be needed, so as to allow the green bodies a greater say in water management. One specific option is outlined for the Israeli case.

The discussion of this proposal in the Israeli case raises, however, several concerns and limitations. One, it pertains only to a limited set of aquifers, where there are clear well-known relationships between water tables and highly valued GDEs. Two, there is a need for an alternative source of water for domestic consumption, if domestic consumption is to be affected. In Israel the shift toward desalinization allows for such an idea to be put forward. It should be noted, however, that one cheaper option for additional water is water conservation, in which case this idea has to be advanced in tandem with water conservation measures. Three, it is questionable whether this approach is suitable for shared aquifers, especially if the GDEs lie primarily on one side of the border (the side of the lower riparian).

This paper argues that the main factors driving aquifer over-exploitation are grounded in the political economy of groundwater exploitation. Hence, the measures needed to constrain excessive pumpage need also to be discussed within such a framework. In particular, this paper explores the option of cobbling a coalition between environmentalists and water managers for advancing and maintaining *red lines* in aquifers, and identified some of the issues that need to be discussed and analyzed before this idea is further advanced. The answers to these issues will clearly differ by place and over time. Thus, while the proposal made here may not be applicable everywhere, it may be worthwhile to consider it with regard to aquifers in semi-arid countries with relatively high capacities and few trans-boundary problems.

REFERENCES

Arlosoroff, S. (2001). Water resource management in Israel, in: E. Feitelson and M. Haddad (eds) *Management of*

Shared Groundwater Resources: The Israeli-Palestinian Case with an International Perspective, Kluwer, 57–74.

Bar-Or, Y. (2000). Restoration of the rivers in Israel's coastal plain, *Water, Air and Soil Pollution* 123, pp. 311–321.

Berkes, F. (1996). Social systems, ecological systems and property rights, in: S. Hanna, C. Folke and K-G Maler (eds.) *Rights to Nature: Ecological, Economic, Cultural and Political Principles of Institutions for the Environment*, Island Press, 87–110.

Bromley, D. (1991). *Environment and Economy: Property Rights and Public Policy*, Blackwell, Oxford, UK.

Dery, D. and Salomon I. (1997). After me the deluge: uncertainty and water policy in Israel, *Water Resources Development* 13, pp. 93–110

Haddad, M.; Feitelson, E. and Arlosoroff, S. (2001a). The management of shared aquifers, in: E. Feitelson and M. Haddad (eds) *Management of Shared Groundwater Resources: The Israeli-Palestinian Case with an International Perspective*, Kluwer, 3–24.

Haddad, M.; Feitelson, E.; Arlosoroff, S. and Nasseredin, T. (2001b). A proposed agenda for joint Israeli-Palestinian management of shared groundwater, in: E. Feitelson and M. Haddad (eds) *Management of Shared Groundwater Resources: The Israeli-Palestinian Case with an International Perspective*, Kluwer, 475–493.

Harpaz, Y.; Haddad, M. and Arlosoroff, S. (2001). Overview of the mountain aquifer, in: E. Feitelson and M. Haddad (eds) *Management of Shared Groundwater Resources: The Israeli-Palestinian Case with an International Perspective*, Kluwer, 43–56.

Kahneman, D. and Tversky, A. (1979). Prospect theory: an analysis of decision under risk, *Econometrica* 47, pp. 263–291.

Ortal, R. (2001). The scope of damage to aquatic ecosystems in Israel following the "enforced drought policy" in Israel, Science and Conservation Division, Aquatic Ecosystems Branch, Nature Reserves and national Parks Authority (in Hebrew).

Ostrom, E.(1990). *Governing the Commons: The Evolution of Institutions for Collective Action*, Cambridge University Press.

Reisner, M. (1986). *Cadillac Desert: The American West and its Disappearing Water*, Penguin.

Shacham, G.; Rosenthal, G.; Amit, Z.; Gafni, S. and Shacham, G. (2002). *The Right of nature for Water*, Nature Reserves and National Parks Authority (in Hebrew).

Shah, T.; Molden, D.; Sakthivdivel, R. and Seckler, D. (2000). *The Global Situation: Overview of Opportunities and Challenges*, International Water management Innstitute, (available at: www.cgiar.org/iwmi/pubs/WWVisn/GrWater.htm, accessed November 15, 2002).

Sophocleous, M. (2003). Environmental implications of intensive use with special regard to streams and wetlands, in: R. Llamas and E. Custodio (eds.) *Intensive Use of Groundwater: Challenges and Opportunities*, A.A. Balkema, Lisse/Abingdon/Exton/Tokyo, 93–112.

State Comptroller (1990). *A Report on the Management of the Water Market in Israel*, Jerusalem (in Hebrew).

Water Commissioner (2002). *Strategic Plan for the Water Sector 2002–2010*, Final Report, Planning Wing, Water Commissioner, Ministry of National Infrastructure (in Hebrew).

Effects of localised intensive aquifer exploitation on the Doñana wetlands (SW Spain)

Marisol Manzano
Universidad Politécnica de Cartagena, Murcia, Spain
marisol.manzano@upct.es

Emilio Custodio
Instituto Geológico y Minero de España, Madrid, Spain
Universidad Politécnica de Cataluña, Barcelona, Spain
e.custodio@igme.es

Carlos Mediavilla
Instituto Geológico y Minero de España, Sevilla, Spain
oficina.sevilla@igme.es

Carlos Montes
Universidad Autónoma de Madrid, Madrid, Spain
c.montes@uam.es

ABSTRACT

The Doñana region (SW Spain) hosts the Guadalquivir river marshes and two widespread protected natural areas of international relevance. There are diverse wetlands types, most of which are groundwater-dependant. A large number of permanent and temporal phreatic shallow lagoons dominate both in the regional aquifer recharge areas and in the aquifer discharge areas fringing the marshes. Wetlands have very different geomorphic origin and hydrologic regime, and they are greatly responsible for most of the internationally appreciated ecological values of the area. However, because of their close relationship to the aquifer the original hydrology of most wetlands is modified, as it is the aquifer original regime. Intensive groundwater pumping in the aquifer discharge areas nearby the marshes since the end of the 1970s has changed groundwater flow and regime in many areas, and consequently many wetlands have been affected. Accumulated inter-annual lowering of the most exploited deep aquifer piezometric levels has produced a water table drawdown, which results in a decrease of natural discharge through seepage and phreatic evapotranspiration and has modified wetland water regime. Other human activities disturbing the aquifer natural regime and that of the related wetlands are sediment relocation due to deforestation, introduction of foreign plant species (e.g. eucalyptus) with greater annual water consumption than native Mediterranean species, and groundwater pollution by urban, industrial and agricultural wastes.

Keywords: *wetlands, intensive pumping, contamination, drawdown, groundwater discharge, Doñana.*

1 INTRODUCTION

The Doñana natural region and aquifer are placed in the SW Atlantic coast of Spain, partly occupying the ancient estuary of the Guadalquivir river but also extending to the north and west of it (Figure 1). It is a singular and fluctuating space where both geographical

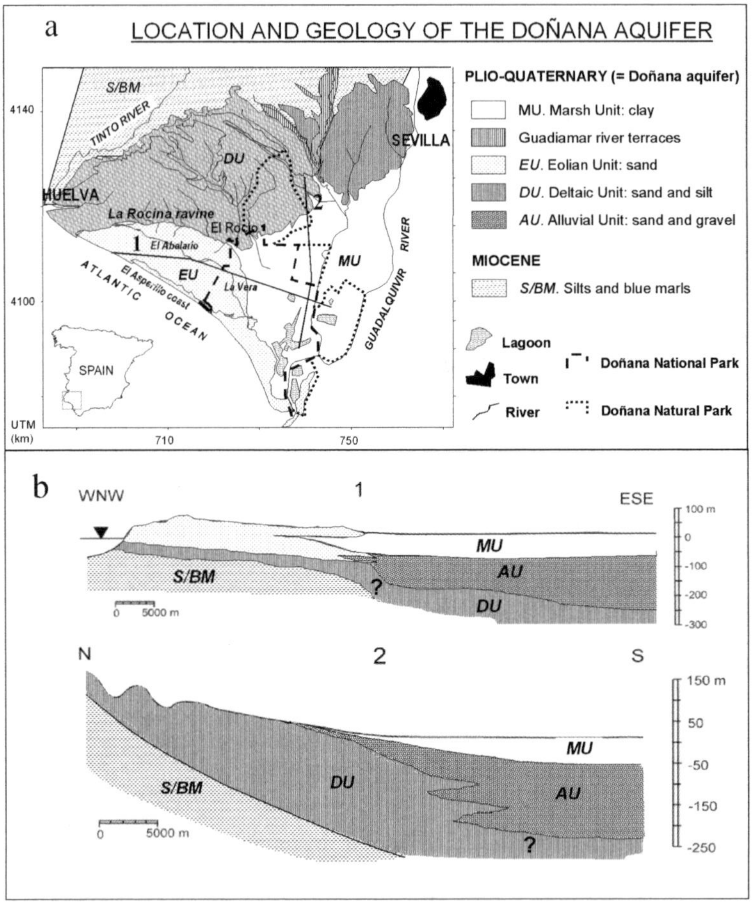

Figure 1. a) Location of the Doñana region, aquifer extent and regional geology. b) Schematic geological cross sections of the Doñana aquifer system. The confined unit AU (Alluvial Unit) does not crop out in a).

and biological factors coincide to produce high value wild fauna and vegetation habitats. Both surface and groundwater play an essential role.

The region known as Doñana, shared by the provinces of Huelva, Sevilla and Cádiz, covers around 2,500 km^2. The closest relevant cities are Sevilla to the NE, Cádiz to the SE, and Huelva to the NW. Agriculture and tourism, the two main economical activities, compete for water with environmental preservation. Part of this area is under the two highest protection figures for natural spaces existing in the Spanish law: the Doñana National Park, managed by the Spanish Ministry of the Environment, and the Doñana Natural Park, managed by the Regional Government of Andalusia. Altogether they cover more than 1,000 km^2. Most of the Parks area is uninhabited, except for people in charge of the protection, the occasional developers of some natural resources, and the visitors.

Within the National Park boundaries human activity is greatly restricted, while across the Natural Park some traditional activities like pinecones collection and old-style charcoal preparation are allowed. Cattle's breeding existed until very recently but now it is being cut down. Outside the Parks agriculture and sun and beach tourism are still the main activities, although officially promoted environmental tourism is also growing up fast. Agriculture concentrates in three areas: a central one surrounding El Rocío village; a second one close to the E of Huelva city, between the Atlantic ocean and the Tinto river, and a third one across the marshes NE boundary. In the two first areas strawberries and citrics are the main cultives, and groundwater is the only source for irrigation. In the third area rice and cotton are the main products; in the large eastern sector irrigation is done with river water but a sector in the west uses groundwater, and intensive groundwater pumping has led to serious environmental problems.

Groundwater is also abstracted to supply the towns and the large touristic resorts of Matalascañas and Mazagón, which also affect wetlands. But the impact is local and lesser than the generated by abstractions to irrigate agriculture and to keep artificially some wetlands. Thus, the urban and tourism effect will not be considered in what follows.

2 ORIGIN AND CHARACTERISTICS OF THE DOÑANA AQUIFER SYSTEM

The Doñana aquifer system consists of detrital Plio-Quaternary sediments lying above a thick (>2,000 m) sequence of Miocene marls. Pliocene silts and sands (Deltaic Unit in Figure 1) and Quaternary sands of the Eolian Unit crop out to the N and W of the marshes respectively, giving way to extensive water table areas. The Eolian Unit disappears to the N and E, while the Deltaic Unit plunges to the SE and become overlapped by the more permeable Plio-Quaternary sands and gravels of the Alluvial Unit (Figure 1), which develops under the clay layers of the Marsh Unit. The Quaternary lithology and thickness are quite variable under the marshes: Pleistocene and Holocene sand and gravel layers thicken from N to S, with interlayers of contemporaneous marine clays, but more to the S (not shown in Figure 1) they change progressively into silts, sands and clays, overlapping the Miocene and Pliocene marls and silts and giving rise altogether to a thick (>300 m) sequence of fine sediments. The Miocene top is at a depth between 50 to more than 200 m under the marshes.

The aquifer surface area is around 3,400 km^2 and the confined area is about 1,800 km^2 (Figure 1a). Due to tectonic sinking and subsidence the maximum thicknesses appear to the SE (>300 m), while to the SW the aquifer is <100 m thick, and to the N its thickness is 50–100 m (Salvany and Custodio, 1995).

Recharge takes place by rainfall infiltration in the cropping out area to the W and N of the marshes. The regional water table follows closely the land topography in most of the area (Figure 2a), so groundwater discharges locally to many ravines that are aquifer drains temporarily or all along the year. At the regional scale groundwater flows southwards and eastwards, that is, towards the confined area and to the sea. Permanent discharge to the Atlantic Ocean along the cliffy western coast of El Asperillo-El Arenosillo is conspicuous in the western water table area. In the SE confined area most of the aquifer is filled with saline water; fresh water outflow can be only produced in a relatively short tract that is now under study. Under undisturbed conditions discharge also occurs as springs, seepage and evapotranspiration by phreatophytes along the contact between the water table and the marshes. Finally, some discharge also occurs to the marsh surface as small upward flows throughout the clays in the NW part of the marshes.

This scheme seems to have been operating since the latest sea level stabilisation, some 6,000 yr BP (Zazo *et al.*, 1996), until some 30 yr ago. But the natural flow pattern has been modified during the past three decades because of localised intensive pumping for irrigation. Pumping concentrates in three sites: close to the marshes in the northern water table area, in the W corner of the aquifer, and around El Rocío village, in the centre of the aquifer (Figure 2b). Therefore, at present most of the recharged water is pumped out from the unconfined areas close to the marshes, which means close to or in the natural discharge areas.

Intensive and localised pumping has produced a piezometric and phreatic level drawdown, and the consequent reduction or cancellation of natural discharge. It has also produced a local groundwater flow reversal between the confined permeable layers and the overlaying clays to the N of the marshes, so as very low upward flows seems to exist currently across the confined area.

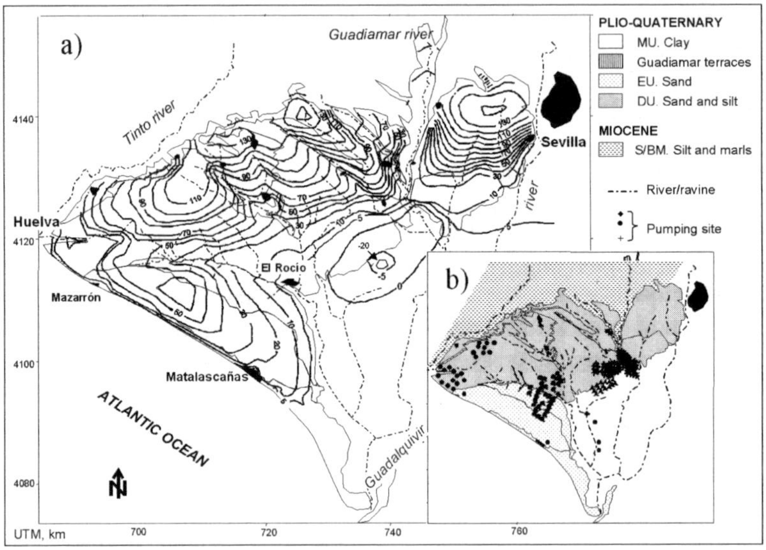

Figure 2. a) Regional piezometry in 1997. b) Location of the main groundwater abstraction sites. (Modified after UPC, 1999).

3 HYDROCHEMISTRY AND GROUNDWATER QUALITY

The water table area contains very low to medium mineralised groundwater of two different chemical types:

1) In the western water table area it is of the Na-Cl type due to the marine rainwater chemical signature and a practical absence of reactive minerals, either in the siliceous sands of the Eolian Unit, which forms most of the aquifer, and in the sands and gravels of the deep Deltaic Unit. Calcium carbonate only appears deep to the E and NE of this western sector as shell deposits, increasing both the alkalinity and Ca contents of the deep groundwater flowing upwards along the W border of the marshes.
2) In the northern water table area both shallow and deep groundwater are of the CaNa-HCO_3Cl type due to the widespread presence of calcium carbonate in the whole aquifer thickness (Manzano *et al.*, 1991; Iglesias, 1999; Iglesias *et al.*, 1996; Lozano *et al.*, 2002).

Groundwater mineralisation increases conspicuously from the water table areas to the confined part. Under the marshes groundwater mixes with a body of saline water which extents roughly through the SE half of the system, filling both the permeable layers and the confining clays. This salinity has traditionally been attributed to connate marine water trapped during sediment settling, which has not been flushed out to the sea due to the low hydraulic head existing since the latest Holocene sea level stabilisation (FAO, 1972; IRYDA, 1976; IGME, 1982; Custodio, 1992). ^{14}C ages support this hypothesis: the more saline waters are older than 10–15 ka (Baonza *et al.*, 1984, Manzano *et al.*, 2001). Sulphate reduction and Na/Ca exchange in the mixing front, when it is displaced, seem to be the main (at least the most visible) chemical reactions modifying the mixed waters, but the role of eventual calcite dissolution or precipitation, as well as organic matter oxidation and CH_4 generation is not yet well known (Manzano *et al.*, 2001).

Following other studies, salinity in the shallower part of the clays has been explained as due to evaporation of ponding surface marine water, probably when the area behaved as a tidal marsh. The concentrated marine salts salinized pore-water in the upper 40–50 m of clays, mainly by molecular diffusion (Rodríguez-Arévalo, 1988; Konikow and Rodríguez-Arévalo, 1993).

Recently drilled boreholes along the coastal sand barrier showed a thick (>300 m) column of sands, silts and clays with pore-water more saline than sea-water (>60 µS/cm). This puzzled the existing conceptual model, as until now the littoral barrier was supposed to be the area through where fresh groundwater in the confined aquifer discharged to the sea. New studies are currently developing to ascertain the extension, hydraulic situation and age distribution of this saline body.

Available data on groundwater quality status consists of a few hundred systematic analyses performed in the agricultural areas during the 1980s by public institutions (IARA and IGME), and several hundreds of analyses from different research projects carried out by different universities (Iglesias, 1999; Manzano *et al.*, 2001; Romero *et al.*, 2001). According with these data, agriculture-derived pollution is the only significant process affecting groundwater quality. Nitrate contamination is clearly observed at 40-50 m depth in the NE agricultural area, although it is difficult to know if NO_3 is at this depth in the porous medium by groundwater flow or it goes down through the wells during pumping, induced by the local water table depression. SO_4 data show concentrations not higher than expected from natural sources (rain water and lithology), and even lower in the confined area due to reducing conditions.

A two-year sampling in a especially designed nested borehole placed within the agricultural area W to El Rocío illustrated that very little NO_3 and pesticides have already reached the uppermost part (<20 m) of the sandy saturated zone in this area. This would mean that most of those products are still in the 2–6 m thick unsaturated zone (Custodio, 1994), but there are no data from other areas (Iglesias, 1999). Although the available information is scarce, unpublished data referring to excess irrigation water show elevated concentrations of NO_3, SO_4 and pesticides in surface water of the small creeks crossing the cultivated area. Those waters flow to La Rocina ravine and to the marshes NW boundary, a vegetated fringe known as La Vera. Both of them have a significant ecological value, mainly due to the large flow and very good quality of groundwater discharge.

With respect to other quality problems, pollution of shallow phreatic water by sewage around some small settlements has also been documented. Surface water pollution by heavy metals and organics derived from food and agriculture-related industry has been reported in the creeks and ravines nearby Almonte, Hinojos and Pilas villages, but data on groundwater are not available. These surface waters flow southwards into the marshes, where they introduce those pollutants.

4 AQUIFER-WETLAND RELATIONSHIP IN DOÑANA

Wetlands of different origin and hydrology are abundant in Doñana. Proposals to characterise and classify them using a hydromorphic approach have been published elsewhere and will not be reproduced here (Manzano, 2001; Manzano *et al.*, 2002; CMAJA, 2002). Hydrology is the main driving factor of wetland ecology (Custodio, 2000), and wetland location on the groundwater flow system is the main factor controlling their hydrology (water source, mode of drainage, inundation frequency and lasting, mineralisation and hydrochemical type):

- In the aquifer *recharge area* small depressions of eolian and erosive origin holding permanent or temporary water bodies are abundant. Frequently they are phreatic wetlands with visible water only during the wet season, while during the dry period rain-contributed water may become perched above the water table. Other seasonal wetlands in this area get their water from shallow vadose flows generated after rainfall episodes. Those flows discharge to small topographic depressions when, in their path down to the water table, they find shallow horizontal layers of low vertical permeability, typically old lagoon bottom layers or hydromorphic horizons now covered by eolian sands (Custodio *et al.*, 1995) (Figure 3). Whichever the water source and the inundation lasting, all of them hold aquatic and phreatophyte species that contributed to retain eolian sands during thousands of years and helped to create a thick (up to 20 m) and large (several hundred km^2) vegetated eolian cover which is one of the most characteristic features of the area.

– Most wetlands in the aquifer *discharge area* are located in the densely vegetated lower reaches of the ravine beds. Under natural conditions the vegetation belt (locally called *algaida*) maintain a permanent phreatic discharge that allows aquatic and phreatophytic vegetation to last around the year. This is the case of small phreatic watercourses discharging to La Rocina ravine and to the marshes along La Vera fringe (Cañada Mayor, Soto Chico, Soto Grande,...), and also that of some permanent ravines and creeks discharging to the sea in the western cliffy coast of El Asperillo. La Rocina ravine has many shallow-ponding areas along its course and support well-developed riverine vegetation.

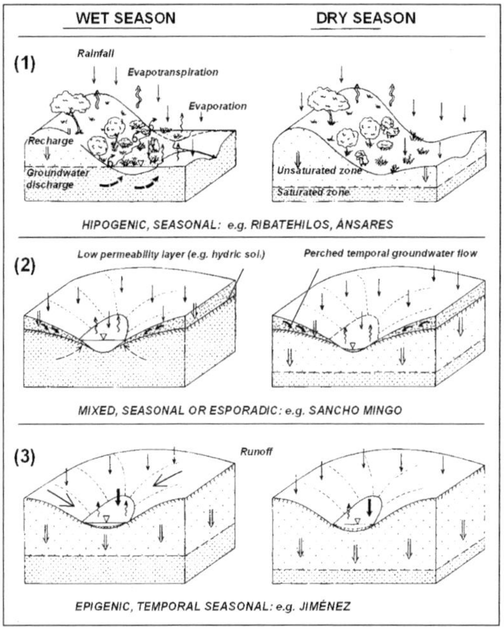

Figure 3. Some hydrological wetland types in the water-table area of El Abalario (see Figure 1), as a representative of wetland's diversity in Doñana. Though being an area of shallow phreatic level, not all the wetlands are purely phreatic like type (1), which stand water only during high water-table stages. Some wetlands are phreatic only in wet periods, while in dry periods they use to get water, but mostly supplied by vadose groundwater flows (type 2). Finally, some formerly phreatic wetlands are nowadays perched and get their water mostly by direct rainfall and runoff (type 3). Thicker arrows identify the main water source in each case. (Modified from Manzano, 2001).

The Guadalquivir river *marshes,* formerly tidal and nowadays fluvial, get their water from several ravines entering through its N and NE boundaries. Only a small tidal lagoon remains in the Guadalquivir river right bank, close to the river mouth. Marsh related wetlands are of three types: a) seasonal fresh to saline water ponds remnants from the wet season; b) permanent saline quicksands rounded spots (locally called *ojos* or eyes) fed by upward groundwater flows from confined sand layers; and c) permanent fresh to saline lagoons nearby the inner limit of the littoral sand bar (locally called *lucios*), which receive a mixture of local phreatic water and surface water. Some formerly permanent lagoons, also called *lucios* but located far from the coast, are nowadays artificially fed with pumped groundwater but in the past they were supplied by groundwater upward flows discharging into the marsh surface.

The combined result of the many hydrological situations, water salinity and hydrochemical types of these wetlands is the main cause of the biodiversity for which the Doñana region is worldwide known.

5 THE IMPACT OF LOCALISED AND INTENSIVE GROUNDWATER ABSTRACTION ON WETLANDS

Intensive groundwater pumping in or close to areas of natural discharge is probably the most relevant factor causing adverse impacts to the Doñana wetlands, although it is not the only one. Also significant are the impacts of soil use changes during the last 50 years, mostly the substitution of native vegetation by eucalyptus trees (Sousa-Martín and García-Murillo, 1999), and soil, water and air contamination by agricultural and industrial activities.

Following the FAO and Spanish Government irrigation development programme *Plan Regable Almonte-Marismas*, groundwater pumping for irrigation started in 1973–1974 to the N and W of El Rocío village and across the NE reaches of the marshes, which had been partially drained to allow cultivation. Groundwater pumping started to be intensive around 1980. The cropped surface spread until the middle 1980s, but after that date large sectors around El Rocío village started to be abandoned due to different reasons. The remaining crops in this location were concentrated along La Rocina ravine lower course and close to La Vera fringe, that is, in the area of the aquifer natural discharge. In the meanwhile the NE cultivation area specialised in cotton and rice, most of them irrigated with groundwater. Wells drilled further into the marshes were progressively set in work, inducing the fresh-saline water mixing zone displacement towards irrigation wells located to the N (Mediavilla *et al.*, 1995).

Recorded groundwater level temporal evolution in different wells shows some different patterns (Figure 4):

- To the NE of El Abalario village, some 15–20 km away from the pumping and cultivation sites, seasonal oscillations controlled by the irrigation/pumping pattern seem to dominate, but a superposed continuous interannual water table drawdown can also be seen. In this area (El Abalario, Peladillo, La Mediana) a small lowering of the water table seems to have occurred, converting some phreatic and formerly permanent or seasonal wetlands into sporadic ones. This seems to be the combined effect of groundwater pumping to the NE and of local increase of phreatic transpiration after substituting native vegetation by eucalyptus trees some 50 years ago. As eucalyptus may develop rapidly deeper roots than native plants, and they were placed in an area of very shallow (0.5 to < 2 m) water table depth, after several decades the increase of evapotranspirated phreatic water led to a significant water table drawdown. Even being small, this sufficed to eliminate most of the local phreatic discharges to erosive and eolian depression, such as the well-documented lagoons of El Abalario, Ribetehilos and La Mediana (Sousa-Martín and García-Murillo, 1999). Although there is little quantitative information related to that aspect, Trick's work (Trick, 1998; Trick and Custodio, 2003) is definitive to confirm this effect (Figure 5).
- In the two main pumping sites of El Rocío and NE of the marshes, well-developed water table depressions are nowadays clearly visible. Comparing the 1996 piezometric surface to that of 1972, before the beginning of intensive pumping, a large piezometric drawdown cone developed to the NE of the marshes, with water levels down to 20 m below sea level (see Figure 2a). This large cone triggered some local flow reversals between the confined

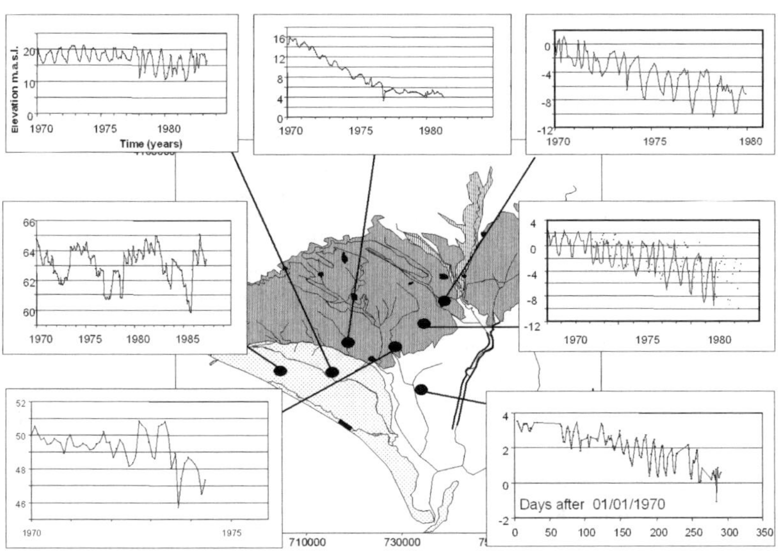

Figure 4. Temporal evolution of phreatic and piezometric levels in different areas showing the impact of groundwater abstraction. (Modified after UPC, 1999).

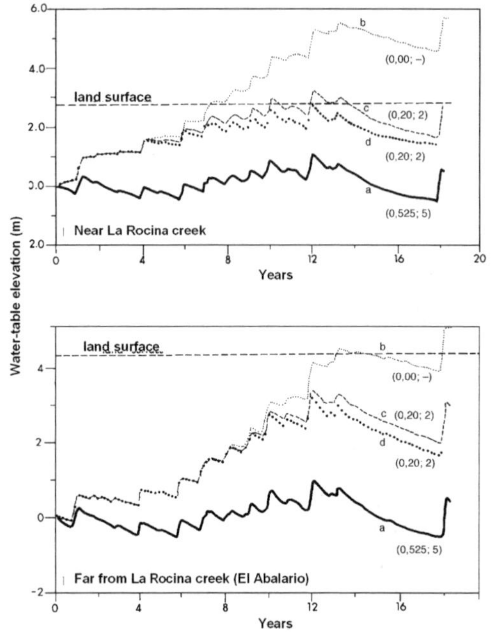

Figure 5. Modelled water-table evolution in El Abalario area following different vegetation-management alternatives. Pumping volumes maintain as nowadays. φ is the maximum phreatic evaporation (m/yr) and d is the maximum depth available for phreatic evaporation (m). The simulated scenarios are:

$\varphi = 0.525$ *m/yr and* $d=5$ *m* is actual calibrated situation;

$\varphi = 0.2$ *m/yr and* $d=2$ *m* is forecasted evolution if eucalyptus were substituted by native vegetation with *d* around 2 m;

$\varphi = 0.2$ *m/yr and d = "no"* is forecasted evolution if eucalyptus were substituted by a native vegetation able to adapt to any "d" (d is variable), and

$\varphi = 0.0$ *m/yr* is evolution if eucalyptus were eliminated but not substituted by any other vegetation.

The figure suggest that: 1) eucalyptus elimination without reforesting would lead to a phreatic level rise above soil surface in less than 8 years close to La Rocina creek, and in some 15 years far to it; 2) eucalyptus substitution by a vegetation able to reach whichever water-table depth would rise the phreatic level some 2 m close to La Rocina, and between 2 and 3 m far to it; 3) the impact of eucalyptus introduction some 50 years ago in El Abalario area was water-table lowering between 0.5 and a few metres. (Modified after Trick, 1998, in Custodio, 2000).

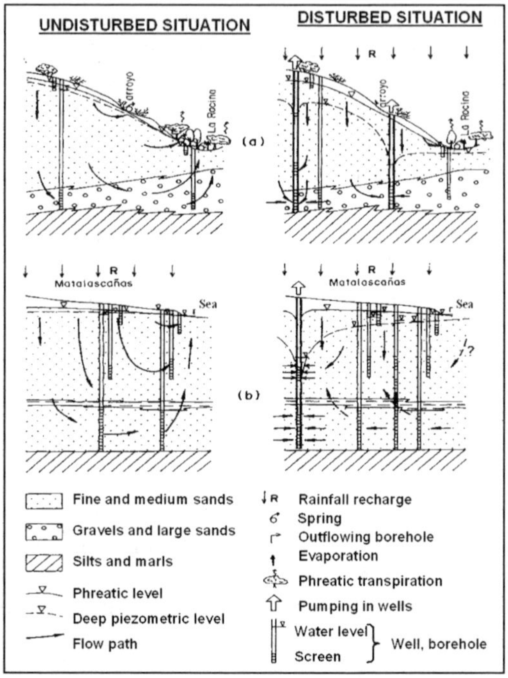

Figure 6. Illustration of groundwater flow reversing between the water-table and the deep piezometric levels due to pumping in deep layers. The main impacts are: (a) decrease of groundwater discharge to ecotones and to phreatic lagoons nearby El Rocío; (b) disturbance of freshwater-saltwater equilibrium nearby the coast. (Modified from Custodio *et al.*, 1995).

permeable layers and the overlaying clays, causing salinization of some agricultural wells by mixing with saline water existing in the confined layers (Suso and Llamas, 1990; Llamas, 1988, 1992; Custodio and Palancar, 1995; Mediavilla *et al.*, 1995; Trick, 1998; UPC, 1999; Manzano *et al.*, 2001). It also caused that some permanent and temporal phreatic lagoons in this area become sporadic or even erratic, although deforestation in the Guadiamar and Cigüeña rivers basins had also contributed to this effect by filling with sediments some of the depressions.

Some formerly permanent phreatic wetlands (lagoons and *algaidas*) around El Rocío also turned seasonal or even sporadic, eventually holding water only in very wet periods due to accumulated water table lowering during the last decades. Even though the exploited permeable layers are somewhat deep (45–70 m), the accumulated phreatic drawdown in the overlying fine sands layer led to water table lowering (Figure 6) and to natural discharge decrease to local lagoons and ravines like La Rocina, Cañada Mayor, Soto Chico or Soto Grande, which in the past were the main sources supplying freshwater

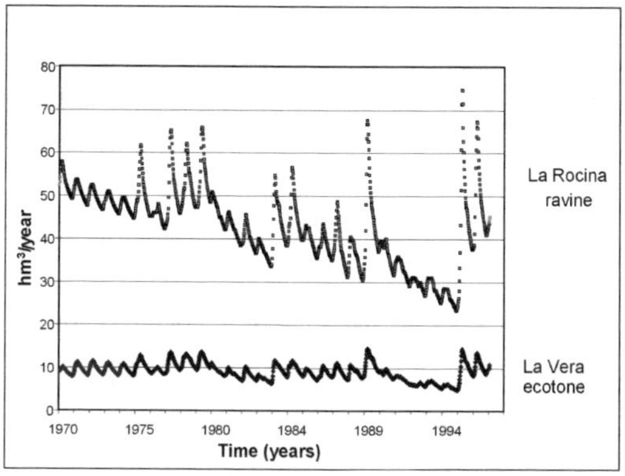

Figure 7. Groundwater discharge evolution to La Rocina ravine and La Vera ecotone (see location in Figure 1a) as calculated by groundwater flow numerical modeling. Note that the extreme situation of 1994 was due to the combined effect of 1) accumulated inter-annual water level lowering due to pumping, and 2) a 4 years long drought. (Modified after UPC, 1999).

and supporting the National Park wild fauna during the dry season. Moreover water quality deteriorated because most of the ravines are currently agricultural drains. Nutrient excess and the presence of other agrochemical products can change the types and/or the relative abundance of characteristic plant and fauna species (Muñoz-Reinoso, 2001), thus altering wetlands ecology and even reducing biodiversity.

Groundwater flow numerical models show that the groundwater discharge reduction to La Rocina ravine and La Vera ecotone for the period 1970–1996 has been quantitatively important (Martín-Machuca and Virgós, 1995; UPC, 1999). In addition, the eventual coincidence of groundwater pumping and a severe drought led to especially adverse situations like that of 1994, when groundwater discharge was only around 50% of that in 1970 (Figure 7). The transient period until the effect of a water balance change fades out is more than 15 years.

6 CONCLUDING REMARKS

Most wetlands in the Doñana region are groundwater-dependant, from those in the water table aquifer area to those placed on top of the clay marshes and receiving phreatic water either as seepage along the marshes NW and W boundaries or through the groundwater fed ravines entering the area. Thus, wetlands in the aquifer discharge area surrounding the marshes are the most vulnerable to localised groundwater abstraction for agriculture, but some wetlands on the marshes can also be affected by intensive pumping in the nearby water table area. Both situations can be already perceived.

Intensive groundwater withdrawal during two decades, localised in natural discharge areas nearby the marshes, led to a more that 15 years transient period until a new equilibrium is attained. Accumulated drawdown of the more permeable deep aquifer piezometric levels induced a water table lowering. Therefore, seepage along the ecotones decreased, flowing deep wells and boreholes ceased to flow, deep groundwater discharge through the marsh clay diminished, and a kilometric-size piezometric cone developed in the NE part of the confined aquifer, which induced saline groundwater to displace from S to N salinizing some agricultural wells.

This resulted in severe hydrological damage to most wetlands through decreasing the inundation and/or the soil saturation frequency and lasting. Water table lowering also impedes phreatophyte roots to reach the water table level. Consequently, many small phreatic lagoons in the water table around the cultivated areas (El Acebuche; El Alamillo; Peladillo; La Mediana, and La Rocina, to the W; Villamanrique, and Guadiamar, to the N), which were permanent 20–30 years ago, are nowadays seasonal or even sporadic, holding water only in very wet years. Also the formerly permanent ravines discharging to La Rocina creek or to the marshes along La Vera fringe are nowadays seasonal and fed mostly with excess irrigation water from the nearby crop fields.

The introduction of eucalyptus trees in most of the western water table area (Moguer, El Abalario, La Mediana, La Rocina) some 50 years ago also impacted local phreatic wetlands. The increase of evapotranspirated volumes led to a water table drawdown large enough to reduce or eliminate local phreatic discharges to many lagoons between El Abalario height and La Rocina creek. This seems to have been the case of the formerly permanent lagoon-complexes of Ribetehilos and La Mediana.

Finally, surface and groundwater pollution by agrochemicals (nutrients, pesticides), industrial wastes (oil-derived wastes, heavy metals) and urban sewage around small settlements is also jeopardising wetland performance and values in Doñana.

To restore the Doñana wetlands natural role, first of all the water balance terms of the aquifer has to be changed and enough time has to be wait for transient evolution to fade out. This can be attained by reducing total abstraction and by well relocation. Secondly, effectiveness of irrigation water management has to be improved, and agrochemicals uses have to move to less aggressive methods, like the biological ones. Although urban sewage treatment is improving fast in the area, little or no treatment at all is applied to industrial and agricultural wastes in the tributaries to the area. Moreover, attention should also be paid to urban sewage network losses. Given the reasonable knowledge acquired in the last 15–20 years about the aquifer and the impact of aquifer disturbance on wetlands performance, most of these measures could start to work in a short period of time.

ACKNOWLEDGEMENTS

This work comes out from several research projects developed by the authors in Doñana during the last 15 years, following those started in the late 1980s by Dr. M.R. Llamas and co-workers at the Complutense University of Madrid. Among the more recent ones are the Spanish-founded projects "Procesos de recarga en el acuífero de Doñana" (CICYT-AMB95-0372), "Sistemas de apoyo a la gestión de los recursos hídricos subterráneos del ecosistema litoral de Doñana mediante técnicas de evaluación funcional de humedales" (CICYT HYD97-0321-C02-01), and "Gestión de recursos hídricos y conservación de los humedales hipogénicos del manto eólico litoral de Doñana" (CICYT REN2001-1293-CO2-01/HID), and the European Union-founded projects "PALAEAUX" (ENV4-CT95-0156), and "BASELINE" (EVK1- 1999-0006). Several agreements between the Geological Institute of Spain (IGME) and the Technical Universities of Catalonia and Cartagena allowed detailed studies or some hydrogeological aspects. Also many studies related to wetlands hydrology were supported by the Environment Department of the Regional Government of Andalucía.

REFERENCES

Baonza, E.; Plata, A. and Silgado, A. (1984). Hidrología isotópica de las aguas subterráneas del Parque Nacional de Doñana y zona de influencia. Centro de Estudios y Experimentación de Obras Públicas (CEDEX), Madrid, Cuadernos de Investigación C7: 1–139.

CMAJA (2002). Plan Andaluz de Humedales. Consejería de Medio Ambiente, Junta de Andalucía. Sevilla.

Custodio, E. (1992). Preliminary outlook of saltwater intrusion conditions in the Doñana National Park (Southern Spain). Study and Modelling of Salt Water Intrusion. CIMNE-UPC. Barcelona: 295–315.

Custodio, E. (1994). Posibles procesos de contaminación agrícola de aguas subterráneas en el área de Doñana (Huelva). Análisis y Evolución de la Contaminación de las Aguas Subterráneas en España . (Alcalá de Henares Nat. Congr.). Asoc. Intern. Hydrogeol.-Grupo Español. II: 283–308.

Custodio, E. (2000). Groundwater-dependent wetlands. Acta Geologica Hungarica, 43(2): 173–202.

Custodio, E. and Palancar, M. (1995). Las aguas subterráneas en Doñana. Revista de Obras Públicas, Madrid, 142 (3340): 31–53.

Custodio, E.; Manzano, M.; Iglesias, M. and Giráldez, J.V. (1995). Estudio hidrogeológico de El Abalario (Huelva). Bases para la restauración de la zona húmeda. (Unpublished report elaborated for the Natural Park of Doñana, Environmental Agency, Junta de Andalucía; Sevilla, Spain).

FAO (1972). Proyecto de utilización de aguas subterráneas para el desarrollo agrícola de la cuenca del Guadalquivir; Anteproyecto de transformación en regadío de la zona Almonte-Marismas (margen derecha). Programa de las Naciones Unidas para el Desarrollo, Organización de las Naciones Unidas para la Agricultura y la Alimentación. Technical Report I, AGL: SF/SPA 16. Roma, 2 volumes.

Iglesias, M. (1999). Caracterización hidrogeoquímica del flujo del agua subterránea en El Abalario, Doñana, Huelva. Doctoral Thesis. Civil Eng. School, Tech. Univ. of Catalonia, Barcelona, Spain.

Iglesias, M.; Custodio, E.; Giráldez, J.V.; Manzano, M. and Ordóñez, R. (1996). Caracterización química de la lluvia y estimación de la recarga en el área de El Abalario, Doñana, Huelva. El Agua en Andalucía. Proc. IV Symp. Instituto Tecnológico Geominero de España (ed.), Madrid: 99–121.

IGME (1982). Hidrogeología del Parque Nacional de Doñana y su entorno. Instituto Geológico y Minero de España. Madrid, 1–120 + maps.

IRYDA (1976). Informe final de sondeos de la zona regable de Almonte-Marismas (Huelva-Sevilla). Instituto para la Reforma y el Desarrollo Agrario, Sevilla, 1–110 + anexes and maps.

Konikov, L.F. and Rodríguez-Arévalo, J. (1993). Advection and diffusion in a variable-salinity confining layer. Water Resources Research, 29 (8): 2747–2761.

Llamas, M.R. (1988). Conflicts between wetlands: new constraints in groundwater management. Management: Quantity and Quality. Inter. Assoc. Hydrol. Sciences. Publ. 188: 295–304.

Llamas, M.R. (1992). Wetlands: an important issue in hydrogeology. Aquifer Overexploitation. Intern. Assoc. Hydrogeologists, Heise, Hannover, Selected Papers 3: 69–86.

Lozano, E.; Delgado, F.; Manzano, M. and Custodio, E. (2002). Caracterización hidrogeoquímica de las aguas subterráneas y superficiales de la Vera del Parque Nacional de Doñana (SW de España). Groundwater and Human Development (Bocanegra, Martínez and Massone, eds.). Univ. Mar del Plata, Argentina: 1348–1358 (CD).

Manzano, M. (2001). Clasificación de los humedales de Doñana atendiendo a su funcionamiento hidrológico. Hidrogeología y Recursos Hidráulicos, Madrid, XXIV: 57–75.

Manzano, M.; Custodio, E. and Poncela, R. (1991). Contribución de la hidrogeoquímica al conocimiento de la hidrodinámica de los acuíferos del área de Doñana. Agua en Andalucía. Proc. III Symp. ITGE , Madrid: 2: 475–486.

Manzano, M.; Custodio, E.; Loosli, H.H.; Cabrera, M.C.; Riera, X. and Custodio, J. (2001). Palaeowater in coastal aquifers of Spain. Palaeowaters in Coastal Europe: Evolution of Groundwater since the late Pleistocene. (Edmunds, W.M. y Milne, C.J., eds.). Geological Society London, Sp. Publ. 189: 107–138.

Manzano, M.; Borja, F. and Montes, C. (2002). Metodología de tipificación hidrológica de los humedales españoles con vistas a su valoración funcional y a su gestión: Aplicación a los humedales de Doñana. Boletín Geológico y Minero, 113 (3): 313–330.

Martín-Machuca, M. and Virgós, L.I. (1995). Modelo matemático del acuífero de Almonte-Marismas. Hidrogeología y Recursos Hidráulicos, XIX. Madrid: 639–660.

Mediavilla, C.; Martín-Machuca, M.; Cantos, R.; Mantecón, R.; Coleto, I. and Tenajas, J. (1995). Análisis de la situación hidrodinámica y de la posición de la zona de tránsito agua dulce-agua-salada en el extremo oriental de la unidad hidrogeológica Almonte-Marismas. Hidrogeología y Recursos Hidráulicos, XIX. Madrid: 463–471.

Muñoz-Reinoso, J.C., (2001). Vegetation changes and groundwater abstraction in SW Doñana, Spain. Journal of Hydrology, 242:197–209.

Rodríguez-Arévalo, F.J. (1988). Origen del movimiento del agua intersticial en el acuitardo arcilloso de las marismas del Guadalquivir. Doctoral Thesis. Geol. Sci. Fac., Complutense Univ., Madrid, 1–300 + anexes.

Romero, E.; González, A.; Garrido, R.; Orihuela, D.L.; Fidelibus, M.D. and Tulipano, L. (2001). Procesos de salinización del acuífero Almonte-Marismas en el preparque Norte del Parque Nacional de Doñana en base al estudio de los iones mayoritarios. (Pulido Bosch, A.; Pulido Leboeuf, P. Vallejos Izquierdo, A., eds.). El Agua en Andalucía, Proc. V Symp., Almería, II: 261–270.

Salvany, J.M. and Custodio, E. (1995). Características litológicas de los depósitos pliocuaternarios del Bajo Guadalquivir en el área de Doñana: implicaciones hidrogeológicas. Rev. Soc. Geol. De España 8 (1–2): 21–31.

Sousa-Martín, A. and García-Murillo, P. (1999). Historical evolution of the Abalario lagoon complexes (Doñana Natural Park, SW Spain). Limnetica, 16: 85–98.

Suso, J.M. and Llamas, M. (1990). El impacto de la extracción de aguas subterráneas en el Parque Nacional de Doñana. Estudios Geológicos, 46: 317–345.

Trick, T. (1998). Impacto de las extracciones de agua subterránea en Doñana: aplicación de un modelo numérico con consideración de la variabilidad de la recarga. Doctoral Thesis. Civil Eng. School, Tech. Univ. of Catalonia, Barcelona, Spain.

Trick, T. and Custodio, E. (2003). Hydrodynamic characteristics of the wester Doñana Region (area of El Abalario), Huelva, Spain. Hydrogeology Journal (in press).

UPC (1999): Modelo regional de flujo subterráneo del sistema acuífero Almonte-Marismas y su entorno. Grupo de Hidrología Subterránea (UPC, Barcelona) – Instituto Tecnológico Geominero de España, Madrid: 114 + anex. (Unpublished report).

Zazo, C.; Goy, J.L.; Lario, J. and Silva, P.G. (1996). Littoral zone and rapid climatic changes during the last 20,000 years: the Iberian study case. Z. Geomorph. N.F.; Berlin–Stuttgart, Suppl. 102, 119–134.

Poverty alleviation *versus* mass poisoning: the dilemma of groundwater irrigation in Bangladesh

Mohammed Mainuddin
International Water Management Institute, Southeast Asia Regional Office, Bangkok, Thailand
m.mainuddin@cgiar.org

ABSTRACT

Although many of the countries in Asia have tapped their underground aquifers to supplement their surface water supplies, no country in the region is as dependent upon groundwater irrigation as Bangladesh. The contribution of groundwater irrigation to the overall irrigated area was 70% in 1999 compared to only 4% in 1972. Groundwater development has enabled Bangladesh to attain self sufficiency in staple food production together with significant reductions on poverty. However, this good run with groundwater irrigation is under serious threat because of arsenic contamination. Millions of people are at the risk of being poisoned by consuming food produced with arsenic contaminated groundwater besides drinking water directly. Many analysts allege that abstraction of groundwater for irrigation is the cause of contamination, whereas others suggest that contamination of groundwater is a natural phenomenon. This paper evaluates the contribution of groundwater on poverty alleviation and the impending threat to its sustainability due to arsenic contamination. In this context, the overall water management strategies over the past are briefly discussed and the need for an integrated water resources management strategy is proposed.

Keywords: *groundwater, arsenic contamination, poverty alleviation, mass poisoning.*

1 INTRODUCTION

The growth of groundwater irrigation has been the second most dramatic episode in Asian agriculture of the last two decades, after the spread of green-revolution technology (Siamwalla, 2001). Although many countries in Asia have exploited their underground aquifers in order to supplement their surface water supplies, no country in the region is as dependent upon groundwater irrigation as Bangladesh. Groundwater irrigation plays a crucial role in Bangladesh's agriculture and, therefore, in the national economy. After independence in 1971, of the 8.25 million ha of net cultivated area, only 1.05 million ha was irrigated and the contribution of groundwater to total irrigated area was only 3% (BBS, 1976). In 1999 (years, in these documents, refer to the financial year, July to June, e.g., 1999 refers to 1998–1999), of the 3.99 million ha of irrigated area, approximately 70% of the irrigated water was from groundwater (BBS, 2002). Groundwater development and the green revolution have gradually enabled Bangladesh to emerge from being a *basket case* to partial self sufficiency in staple food production together with significant reductions on poverty. Hence, groundwater irrigation has become a formidable tool for poverty alleviation in Bangladesh.

The advantages of exploiting groundwater irrigation sources are under serious threat due to arsenic contamination. Recent evidence has shown that groundwater sources of 61 districts out of total 64, is contaminated with arsenic. Some researchers attribute groundwater abstraction for irrigation as the cause of arsenic contamination whereas others suggest that the origin of arsenic rich groundwater is a natural phenomenon that has no relationship with excessive groundwater abstraction (BGS, 2000; Smedley and Kinniburgh, 2002; Nickson *et al.*, 1998; Harvey *et al.*, 2002). Since the detection of arsenic in drinking water, a great deal of effort has been diverted towards the determination of the cause of contamination and the removal of arsenic from drinking water. It is estimated that of the 125 million of inhabitants of Bangladesh between 35 million and 77 million people are potentially exposed to arsenic poisoning through drinking water (Smith *et al.*, 2000). It is predicted that 200,000–270,000 people will die of cancer from drinking arsenic contaminated water in Bangladesh (WHO, 2001). This has been described as the greatest mass poisoning in human history (Smith *et al.*, 2000). However, these figures may be conservative if arsenic is shown to be entering the food chain through the consumption of crops irrigated by contaminated water. A recent study undertaken by Mehrag and Rahman (2003) suggests that arsenic from contaminated groundwater is being taken up by rice the staple food for the region. However, no comprehensive studies have been undertaken to assess the consequence of arsenic entering the food chain through irrigation with contaminated groundwater. In this respect drinking water is a small proportion of the total groundwater consumed when compared to that used in the production of irrigated crops. While a great deal of attention has been given to find alternative sources of drinking water such as rain water harvesting, community ponds, and developing low-cost filters to remove the arsenic from contaminated water, scant attention has been given to alternatives to groundwater irrigation or reducing the dependence on contaminated groundwater for irrigation. If a safe drinking water source is successfully introduced to Bangladesh, such as rainwater harvesting or treatment of tubewell water, it is likely that such technologies will not provide the volume, and/or will be too expensive, to be used to produce uncontaminated irrigation water.

Whatever the cause, there are serious implications of arsenic entering the food chain through crops that are irrigated with contaminated water, and then consumed by humans. Due to continued population growth, pressure on agricultural lands thereby on the groundwater is likely to stay on or even grow for several decades for producing food as well as supporting rural livelihoods if the present trend continues (World Bank, 2000). Therefore, developing and managing groundwater resource in a sustainable manner poses many challenges. This paper discusses the contribution of groundwater irrigation to food security as well as poverty alleviation since its inception and the imminent threats to its sustainability due to arsenic contamination and continued expansion of groundwater irrigation. In this context, the overall water management strategies over the past are briefly discussed and the need for an integrated water resources management strategy is proposed. In addition, the paper outlines some strategically important areas for the sustainability of groundwater that require careful attention by the agencies concerned.

2 EVOLUTION OF GROUNDWATER IRRIGATION IN BANGLADESH

Groundwater abstraction technology was introduced into Bangladesh in the 1960s by a government agency, later called the Bangladesh Agricultural Development Corporation (BADC). Prior to this, limited groundwater exploitation for drinking and irrigation purposes was undertaken and this was confined to dug wells accessing shallow aquifers. The BADC initiated a heavily subsidized deep-tubewell (DTW) program, installing and renting wells to cooperatives, which contributed nominally to the costs of operation and maintenance (O&M). The BADC also rented shallow tubewells (STWs) to farmers, who paid a larger share of the O&M costs themselves. From the late 1970s through to the early 1980s there were a decrease in public sector involvement in minor irrigation and a gathering momentum in privatizing this sector activity. As a result, in the late 1970s, the government began to ease the BADC out of its role as sole provider of equipment to farmers. There were simultaneous moves to discontinue the DTW rental program and sell both new and old DTWs to groups in the private sector. Subsidies on STWs were reduced and in its place credit for the purchase and installation of tubewells was provided. Import duties were reduced and the private sector was, for the first time, allowed to import the equipment. Consequently, the number of STWs increased from 2,200 in 1976 to 120,000 in 1984 (NMIDP, 1997).

The expansion of STW slowed in 1984 and practically stopped during 1985 through 1987 due to the imposition of multiple government controls, which were triggered because of greater than expected drawdown of groundwater in a number of northern districts. The slow growth of the minor irrigation sector of the previous years prompted the government to remove the restrictions imposed in 1989 (Mandal and Parker, 1995). Since then, private importation and sales of STW increased sharply (Figure 1). By 2000, the number of STWs had reached 819,137 (NMIDP, 2001). Some observer dubbed it as the *quiet revolution* (Chakravorty, 2001) or *pump revolution* (Shah *et al.*, 2000). DTW growth, however, has proven to be unsustainable in an unsubsidized environment. At present, there are no subsidies on DTWs and the BADC has fully withdrawn from DTW support.

However, initial investment for installing STW and the capacity of these pumps were too large for the multitude of marginal farmers. To assist these farmers, the International Development Enterprises (IDE) introduced the treadle pump, a foot-operated device that uses a bamboo, or a PVC pipe to pump water from shallow aquifers, as an ideal technology during the

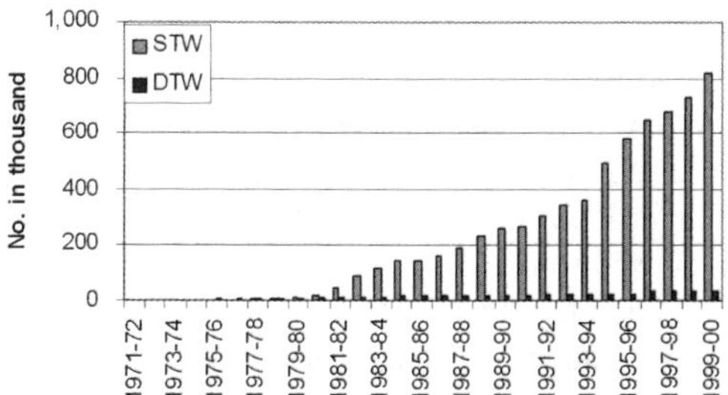

Figure 1. Growth of shallow tubewell (STW) and deep tubewell (DTW) in Bangladesh.
Source: NMIDP (1997), NMIDP (2001), Aktheruzzaman and Jaim (1999).

mid-eighties. Treadle pumps gained rapid popularity because of its cost-effectiveness and affordability. According to Downing and Polak (2000), an estimated 1.25 million treadle pumps were in the field at the end of 1999.

3 GROUNDWATER, AGRICULTURE AND POVERTY ALLEVIATION

The economy of Bangladesh is primarily dependent on agriculture. The agriculture sector is the single largest contributor to income and employment generation and a vital element in the country's challenge to achieve self-sufficiency in food production, reduce rural poverty and foster sustainable economic development. The agriculture sector (crops, forest, fisheries and livestock) contributes about 36% of GDP, of which the crop sub-sector contributes 71%. The sector generates 63.2% of total national employment, of which the crop sectors share is nearly 55% (SDNP, 2002).

Bangladesh's agriculture has made considerable strides in the last decades. Both production and mean yields of rice, the major crop, have risen constantly (Figure 2). Rice occupies 77% of the total cropped area (Bhuiyan *et al.*, 2002) and uses 82% of the irrigated water (BBS, 2001). The total production and average yield of rice in 1972 was 9.8 million tons and 1.05 ton/ha, respectively. In 1999 the total production was approximately 20 million tons and the average yield had increased to 1.97 ton/ha. This growth has resulted in food-grain self-sufficiency being reached – an objective few had thought achievable. In 1990, food grain deficit was 1.66 million ton (Hamid, 1991).

Production increases have resulted from a substantial intensification of agriculture rather than from increases in cultivated area; indeed, the area available for cultivation has fallen slightly in the last decades. Cropping intensity has grown substantially to 175% in 1999 from 145% in 1975, with an increasing proportion of land being double- or triple- cropped. Double cropping is practiced on 57% of the land and the triple cropped area is 21% (BBS, 2002). This growth in intensity was driven by increased cultivation during the dry season, made possible by

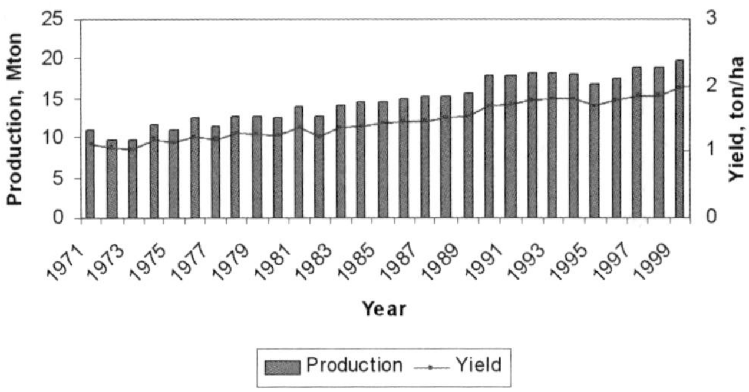

Figure 2. Yield and production of rice in Bangladesh.
Source: Statistical Year Books of Bangladesh, Bangladesh Bureau of Statistics.

Thus groundwater irrigation has significantly improved the socio-economics of tubewell owners as well as neighboring farmers who buy water from these owners. Through the introduction of groundwater irrigation systems, farmers have been able to cultivate high yielding *boro* (winter) rice, wheat, maize, potato, mustard, vegetables etc. Therefore, groundwater development and the green revolution have gradually enabled Bangladesh to achieve self-sufficiency in food production which significantly contributes to poverty alleviation. This is despite an extremely high population density, rapid population growth, and extreme vulnerability to floods and droughts, with consequent loss of crops.

the availability of irrigation by groundwater through the rapid increase in adoption of STWs. Figure 3 shows the percentage of the *Thana* (sub-district administrative area) net command area (NCA) irrigated by groundwater. The figure clearly indicates the extensive use of groundwater in Bangladesh. The contribution of different irrigation technologies to total irrigated area has changed considerably over time (Figure 4). The steep rise of tubewells irrigated area is associated with the use of STW as there has been no growth in DTW. STWs have increased in importance, from 24.4% of total irrigated area in 1983 to 52.1% in 1999 (NMIDP, 2001). The contribution of groundwater to total irrigated area has increased from 4% in 1971 to 70% in 1999 (Figure 4).

Use of high-yielding seed varieties (HYV) from 11% to 55% of the area during the past 25 years, and an increase in fertilizer consumption have also contributed to this growth in production (FAO, 1998). However, using HYV seed varieties and increasing rates of fertilizer have been possible because of the availability of irrigation from groundwater. Roy and Mainuddin (2003) reported from a tubewell owners' survey carried out in 40 villages throughout Bangladesh that 100% of tubewell owners in the North-Western region and 88% of tubewell owners in the rest of Bangladesh have expressed that it has contributed greatly to the village economy. If there were no groundwater irrigation, farmers would have to grow non-irrigated crops such as oilseeds or pulses, yields of which are very low.

To put the poverty profile in Bangladesh in brief, extreme poverty prevails among 22.7% of rural households and moderate poverty among 29.2%. Besides these, another class of the poor with vulnerability to income erosion comprises about 21%. This poverty situation has followed more or less a trend of decline over the past. Another characteristics development is the decline of malnutrition, which reached the lowest in 1996. Groundwater irrigation development has played a significant role in causing this decline over time. The most recent estimate of the Human Poverty Index (HPI) has dropped more than 20% during last 15 years (1981 to 1997) (Chakravorty, 2001).

The development of groundwater irrigation has increased livelihood of the country. In one irrigation season, 3 labors can get employment for 3 months per hectare of land. According to FAO (1999) groundwater irrigation in Bangladesh has increased the employment in agriculture since 1985 by 250% (cited in Karkkainen, 2001). Besides improving livelihoods, intensive groundwater irrigation is also contributing to alleviate the acute flood-proneness and water logging of the country. The depth and severity of flood would have been much more should there be no use of groundwater. Treadle pumps, which self select the poor, also

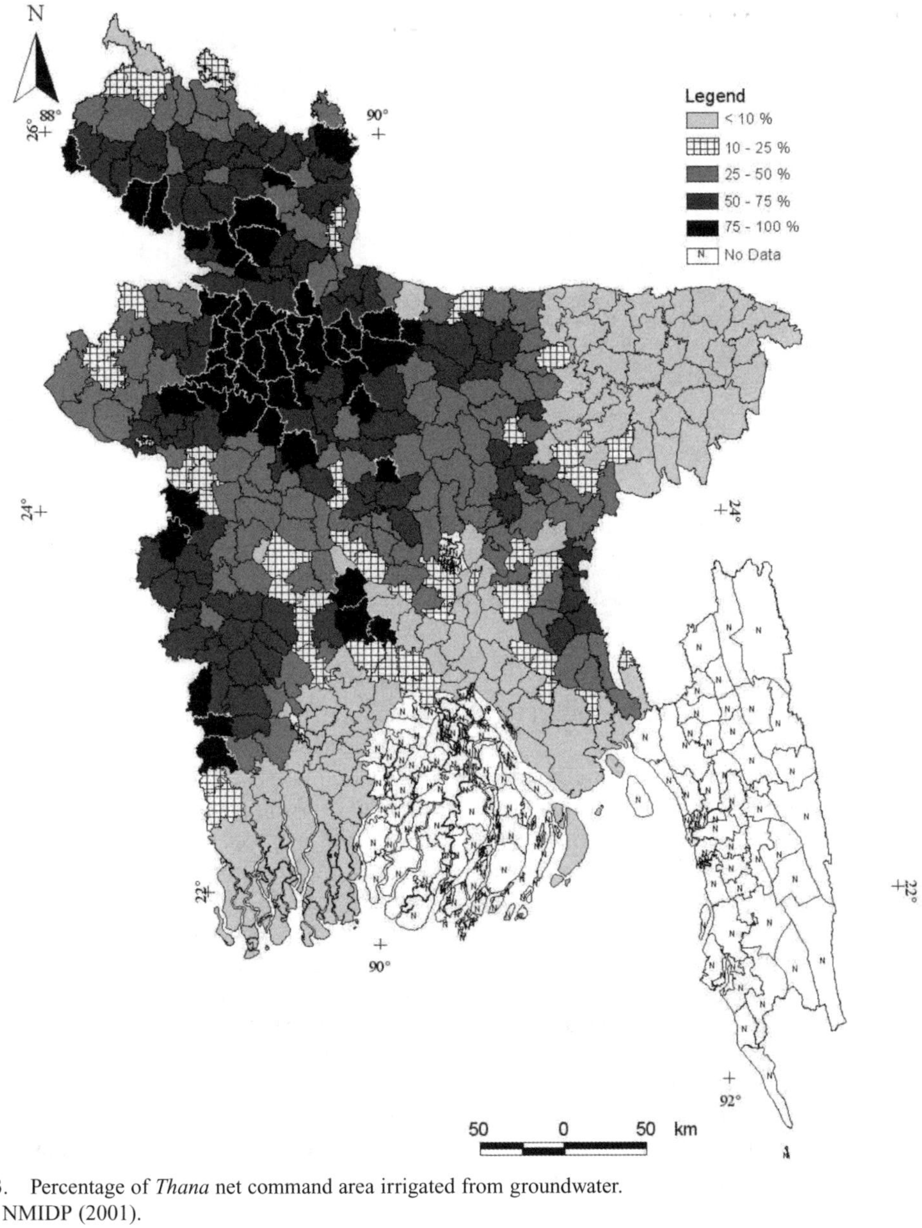

Figure 3. Percentage of *Thana* net command area irrigated from groundwater.
Source: NMIDP (2001).

contributing significantly to the poverty alleviation programs of Bangladesh. An impact study found that farmers conservatively earn US$ 100 annually on their US$ 24 investment in the installed pumps (Downing and Polak, 2000). More enterprising farmers earn far more. In a country where per capita income is estimated at US$ 220 (1999 estimate), even the most conservative figure is indeed substantial. Shah *et al.* (2000) rightly dubbed this as "pedaling out of poverty". Thus groundwater irrigation has emerged as a formidable tool for livelihood improvement and poverty alleviation in Bangladesh.

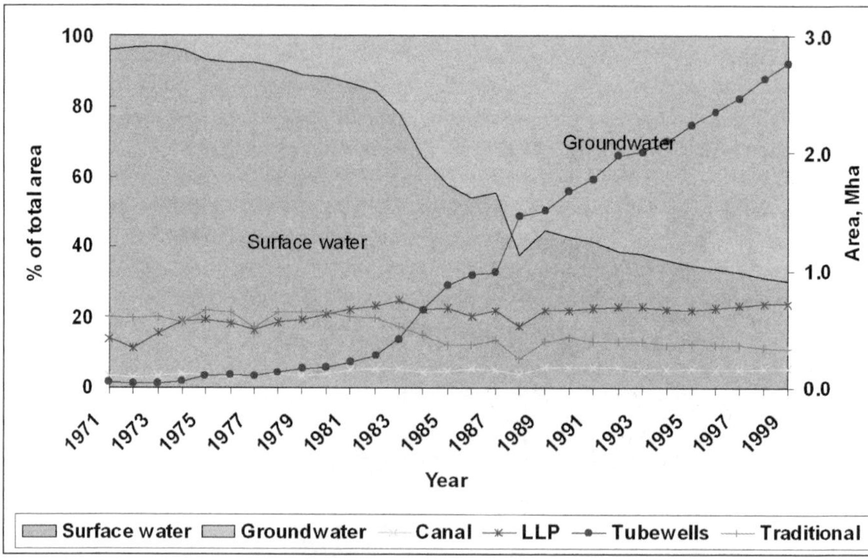

Figure 4. Percentage of total irrigated area by source of water and area irrigated by different technology. Source: Statistical Year Books of Bangladesh. Bangladesh Bureau of Statistics.

4 OVEREXPLOITATION AND ARSENIC CONTAMINATION OF GROUNDWATER

Concern has been expressed in some quarters regarding the possible negative impact of increased groundwater extraction by increasing numbers of private sector tubewells. Unregulated extraction of groundwater may lead to land subsidence, lowering of underground water levels, and drying up of drinking water pumps and surface water sources (Mandal and Parker, 1995). The extent and duration of such damage is in dispute. According to the Main Report of the National Water Management Plan (WARPO, 2001), in most areas groundwater levels continue to return each year to the same level except Dhaka city where water levels are falling due to heavy abstractions. There were some seasonal fluctuations such as in 1983 as reported earlier but evidence of long term declines in groundwater level is yet to be observed in other areas (SWMC, 2000). Salinization a problem often encountered in irrigated areas, also does not appear to be an issue in Bangladesh (Pagiola, 1995).

Arsenic was first detected in drinking water in 1993, and the issue came to the fore at the beginning of 1995. Recent findings of the British Geological Survey (BGS, 1999) show that groundwater of 61 surveyed districts, out of total 64, are contaminated with arsenic (Figure 5). The population exposed to the arsenic poisoning through drinking water is more than 35 million. Thousands of people are suffering from arsenic diseases ranging from melanosis to skin cancer and gangrene. In 241 villages where arsenic poisoning was suspected, thousands of arsenic patients and a total of 40 deaths due to arsenic-related diseases were identified by the School of Environment Studies (SOES) and Dhaka Community Hospital (DCH) survey (SOES-DCH, 2000). Although the presence and attendant risks of arsenic in the shallow aquifer are relatively well recorded, there is much less information about the safety of the deeper aquifer in this regard. Current evidence points to it probably being generally safe, a view supported by the Government's Technical Committee, but contradictory views are also expressed. As per tests by DPHE, water supplied in 28 municipalities from DTW are contaminated with arsenic, some having levels beyond the WHO-recommended (0.01 mg/L) safety limit (The Daily Star, 2002a).

A wide range of explanations for the origin of arsenic in groundwater have been put forward. Of all the natural and anthropogenic theories

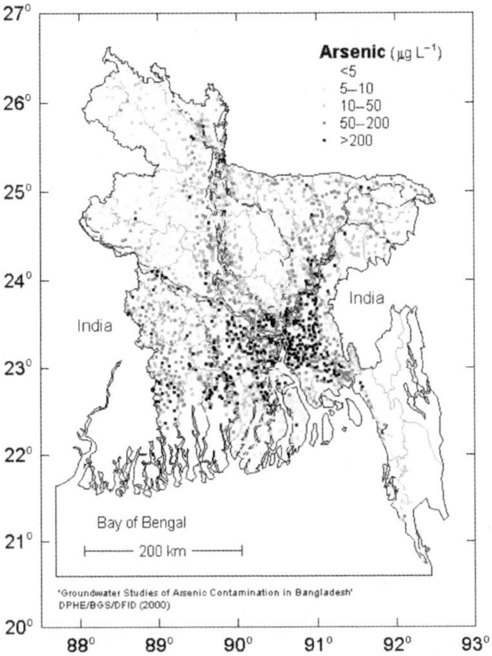

Figure 5. Occurrence of arsenic in groundwater in Bangladesh.
Source: BGS (2000).

investigated, only those proposing a geological origin can account for the widespread distribution of arsenic though Anwar (2000) argued that arsenic contamination is anthropogenic in Bangladesh i.e. it is caused by or influenced by human activities: more likely the result of indiscriminate use of substandard agrochemicals in the soils for some three to four decades from the start of the *Green Revolution* to this day. Of the geological explanation, two hypotheses are prevailing (Fazal *et al.*, 2001a) namely: a) Pyrite oxidation and b) oxy-hydroxide reduction. It is generally thought excessive withdrawal of groundwater for irrigation as the main reason for arsenic contamination linked with the first hypothesis.

a) Pyrite oxidation: Arsenic is assumed to be present in certain sulphide minerals (pyrites) that are deposited within aquifer sediments. Due to the lowering of water table below these deposits as a result of groundwater pumping or any other human activity, arseno-pyrite oxidizes in the vadose zone releasing arsenic as arsenic absorbed on iron hydroxide. During the subsequent recharge period, iron hydroxide releases arsenic into groundwater. According to this hypothesis, the origin of arsenic rich groundwater is man-made, which is a recent phenomenon. The lack of prior reported arsenic patients before irrigation development based on groundwater extraction in the country also supports the pyrite oxidation hypothesis, although a significant number of people were using groundwater through tubewells or dug wells as a drinking source at that time. Until only two years ago, groundwaters in Rangpur, Jamalpur and Bogra (heavy groundwater abstraction areas) were considered as safe. Now they are not only contaminated but also these areas contain many identified arsenic patients (SOES-DCH, 2000). Pyrite oxidation hypothesis was first proposed by Nickson *et al.* (1998).

b) Oxy-hydroxide Reduction: BGS (2000) rejected the 'pyrite-oxidation' hypothesis and accepted the 'oxy-hydroxide reduction' hypothesis. They mentioned that groundwater arsenic contamination in Bangladesh is a natural process that has no relationship with excessive groundwater withdrawal. Arsenic is assumed to be present in alluvial sediments with high concentrations in sand grains as a coating with iron hydroxide. The sediments were deposited in valleys eroded in the delta when the stream base level was lowered with the lowered sea stand during the last glacial advance. The organic matter deposited with the sediments reduces the arsenic-bearing iron hydroxide and releases arsenic into groundwater. Therefore, the origin of arsenic rich groundwater is due to a natural process, which implies that the arsenic in groundwater has been presents for thousands of years without being flushed from the delta.

A consensus on the cause of arsenic contamination of groundwater is yet to be found in the scientific community. Fazal *et al.* (2001b) investigated the correlation between different hydro-geological parameters and arsenic contamination and compared the results with the BGS (1999) and Nickson *et al.* (1998) findings. They stated that neither pyrite oxidation nor the oxy-hydroxide reduction hypotheses using existing data are enough to draw any final conclusion. Final conclusion could be made only by the time-dependent relationships using reliable field data. However, Harvey *et al.* (2002), based on their pumping test conducted in the Munshiganj district, 30 km from south of Dhaka, concluded that act of pumping water for irrigation can raise the arsenic levels in water. Therefore, the debate whether groundwater irrigation is responsible for arsenic contamination is yet to be settled.

Whatever the cause there are serious implications of arsenic contamination as recent studies have found that arsenic is entering the food chain through crops that are irrigated with arsenic contaminated water. Approximately one fifth of the population in most of the arsenic affected areas in the country is at risk of consuming arsenic-contaminated rice, wheat and vegetables (arum, tomato, papaya, cauliflower, cabbage, onion, bean, and leafy vegetables) (Haq, 2002a). A study by the Department of Soil, Water and Environment of Dhaka University and the Commonwealth Scientific and Industrial Research Organization (CSIRO) detected significant amounts of arsenic transferred from groundwater to crops although many crops are still safe (Haq, 2002a). A recent study by Mehrag and Rahman (2003) found that arsenic is entering rice, through irrigation water pumped from contaminated aquifers. In Bangladesh, where rice is the staple food, the contribution of arsenic to dietary exposure, including contaminated drinking water, is considerable. Arsenic intake from rice would account for 17.3 and 29.6% of arsenic consumed if rice contained 0.1 and 0.2 μg/g of arsenic, respectively (Mehrag and Rahman, 2003). Mehrag and Rahman (2003) found arsenic levels in rice grain ranging from 0.058 to 1.83 μg/g. Their findings suggest that ingestion of rice is a major source of arsenic exposure in Bangladesh. As a consequence of elevated arsenic in crops, consumer confidence is declining (Hasan, undated). This is now becoming an issue as there are levels on bottled water and signs in vegetable markets informing people of arsenic free products. Crops originating in those regions where severe cases of arsenic toxicity have been recorded may not be marketable. This will have significant impact on the economic viability of these farmers.

Irrigation of paddy with arsenic contaminated water is also resulting in elevated arsenic levels in soils. Alam and Sattar (2000) found that arsenic levels in the soil were correlated with local well water concentrations, suggesting that the soils had become contaminated through irrigation with arsenic contaminated water. Mehrag and Rahman (2003) found high levels of arsenic in soils. They showed that soil arsenic levels will rise by 1μg/g/yr if fields are irrigated from tubewells with water containing 0.1 mg/L arsenic. The continued presence of arsenic in soil will have long-term impacts on crop productivity and quality.

Other than arsenic, there are evidences of various chemicals in groundwater at alarming levels, which could cause severe health hazards in long-term consumption. Two joint studies by the BGS and the Department of Public Health Engineering (DPHE) during 1998–1999 have found high concentrations of naturally occurring chemicals like uranium, manganese, boron, sulphur, fluoride and phosphorus in groundwater in several areas (Haq, 2002b). Other hazards include contamination of groundwater from agricultural inputs, improperly constructed sanitation facilities and industrial effluent. Few are prevalent at present but all are in a category where the risks are likely to grow (WARPO, 2001).

All of these findings depict a very grim future of groundwater irrigation and the socio-economic development of Bangladesh. The consequent adverse impact of arsenic ingestion to human and animal health may lead to significant economic loss through human loss, health and decline in crop productivity. There is a serious risk that we may face a bigger catastrophe than we can visualize. If steps are not taken in the right direction, arsenic contamination could threaten the very existence of our civilization.

There is clearly evidence that groundwater irrigation is a formidable tool for poverty alleviation. Nevertheless, it could also be the cause of mass poisoning. If so, then, those who today benefit from groundwater development may be ill-served by allowing unsustainable exploitation that will tomorrow leave them more impoverished.

5 OTHER THREATS TO THE SUSTAINABILITY OF GROUNDWATER

In addition to contamination by arsenic and other chemicals there are other threats in the development and management of groundwater resources.

According to the World Resources Institute's (1998) projection, the population of Bangladesh could reach 190 million in 2020 and 218 million in 2050. A projection of food grain demand and supply has shown that even with modest economic growth rate of 4%, demand for food grains will increase to 30 million tons in 2010 (present production is approximately 20 million tons). Assuming annual growth rates of 3% in irrigated area and 3.5% in rice yields, supply would only grow to 28 million tons in the year 2010. The shortfall would amount to 2 million tons by 2010. A large portion of this increased production is expected to come from expansion of irrigation. Currently 7.6 million ha out of the total cultivable land of 8.90 million ha used in agriculture are suitable for irrigation, and about 3.99 million ha is irrigated. MPO (1991) estimated that the irrigated area would reach 5.5 million ha by 2005 and 6.9 million ha by 2020. Currently Bangladesh withdraws 22,500 Mm^3 of water (WRI, 1998). According to the MPO (1991), the total requirement for water consumption in 2020 will be 24,370 Mm^3, and supply will be 23,490 Mm^3. Thus, there would be a shortage of 880 Mm^3.

If the present trend of irrigation development continues, where there is no growth of surface irrigation technology, it is obvious that all of this increased withdrawal will come from groundwater (Figure 4). This would make the already fragile groundwater resources highly unsustainable. Figure 6 shows the increased drawdown in dry years at full irrigation development in the groundwater irrigable *thanas* (GIT, sub-district administrative area) of Bangladesh. In many areas drawdown would increase further up to 6 m in dry season. However, the available resources will not be able to meet the demand. At present, during the dry season, when most of the irrigation occurs, heavy pumping causes the groundwater tables to drop below the suction level of STWs and treadle pumps. This happens over a third of the irrigated area. Poor farmer who cannot afford to invest in more efficient pumping devices suffer the most. Increased drawdown will further aggravate this situation.

The above estimate does not take into consideration the possible impact of climate change. The most recent projections (World Bank, 2000) indicate the following for Bangladesh:

- A maximum rise in sea level in the order of 300 mm by the year 2030 and 500 mm by 2050.
- An increase in monsoon rainfall of about 11% by the year 2030 and 28% by 2050. Dry season rainfall is expected to decrease 3% by 2030 and 37% by 2050.
- A rise in average annual temperature between 0.7°C in monsoon and 1.3°C in winter (dry season) by 2030 and 1.1°C in monsoon and 1.8°C in winter by 2050.

Global climate change would, through higher temperatures, further increase crop water requirements. Changes in climate may affect irrigation requirements for all the three cropping seasons. Increase in temperature will lead to escalating irrigation demands by 200 Mm^3 for March only (Brammer *et al.*, 1996). Additional irrigation demand would increase abstraction of groundwater significantly. This would be further augmented by the reduced trans-boundary surface water inflows into Bangladesh due to increased demand in upper riparian regions experiencing lower rainfall and higher temperatures. All these would seriously change the balance of recharge and demand on aquifers within the groundwater irrigable area of Bangladesh, and make groundwater use highly unsustainable.

6 TOWARDS SUSTAINABLE GROUNDWATER USE

The development of water resources in Bangladesh has a 40-year history, spanning the period 1957 to 1998. During this era, under various governments in the erstwhile Pakistan and Bangladesh, water resources development shifted noticeably. These shifts have affected the natural water regime in Bangladesh that has adversely affected its people. Changes made to land and environment through implementation of the various policies over time have complicated management of water resources in Bangladesh (World Bank, 2000). The orientation of all water sector development to this time was almost exclusively focused on achieving the goal of

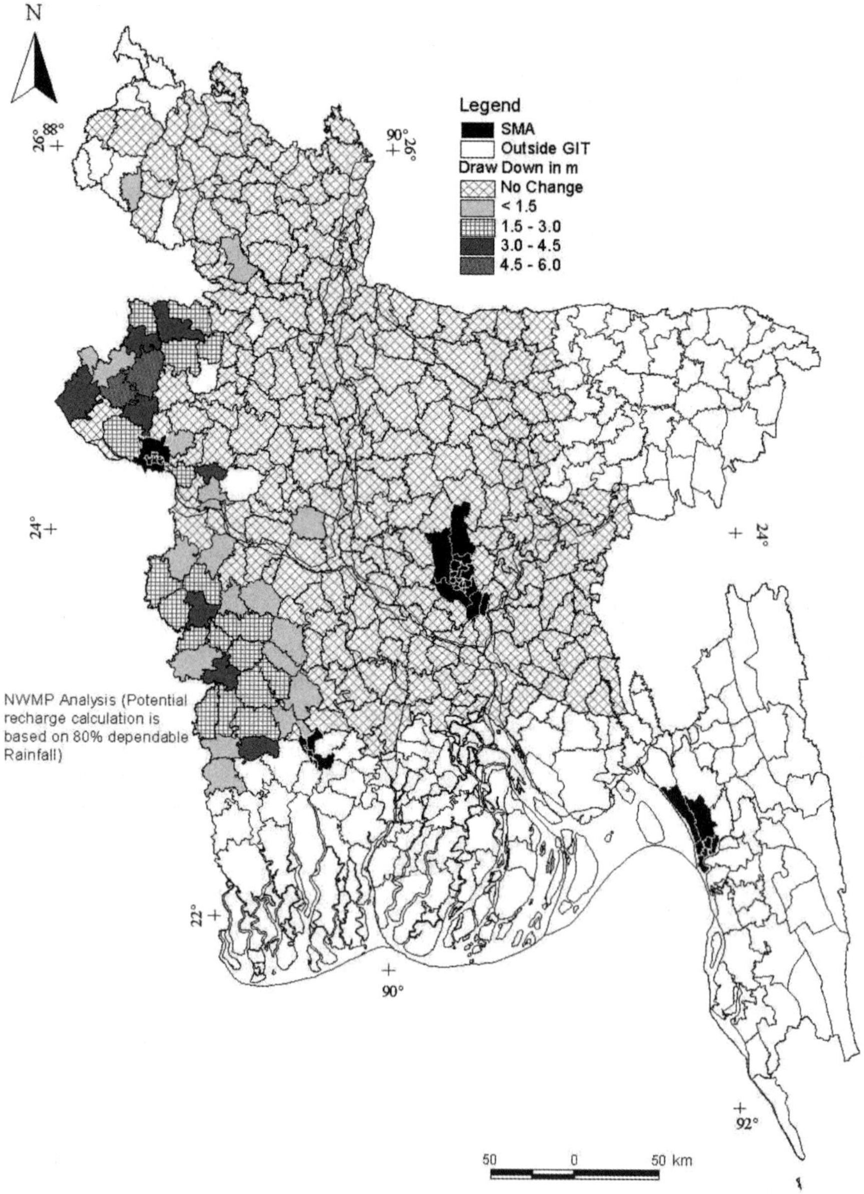

Figure 6. Increased drawdown in dry year at full irrigation development.
Source: WARPO (2001).

increasing agricultural production to achieve national self sufficiency. Development during this time was strictly sectoral with very little inter-sectoral communication. Water management was not considered in a comprehensive and integrated manner. Social and environmental impacts of water resource development were not addressed.

To address the impending challenges for sustainable use of water, particularly groundwater, there is a need to move away from a sectoral approach and towards comprehensive and integrated approaches. Integrated water resources management is a proper mechanism to account for the many aspects of water supply and demand, and to properly consider long term issues of sustainable resource management. Through integrated management, different sectoral interests are coordinated and important links are made with land use and environmental planning and management. Therefore, if productivity gains are to become instruments of poverty alleviation, it is necessary to implement an integrated approach to water management, which is mainly an integration of surface water and groundwater in a river basin context in the case of Bangladesh.

Bangladesh occupies a delta plain formed by the three great rivers Ganges, Brahmaputra and Meghna (GBM). A total of 700 rivers flow throughout the country, having a total length of 22,155 km eventually discharging into the Bay of Bengal (BBS, 2001). These give Bangladesh the highest per capita availability of surface water (20,000 m^3/capita according to an estimate in 1995) (SEI, 1997; The Daily Star, 2002b). A lack of suitable reservoir locations and the complexity of the flat deltaic river plain, make large-scale surface irrigation difficult to develop and maintain. However, the surface water sector has developed very little when compared to its actual potential. Among the minor irrigation technologies comprising STWs, DTWs, low-lift pumps (LLPs), and manually operated pumps, only LLPs are based on surface water. While growth of STWs irrigated area has picked up sharply, (460% from 1983 to 1999), LLP shows a long-term relative decline of 4.2% during this period. Traditional systems, also based on surface source, have declined considerably, from 28.6% or total irrigated area in 1983 to 8.3% in 1999 (Figure 4). Many researchers suggest that lack of surface water utilization due to upstream withdrawal and conflict between fish versus LLP as the major cause of declining for this form of irrigation. There are other causes, apart from this, which also contributed significantly to this decline. These are:

i) Lengthy bureaucratic procedure to rent or to buy LLPs from BADC.
ii) The discharge capacity of LLPs is usually higher than that of STWs which complicate the management and reduce the performance.
iii) In comparison to LLPs, STWs are easy to operate, need less initial investment, widely available and less expensive for repair and maintenance.

Due to ease of using groundwater, farmers have shifted to the use of STWs even in the areas where there is surface water potential. The rapid expansion of groundwater irrigation has driven away the attention of the farmers from the surface water sources such as rivers and natural water bodies. The importance of rivers which was once the main route for rural transport, has further declined due to the massive development of rural road networks. All these developments left the rivers totally unattended that resulted in the encroachment of the land for cultivation by gradually filling up the river beds and natural water bodies. Nevertheless, there is still scope to develop surface water resources. The first step towards the sustainability of groundwater is, therefore, to give priority to the development of surface water resources through river basin management.

The first step towards river basin management is to undertake a water balance or water accounting for the basin. River basin water accounting provides an insight into where institutional change need to be targeted thereby resulting in institutional transformation to use the basin water efficiently, more production per drop and sustain the productivity and environment of the system (Sakthivadivel and Molden, 2001). In a riverine Bangladesh, many rivers are in direct contact with the aquifer, and generally river contributes to aquifer from the month of March to November (SWMC, 2000). The usable capacity of groundwater aquifers should be developed by planned extraction of groundwater during periods of low precipitation while subsequent replenishment can be made during periods of surplus surface supply. The concept of river basin planning and water accounting can play a significant role for optimal planning for the conjunctive use of surface water and groundwater.

There are several different technical issues to address in improving water resources management in Bangladesh, following are the strategically important areas where major challenges exist that require careful attention by the agencies concerned.

Promoting the Expansion of LLP Irrigation: There are currently some limited opportunities to develop locally available surface water resources in a manner that will promote further expansion of LLP irrigation. Government should take steps to support expansion of LLP irrigation. The encroachment and filling of the natural water bodies and the small tributaries of the river should be stopped. The storage and navigational capacity of the rivers should be enhanced by cost-effective dredging.

Water Saving Irrigation Practices: At present, "water saving irrigation (WSI)" to produce *more crop per drop* is a popular issue and considerable research is being carried out towards that direction. Several reports from China claim that the WSI techniques could increase on-farm water productivity by 20 to 35% compared with the traditional irrigation practices (Li, 2001). If this claim is to be believed, significant amount of water can be saved and diverted to non-irrigated areas or pressure on groundwater can be eased by practicing WSI techniques for rice as more than 80% of irrigated area are covered by rice crop in Bangladesh.

Conjunctive Management of Groundwater and Rainwater: Conjunctive management of groundwater and rainwater is also very important during the dry season. Groundwater is predominantly used for dry season irrigation in Bangladesh when 22% of the annual rainfall occurs. Maximum use of rainwater can decrease the pumping of groundwater. Existing cropping pattern should be compared with the long-term average rainfall to investigate whether changes in the cropping calendar can maximize the rainfall use.

Increasing the Efficiency of Public Irrigation Schemes: There is a total of 15 existing major (over 2,000 ha) surface irrigation schemes, covering 480,000 ha net command area (NCA). Present irrigation intensities of these schemes are very low. In the 1996–1998 period only 46% of the 15 major schemes NCA was irrigated in the main irrigation season, the *rabi* (winter), and only another 7% in the *Kharif I* (pre-monsoon) season. Other areas are covered with groundwater irrigation (WARPO, 2001). Action is required to improve the performance and management of existing surface irrigation schemes, which eventually will decrease the pumping of groundwater.

7 CONCLUSION

In this paper, the evolution of groundwater irrigation, its rapid growth over the last two decades, and its role in attaining food grain self-sufficiency in Bangladesh is discussed. The paper shows how groundwater development has become a formidable tool for poverty alleviation. Alongside, the risk of using this resource due to arsenic contamination and the challenges ahead for its sustainable use are outlined.

Sustainable groundwater management is a very complex issue, particularly in Bangladesh, where agricultural production is still the mainstay of the rural population's livelihoods. A vast majority of the people's livelihoods are still inextricably linked to the use of groundwater for irrigation. Groundwater is the raw material that supports a population with complex, interwoven and overlapping livelihood strategies. However, millions of people are at risk of being poisoned by consuming food irrigated with arsenic contaminated groundwater besides drinking contaminated water directly. This makes the issue of groundwater management even more complex. A strategy for the management of groundwater resources is suggested as a component of an integrated water resources management (IWRM), which promotes the coordinated development of surface water and groundwater in order to maximize the resultant economic and social welfare without compromising sustainability. Increased productivity of water, increased food security for smallholders and marginal farmers, and poverty reduction will remain elusive without an integrated approach to water resources management.

Realizing the need for a coherent water strategy for water resource exploration, use and management, the Government of Bangladesh began the process of drafting a National Water Policy (NWPo) for the first time in the country's history, which was published in 1999. The goal of the NWPo is "to ensure progress towards fulfilling national goals of economic development, poverty alleviation, food security, public health and safety, a decent standard of living for the people and protection of the natural environment" (MoWR, 1999).

The importance of IWRM has been recognized in the *National Water Policy.* The NWPo seeks a remedy to this chaotic situation by bringing order and discipline in the exploration, management and use of water resources in Bangladesh. It clearly and unequivocally declares that all necessary means and measures will be taken to manage the water resources of the country in a comprehensive, integrated and equitable manner. The NWPo provides an extensive framework for management of the water sector. However, in setting priorities for allocating water during critical periods, the NWPo gives this sector a relatively low priority though the largest demand for both surface and groundwater is to support irrigation in the dry months and sets the following order: domestic and municipal uses, non-consumptive uses (e.g., navigation, fisheries and wild life), sustenance of the river regime, and other consumptive and non-consumptive uses including irrigation, industry, environment, salinity management, and recreation (MoWR, 1999).

However, the declaration of the NWPo is a bold step towards integrated water management in Bangladesh. Many adverse and counter-productive situations have been created due to lack of coordination in development programs and use of water resources (MoWR, 1999). For a water-dependent country like Bangladesh, this situation is highly detrimental to its overall development and this need to be addressed. The *water policy* lays down the broad principles of development of water resources and their rational utilization. It will help guide both public and private actions in the future for ensuring optimal development and management of water that benefit both individuals and the society at large. However, future concerns prevail regarding the implementation of national policies, due to lack of institutional capability and awareness to properly address the policy objectives and goals. In practice, there are very few cases of integrated water resources management *on the ground*, and it may still take a long time to materialize.

ACKNOWLEDGEMENT

The author is thankful to Andrew Noble, Principal Researcher of IWMI Southeast Asia Office, for his comments and editorial correction on an earlier draft of this paper. The views expressed in this paper are not necessarily the views of IWMI.

REFERENCES

Aktheruzzaman, M. and Jaim, W.M.H. (1999). Market for groundwater irrigation in Bangladesh: the supply side. *Journal of Applied Irrigation Science*, 34(1): 23–39.

Alam, M.B. and Sattar, M.A. (2000). Assessment of arsenic contamination in soils and waters in some areas of Bangladesh. *Water Science and Technology*, 42: 185–192.

Anwar, J. (2000). *Arsenic Poisoning in Bangladesh: End of a Civilization.* Palash Media, Dhaka, Bangladesh, 336 pp.

Bangladesh Bureau of Statistics (BBS) (1976). *Statistical Yearbook of Bangladesh.* Bangladesh Bureau of Statistics, Dhaka, Bangladesh.

Bangladesh Bureau of Statistics (BBS) (2001). *Statistical Yearbook of Bangladesh.* Bangladesh Bureau of Statistics, Dhaka, Bangladesh.

Bangladesh Bureau of Statistics (BBS) (2002). *Statistical Yearbook of Bangladesh.* Bangladesh Bureau of Statistics, Dhaka, Bangladesh.

Bhuiyan, N.I.; Paul, D.N.R. and Jabber, M.A. (2002). Feeding the extra millions by 2025 – challenges for rice research and extension in Bangladesh. Keynote paper presented in the National Workshop on *Rice Research and Extension – 2002* held at Bangladesh Rice Research Institute, Joydebpur, 29–31 January.

Brammer, H.; Asaduzzaman, M. and Sultana, P. (1996). Effects of climate and sea-level changes on the natural resources of Bangladesh. In R.A. Warrick and Q.K. Ahmad (eds.), *The Implications of Climate and Sea-Level Change for Bangladesh.* Kluwer Academic Publishers, Dordrecht, The Netherlands.

British Geological Survey (BGS) (1999). *Groundwater Studies for Arsenic Contamination in Bangladesh.* Main Report and Supplemental Volumes 1–3. Dhaka, Bangladesh.

British Geological Survey (BGS) (2000). *Groundwater Studies for Arsenic Contamination in Bangladesh.* Main Report and Supplemental Volumes 1–3. Dhaka, Bangladesh.

Chakravorty, N. (2001). An overview of irrigation and rural poverty issues in Bangladesh. In I. Hossain, and E. Biltonen (eds.), *Managing Water for the Poor. Proceedings of the Regional Workshop on Pro-Poor Intervention Strategies in Irrigated Agriculture in Asia.* International Water Management Institute, Colombo, Sri Lanka, 220pp.

Downing, J. and Polak, P. (2000). *The Development and Commercialization of the Treadle Pump in Bangladesh: A Case of Product Marketing on a Mass Scale*. United States Agency for International Development, Bethesda, USA, 33pp.

(Food and Agriculture Organization) FAO. (1998). *FAO/WFP Crop and Food Supply Assessment Mission to Bangladesh.* Special Report, FAO, Rome.

(Food and Agriculture Organization) FAO. (1999). *On World Water Day, FAO Calls for Irrigation for Small Farmers*. http://www.fao.org/news/1999/990306-e.htm

Fazal, M.A.; Kawachi, T. and Ichion, E. (2001a). Extent and severity of groundwater arsenic contamination in Bangladesh. *Water International*, 26(3): 370–379.

Fazal, M.A.; Kawachi, T. and Ichion, E. (2001b). Validity of the latest research findings on causes of groundwater arsenic contamination in Bangladesh. *Water International*, 26(3): 380–389.

Hamid, M.A. (1991). *A Data Base on Agriculture and Foodgrains in Bangladesh*. Bangladesh Agricultural Research Council, Dhaka, Bangladesh.

Haq, N. (2002a). Arsenic Creeps into Food Chain: Impact on Health still Unknown. *The Daily Star*, 3(974), Thursday, June 6, 2002. Bangladesh.

Haq, N. (2002b). Drinking Death in Groundwater. *The Daily Star*, 3(877), Wednesday, February 20, 2002, Bangladesh.

Harvey, C.F.; Swartz, C.H.; Badruzzaman, A.B.M.; Keon-Blute, N.; Yu, W.; Ali, M.A.; Jay, J.; Beckie, R.; Niedan, V.; Brabander, D.; Oates, P.M.; Ashfaque, K.N.; Islam, S.; Hemond, H.F. and Ahmed, M.F. (2002). Arsenic mobility and groundwater extraction in Bangladesh. *Science*, 298: 1602–1606.

Hasan, M.A. (undated). Contamination of Soil Due to Irrigation with Arsenic Laden Water and Its Impact on Phosphorus Leading to Crop Production in Bangladesh. http://www.eng-consult.com/arsenic/articles/DU-ACIAR_Project.html

Karkkainen, T. (2001). *Irrigation in South Asia (with a focus on drip irrigation)*. Research Report, Laboratory of Water Resources, Helsinki University of Technology.

Li, Y.H. (2001). Research and practice of water-saving irrigation for rice in China. In R. Barker, R. Loeve, Y. H. Li, and T. P. Toung (eds), *Water-Saving Irrigation for Rice, Proceedings of an International Workshop*, International Water Management Institute, Colombo, Sri Lanka, 123pp.

Mandal, M.A.S. and Parker, D.E. (1995). *Evolution and Implications of Decreased Public Involvement in Minor Irrigation Management in Bangladesh*. Short report series on locally managed irrigation, International Water Management Institute, Colombo, Sri Lanka, 22pp.

Master Plan Organization (MPO). (1991). *National Water Plan-Phase II, Vol. I and Vol. II*. Master Plan Organization, Dhaka, Bangladesh.

Mehrag, A.A. and Rahman, M.M. (2003). Arsenic contamination of Bangladesh paddy field soils: implications for rice contribution to arsenic consumption. *Environment Science and Technology*, 37(2): 229–234.

Ministry of Water Resources (MoWR). (1999). *National Water Policy*. Ministry of Water Resources, Government of the People's Republic of Bangladesh, Bangladesh.

National Minor Irrigation Development Project (NMIDP). (1997). *National Minor Irrigation Census, Bangladesh: 1995-96 Irrigation Season*. Ministry of Agriculture, Government of Bangladesh.

National Minor Irrigation Development Project (NMIDP). (2001). *National Minor Irrigation Census, Bangladesh: 1999–2000 Irrigation Season*. Ministry of Agriculture, Government of Bangladesh.

Nickson, R.; McArther, J.; Burgess, W.; Ahmed, K. M.; Ravenscroft, P. and Rahman, M. (1998). Arsenic poisoning of Bangladesh groundwater. *Nature* 395.

Pagiola, S. (1995). *Environmental and Natural Resources Degradation in Intensive Agriculture in Bangladesh*. The World Bank, 32pp.

Roy, K.C. and Mainuddin, M. (2003). Socio-Ecology of Groundwater Irrigation in Bangladesh. Workshop on *Groundwater Socio-Ecology of Asia: Governing a Colossal Anarchy*, held on 26 January 2003, at Anand, India.

Sakthivadivel, R. and Molden, D. (2001). Linking water accounting analysis to institutions: synthesis of five country studies. In C. L. Abernethy (ed), *Intersectoral Management of River Basins*, International Water Management Institute, Colombo, Sri Lanka, 416pp.

Shah, T.; Alam, M.; Kumar, M.D.; Nagar, R.K. and Singh, M. (2000). *Pedaling Out of Poverty: Social Impact of a Manual Irrigation Technology in South Asia*. International Water Management Institute, Sri Lanka.

Siamwalla, A. (2001). *The Evolving Roles of the State, Private, and Local Actors in Rural Asia*. Asian Development Bank, Manila, Philippines.

Smedley, P.L. and Kinniburgh, D.G. (2002). A review of the source, behaviour and distribution of arsenic in natural waters. *Applied Geochemistry*, 17(5): 517–568.

Smith, A.H.; Lingas, E.O. and Rahman, M. (2000). Contamination of drinking-water by arsenic in Bangladesh: a public health emergency. *Monthly Bulletin, World Health Organization*, 78(9): 1093–1103.

SOES-DCH. (2000). *Groundwater Arsenic Contamination in Bangladesh*. School of Environment Studies, Jadavpur University, Kolkata, India and Dhaka Community Hospital, Dhaka, Bangladesh.

Stockholm Environment Institute (SEI). (1997). *Comprehensive Assessment of the Freshwater Resources of the World Water Futures: Assessment of Long-Range Patterns and Problems*. Stockholm Environment Institute.

Surface Water Modeling Center (SWMC). (2000). *Surface Water Groundwater Interaction Model Study*. Technical Report No. 6, SWMC, Dhaka, Bangladesh.

Sustainable Development Networking Program (SDNP). (2002). www.sdnbd.org/sdi/issues/agriculture/main.htm

The Daily Star (2002a). Develop Surface Water as Main Source of Drinking Water. *The Daily Star*, 3(902), Thursday March 21 2002, Bangladesh.

The Daily Star (2002b). Use surface water as much as possible: PM, *The Daily Star*, 3(1107), Thursday, October 17, 2002, Bangladesh.

WARPO (2001). *National Water Management Plan*. Vol. 1–5, WARPO, Dhaka, Bangladesh.

World Bank (2000). Bangladesh: *Climate Change and Sustainable Development*. Report No. 21104 BD, The World Bank, Dhaka, Bangladesh.

World Health Organization (WHO) (2001). http://www.who.int/inf-fs/en/fact210.html

World Resources Institute (WRI) (1998) *World Resources: 1998-99*. World Resources Institute. New York: Oxford University Press.

Fluoride contamination of groundwater in India – country update

Hema T. Chaturvedi[(1)], D. Chandrasekharam[(2)] and A.A. Jalihal
Department of Earth Sciences, Indian Institute of Technology, Bombay, India
[(1)] *hematc@iitb.ac.in*
[(2)] *dchandra@geos.iitb.ac.in*

ABSTRACT

About 63 million people in rural India suffer from dental, skeletal and or non-skeletal fluorosis due to consumption of groundwater with high fluoride content. The problem of fluorosis is more severe than the arsenic poisoning reported from the West Bengal Delta plains. The source of fluorine, unlike arsenic, is geogenic. Arid climate, low rainfall coupled with high evapotranspiration, low recharge and high residence time of groundwater are some of the factors enhancing the fluoride concentration in groundwater. Though several defluoridation techniques are available, cost factor is a deterrent in implementing such technologies in rural areas. It is time that cost effective defluoridation methods are evolved using locally available natural products. Preliminary attempts indicate that such methods are possible to develop and implement in rural areas in India.

Keywords: *groundwater, fluoride, fluorosis, defluoridation.*

1 INTRODUCTION

In many parts of the world, high concentration of fluoride in groundwater is causing widespread fluorosis. Fluorine exists in a variety of rocks and minerals and can enter groundwater due to water-rock interaction processes. Nearly 70% of rural world population suffering from fluorosis depends on groundwater. According to WHO (1984) guidelines, in areas with a warm climate, the optimal fluoride concentration in drinking water should be below 1 ppm, while in cooler climatic regions it could be up to 1.2 ppm. Since the threshold limit of fluoride to cause dental fluorosis is 1.5 ppm, WHO sets the upper limit of fluoride in drinking water at 1.5 ppm. However this guideline is not universal. For example in India this value is set at 1 ppm (WHO, 1984).

India is one among such countries where over 63 million people depend on groundwater with fluoride concentration varying from 1 ppm to 20 ppm (Teotia *et al.*, 1981). Almost all the states in India are facing this problem (Figure 1, Table 1). Fluorosis in India is much more severe than arsenic poisoning in west Bengal. According to recent statistics the number of people subjected to arsenic poisoning in India is 38 million and confined to West Bengal (Bhattacharya, 2002) and is far less than those affected by fluorosis. While the source of arsenic in west Bengal is not well understood (Chandrasekharam, 2002; Stueben *et al.*, 2003) the source of fluoride is well defined and is geogenic. In this paper the present status of fluorosis in India is discussed in the light of the use of groundwater, climate, geology and source parameters.

2 HIGHLY ENDEMIC STATES

Rajasthan, Gujarat and Andhra Pradesh are highly endemic fluorosis states (Figure 1), where 65%

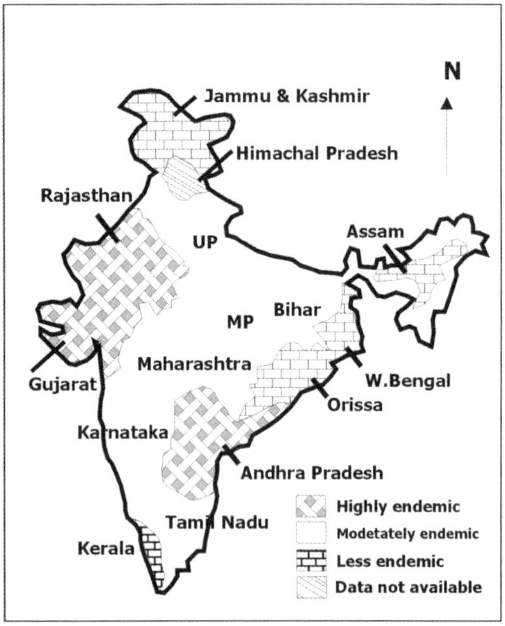

Figure 1. Status of fluorosis in India. State boundaries are not shown for clarity. Map not to scale. (Modified after Susheela, 1999).

of population depending on groundwater with fluoride varying from 0.4 ppm to >20 ppm (Table 1). The annual rainfall in these states is low (69–250 cm; Table 1) and annual evapo-transpiration is extremely high (1,800–2,100 mm/yr; www.drought.unl.edu). In Rajasthan, groundwater occurs at depths of about 25 m to 80 m and is confined to crystalline igneous, metamorphic and sedimentary rocks while in Gujarat it occurs at a depth greater than 100 m (hard rock aquifers in the Gujarat mainland) while in Kutch it occurs in shallow sedimentary aquifers. Limestone, shale and sandstone which form the sedimentary aquifers contain 200 ppm to 940 ppm of fluorine (Fleischer and Robinson, 1963). The crystalline rocks of Rajasthan host a variety of fluorine bearing minerals and tourmaline (with ~1% F) bearing pegmatite are found extensively in Rajasthan (Balasubramanium and Kumar, 1974; Sengupta, 2002).

One of the largest fluorspar deposits in India is located in Ambadongar, Gujarat. It forms veins of varying dimensions within brecciated and fractured carbonatite and associated host rocks, mostly as replacement and cavity fillings (Sukeshwala and Udas, 1963, 1964; Bose and Banerjee, 1974). Groundwater occurring in and around Ambadongar has fluoride content of 12.9 ppm (Gupta and Deshpande, 1998).

In Andhra Pradesh, groundwater along the costal zones occurs in laterites and sandy alluvium while in the interior parts, hard rocks form the main aquifer. Groundwater in the hard rocks, consisting of granite, gneiss, khondalites and charnockites, occur at depths of about 100 m. These rocks form a part of the well known eastern ghat khondalite belt. All the igneous suite of rocks contains fluorite, hornblende and biotite, which are the main source of fluoride (Ramamohan *et al.*, 1993; Prasad Rao *et al.*, 1974).

3 MODERATELY ENDEMIC STATES

Fluorosis is moderately endemic in Orissa, Uttar Pradesh, Karnataka, Punjab, Haryana, Tamil Nadu, Madhya Pradesh and Kerala (Figure 1). About 65–86% population in these states drink groundwater with 16 ppm of fluoride. Except for Kerala (which receives uniform rainfall of ~300 cm); the rainfall in the remaining states is far below this value and varies widely (Table 1).

In Orissa, a large population depends on groundwater occurring in crystalline rocks (e.g. charnockites, khondalites and granite gneisses) at depths of 200 m. Like Rajasthan, these rocks are traversed invariably by pegmatite veins. Fluorine bearing minerals like fluorite, fluorapatite, biotite, hornblende and tourmaline occur in these rocks which form the main source of fluoride in groundwater (Das *et al.*, 2000). On the contrary, the groundwater conditions in Uttar Pradesh is quite different from Orissa, in the sense that a large part of this state in covered by Quaternary Gangetic alluvium (gravels, sands and clays) bounded on the north by the Himalayan mountain belt (which contains granites, shales, limestone and quartzites) and on the south by the Vindhyan chain of mountains (represented by sandstones and quartzites). The river Ganges originates from the above said rock types and forms the main recharge source for alluvial aquifers.

In the interior parts of Karnataka which fall under the rain shadow region, groundwater is the main source for drinking. Groundwater with

Table 1. Parameters related to fluoride in groundwater of India.

States	F^- (ppm)	Climate	Rainfall (cm)	Lithology	% population affected
Rajasthan	0.1-16	Arid to semi-arid	70–160	Gneisse, schists, phyllites	77
Gujarat	0.4–12.9	Humid-arid	57–94	Gneisse, schists, phyllites	65
Andhra Pradesh	0.5–20	Humid-Semi-arid	69–250	Granites, gneisses, Charnockites and Khondalites	73
Orissa	3.5–16.4	Equable	50–150	Granites, khondalites and gneisses, of shale, coal seams, and limestone	86
Uttar Pradesh & Delhi	1–8.9	Temperate	91–100	Granites, gneiss, shale and limestone, alluvium	80 & 9
Karnataka	0.8–12.8	Semi-tropical to semi-arid	70–350	Gneisses, granulites, granites, schists, sandstone, shale and limestone	68
Punjab & Haryana	0.64–13	Temperate	35–125	Alluvium	70 & 75
Tamil Nadu	0–5	Tropical	33–99	Charnockites, granites, gneisses, granulites	65
Madhya Pradesh	0–19	Temperate to humid	100–135	Granites and gneisses	
Kerala	N.A.	Tropical	~300	Laterite, sand and clay.	
Maharashtra	0.7–10	Tropical monsoon	70–350	Gneiss and granites and Gondwana sedimentary	61
Bihar	10–17	Sub-tropical	35–121	Schist, quartzite, granite and granulite.	86
Assam	1–20	Tropical	170–305	N.A.	88
West Bengal	N.A.	Tropical	150–300	N.A.	72
Jammu & Kashmir	N.A.	Alpine to sub-tropical	N.A.	N.A.	76

Sources: Alvi and Koteswaram, 1985; Gupta *et al.*, 1993; Gupta and Deshpande, 1998; Rammohan Rao *et al.*, 1993; Das *et al.*, 2000; Gupta *et al.*, 1994; Faruqi *et al.*, 1997; Jalihal, 2002; Singh, 1998; Adyalkar and Radhakrishna, 1974; Chakraborti *et al.*, 2000; Burman, Dev *et al.*, 1995; Chowdhary *et al.*, 1964; Saxena and Ahmed, 2003.
N.A.: Data not available.

high fluoride content occurs in hard-rock aquifers (30–90 m depth) and supports a population of 68% (Table 1). Due to semi-arid and high evapotranspiration conditions (>1000 mm/yr), a major part of the litho-units are highly weathered promoting high degree of water-rock interaction (Sumalatha *et al.*, 1999). Fluorite and other fluorine bearing minerals occur extensively as joint fillings and also as primary minerals in granites and gneisses. The Arsekere granite, for example, contain large crystals of apatite and fluorite (~5 mm; Jalihal, 2002). The cleavages of certain minerals like muscovite and biotite, occurring in the hard rocks, host elongate blebs of blue fluorite. Pegmatite veins containing large grains of fluorite and tourmaline occur widely in granites and gneisses. The well-known Kolar schist belt with gold mineralization contain large amount of fluorapatite and fluorite (Sumalatha *et al.*, 1999). The volume of occurrence of granites in the state is very large which has given way to large-scale mining of granite for building purpose. Thus hundreds of small and big granite cutting and polishing industries are thriving in the state. Granite industry has a serious impact on the fluoride problem in this state (Jalihal, 2002).

In the case of Punjab, located between the Indo-Gangetic alluvial plains, the southern part of the state adjacent to Rajasthan, 75% of the population drink groundwater with fluoride content varying from 0.64 ppm to 13 ppm (Table 1). Aeolian derived sand from Rajasthan desert containing mineral grains like muscovite and biotite with high fluorine content (Faruqi *et al.*, 1997) appears to be the main source contributing fluoride to the aquifer.

In Tamil Nadu, 65% of the population drink groundwater with fluoride varying from 0–5 ppm (Table 1). Tamil Nadu consists of a central arid plain and is bounded on the west by the Western Ghats and isolated rocky hills on the east. This state is drained by the river Cauvery which provides drinking water to those communities (~35%) located along the river bank. Due to repeated failure of monsoon (annual rainfall is <900 mm) and drought conditions (with summer temperature exceeding 45°C) this part of state faces acute shortage of drinking water. The depth to water table varies from 3–25 m in the state (www.twadboard.com). In addition to this, over-exploitation of groundwater in 1980s under the UNDP programme has lowered the water table permanently (~3,383 Mm^3 of water has been extracted from static storage; CGWB, 1975) in the entire state. Groundwater occurs under confined conditions in charnockites, granite and gneisses aquifers. Both the hard rocks and the river alluvial aquifers contain fluorine bearing minerals. Fluorine bearing phosphatic nodules in the sedimentary formations (known as the Cretaceous Formations of Tiruchirapalli; Murthy, 1974) and other fluorine bearing minerals mentioned above are the main source contributing fluoride to the groundwater.

In Madhya Pradesh a variety of rock units extending in age from the Precambrian to Recent (Krishnan, 1985; Banerjee and Sathyanarayana, 1964; Bose and Banerjee, 1974; Adyalkar and Radhakrishna, 1974) occur. The fluoride concentration in groundwater varies from 0.1–3.4 ppm (Saxena and Ahmed, 2003) with depth to water level varying from 3.8 m to 20 m. However much higher levels of fluorine are reported in thermal spring waters (>39 ppm; Minissale *et al.*, 2000) in the Narmada-Tapi valley. Both the litho-units as well as the thermal springs are sources of fluorine in this state.

The state of Kerala has three distinct physiographic divisions – the highlands containing igneous and metamorphic rocks similar to those described above, the midlands consisting mainly of 10 m to 15 m of laterite derived from the weathering of these rocks and the coastal alluvium consisting of sand and clay. The laterite is the main aquifer in the state and groundwater extracted through open dug wells is extensively used for irrigation and drinking. The density of open dug wells is highest in this state amounting to about 100 wells in 1 km^2. Though there are thirteen rivers in the state, due to high degree slope, the run-off is very high. About 90% of the state's population live in the midlands (Chandrasekharam, 1989). The problem of high fluoride in groundwater in Kerala has been reported so far in Allepy and Palaghat districts (Susheela, 1999). Data on other districts are not available.

4 LESS ENDEMIC STATES

In the earlier published reports (Susheela, 1999) Maharashtra, Bihar, Assam, West Bengal, Jammu and Kashmir are grouped under less endemic fluorosis affected states. However, recent investigations show that parts of Assam and Maharashtra are also severely affected by fluorosis and hence cannot be completely considered as less endemic fluorosis states. For example 88% of population in Karbi-Anglon district of Assam drink groundwater with 20 ppm of fluoride (Saji, pers. comm.) while Nanded district in Maharashtra fluoride content in groundwater is over 10 ppm (Shukla, pers. comm.). Thus detailed investigation related to the problem of fluorosis in Assam and other eastern states (Mizoram, Meghalaya, Manipur etc.) is lacking. These states have contrasting geology and rainfall pattern. Assam and West Bengal are located over thick alluvial formations and receive 1,700–3,500 mm rain annually while interior Maharashtra falls under rain shadow region (eastern part of the Western Ghat ridge) and receives only 700 mm of rain annually. Drought conditions also prevail in certain parts of Maharashtra. Except for small areas north and south of Nagpur, the remaining part of Maharashtra is covered with volcanic rocks. Though the volcanic rocks contain 450 ppm of fluorine (Seraphim, 1951; 3.25% of F is reported from the giant plagioclase feldspar in volcanic rocks; Sahasrabudhe, 1974), fluorosis is reported only in Nanded district in granites and gneiss aquifers.

Bihar (including Jharkhand, the recently formed state) has typically two distinct physiographic units: SE sloping plains in the north, and plateau towards south. In the Singbhum district of south Bihar (Jharkhand state), Archeans are exposed which host a variety of minerals of economic importance. The rock types include schist, quartzite, granite and granulite. Pegmatites occurring in the granites contain tourmaline. Apatite occurs abundantly as fissure, breccia and shear fillings and also as tabular cryatals in the pegmatite veins (Dhruva Rao and Balmiki, 1963). For example the Hazaribagh mica peridotite dykes contains 12.5% of apatite (Bose and Banerjee, 1974).

5 DISCUSSIONS

From the forgone discussion it is apparent that the hard rock aquifers in all the states invariably contain fluorine bearing minerals, with fluorine content varying from 0.10% to 2.38% in biotite, 0.14–0.22% in muscovite, 0.23–1.50% in hornblende, 3.5% F in fluor-apatite, 48.7% F in fluorite and 0.07–1.27% in tourmaline (Deer *et al.*, 1985; Deshmukh *et al.*, 1995).

From the Table 1 the relationship between rainfall, groundwater availability and population affected by fluorosis is quite clear. There is a wide gap between the available groundwater and population depending on this source (Figure 2). In addition to this, excess withdrawal of groundwater, as exemplified from the UNDP project of Tamil Nadu, is also another factor which has enhanced fluoride levels in groundwater. The people affected by fluorosis thus are those depending only on groundwater source in all the states. Direct example can be drawn from Rajasthan and Orissa where the fluoride level is more or less similar (16 ppm) but Rajasthan is affected severely compared to Orissa for the simple reason that Orissa has alternate water supply source like rivers and hydro-electric projects while Rajasthan has to depend entirely on rainfall and groundwater. In fact world over, fluorosis is directly related to the source rocks, especially where mica minerals and hornblende are found abundant in the hard rock aquifer (Ren and Shugin, 1988; Baoshan and Yetang, 1988; Yong and Hua, 1991; Hamamoto, 1957; Boyle and Chagnon, 1995; Morris, 1965; Manji *et al.*,

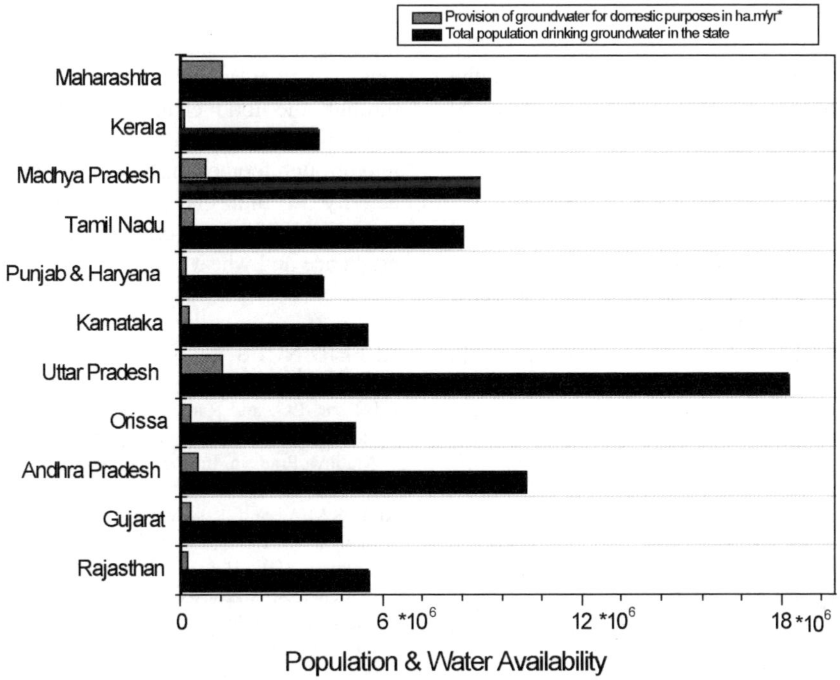

Figure 2. Population and water availability in fluoride endemic states of India *ha.m is a volume measure of a hectare surface area that is 1 m high.

1986; Smet, 1992; Brouwer *et al.*, 1998; Saether *et al.*, 1995). For example high fluoride water in China occurs in biotite/muscovite and hornblende (F content 1–7%) bearing granitoid rather than in rocks containing fluorite bearing granitoid (Yong and Hua, 1991). Similarly the Bongo granite aquifer in Ghana contains 792 ppm of fluorine and is the main source of fluoride in groundwater (Apambire *et al.*, 1997). Since the solubility of fluorite compared to biotite and hornblende is low (Appelo and Postma, 1993; Handa, 1977) these ferro-magnesian minerals dissolve faster compared to fluorite.

Besides the hard rock aquifer, thermal springs, as reported above from the Narmada-Tapi region, are also a source of fluoride.

Thus it is apparent that fluorosis in India is geogenic and is related to the rainfall, climate and the degree of water-rock interaction. In addition to this, mining activity in states like Karnataka is an additional source enhancing fluoride levels in groundwater.

Most of the rural population, affected by fluorosis, depend on agricultural income and thus represent economically poorest section of the society. Though modern technological development could provide several defluoridation techniques (Bower and Hatcher 1967; Killedar and Bhargava, 1988a; Killedar and Bhargava, 1988b; Maruthamuthu and Venkatanarayana 1987; Singano *et al.*, 1998; Bjorvartn *et al.*, 1998; Garmes *et al.*, 2002; Mahramanlioglu *et al.*, 2002; Cengelog¢lu *et al.*, 2002), economic consideration is a major hurdle in effectively implementing such technologies. The Nalgonda defluoridation method using alum is considered to be the most effective (Nawalkhe *et al.*, 1975). Minimum cost of this method is around US$ 600 for a community hand-pump. In India, a maximum of two or three hand-pumps serve an entire village community. Assuming that initial expenditure is borne by the local governments, wherever community rural water supply scheme exists, subsequent maintenance of the systems rests on individual community and becomes too expensive. In the absence of rural community water supply, dug/bore wells are the main source of drinking water. Installing Nalgonda method at all such locations is not feasible.

The alternate method for the rural population is either to use the river water or rain water since in both these sources the amount of fluoride is well below the limit of 1.5 ppm. The Indian rivers contain about 0.013–1.7 ppm of fluoride and thus are the best source of drinking water for rural population (Madhavan and Subramanium, 2001). But many states in India, like Rajasthan, fall under dry arid climatic zone and groundwater replenishment is poor due to low rainfall and high evapotranspiration rates.

Modern technology should focus on developing defluoridation methods using locally available natural material (e.g. plant roots, vegetable seed, etc.). Preliminary attempts have been made to reduce fluoride concentration in groundwater occurring in Hungund, Karnataka using fried tamarind seed powder (Jalihal, 2002). This method is able to reduce the fluoride concentration from 8 ppm to 6.6 ppm. Such methods will be cost effective and affordable by millions of rural population.

The investigation over the last two decades, focused on spatial distribution of fluoride in groundwater and no systematic investigation related to the aquifer parameters, depth distribution *vs.* fluoride content in groundwater has been carried out. Considering the regions affected by fluorosis (Table 1) the problem will become more serious if appropriate measures are not taken immediately. Recent investigation indicates that arsenic in west Bengal is entering the human body through the food chain also (Paramita, 2003). Similarly fluoride can also enter the human system through food chain. Since 70% of agricultural activity in India is through irrigation, future research should focus on fluoride up-take by food crop (rice and wheat).

REFERENCES

Adyalkar, P.G. and Radhakrishna, T.S. (1974). On fluorides in Jhor-Tattapani hot springs of Surguja district, Madhya Pradesh. In: *Proceed. symposium on fluorosis*, Hyderabad. 297–299.

Alvi, S.M.A. and Koteswaram, P. (1985). Time series analysis of Annual Rainfall over India. *Mausam*. 36, 4.

Apambire, W.B.; Boyle, D.R. and Michel, F.A. (1997). Geochemistry, genesis, and health implications of fluoriferous groundwaters in the upper regions of Ghana. *Environmental Geology*, 33, 13–24.

Appelo, C.A.J. and Postma, D. (1993). Geochemistry, groundwater and pollution. A.A. Balkema Pub. Co., The Netherlands, 203p.

Balasubramanium, K.S and Kumar, S.S. (1974). The mineralogy and geochemistry of fluorite deposits of Kahela- Mando Ki Pal, Rajasthan. In: *Proceed. symposium on fluorosis*, Hyderabad, India. 19–28.

Banerjee, S. and Satyanarayana (1964). Report on the investigation of the fluorite-lead-zinc occurrences in the Chandidongari area, Durg district, Madhya Pradesh. Unpublished Report, *Geological Survey of India.*

Baoshan, Z. and Yetang, H. (1988). Geochemical environment related to human endemic fluorosis in China. In: Thornton (ed.) *Geochemistry and Health.* Science Reviews, Northwood, 93–96.

Bhattacharya, P. (2002). Arsenic contaminated groundwater from the sedimentary aquifers of South-East Asia. *XXXII IAH & VI ALHSUD Congress* 2002. 357–365.

Bose, S.K. and Banerjee, S. (1974). Distribution of fluorine rich rocks in India and their relationship with tectonic lineaments. In: *Proceedings of the symposium on fluorosis*, Hyderabad, India. 93–114.

Bower, C.A. and Hatcher, J.T. (1967). Adsorption of fluoride by soils and minerals. *Soil Science*, 103(3), 151–154.

Boyle, D. R and Chagnon, M. (1995). An incidence of skeletal fluorosis associated with groundwaters of the Maritime Carboniferous basin, Gaspe Region, Quebec, Canada. *Environmental Geochemistry and Health.* 17, 5–12.

Brouwer, I.D.; Backer, D.O.; DeBruin A. and Hautvast J.G.A.J. (1988). Unsuitability of WHO guidelines for fluoride concentration in drinking water in Senegal. *Lancet.* 130, 223–225.

Burman Dev, G.K.; Singh, B. and Khatri, P. (1995). Hydrogeochemical studies of groundwater having high F content in Chandrapur district of Vidharbha region Maharashtra. *Gondwana Geological Magazine*, 9, 71–80.

Cengelog¢lu, Y.; Kýr, E. and Ersoz, M. (2002). Removal of fluoride from aqueous solution by using red Mud. Separation and Purification Technology 28, 81–86.

CGWB (1975). Hydrogeology of 56 G (East) and 56 K (West), India, Technical Report, *Central Ground Water Board*, 1–24.

Chakraborti, D.; Chanda, C.R.; Samanta, G.; Chowdhary, U.K. and Singh, B. (2000). Fluorosis in assam, India. *Current Science*, 78, 1421–1423.

Chandrasekharam, D. (1989). Anomalous SO_4 – Cl groundwater in the coastal aquifer, Kerala. *Indian Academy of Science*, 98, 287–295.

Chandrasekharam, D. (2002). Geogenic arsenic pollution in groundwater. *XXXII IAH & ALHSUD Congress, Argentina,* 2002. 373–379.

Chowdhary, A.N.; Bose B.B. and Banerjee, G. (1964). Studies in geochemistry of thermal springs at Bakreshwar. In: *Proceedings of the 22nd I G C*, New Delhi, India. 12, 143–160.

Das, S.; Mehta, B.C.; Samanta, S.K.; Das, P.K. and Srivastava, S.K. (2000). Fluoride hazards in groundwaters of Orissa, India. *Indian Journal of Environ. Hlth.* 1, 40–46.

Deer, W.A.; Howie, R.A. and Zussman, J. (1985). An introduction to rock forming minerals. John Wiley and Sons, New York, 528p.

Deshmukh, A.N.; Wadaskar, P.M. and Malpe, D.B. (1995). Fluorine in the Environment: A review. *Gondwana geological Society*. 9, 1–21.

Dhruva Rao, B.K. and Balmiki, P.A. (1963). Report on the exploration of apatite deposits in Singhbhum district, Bihar. *Progress Report, Indian Bureau of mines* (unpublished).

Faruqi, N.H.; Khan, A.U. and Joshi, K.C. (1997). Incidence of fluoride in groundwater in the semi-desertic terrain of Punjab, Haryana and adjoining parts of Rajasthan. *Geol. Surv. India.* Spl. Pub., 48, 111–114.

Fleischer, M. and Robinson, W.O. (1963). Some problems of the geochemistry of fluorine. In: Shaw, D.M.(ed), *Studies in Analytical Geochemistry*. Toronto, University of Toronto Press.

Garmes, H.; Persinb, F. G.; Sandeauxb, J.; Pourcellyb, G. and Mountadar, M. (2002). Defluoridation of groundwater by a hybrid process combining adsorption and Donnan dialysis. Desalination 145: 287–291.

Gupta, S.K. and Deshpande, R.D. (1998). Depleting groundwater levels and increasing fluoride levels in Mehsana district Gujarat. www.globenet.org/preceup/pages/ang/chapitre/capitali/cas/indmehs.htm.

Gupta, S.C.; Rathore, G.S. and Doshi, C.S. (1993). Fluoride distribution in groundwaters of South-Eastern Rajasthan. *Indian Jour. Environ. Hlth.* 35, 97–109.

Gupta, M.K.; Singh, V.; Rajwanshi, P.; Srivastava, S. and Das, S. (1994). Fluoride in groundwater at Agra, *Indian J. Environ Hlth.* 36, 43–46.

Hamamoto, E. (1957). On bone changes observed in residents of a high fluorine zone. In: Utzino S. (ed.) *Medico-dental researches on fluorides.* Japan Society for the promotion of Science, Tokyo, 118–130.

Handa, B.K. (1977). Presentation and interpretation of fluorine ion concentrations in natural waters. In Proc. Sym. Fluorosis, Hyderabad, 1977, 317–347.

Jalihal, A.A. (2002). High fluoride concentration and high salinity problems in groundwater around Hungud Ilkal area, Bagalkot district, Karnataka. *Unpublished Thesis.*

Killedar, D.J. and Bhargava, D.S. (1988a). An overview of defluoridation methods (part 1). *Jour. IPHE Ind.*, 1, 6–13.

Killedar, D.J. and Bhargava, D.S. (1988b). An overview of defluoridation methods (part 2). *Jour. IPHE Ind.*, 2, 37–44.

Krishnan (1985). *Geology of India and Burma*, 6th Edition, CBS publishers.

Madhavan, N. and Subramanium, V. (2001): Fluoride concentrations in river waters in South Asia. *Current Science*.80, 10, 1312–1319.

Mahramanlioglu, M.; I. Kizilcikli, I. and Bicer, I.O. (2002). Adsorption of fluoride from aqueous solution by acid treated spent bleaching earth. Jour. Fluorine Chemistry 115, 41–47.

Manji, F.; Baelum, V. and Fejerskov, O. (1986). Dental fluorosis in an area of Kenya with 2mg/L fluoride in the drinking water. *J Dent. Res.*, 65, 659–662.

Maruthamuthu, M. and Venkatanarayana, R. (1987). A native index of defluoridation by serpentine. *Fluoride*, 20(2), 64–67.

Minissale, A.; Vaselli, O.; Chandrasekharam, D.; Magro, G.; Tassi, F. and Casiglia, A. (2000). Origin and evolution of "intracratonic" thermal fluids from central-western Peninsular India. *Earth. Planet. Sci. Lett.*, 181, 377–394.

Morris, J.W. (1965). Skeletal fluorosis amongst Indians of the American southwest. *Am. J. Roentgeonal Radium Ther Nucl Med.*, 94, 608–615.

Murthy, S.R.N. (1974). Serpentine and Magnesite from Chalk hills, Salem, Tamil Nadu, India. *Proceedings in the symposium on fluorosis*, Hyderabad, India, 67–79.

Nawalkhe, W.G.; Kulkarni, D.N.; Pathak, B.N. and Bulusu, K.R. (1975). Defluoridation of water by Nalgonda technique. *Indian Journ. Environ. Health.*, vol.17, no.1, pp 26–65.

Paramita, A. (2003). Arsenic contamination in groundwater, surface water, soils and uptake by plants in Malda district, west Bengal, India. M.Tech., Thesis, IIT, Bombay (unpublished).

Prasad Rao, A.D.; Ahluwalia, A.D. and Kurien, T.K. (1974). Geology of the fluorosis affected areas in Podili, Darsi and Kanigiri Taluks, Prakasam district. *Proceedings of the symposium on fluorosis,* Hyderabad, India, 79–84.

Ramamohan Rao, N.V.; Rao, N.; SuryaPrakash Rao, K. and Schuling, R.D. (1993). Fluorine Distribution in waters of Nalgonda District, Andhra Pradesh, India. *Environmental Geology,* 24, 84–89.

Ren, F.R. and Shugin, J. (1988). Distribution and formation of high fluorine groundwater in China. *Environ.Geol. Water Sci.* 12, 3–10.

Saether, O.M.; Reimann, C.; Hilmo, B.O. and Taushani, E. (1995). Chemical composition of hard and soft rock groundwaters from Central Norway with special consideration of fluoride and Norwegian drinking water limits. *Environ. Geol.* 26, 147–156.

Sahasrabudhe, Y.S. (1974). Fluorine-rich plagioclase in the Deccan Traps and its influence on groundwater. *Proceedings of the symposium on fluorosis*, Hyderabad. 29–37.

Saxena, V.K. and Ahmed, S. (2003). Inferring the chemical parameters for the dissolution of fluoride in groundwater. *Environ. Geol.* 43, 731–736.

Sengupta, S. (2002). Hydrogeological study in and around Bilwara district, Rajasthan, India. M.Tech., Thesis, IIT, Bombay (unpublished).

Seraphim, R.H. (1951). Some aspects of the geochemistry of Fluorine. *Thesis.* Mass. Inst. of Tech.

Singano, J.J.; Mashauri, D.A.; Dahi, E. and Mtalo, F.W. (1997). Effect of pH on Defluoridation of Water by Magnesite. In: *Proceedings of the First International Workshop on Fluorosis and Defluoridation of Water,* 18–22 October 1995, Tanzania, p.30–34, The International Society for Fluoride Research, Auckland.

Singh, Y. (1998). Agricultural utility of groundwater from parts of Rewa district Madhya Pradesh. *Indian J. Environ. Hlth.* 40, 261–270.

Smet, J. (1992). Fluoride in drinking water. In Frencken J.E. (ed). Endemic fluorosis in developing countries. *Report of a symposium held in Deft, Netherlands.* Leiden TNO Institute of Preventive Health Care. 10–19.

Stueben, D.; Berner, Z.; Chandrasekharam, D. and Julie Karmakar (2003). Arsenic pollution in groundwater of West Bengal, India: Geochemical evidences for mobilization of As under reducing conditions. *App. Geochem.* 18, 1417–1437.

Sukheswala, R.N. and Udas, G.R. (1963). Note on the Carbonatite of Amba Dongar (Gujarat State) and its economic potentialities. *Science and Culture.* 29, 563–568.

Sukheswala, R.N. and Udas, G.R. (1964). The carbonatite of amba Dongar India-Some structural considerations. *Proceedings of the 22nd International Geological congress*, New Delhi, India.7,1–13.

Sumalatha, S.; Ambika, S.R. and Prasad, S.J. (1999). Fluoride status of groundwater contamination of Karnataka. *Current Science.*76, 730–888.

Susheela, A.K. (1999). Fluorosis management programme in India. *Current Science.* 77, 1250–1255.

Teotia, S.P.S.; Teotia, M. and Singh, R.K. (1981). Hydrogeochemical aspects of endemic skeletal fluorosis in India: an epidomological study. Fluoride, 14: 69–74.

WHO (1984). *Guidelines for drinking water quality*, 1,139.

Yong, L. and Hua, Z.W. (1991). Environmental characterization of regional groundwater in relation to fluoride poisoning in North China. *Environ. Geol. Water Sci.* 1 18, 3–10.

Impacts of increasing abstraction on groundwater quality in two areas of Southern Yemen

Jane Dottridge[1] and Paul Hardisty[2]
Komex, London, UK
[1] *jdottridge@london.komex.com*
[2] *phardisty@komex.com*

ABSTRACT

The impacts of increasing abstraction on groundwater quality may be more critical than drawdown. Recent water resources studies in two areas of southern Yemen show contrasting impacts of abstraction and implications for over-exploitation. The differences are explained by the geology and natural hydrological regime.

In Wadi Hadramaut, a deep valley incised in a desert plateau, pumped abstraction has replaced natural discharge from an extensive sandstone aquifer. Historical discharge by seepage and evaporation resulted in poor water quality in the shallow sediments. Despite very low recharge, rapid increases in abstraction caused little drawdown at first, due to vast storage. Water levels are declining locally in areas of intensive pumping, with some deterioration in water quality due to urbanisation and cross-completed boreholes.

In the Tuban coastal plain, the alluvial aquifer receives regular recharge from floods and supplies water for agriculture and the city of Aden. Although the groundwater levels are declining in the central zones with concentrated abstraction, overall balances show only a gradual depletion of storage. This is due to increasing inflow of seawater from the coast. Combined with the high vulnerability of the alluvium, this results in widespread deterioration in water quality.

Keywords: *groundwater, over-exploitation, abstraction, safe yield, water quality.*

1 INTRODUCTION

The concepts of overexploitation of groundwater and *safe yield* of aquifers have been used since the 1920s, when safe yield was defined as "the water that can be abstracted permanently *without producing undesirable results*" (Meinzer, 1923). As international understanding of groundwater resources has developed, the concept of over-exploitation has broadened to consider the economics of groundwater development, protection of water quality and environmental impacts. The definitions of safe yield, undesirable results and overexploitation are sufficiently vague to generate a diversity of opinions among those involved in groundwater development. Overall, there is a consensus that it is essential to understand the consequences of abstraction and to realise that "groundwater mining is *conscious and planned* abstraction at a rate greatly exceeding aquifer recharge" (UN, 1992). Effective resource management is only possible when there is a clear conceptual understanding of the hydrogeological system, and sustained monitoring of water levels, water quality, abstraction and hydrological conditions. This paper explores the implications of these requirements, illustrated by case studies from two contrasting areas in southern Yemen.

Yemen is an arid country, with an expanding population and rapidly growing demand for water, and thus faces a serious water management challenge. Water issues have received most attention in the major urban centres, such as the Sana'a basin where aquifer levels are steadily declining (Alderwish and Dottridge, 1999) and Ta'iz, which has experienced chronic shortages in water supply for 10 years (Handley, 2000). However, effective water management is vitally important in other areas, with high population density and growing competition for scarce resources among different water users. Recognizing this situation, the UN and the government of Yemen implemented a coordinated programme to develop an integrated water resources management strategy for four important water regions of the country, including the Hadramaut and Tuban-Abyan regions described below.

2 EXTENSIVE AQUIFER WITH RISING ABSTRACTION-HADRAMAUT REGION

Wadi Hadramaut is a spectacular valley, deeply incised into the Jol Plateau in southern Yemen (Figure 1). The valley, which runs west-east for 100 km, is approximately 300 m deep and up to 5 km wide (Figure 1). The southern plateau has an average annual rainfall of over 200 mm, which declines to less than 50 mm in the north. Recharge through infiltration on the plateau is supplemented by major flood events in the valley, every 3 to 5 years. Estimated natural recharge, excluding return flows from irrigation and domestic water use, averages 115 Mm^3/yr. The sheltered Hadramaut valley is an important feature in the water resources of southern Yemen, and supports the main agricultural zone and centres of population in this region. Groundwater is heavily utilised, with a large component to supply irrigation water, and a smaller proportion for domestic use.

The plateau is underlain by the thick Mukalla sandstone aquifer, topped by outcropping Tertiary limestones and marls. In the valley, the layers above the Mukalla sandstone have been removed by erosion, but the valley is backfilled by a limestone conglomerate, overlain by Quaternary alluvium. This results in a sequence of three aquifers, which are interconnected by both lateral and vertical flow, but have distinct hydraulic properties and water quality.

Historically, groundwater was only abstracted from the alluvium by shallow dug wells. Since

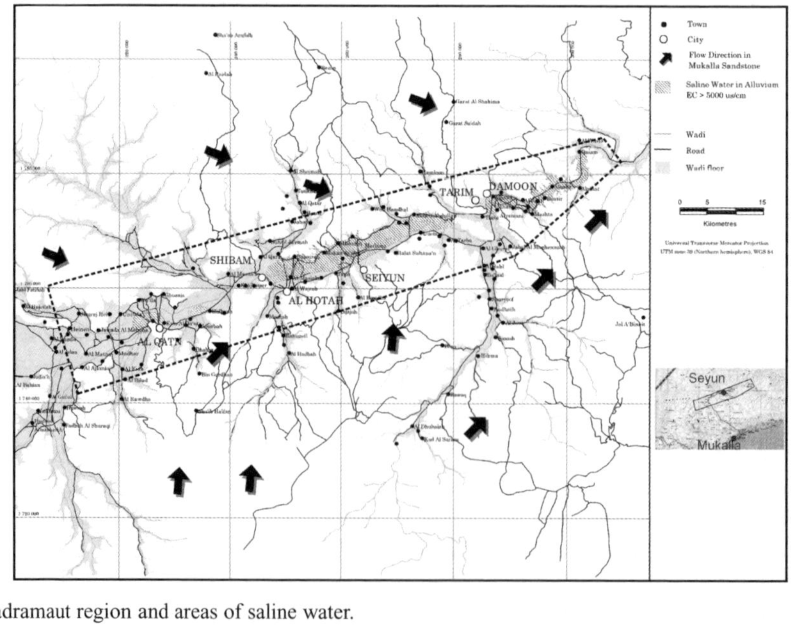

Figure 1. Hadramaut region and areas of saline water.

the 1950s, the number of wells has increased, supplemented since the 1970s by drilled boreholes and dug wells deepened by drilling (Figure 2). The rate of borehole construction accelerated following the unification of Yemen in 1990. Based on historical data from previous studies and a complete well inventory in 2001 (Komex, 2002a), the increase in groundwater abstraction with time has been calculated, as shown in Figure 3. The resulting trends in water levels are illustrated by two typical hydrographs in Figure 4.

Water quality in the Mukalla sandstone is generally good, with EC values ranging from 600 μS/cm to 1,000 μS/cm and up to 1,500 μS/cm in the centre of the main valley. The alluvial and conglomerate aquifers show similar distributions of salinity with higher salinity groundwater (EC from 5,000 μS/cm to 8,770 μS/cm) in the centre of the wadi (Figure 1). Although water quality data are sparse, this pattern appears to be fairly stable with time, and there is some evidence that it precedes modern groundwater development. It is thought that naturally occurring saline water has become concentrated by evaporation of groundwater discharge and recycling of evaporation water. Comparison of analyses made in 1985/86 (MMP, 1987) and 2001 (Komex, 2002a) shows that there has been a slight increase in groundwater salinity in the more heavily populated areas, which appears to be associated with urban and agricultural development rather than abstraction rates. Nitrate and chloride values are higher within the towns, apparently due to groundwater contamination from sewage, which is discharged via cesspits.

The main impacts of increasing abstraction are that:

- Natural outflows such as seepage and baseflow have ceased. Perennial water is now only found at one location in the valley;
- Water has been drawn from storage, especially in the areas of concentrated abstraction (Figure 4), but rates of drawdown are low;
- In a few locations with high abstraction, the water quality in the Mukalla sandstone has deteriorated due to reversal of the natural upward gradients by heavy pumping combined with cross-completed boreholes acting as conduits;
- Elsewhere there has been little measurable impact, apart from a slight general decline in water levels.

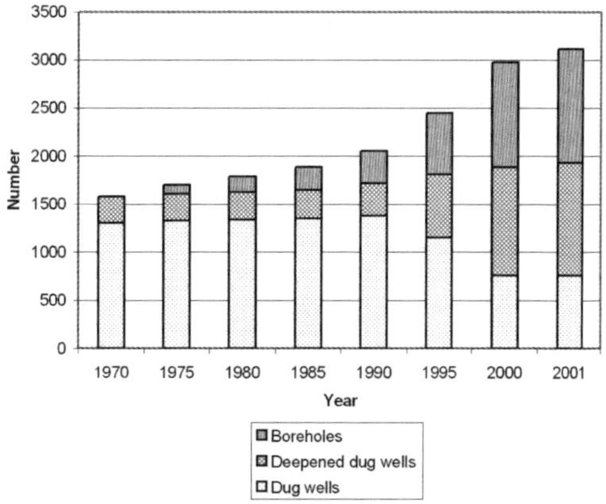

Figure 2. Number of new wells and boreholes, 1970–2001.

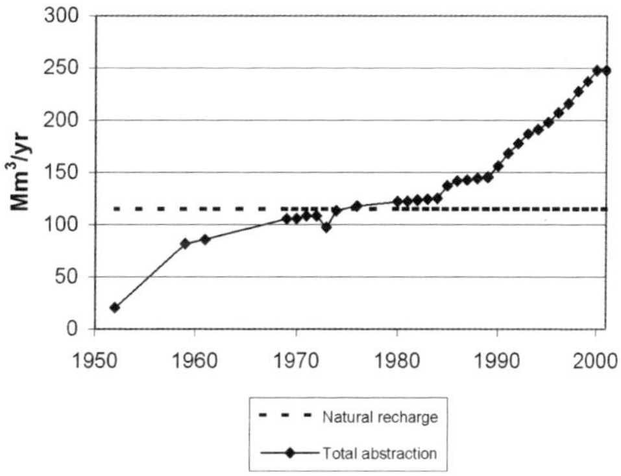

Figure 3. Historical groundwater abstraction.

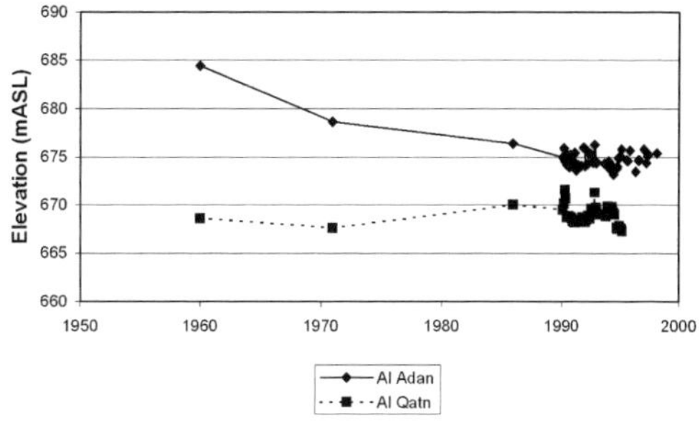

Figure 4. Long term groundwater hydrographs.

3 COASTAL PLAIN AQUIFER WITH FLUCTUATING ABSTRACTION – WADI TUBAN

The coastal plain of wadi Tuban covers an area of approximately 700 km^2, sloping gently from the mountain front to the coast at Aden. (Figure 5). Runoff from a large catchment in the mountains, which consist of volcanic and basement rocks, supplies water to wadi Tuban. The regular flood flows, which average 123 Mm^3/yr, are diverted for spate irrigation, with water only reaching the coast under extremely wet conditions. An estimated 58% of the inflow recharges the alluvial aquifer through wadi bed infiltration and return flows from spate irrigation. Direct recharge from local rainfall is negligible.

The coastal plain is underlain by a wedge of Quaternary and recent sediments, mostly alluvium and outwash plain deposits, which are cemented towards the base. The alluvial deposits

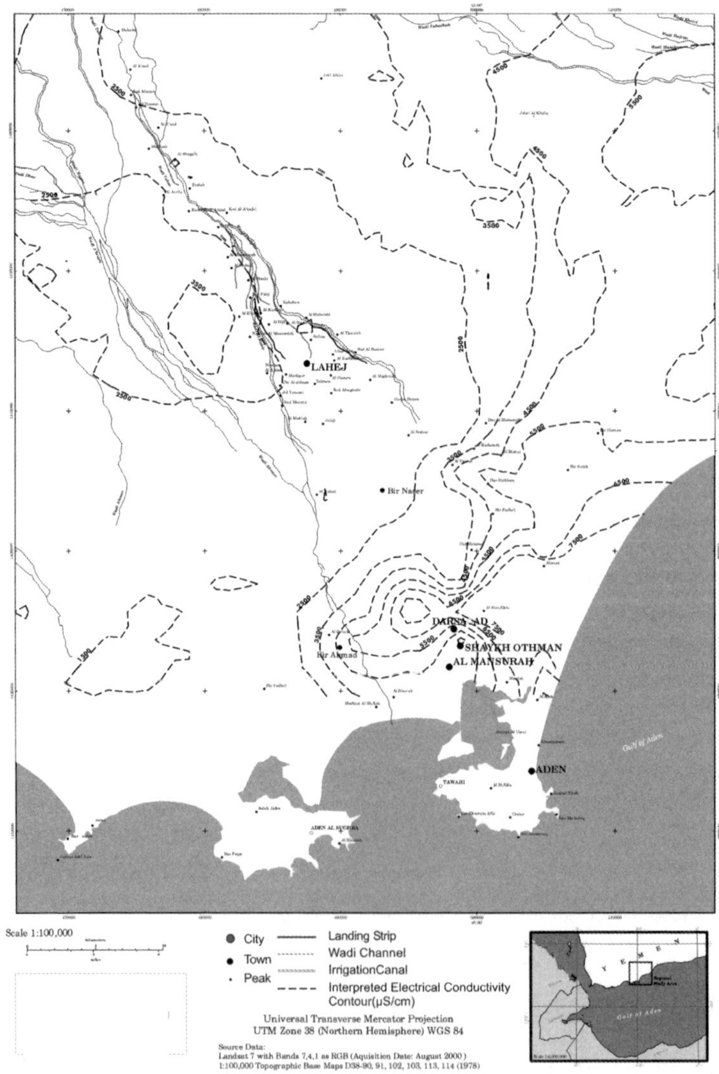

Figure 5. Tuban Area with contours of EC.

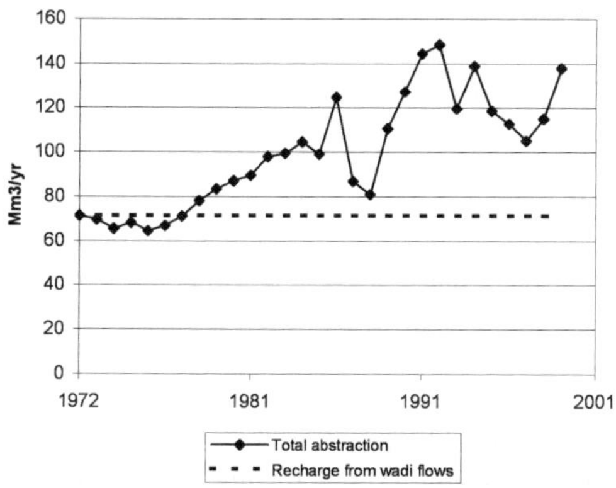

Figure 6. Estimated groundwater abstraction from Tuban aquifer.

thicken towards the coast, reaching over 700 m near Aden. Although the alluvium generally becomes finer towards the coast, the heterogeneity of the sediments generates highly variable hydraulic properties with no systematic pattern.

The bulk of groundwater abstraction is used for irrigation, both to supplement spate flows and for areas beyond the command of the surface water. The annual volume used for irrigation is estimated to have doubled between the late 1970s and 1990s, with peaks in dry years when flood volumes are low. This water also provides domestic supplies for much of the rural population. Urban supplies are abstracted from six wellfields, three for local supplies and the three largest for the city of Aden, which also imports water from the nearby Abyan aquifer. Figure 6 shows the estimated annual abstraction from 1972 to 2000. As losses are high, particularly in urban areas, the total recharge to groundwater is considerably in excess of the recharge derived solely from wadi flows. In 2000, total abstraction was estimated to be 138 Mm^3, exceeding the total recharge of 112 Mm^3 by 26 Mm^3. The resulting trends in water levels are shown in Figure 7.

A comprehensive survey of water quality in 2001 (Komex, 2002b) concluded that the aquifer contains a highly mineralized groundwater, with total dissolved solids (TDS) averaging 1,950 mg/L, considerably above WHO drinking water guidelines of 1,000 mg/L. TDS and electrical conductivity are, lowest close to the wadi channels and canal systems, a direct result of

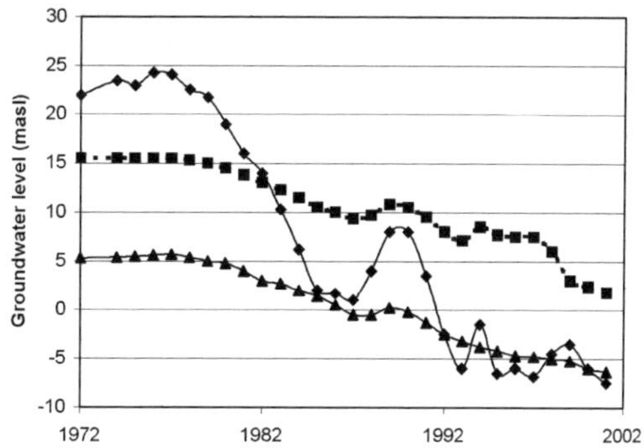

Figure 7. Changes in groundwater elevation with time.

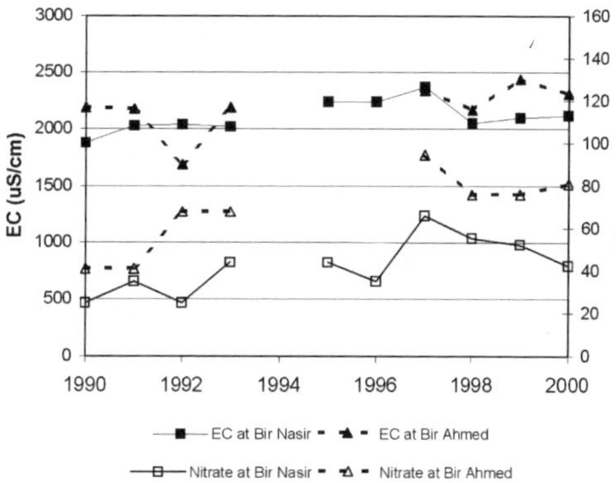

Figure 8. Changes in water quality with time.

fresh water recharge being concentrated in these areas (Figure 5). Elevated groundwater conductivity near the coast and a steep conductivity gradient normal to the coastline (Figure 5) indicate seawater intrusion. The wellfield closest to the coast was closed in 1984 due to deteriorating water quality caused by seawater intrusion. The key parameter indicating health risk and likely pollution was nitrate, with an average value of 64 mg/L and half of the samples elevated beyond the WHO guidelines of 50 mg/L. Monitoring of the main wellfields shows fluctuations in both nitrate and conductivity superimposed on a rising trend (Figure 8).

4 SUSTAINABILITY AND OVER-EXPLOITATION

Current abstraction from the Hadramaut aquifers is estimated to be approximately

250 Mm^3/yr (in 2000/01), considerably in excess of the total recharge, which averages 150 Mm^3/yr. The abstraction from the Tuban aquifer of 138 Mm^3/yr in 2000 also exceeded the estimated recharge of 112 Mm^3/yr, but the discrepancy is much smaller. Using the conservative definitions, abstraction exceeds the safe yield and both aquifers are over-exploited. However, for a limited period, effective use can be made of groundwater from aquifer storage, and the safe yield is only exceeded if this produces undesirable results, or rapidly becomes unsustainable.

The most noticeable impact of the recent increases in abstraction in Wadi Hadramaut is the cessation of seepage and baseflow, but as little use was made of this water and the quality was poor, the impact is not significant. In addition, there is a sustained decline in groundwater levels at a rate of 0.1 m/yr to 0.25 m/yr, indicating withdrawal of water from storage. Drawdown and interference effects, which have caused many dug wells to dry up or require deepening, are concentrated in the areas with most intense abstraction. In general, there has been no detectable deterioration in water quality, apart from an increase in nitrate and salinity in urban areas and localised increases in the salinity of the sandstone groundwater. This apparent lack of impacts is partly due to inadequate monitoring of both water levels and quality, and it would be difficult to identify regional trends while drilling of new boreholes continues at an accelerating rate.

In Tuban, the impacts of a smaller imbalance between abstraction and recharge appear to be more severe, with declining water levels in the wellfield areas and evidence of seawater intrusion in the coastal area. These observations and results of modelling the aquifer suggest that the current groundwater usage practices in the Tuban delta are unsustainable. This is partly due to the overall imbalance between abstraction and recharge, but the location of the high abstractions close to the coast is more critical. Unless the rate of abstraction in the southern zone is reduced, then the observed intrusion of saline water at depth will continue with subsequent long term damage to the coastal aquifer and its users.

5 CONCLUSIONS

Using the conservative definition of safe yield, both aquifers are over-exploited as abstraction exceeds recharge and thus present abstraction rates cannot be maintained indefinitely. A contrasting opinion is that effective use of groundwater resources requires use of storage, but in a conscious and planned way, without undesirable results. Despite concerns about the effects of over-abstraction on the aquifer, pumping from the Hadramaut aquifers has only caused to date a localised deterioration of water quality and has caused only gradual drawdown along the valley. Thus, provided that the use of storage is monitored and future abstraction is managed, the aquifer may not be considered to be over-exploited. In the Tuban coastal plain, the location of abstractions close to the coast causes saline intrusion rather than use of aquifer storage and a consequent deterioration in water quality. These undesirable impacts therefore lead to the conclusion that the current abstraction exceeds the safe yield of the Tuban aquifer, using any definition of safe yield.

Despite the contrasting conclusions on over-exploitation, both areas suffer from several similar and fundamental problems, which jeopardise the future sustainability of the groundwater resources. The aquifers, particularly the Tuban alluvium, are vulnerable to pollution from diffuse and point sources. The most significant threat is from ubiquitous disposal of untreated sewage to ground, as shown by the high and rising nitrate values. In addition, monitoring of water quality and water levels is patchy, both spatially and temporally, and estimates of abstraction rates are made infrequently. Regular and systematic collection of reliable monitoring data is an essential step in protection and management of the aquifers. More proactive management may be advisable in future, including controls on drilling of new boreholes and the rates and locations of groundwater abstraction.

ACKNOWLEDGEMENTS

The co-operation and generous assistance of the chairman and staff of the National Water

Resources Authority and other agencies in Yemen made this work possible. This work formed part of project UNDP/Netherlands water programme Yem/97/200, a joint programme of the UN, the Netherlands and the government of Yemen. The views expressed in this paper are those of the authors alone and do not represent official policy of the UN or government of Yemen.

REFERENCES

Alderwish, A. and Dottridge, J. (1999). Urban recharge and its influence on groundwater quality in Sana'a, Yemen In: *Groundwater in the urban environment*, Volume 2: City case studies (ed. Chilton, P.J), p. 85–90, Balkema, Rotterdam.

Handley. C.D. (2001). *Water Stress: Some symptoms and causes, A case study of Ta'iz, Yemen*, SOAS Studies in Development Geography, Ashgate Press, Aldershot.

Komex (2002a). *Water Resources Management Studies in the Hadramaut region*, Draft final report for UN and NWRA.

Komex (2002b). *Water Resources Management Studies in the Tuban-Abyan region*, Draft final report for UN and NWRA.

Meinzer, O.E. (1923). The occurrence of groundwater in the United States, with a discussion of principles, *US Geological Survey Water Supply, Paper* 489, US Geological Survey, Reston, Virginia.

MMP (Sir M. Macdonald and Partners) (1987). *Wadi Hadhramaut Agricultural Development Project, Phase 2*, Borehole construction and groundwater studies.

UN (1992). *Inter-regional workshop on groundwater overexploitation in developing countries*, United Nations Department of Technical Cooperation for Development, report INT/90/R43, UN, New York.

When intensive exploitation is a blessing: effects of ceasing intensive exploitation in the city of Barcelona

Enric Vázquez-Suñé[(1)], Xavier Sánchez-Vila[(2)] and Jesús Carrera[(3)]
Hydrogeology Group. School of Civil Engineering, Technical University of Catalonia (UPC), Barcelona, Spain
[(1)] *enric.vazquez-sune@upc.es*
[(2)] *xavier.sanchez-vila@upc.es*
[(3)] *jesus.carrera@upc.es*

Ramón Arandes
Ajuntament de Barcelona, Spain
rarandes@mail.bcn.es

ABSTRACT

Several aquifers underlie the city of Barcelona. They have been heavily exploited since the 19th century. Exploitation gradually increased throughout the 20th century, mostly for industrial purposes. This strong sustained exploitation produced marked decreases in the piezometric level. In some areas, the water table was continuously below sea level, leading to salinization and loss of groundwater quality. Starting in the 1970s, the economic crisis and the urban pressure caused many industries to disappear or to move outside the city. This led to a decrease in extractions and a sustained recovery of the piezometric levels with a decrease in the seawater intrusion and a progressive recovery of the water quality till the current level. The immediate consequence of the reduction in groundwater pumping is the flooding or, at least, an increase of seepage to underground urban structures, particularly the subway system, basements and underground parking areas. Because of the strong economic implications, the Municipality has developed a project of re-exploitation of the aquifer in selected zones for municipal uses.

In this context, the objective of our work is to discuss not only the effects of ceasing intensive exploitation in the city of Barcelona, but also the initiatives to reduce some of them. In order to assess the amount of pumping that would be needed to maintain the water table at an acceptable level; we evaluated the recharge in the present conditions. This has been done through water balances and hydrochemical studies. It allowed us to develop a zonation of hydrochemical characteristics of groundwater according to the source of recharge. The evaluations indicate that leakage in supply water pipes is probably the main contribution of recharge to the aquifers. Moreover, urban groundwater is of sufficient quality for most uses other than domestic supply. This implies that pumping groundwater, which can then be used to satisfy many municipal water needs at a very low cost, can fight most of the negative impacts of water level rise. This has started to become implemented in the city with a secondary distribution network of phreatic water.

Keywords: *Structure drainage, urban area, hydrogeology, parking lot, administration.*

1 INTRODUCTION

Many aquifers underlying the city of Barcelona have been exploited since old times, but until the 19th century there has not been intensive exploitation. Groundwater abstraction increased throughout the 20th century, mainly for industrial and supply uses. This strong and continuous exploitation caused an important decrease of piezometric levels. In some areas the levels were permanently below the sea level, causing a marine intrusion and thus a loss in water quality. Since the 1970s the economic crisis and the urbanistic pressure were the reason for many factories to close or to move to the outskirts of the city. This cause during the last 30 years a decrease in groundwater exploitation. This situation led to a backwards marine intrusion and thus, to an improvement of the groundwater quality. On the other hand, there was also an increase in groundwater infiltration in buildings and infrastructures, most of them built when piezometric levels were at a historical minimum, such as the subway system, underground parking lots, basements, etc. (UPC *et al.*, 1997).

In general, the essential processes affecting underground water in any urban area are not very different from those in rural areas, although there are some significant peculiarities. Some of these processes are the changes that take place on the quality and amount of water, which are related to: changes in the hydrogeological cycle (due to urbanization processes), variations in the piezometric levels and underground pollution levels (due to human activity and the presence of underground structures that interfere with aquifers).

The aim of this paper is to show, the aspects to be considered in hydrogeological studies in urban areas, the results extracted from the studies of the Barcelona area, and finally to illustrate some of the initiatives addressed to minimize potentially negative effects.

2 SIGNIFICANT ASPECTS IN URBAN HYDROGEOLOGY STUDIES. APPLICATION TO THE BARCELONA AREA

Generally the aspects that should be taken into account in urban hydrogeology are, firstly to identify the problems, determine their causes, see the effects caused by the changes in the hydrogeological cycle in the area and, secondly, to develop and apply methodologies to quantify and control these effects. These aspects will be detailed below, starting with some general features, and following with their application to Barcelona.

2.1 *Fluctuation of piezometric levels. Problems on underground structures*

In many modern cities, early development phases were associated to industrial activities. Industry is a great water consumer, and that is why the industry proliferation in urban areas causes an important decrease of piezometric levels (water mining). Additionally, industry can be a source of pollution for aquifers (Cheney *et al.*, 1999). In the most advanced phases of urban development, industry tends to disappear or to move outside the urban area. This situation tends to a rise of piezometric levels. This increase can be due to the recovery of levels before pumping and/or an increase of recharge due to losses in the supply and sewage systems, both favored by a population that tends to increase significantly and to a generally low degree of conservation of the piping systems.

The interaction between groundwater and underground structures could be seen from two points of views. On the one hand, the structures alter groundwater flow (Marinos and Kavvadas, 1998), while on the other hand water can cause problems or condition those same structures (Marinos and Kavvadas, 1997). In the latter case it is necessary to differentiate between the effects that will cause the rising and the lowering of the groundwater levels. The two problems are absolutely different.

Water level lowering can cause subsidence or consolidation problems, with its consequent risk for structures, and also marine intrusion problems in coastal cities. Furthermore sea water, salty and sulphated, can cause serious corrosion problems in contact with metallic and concrete structures.

The evolution of the piezometric levels in the city of Barcelona shows the typical development of an industrial city. Groundwater abstraction for industrial and supply uses increased much since the beginning of the 1970s. Drawdowns in many parts of the city ranged from a few to 15 m. Since then, the abstraction kept decreasing rapidly due to pollution, relocation of the industries, improvement of production systems, etc. Currently in most of the areas, the levels have recovered to values similar to those of the beginning of the 20th century.

As the levels increased during the last years, a progressive increase of water infiltration has been observed in several urban underground infrastructures, public or private (subway system, basements, sewer system, underground parking lots, etc.). A significant part of the subway system and many other constructions in Barcelona were built during the period 1950–1975, when the maximum depression of the levels was attained. During this time, neither the designers, nor constructors thought that the levels could increase or recover the original heights. In order to understand the magnitude of the problem, it should be noted that the water that has to be pumped in 1996 from the tunnels in the subway system is equivalent to 12 Mm^3/yr. At least two thirds of this water volume (about 8 Mm^3/yr) comes from the inflow of phreatic water. In the railway system, 5 Mm^3/yr are drained. In the Sant Adrià del Besós area, in just one of the parking lots, some 8 Mm^3/yr are drained. Just for comparison, the total amount extracted is around 12% of the water supply to the city. Furthermore, the problem of infiltrations is not local but rather generalized throughout the city, even in areas with an elevated topography (Figure 1).

Currently, the economic impact derived from this situation is very high, due to the cost of drainage, impermeabilization, pumping installation, evacuation of pumped water, and, of course, to energetic costs. Furthermore, it is important to note that it seems wrong to use the public sewage system as an evacuation system for this water, as it can affect the correct functioning of the system, reducing its capacity, and increasing the possibility of floods during summer storms seasons, and causing an extra cost in the depuration of sewage, as water volumes reaching the treatment plants increase.

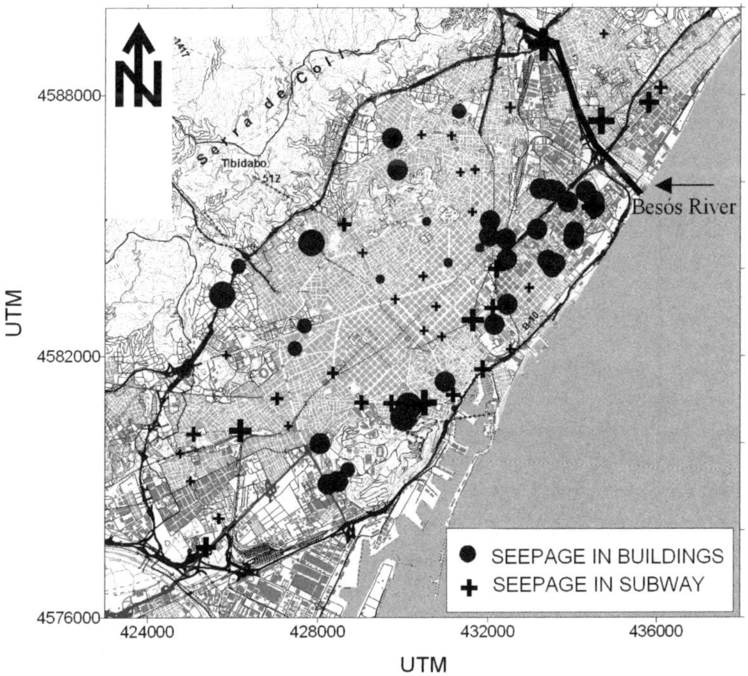

Figure 1. Infiltration problems in different parts of the city of Barcelona. The size of the symbols is proportional to the inflow.

An additional problem is that seepage can jeopardize the stability of the structures themselves, as in most cases infiltrated water drags concrete fines or has great iron content, possibly coming from the degradation of the steel used in concrete reinforcement. This endangers the structural security. An alternative would be keeping the level of underground water under the foundations artificially, but this would also mean an elevated cost (see Ondiviela *et al.*, in this same volume).

2.2 *Pollution of underground water*

Groundwater in urban areas can be polluted from different potential sources, including urban sewage, urban run-off, which lixiviates waste and effluents from vehicles accumulated in the street, waste storage and pollutants in the surface of the area or buried, industries, etc. The mixture of sewage (with a high organic content) with supply water (chlorated) produces organochlorated pollutants. Furthermore, changes in the pH and Eh conditions in the aquifer can mobilize toxic compounds (normally heavy metals).

Fluctuations in piezometric levels could condition pollution. In the case of an increase in the levels, a remobilization of pollution can take place due to two causes: direct lixiviation and changes in chemical conditions which can remobilize some compounds (Navarro *et al.*, 1992).

Other pollution sources could be spills in gas stations and fuel deposits in general, the interaction of aquifers with rivers, channels, etc., direct leaching of polluted water coming from industries, marine intrusion, etc.

In Barcelona, three main sources of underground water pollution were identified: urban waste water infiltration, urban run-off infiltration, and marine intrusion. As discussed previously, the evolution of marine intrusion is conditioned by the evolution of extractions and variations of piezometric levels. In the northeastern part of the city, in the Besós delta area, during the 1970s the intrusion reached some 2 km from the coast. The decrease of extractions and thus the recovery of levels caused again a flow towards the sea and some wells that were salinized regained similar characteristics to those in the 1950s (Figure 2).

2.3 *Evaluation of the recharge and hydrogeological balance*

In order to understand how the aquifer works, it is necessary to know the characteristics of the water flow, that is, its magnitude, relative significance and dependence of the different hydraulic parameters. Furthermore, a detailed analysis of the hydrogeological cycle in urban areas should take into account the peculiarities of each city (Lerner, 1997). For instance, the presence of

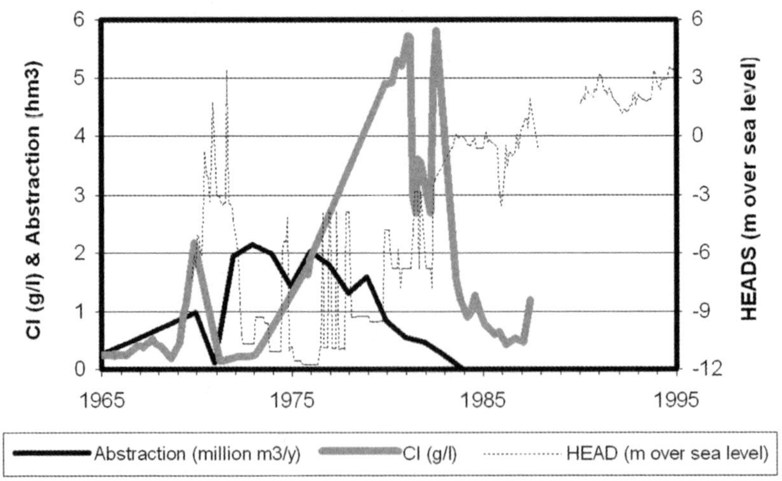

Figure 2. Relationship between abstraction, piezometric level and marine intrusion in Montsolís well. Alluvial aquifer of the Besós River, Barcelona.

rivers, channels, lakes, sea, drains, infrastructures, induced recharge, etc.

Changes in territorial uses imply changes in the hydrological cycle, particularly in the recharge: while rainfall infiltration is normally reduced due to surface impermeabilization, total recharge could increase. This is due to the fact that what infiltrates in the ground (by direct infiltration, drains, discontinuities of the pavement) even in small amounts, does not have a possibility of evapotranspirating. Furthermore, there are new recharge sources, such as the input of water supply systems (normally their losses surpass 15% of the supplied water, with many documented values between 20–25% and a few over 30% to more than 50%) and waste water systems.

Discriminating and quantifying the different recharge sources can take place due to a mass balance containing several conservative chemical species, which are representative of those sources. Castillo (2000) developed a methodology based on the statistic multivariate analysis to estimate the mixture proportion of the different sources in each tested well, which includes the uncertainty in the measures. This methodology was applied to the city of Barcelona (Vázquez-Suñé *et al.*, submitted) to evaluate the amount of water coming from each recharge source and thus the estimation of the loss percentage of the supply and sewer systems, marine intrusion, direct infiltration, etc.

2.4 *Modeling flow and transport*

The quantification of the balance terms can be carried out separately, but it is convenient to make it jointly in order to prove the coherence of the proposed model. Modeling can help achieving this.

The models can play different roles. For instance, (1) the identification of processes, (2) the hydrological characterization and integration, and (3) the management of resources. Each model is specific of each area. It depends on the kind of city, weather, geology and time. In each case, the terms that contribute most significantly to the balance are different.

The application of models in urban areas has certain particularities. First, they should take into account the historical evolution of the city, including the changes in territory uses. Second, the areal recharge depends essentially on specific features (density of supply and distribution systems), directly related to the density of population and with the period when each area of the city developed. Third, the extractions are difficult to evaluate, and normally are underestimated. And fourth, underground structures must be considered when modeling, should they be taken into account independently.

The above-mentioned steps were followed for the hydrological study of an urban area. A hydrological model was built for the city of Barcelona. The main tasks were the characterization of aquifers (geometry and hydraulic parameters, discrimination of recharge sources, limits and boundary conditions and the integration of all this tasks in a conceptual model). The model was calibrated with piezometric data from the period 1900–1999. The Subway and railway systems were explicitly included.

One of the results of the model is the quantification of the balance terms (Figure 3) and their evolution in time. It is verified that the main recharge sources are the losses in the sewer system. A net discharge to the sea is taking place since 1994. Until that moment, marine intrusion was higher than the discharge to the sea. Currently industrial and supply extractions have decreased radically. The greatest extractions take place in the subway and railroad systems and other infrastructures drainage. In spite of this, there is some knowledge on the velocity of these changes in the hydrology of urban areas.

3 SUSTAINABILITY AND MANAGEMENT OF WATER IN URBAN AREAS

For several years, nowadays and probably in the future, many cities will have to face many problems related to groundwater and especially those related to their intensive use. If no correct management policies are applied, the potential problems occurring will be very difficult to solve, leading to significant economic and social impacts. Management should focus on the global water optimization criteria, including consume reduction and reutilization policies. New water

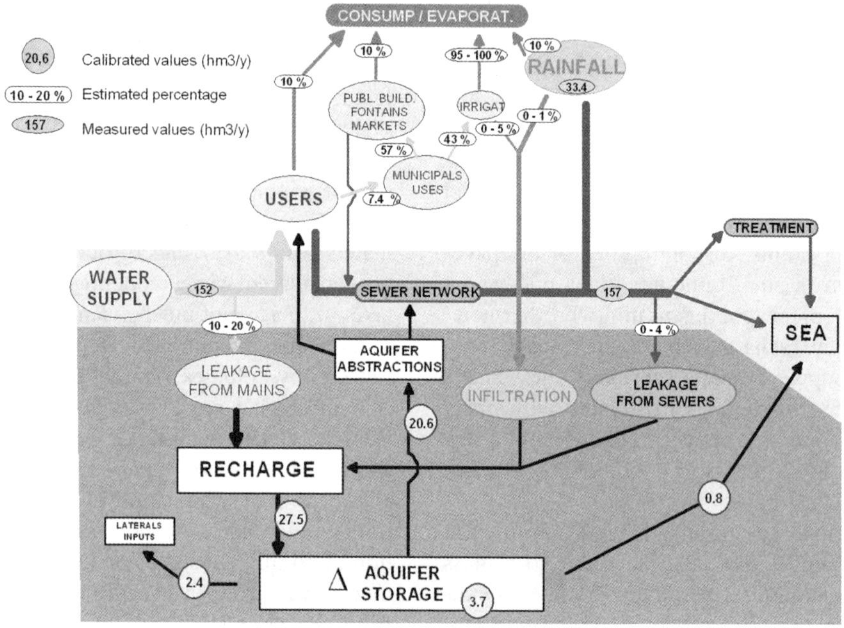

Figure 3. Quantification of underground water balance in Barcelona aquifers.

inputs should be taken into account as new available resources. In any case, the possible uses of underground water are related to their quality and to the potential evolution of recharge sources with time.

Considering the above reflections and due to their great implications, not only social but also economic, the Town Council (in cooperation with CLABSA[1] and the Hydrogeology Group of the Technical University of Catalonia) developed a plan of use of groundwater in Barcelona. The studies carried out and the groundwater flow models were developed in cooperation with the local authorities. One of the aims of the modeling effort was to use it as a management tool. In fact, it has been used to support those decisions related to several projects, especially to determine where it is convenient to exploit groundwater, which water volumes are necessary to maintain piezometric levels in an acceptable situation and what are the potential uses depending on its quality. From these studies, the construction of a secondary supply system was envisaged. The potential use of underexploited groundwater would allow reducing the consumption of drinking water, for uses that do not require high water quality, such as certain local services. In some cases, the exploitation of the aquifer in the vicinity of certain infrastructures helps reducing the phreatic level and thus diminishes the infiltrations. The effects of phreatic level recovery can be fought pumping groundwater, which can be used to satisfy, at a low cost, some of the local demands.

In order to make this additional water available the construction of a secondary supply system was required. This was not envisaged as a unique system, but as a series of multiple use systems placed in the vicinity of the different capture points, for instance, near subway stations. Other systems are located in the vicinity of the deposits to store flow water and primary treatments. Ancient wells and mines were used, but also new wells were drilled, in one of the cases water is pumped from the inside of a building for structural reasons. The use of this water includes watering gardens and trees, filling hydrants, loading trucks for street cleaning.

[1] CLABSA. Clavegueram de Barcelona, S.A., company in charge of the sewage system management in Barcelona that cooperates with the local authorities.

It is believed that untreated groundwater can also be used for the following local services: cleaning-up the sewer system, ornamental fountains, vehicle and installation cleaning in maintenance company stores, air-conditioning of public buildings, regeneration of natural water volumes and humid areas, specially near the Besós river to recover its original environment.

Following the quantification of the resources, the water that has to be pumped to maintain the piezometric level as it is nowadays is of some 10 Mm^3/yr. This volume does not include the part which could potentially be exploited, out of the 8–12 Mm^3/yr that infiltrates annually in the subway and is directly pumped to the sewage system.

In June 1998 a "Pla per l'aprofitament de l'aigua del subsòl de Barcelona" (Plan to exploit the groundwater in Barcelona) (Ajuntament de Barcelona and CLABSA, 1998) was formulated with the aim of using more rationally the hydraulic resources available and introducing environmental sustainability criteria.

The plan will take place in different phases, the first one being that included in the "Plan de Actuación Municipal 2000–2003" for the Barcelona city area, with a previous investment of over 3.5 million €. The Plan suggests fulfilling at least the following tasks:

- Supplying 15% of the 4.2 Mm^3/yr that is consumed by "Parques y Jardines" (local organism in charge of the Barcelona green areas) for watering.
- Supplying 20% of the 0.5 Mm^3/yr to ornamental fountains.
- Use 0.025 Mm^3/yr of water to clean the sewer system, basically regulating deposits.

Summing up all this, we would reach a total consumption of 1.10 Mm^3/yr, which is 11% of the resources that could be used, and 13% of the Barcelona municipal services water consumption (8.14 Mm^3/yr).

On the other hand, regarding the environmental uses, the Plan envisages capturing water in wells near the subway station of Baró de Viver some 1.5 Mm^3/yr and pouring it in the Besós river to improve the quality of the surrounding water volume and the watering of the riverside park. At the same time the levels decrease near the subway station of Poble Sec by pumping 10–20 L/s to the of Montjuic Botanic Garden (some 1.2 Mm^3/yr at the end of the plan, and 0.25 Mm^3/yr, in this first phase).

During the year 2001 the total consumption of groundwater for local uses was 429,408 m^3.

4 CONCLUSIONS

The decrease of intensive pumping that occurred in the Barcelona aquifers since the 1970s caused a recovery of piezometric levels, in some cases up to 15 m. This recovery caused infiltration or, even flooding, of some structures and underground spaces in the city (subway and railroad tunnels, parking lots, basements, etc.).

It is necessary to focus the study of these effects and their causes on the singularities and conditions that an urban environment imposes on a hydrogeological environment.

Piezometric levels in the city of Barcelona display the typical evolution of an industrial city, clearly associated to the changes in groundwater use. In early times (late 19th century) continuous abstraction caused important groundwater drawdowns. Later (since 1970s) this extraction was reduced gradually until the present time. Levels recovered to those of the first half of the 20th century.

Different recharge sources and its quality were identified and evaluated. These include water coming from the supply and sewage systems, direct infiltration, marine intrusion, etc. In the case of seawater intrusion the diminution of abstractions and thus the recovery of water levels caused a new flow towards the sea and in some salinized wells the original water quality was gradually recovered.

The numerical modeling of Barcelona aquifers allowed identifying the processes (validation of the conceptual model), the hydrogeological characterization and integration (quantification of the terms implicated), and the management of the resources. It was calculated that in order to maintain the levels as they are at present, an additional pumping of 10 Mm^3/yr should take place.

The potential use of underexploited groundwater would allow reducing the consumption of drinking water for uses in which there is no need of a high quality, such as certain local services.

The local authorities developed a plan for groundwater use in Barcelona. This plan is based on the fact that water management should focus on the global optimization of water following savings and reuse policies, while the impact on groundwater due to an increase in the levels hopefully reduces. The use of groundwater is related to its quality and to the possible time evolution of the sources for recharge.

The plan will take place in several phases, the first one being that included in the "Plan de Actuación Municipal 2000–2003". Taking all this into account and regarding local services, the plan suggests attaining a consumption of 1.10 Mm^3/yr, which is 11% of the resources that could be potentially used. During the year 2001, the consumption of underground water for uses of local services was of some 0.43 Mm^3.

REFERENCES

Ajuntament de Barcelona and CLABSA (1998). Pla per a l'aprofitament de l'aigua del subsòl de Barcelona. Ajuntament de Barcelona.

Castillo, O. (2000). Mètode general per a la quantificació de la recàrrega en una àrea urbana. Aplicació al cas de Barcelona (General method for quantifying urban groundwater recharge. The Barcelona city case). Bachelor's thesis, UPC, pp. 92.

Cheney, C.S.; A.M. MacDonald and A. Foster (1999). Groundwater responses to urbanization and changing patterns of industrial development in the Wigan metropolitan area, north-west England. In: Groundwater in the Urban Environment. Selected city profiles. Chilton, J. (ed.). Balkema, 111–117.

Lerner, D.N. (1997). Too much or too little: Recharge in urban areas. In: Groundwater in the Urban Area: Problems, Processes and Management. Chilton *et al.* (ed.), XXVII Congress of the International Association of Hydrogeologists (IAH), Nottingham 21–27 September 1997, pp. 41–47.

Marinos P.G. and M.J. Kavvadas (1997). Rise of the groundwater table when flow is obstructed by shallow tunnels. In: Groundwater in the Urban Area: Problems, Processes and Management. Chilton *et al.* (ed.), XXVII Congress of the International Association of Hydrogeologists (IAH), Nottingham 21–27 September 1997, pp. 49–54.

Marinos P.G. and M.J. Kavvadas (1998). Effects of shallow tunnels on the groundwater table levels. Bulletin of the Int. Assoc. Eng. Geology (IAEG), Vol. 56.

Navarro A.; J. Carrera and X. Sánchez-Vila (1992). Contaminación por lavado piezométrico a partir de vertederos enterrados. Aplicación a un acuífero aluvial. Revista Tecnología del Agua, No. 98, pp 16–34.

Ondiviela, M.; E. Vázquez-Suñé, J.; Nilsson, J.; Carrera, X.; Sánchez-Vila and J. Casas (2002). Effect of intensive pumping of infiltrated water in the "Plaça de la Vila" parking lot in Sant Adrià del Besòs. (Barcelona, Spain). Symposium on intensive use of groundwater. Challenges and opportunities (SINEX). Valencia.

UPC, CLABSA and Ajuntament de Barcelona (1997). Estudi de les aigües subterrànies del Pla de Barcelona. Ajuntament de Barcelona.

Vázquez-Suñé, E.; J. Carrera, O.; Castillo, X.; Sánchez-Vila and A. Soler (2003). Multivariate statistics analysis methodology for the discrimination and quantification of recharge terms in urban areas. Submitted to Ground Water.

Effect of intensive pumping of infiltrated water in the *Plaça de la Vila* parking lot in Sant Adrià del Besòs (Barcelona, Spain)

M. Ondiviela[(1)], E. Vázquez-Suñé[(2)], J. Nilsson, J. Carrera and X. Sánchez-Vila
Hydrogeology Group, Technical University of Catalonia, Barcelona, Spain
[(1)] *monica.ondiviela@upc.es*
[(2)] *enric.vazquez-sune@upc.es*

J. Casas
Ajuntament de Sant Adrià del Besòs (EUSAB), Barcelona, Spain

ABSTRACT

The parking lot located in *Plaça de la Vila* in the city of Sant Adrià del Besòs (northeastern Spain) is facing a huge seepage problem. Groundwater seepage started immediately after the parking lot was constructed, some 30 years ago, and since then total seepage has steadily increased. At present the amount of water evacuated in around 250–300 L/s. The water has to be evacuated to the sewage system, and thus the owner (the City Council) must pay both the cost of electricity for operating the pumps and the cost of dumping the water to the sewage system. This huge abstraction conditions significantly the hydrodynamics and quality of the neighbouring aquifers. To avoid economical, as well as management and maintenance problems, the possible effects of remodeling the parking lot, including ceasing or at least diminishing the abstraction of groundwater, are envisaged. Ceasing pumping would cause a recovery of approximately 4 m in phreatic levels and thus possibly affect other nearby underground urban structures, such as parking lots, basements, and subways tunnels.

Keywords: *Urban hydrogeology, seepage, urban structures, infiltrometer.*

1 INTRODUCTION

The study area is located in Catalonia (northern Spain), more precisely in the Besòs river delta, which includes part of the cities of Barcelona, Santa Coloma de Gramanet and Badalona, and the whole city of Sant Adrià del Besòs.

In 1971 the three-story *Plaça de la Vila* parking lot was opened. A few months later, the parking lot experimented several floods caused by excess groundwater seepage that could not be evacuated by the drainage system. In December of that same year, and after a period of heavy rainstorms, a single flood episode covered one and a half stories, and a year passed before the parking lot dried out naturally (Conde Cabeza, 1971). In any case seepage was observed to increase continuously, and so several pumps were installed in order to drain the groundwater flow that could seep inside the parking lot. Since then pumping has been continuously increasing and currently there are 5 pumps, working 24 hours a day, and pumping at a flow rate of approximately 250–300 L/s.

Only a very small part (<1%) of the water drained is used to water parks and gardens and for the city's ornamental fountains. The rest is poured directly to the sewage system.

The whole situation is not sustainable in economical or environmental terms. In particular there are very high economic costs associated, such as the maintenance of the pumping equipment, a great energetic consumption (50,000 Kwh/month), the maintenance of the sewer system due to the erosion caused by excess water dumping, structural damage to the parking lot due to wedges caused by the dragging of fines, etc.

Given such high costs, the Sant Adrià del Besòs City Council is considering remodeling the parking lot, reducing, and, if possible, stopping groundwater abstraction. A decrease in pumping could cause a recovery of phreatic levels in the vicinity of the parking lot and thus possibly affect other nearby underground urban structures, such as parking lots, basements, and subway tunnels.

In order to quantify these possible effects and to foresee the feasibility of remodeling, the City Council, together with the Hydrology Group of the UPC, are elaborating a complete hydrogeological study and a numerical model of the area. The following sections discuss the most relevant aspects of these studies. The historical evolution, the effects produced by abstractions and the current state of aquifers will be analyzed, and its possible evolution in front of different scenarios determined.

2 HYDROGEOLOGICAL STUDY

The parking lot is placed in the Besòs delta, which stands on a rather impervious base, mainly made of slates, Paleozoic granites and Pliocene clays. On top of this base, the sediments of Quaternary river-delta origin were deposited. In these deposits we can distinguish two permeable levels (sands and thick gravels), which constitute the two main aquifers of the delta. These aquifers are separated by a clayey and silty layer which acts hydrogeologically as a semiconfinement unit (Figure 1).

The parking was excavated in the topmost (phreatic) aquifer, which has a saturated thickness of some 11 m and a hydraulic conductivity of 300 m/d to 500 m/d. The hydraulic conductivity of the lower aquifer (5.5 m thick) is significantly lower, some 50 m/d.

Towards the margins of the delta, the clay layer gets thinner until finally disappearing. Then the two aquifers join together, forming a single one.

Traditionally the Besòs delta aquifers were intensively exploited for industrial and supply purposes. Until 1940, the exploitation of the aquifers of the area was moderate, and mainly addressed to watering and public supply, but the abstraction kept increasing until it reached some 66 Mm^3/yr at the end of the 1960s (MOP, 1966). This overexploitation of the aquifer caused a significant decrease in the phreatic level, with minimum values of the order of 12 m below sea level (Vázquez-Suñé *et al.*, 1999), thus leading to marine intrusion. The possibility of recharging the aquifer with treated residual water was considered in order to stop the progressive salinization that affected the area (Custodio, 1976). Deterioration of the quality and quantity of water from the aquifer, caused by its overexploitation and to waste dumping in the river and subsoil, led many industries to move outside the city and others to cease groundwater abstraction.

The abstractions in the Besòs delta aquifer kept decreasing until now, with an estimated figure of 25 Mm^3/yr (Table 1).

The reduction in abstractions caused the water table to rise (recover). Measured recoveries are as high as 15 m (Figure 2). In its turn, this recovery caused an increase in infiltrations in many underground structures, singularly in the *Plaça de la Vila* parking lot. Present seepage is equivalent to 50% of current abstractions.

The continuous, intense pumping in the parking lot has led to phreatic levels locally very low. A piezometric map shows that water levels are highly conditioned by the presence of the Besòs river, being only some 200 m. The river is then recharging the aquifer in the left margin, a situation that carries on for a few km towards the sea. The right margin becomes effluent only in the final section, close to the sea (Figure 3).

The subway system intersect through the phreatic level of the aquifer. The subway tunnels cause a barrier effect characterized by a rise in the water level upstream, and a decline downstream. The water level variations assuming totally impervious tunnels would be in the order of decimeters. Seepage into the tunnels causes the water levels to decline both sides of the subway system more significantly.

The analysis of the hydrochemical parameters of the water near the *Plaça de la Vila* in Sant Adrià del Besòs allows evaluating qualitatively and quantitatively the groundwater recharge

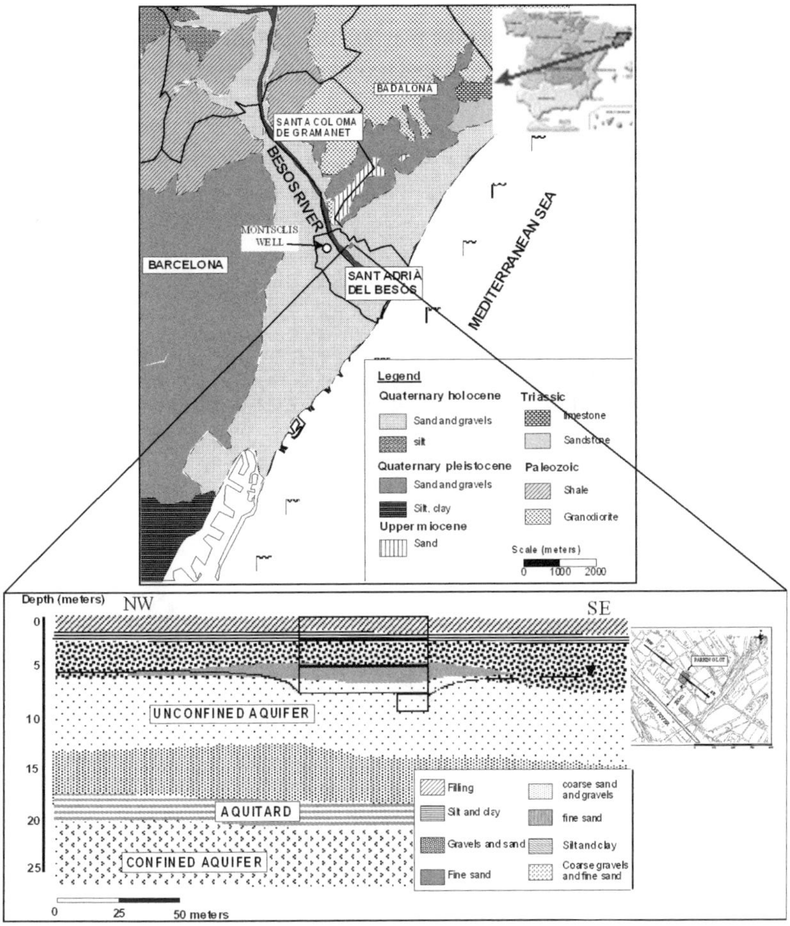

Figure 1. Geological map and local cross section of the studied area.

Table 1. Evolution of abstractions in the Besòs delta aquifer. In others we include 3.5 Mm^3 that are captured within the subway system.

YEAR	ABSTRACTION		
1966	63 Mm^3/yr		
1970	66 Mm^3/yr		
1976	33 Mm^3/yr		
1983	25 Mm^3/yr		
2000	< 20 Mm^3/yr		
2002	< 20 Mm^3/yr	Parking Pça Vila	8-9 Mm^3/yr
		Others	9-12 Mm^3/yr

sources, therefore allowing to study the possible quality evolution and establishing its possible uses and limitations.

Water from the superficial aquifer comes from different sources: the Besòs river, direct recharge from rainfall infiltration, losses in the supply system, losses in the sewage system, and sea water.

The hydrochemical facies distribution in the aquifer depends on which of the recharge terms is more significant in quantitative terms. The

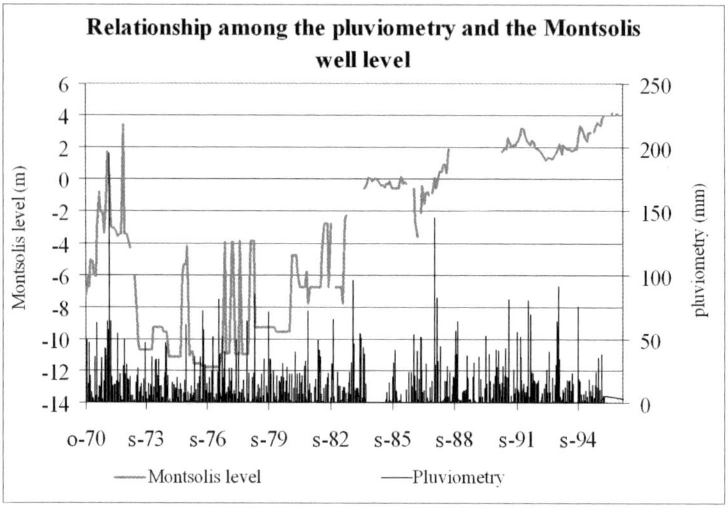

Figure 2. Relationship among the pluviometry and the Montsolís well level (Barcelona well situated right by the river, see geological map).

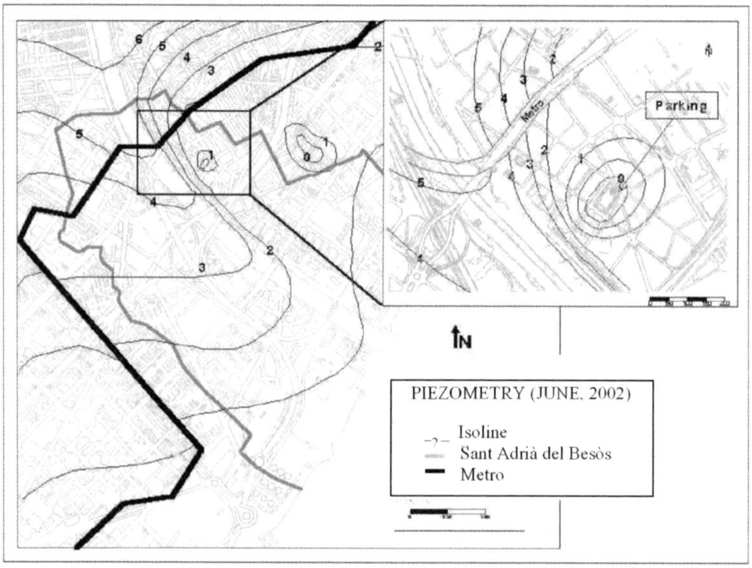

Figure 3. Piezometry of the studied area (June, 2002).

isoconductivity lines map shows that there is marine intrusion in the coast front, which is stressed in the parking lot area due to the heavy pumping (Figure 4). A hydrochemical balance shows that 70%–80% of the water pumped in the parking lot comes from the Besòs river and/or the sewage system.

The chemical parameters of the groundwater, except for the area near the coast, are currently below the values damaging concrete structures, that is, in case a new parking lot is built it would not be necessary for the concrete to have specially resistant characteristics. The mean electrical conductivity is about 1500 μS/cm,

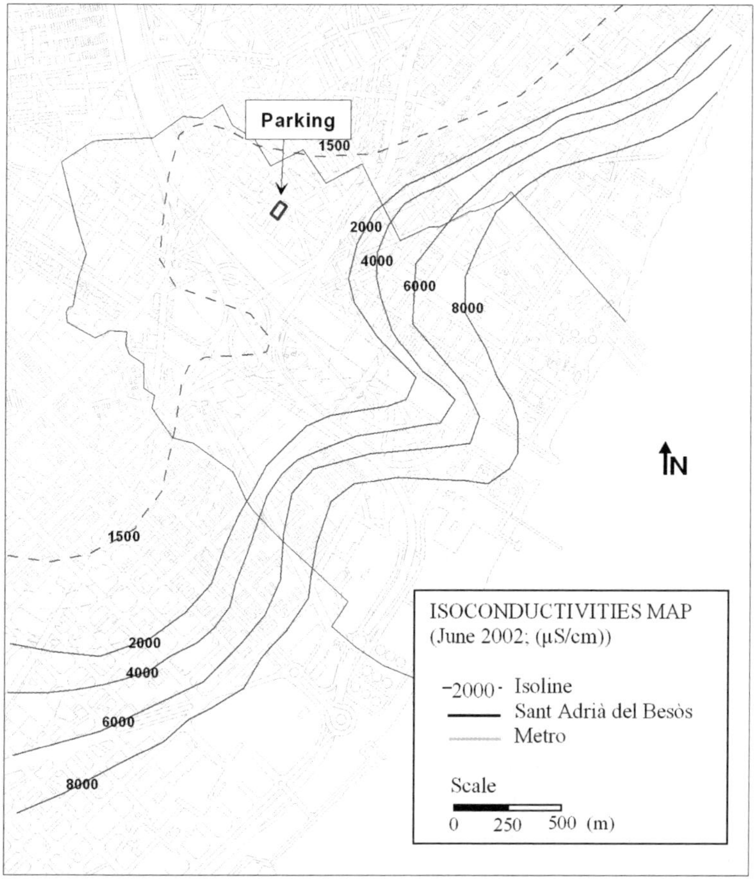

Figure 4. Isoconductivities map (μS/cm) for June 2002.

which is sufficiently low to water parks and gardens if the irrigation system selected is that of intensive watering (using water in excess washes the soil and helps preventing the accumulation of salts in the topmost part of it).

Regarding the hydrogeological balance in the delta aquifer, the most significant terms are abstractions and water infiltration coming from the Besòs river. A total of 13 Mm^3/yr are infiltrated in the river at the final section, out of which 8 Mm^3/yr are actually reaching the parking lot. It is estimated that there is a net water outflow going from the aquifer towards the sea, although there is a clear but very localized marine intrusion wedge (Figures 3 and 4).

Figure 5 shows a summary of the hydrological balance, including the estimated fluxes in Mm^3/yr. A finite element numerical model is being constructed in order to reduce the uncertainly in the balance values.

3 INTERACTION WITH UNDERGROUND STRUCTURES

It is important to establish the water table level with respect to the depth of the underground structures that can be found in the area of interest, such as parking lots, basements, and subway tunnels. The relative location between groundwater levels and structures would determine whether an individual structure might or might not face seepage problems. Notice that the fact that a given structure has no seepage problems could simply be due to the fact that a proper impermeabilization was performed. This does

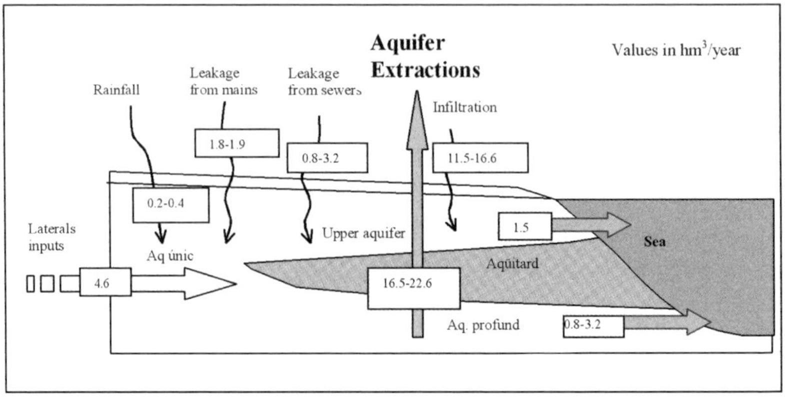

Figure 5. Hydrogeological balance of the Besòs delta.

not mean that structure could not face seepage problems in the future in case the impermeabilization fails over time.

The problem, then is to study what would be the effect of ceasing pumping at the parking lot, thus eliminating the present depression cone. Given that the whole parking lot can be thought as a well with a known radius, we have estimated that the reduction of pumping to half its present value would cause a water table rising of around 2 m in the area. If pumping was totally suppressed the build-up would be of around 3.8 m, placing the level between 2.5 m and 3 m above the sea level in a 300 m radius around the parking lot. This recovery would definitely affect some of the nearby underground structures. The effect on the structures would depend mostly on the building design, their response to an increase of pressures and whether the design included impermeabilization. Note that some of the structures where built after 1970, therefore possibly not accounting for groundwater levels to reach them.

4 CONCLUSIONS

In 1971 the three-story *Plaça de la Vila* parking lot opened. Soon after the inauguration, groundwater seepage was observed, in increasing amount since then, up to the present value of 300 L/s. The City Council has been considering remodeling the parking lot in order to diminishing, and if possible ceasing, the groundwater abstraction.

The parking lot is excavated in the superficial free aquifer of the Besòs River Delta, which has a saturated thickness of some 11 m and a hydraulic conductivity of 300 m/d to 500 m/day.

Abstractions from this aquifer surpassed 65 Mm^3/yr at the end of the 1960s. This overexploitation of the aquifer led to an important decrease of the phreatic level.

Currently abstractions are on the order of 25 Mm^3/yr. This reduction in extractions caused a rebound of the water table, which led to an increase in seepage to underground structures all through the delta. At present seepage accounts for more than 50% of current total abstractions.

Due to the intensity and continuity of pumping in the parking lot, the phreatic levels are locally very low. Water from the superficial aquifer is a mixture from different recharge sources such as the Besòs river, rainfall direct infiltration, losses in the supply and sewage systems and sea water. The water pumped in the parking lot has a 70–80% water content coming from the river (which is only 200 m away and very well connected hydraulically with the parking lot) and the sewage system.

It has been calculated that once pumping ceases, the recovery of phreatic levels would be of around 4 m. This would affect some of the nearby underground structures in a 300 m radius around the parking lot.

5 CURRENT AND FUTURE TASKS

In order to obtain more detailed data on river infiltration, several tests were carried out placing an infiltrometer (Lee, 1977) in the river's bed and measuring the recharge to and the discharge from the aquifer. The results of the tests corroborated, in a qualitative way, that a significant infiltration from the river takes place. These tests are still being carried out in order to obtain the space variation and precise quantification of infiltration.

Another future test would be the total cease of the 5 pumps in the parking lot until water build-up reaches 2 m approximately. Once this level is reached, the pumping will start anew. The variations in level in the piezometers and wells surrounding the area will be measured. This test will allow estimating the hydraulic parameters and the river-aquifer relationship.

All these data will be complemented with the numerical model that is being developed. This model will allow simulating several settings in order to manage groundwater near the parking lot. Different constructive solutions will be considered (barriers, draining, pumping, etc.), as well as other abstractions for public use (street cleaning, environmental arrangement at the Besòs river bed, watering parks and gardens, and so on).

ACKNOWLEDGEMENTS

We wish to thank the *Ajuntament de Sant Adrià del Besòs* for their support both in logistics and in the field work performed.

REFERENCES

Conde Cabeza, M. (1971). Dictamen sobre el aparcamiento subterráneo en la plaza de Calvo Sotelo (Sant Adrià del Besòs). *Internal inform. Ajuntament Sant Adrià del Besòs.*

Custodio, E. *et al.*, (1976). Ensayos para el análisis de la recarga de aguas residuales tratadas en el Delta del Besòs. *II Asamblea Nacional de Geodesia y Geofísica. Sección de ciencias hidrológicas. Barcelona.*

Lee, D.R. (1977). A device for measuring seepage flux in lakes and estuaries. *Limnol. Oceanogr.*, vol 22: 140–147.

MOP (1966). Estudio de los recursos hidráulicos totales de las cuencas de los río Besòs y Bajo Llobregat. *Comisaría de Aguas del Pirineo Oriental and Servicio Geológico de Obras Públicas. 4 vol. Barcelona.*

Vázquez-Suñé, E.; Sánchez-Vila, X. and Carrera J. (1999). Gestión de las aguas subterráneas en zonas urbanas. Conceptualización y modelización: Aplicación a Barcelona (España). *II Congreso Argentino de Hidrogeología, IV Seminario Hispano-Argentino sobre temas actuales de la hidrología subterránea. Santa Fe - Argentina. 28 September -1 October.*

Predicting impacts of abstraction reduction for an aquifer with a century of exploitation

Martin Shepley
Environment Agency. Olton Court, Olton, Solihull, UK
martin.shepley@environment-agency.gov.uk

ABSTRACT

The East Midlands Permo-Triassic Sandstone (EMPTS) is an aquifer with a long history of abstraction that dates back to Victorian times (mid nineteenth century). Since the mid 1960s abstraction has been in excess of precipitation recharge, which has resulted in impacts on the surface water environment. With the forthcoming Water Framework Directive, there will be a requirement to achieve good groundwater status and this could entail modifications in abstraction to achieve ecologically acceptable surface water flows.

A groundwater model of the EMPTS aquifer has been used to predict the improvement of surface water flow by groundwater abstraction reduction. This study has shown that particularly for a low diffusivity aquifer, where the groundwater response to applied stresses is slow, the length of the monitored history of the aquifer is a critical factor. Monitoring of the EMPTS aquifer started in the late sixties post-dating a large part of the development of the aquifer. This represents a challenge for making reliable predictions, particularly if large abstraction reductions are sought. The use of the groundwater model for an *incremental* strategy is advocated, whereby reductions are made according to predictions that are reasonable with respect to the validation period of the model. In this manner, the benefit of groundwater abstraction reduction can be assessed rationally over a time scale that is sensible with respect to the diffusivity of the aquifer.

Keywords: *Prediction simulation, Permo-Triassic sandstone, diffusivity, abstraction history.*

1 INTRODUCTION

In the UK many aquifers have histories of sustained abstraction that are related to the development of the country during the industrial revolution. The East Midlands Permo-Triassic Sandstone (EMPTS) aquifer is such an example where large-scale abstraction started in the nineteenth century. This abstraction has been greater or equal to the precipitation recharge since the mid 1960s with resultant impacts on the surface-water environment. With the introduction of the Water Framework Directive (Council of European Communities, 2000), European environmental legislation has for the first time stipulated an obligation to restore areas where environmental damage has occurred. The directive has an ambitious timescale of implementation whereby the water environment in the member countries should achieve *good status*, in terms of both quality and quantity by 2015. Although this target is currently over a decade away, in aquifers where large-scale long-term abstraction has occurred, the time scale could be difficult to adhere to.

The management of groundwater resources is scientifically a very challenging problem, because, as a general rule, groundwater flows cannot be measured but only inferred. Distributed time-variant numerical groundwater models (referred to throughout this paper as groundwater models) provide the best tool for drawing together the fragmentary information on groundwater flow to give an enhanced and quantified understanding of the groundwater system. However, the robustness of these models is to a certain extent controlled by the amount and reliability of historic data available. Where a large part of the aquifer development has preceded the start of systematic hydrogeological and hydrological monitoring there can be significant uncertainty that flow processes under reduced abstraction regimes are adequately represented in the groundwater model.

The Environment Agency (the water environment regulator for England and Wales) strongly supports the use of groundwater models for groundwater resources management (Hulme *et al.*, 2002). However, if a groundwater model's predictive capability is used then consideration of the history of the aquifer in terms its management is of critical importance. This paper considers the basic hydraulic characteristics of the EMPTS aquifer and reviews the history of its management. With this information the use of a groundwater model is explored to manage the reduction of abstraction in what is a very sluggish (low-diffusivity) aquifer system.

2 THE EAST MIDLANDS PERMO-TRIASSIC SANDSTONE AQUIFER SYSTEM

The EMPTS aquifer extends from Nottingham up to Doncaster (Figure 1). The Permo-Triassic Sandstone does continue further north through Yorkshire to the North Sea, but this part of the Permo-Triassic Sandstone aquifer is not discussed here. The western margin of the aquifer system is formed by the stratigraphically lower Permian Magnesian Limestone Series. Whilst the Magnesian Limestones are considered a major aquifer in their own right, they are separated from the EMPTS aquifer by a number of low-permeability marl beds (the Middle and Upper Permian Marls). In places, the Magnesian Limestones are in contact with the EMPTS due to faulting and the discontinuous nature of these marl beds. Some flow does occur between these two aquifers, however, detailed discussion of this is beyond the scope of this paper.

To the east, the EMPTS is confined by the Triassic Mercia Mudstone Group, a thick sequence of mudstones, with some interbedded sandstones. The base of the Mercia Mudstone consists of the Colwick Formation, a sequence of interbedded mudstones and sandstones that are generally more permeable than the rest of the Mercia Mudstone. Structurally this Permo-Triassic sequence is relatively simple, with a dip of around 2°–3° to the east (Figure 2). The aquifer continues for some distance under the Mercia Mudstone, but the furthest borehole abstracting water from the Permo-Triassic Sandstone is approximately 15 km into the confined zone. The transmissivity (T) of the EMPTS is generally considered in the range of 100–700 m^2/d, with higher values occurring at boreholes where fracturing is locally important (Allen *et al.*, 1997). The specific yield (S) of the aquifer is generally considered to be in the range 0.10–0.15 (Allen *et al.*, 1997). It is therefore unlikely that the diffusivity ($D = T/S$) will be greater than 10,000 m^2/d with more typical values around 3,000 m^2/d. Using a simple calculation assuming two-dimensional flow the response time of this aquifer to a new stress such as a decrease in abstraction can be roughly estimated:

$$t = (l^2/D)$$

where t is the time for aquifer flows to achieve a new equilibrium, and l is a typical distance from the location of stress change to a point of discharge such as a river. Given the distribution of rivers across the EMPTS aquifer, a value for l of 5 km could be considered reasonable. It is therefore improbable that the response times, i.e. the time when full impact of abstraction reduction occurs, would be much less than seven years and could easily be greater than 20 years. This type of calculation is very simplistic, but is good for demonstrating the nature of the problem faced.

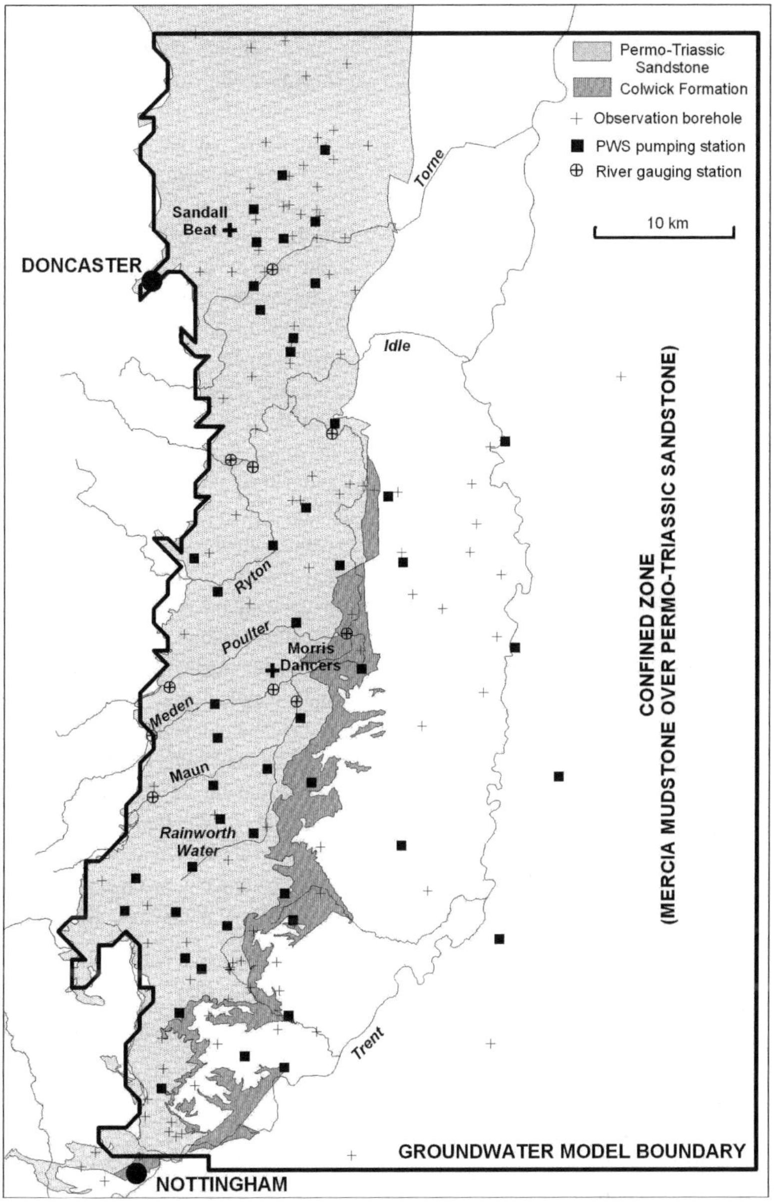

Figure 1. The East Midlands Permo-Triassic Sandstone aquifer.

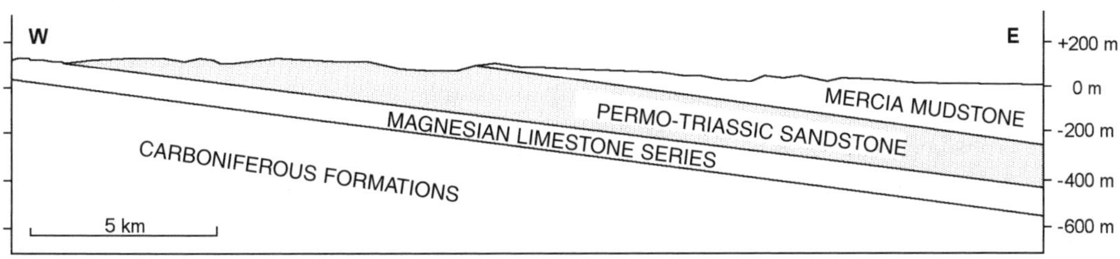

Figure 2. Cross section of the East Midlands Permo-Triassic Sandstone Aquifer system (after Land, 1966).

3 HISTORY OF GROUNDWATER ABSTRACTION AND MANAGEMENT

There has been a long history of groundwater abstraction from the EMPTS aquifer, of which large-scale exploitation started in the mid 19th century. Evidence of this early exploitation exist today in the form of a number of Victorian pumping stations, which are notable architectural features across the outcrop of the aquifer. The degree to which the quantity of abstraction was recorded has largely been driven by legislation. Before the 1945 Water Act of Parliament there are very little data available of measured abstraction. Some data are available from early water supply memoirs (Lamplugh and Smith, 1914), but these give only incomplete records for the time of the survey. Between 1945 and 1965 detailed information is available for a number of years (Downing *et al.*, 1966, 1970) and this shows a steady increase in abstraction from 200,000 m^3/d in 1948 to over 320,000 m^3/d in 1964 (Figure 3). Continuous records of abstraction are available from the late 1960s and these show an increased rise of abstraction that was close to or over 400,000 m^3/d for most of the 1980s and early 1990s (Figure 3).

The current water abstraction-licensing regime was introduced in England and Wales with the 1963 Water Resources Act of Parliament. All abstraction boreholes that were operational were given so-called Licences of Right, which under current legislation are difficult to revoke unless the license holder stops abstraction. At present there are 48 licensed public water supply (PWS) pumping stations (39 of which are Licences of Right), which account for almost 90% of groundwater abstraction from the EMPTS aquifer. Abstraction related to mining has always been exempt from water resources legislation and therefore generally little is known about related abstraction quantities. The Carboniferous Coal Measures beneath the EMPTS (Figure 2) have been extensively mined and related dewatering may have affected the groundwater flows in the EMPTS aquifer.

Considering the long history of groundwater abstraction, in most areas the impacts have not been recognised until relatively recently. In Nottingham long-term falling levels were noted in 1960s reports on groundwater resources (Downing *et al.*, 1966, 1970; Land, 1966), with groundwater levels significantly below the rivers leading to significant loss of surface water to the aquifer, particularly from the River Trent and Leen. There are local reports in Nottingham of falls in wells of 0.15 m/yr since 1927 and 0.60 m/yr since 1938 (Downing *et al.*, 1966; Land 1966). However, with the decline of manufacturing industry in the UK since the 1960's, groundwater abstraction in Nottingham has significantly reduced to such an extent that rising groundwater levels may have caused some local inundation of cellars.

To the north of Nottingham a decline of groundwater levels was not recognised largely due to the lack of systematic monitoring in non-pumped boreholes before the 1960s (Downing *et al.*, 1966, 1970). Perching of some watercourses above the main water table of the EMPTS aquifer, in particular the Rainworth Water (Figure 1), was already noted then (Downing *et al.*, 1966; Land, 1966). However, this was regarded as a

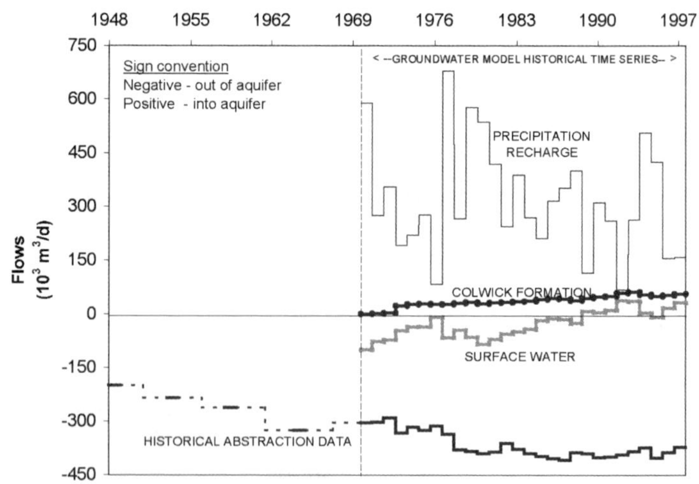

Figure 3. Historical abstraction data and precipitation recharge, and flows to surface water and the Colwick Formation calculated by the Nottingham-Doncaster groundwater model.

natural feature of the aquifer, rather than an impact related to the previous abstraction history.

As a result of 1963 Water Resources Act, the development of extensive water monitoring networks were started throughout England and Wales. Permanent flow gauging stations were constructed on most major rivers crossing the EMPTS outcrop, the first of which were completed in 1965. Particular consideration was given to the hydrogeology, with stations constructed where streams run onto and off the EMPTS outcrop. There are currently ten permanent gauging stations in use on rivers on the aquifer outcrop (Figure 1). A groundwater level monitoring network was also started at the same time, and currently over one hundred observation boreholes are monitored on a monthly or bimonthly basis (Figure 1), both on the outcrop and in the confined zone of the EMPTS. These data have allowed a good data-derived estimation of the water-balance of the EMPTS aquifer for its recent abstraction history, and this has provided a very firm basis for the development of a groundwater model.

4 THE EAST MIDLANDS PERMO-TRIASSIC SANDSTONE GROUNDWATER MODEL

A groundwater model of the EMPTS aquifer was developed during the period 1990–1993 by Professor Ken Rushton and co-workers, School of Civil Engineering, University of Birmingham for the National Rivers Authority (a predecessor organisation of the Environment Agency). The model, often referred to as the Nottingham-Doncaster groundwater model, uses a well-tested finite difference algorithm (Rushton and Redshaw, 1979) to solve the time-variant groundwater flow equation in two dimensions. It extends from the western edge of the EMPTS outcrop to 30 km into the confined zone (Figure 1) and has been validated against a historical time series that spans back to 1969, when the construction of the monitoring network had just started. Some of the work on this groundwater model, particularly relating to river-aquifer interactions has been published (Rushton and Tomlinson, 1995; Rushton *et al.*, 1996).

The model was originally developed to investigate long-term declining groundwater levels in the EMPTS aquifer. It was used in 1994 to predict the impacts of a licence reduction of 13,000 m^3/d in the northern part of the aquifer. In 1998 the Environment Agency updated this groundwater model, extending the historical time series from 1992 to the end of 1997. The predictions made in 1994 have been proved by the model showing a good match with observed data for the extended time series, which includes the licence reduction in the northern part of the aquifer, further demonstrating that this is a robust groundwater model.

The groundwater model has given a firmer appreciation of where abstraction has been balanced by other sources of water apart from precipitation recharge. The long-term precipitation recharge has been calculated at 317,000 m^3/d for the period 1969 to 1997 and thus has probably been on average less than groundwater abstraction since about 1965 (Figure 3). The model calculations show that because of abstraction exceeding precipitation recharge, water has been drawn increasingly from the Colwick Formation, but also from rivers that cross the EMPTS outcrop (Figure 3). It is open to debate to what extent the rivers are acceptable as a source of recharge to supply abstraction. In the aquifer's natural state, it is likely that most if not all rivers were fully gaining along their entire reach across the EMPTS outcrop. Obviously with the onset of abstraction in the mid 19th century the baseflow from the EMPTS aquifer will have been gradually and apparently imperceptibly reduced. The groundwater model shows that the losses to the aquifer have risen such that they are now greater than the baseflow from the aquifer. The Environment Agency currently considers the state of several of the rivers, such as the River Idle, unacceptable and has, in general, allowed no further licensing and progressed a policy of reducing abstraction to the full extent of its current legal powers.

5 PREDICTION SIMULATIONS

In 1999 the groundwater model was used for an extensive series of predictions to investigate the effect of groundwater abstraction reduction on the improvement of surface water flows in the main rivers crossing the outcrop of the EMPTS aquifer. Twenty prediction simulations were performed, examining global abstraction reductions; specific reductions at sources; redistribution of abstraction reductions; and seasonally varying abstractions, mimicking the groundwater component of a conjunctive use scheme. The predictions were performed using a methodology whereby the results were compared with a baseline prediction simulation (Shepley and Taylor, 2003). This baseline prediction simulation used 1993–1997 average abstractions, but with account taken for any anomalous abstraction patterns during that period and programmed future license changes. The baseline prediction simulation therefore represents a licence *status quo*, which should represent the future condition of the aquifer if current use is maintained. The comparison with the baseline prediction simulation gives an objective measure by which the predicted effects of any changes in abstraction can be assessed.

The largest abstraction reduction simulated in the recent prediction exercise was a global reduction of abstraction across the aquifer of 25% relative to the baseline prediction simulation. This equates to a total applied abstraction in this prediction simulation of 277,000 m^3/d, which is similar to the levels of groundwater abstraction from the EMPTS aquifer in the late 1950s. Figure 4 shows a water balance difference plot for the whole groundwater model of the prediction simulation minus the results of the baseline prediction simulation. In this figure a positive difference in flow means that the aquifer, i.e. the EMPTS aquifer is either taking less water from or providing more water to that particular component of the water balance. The overall reduction in abstraction relative to the BPS (approximately 92,000 m^3/d) in the area is balanced by an initial increase in flow to unconfined storage, i.e. a groundwater level rise (Figure 5), which reduces over time as the surface water features start to intercept more flow. The major increase in baseflow to surface water features is predicted to come into effect by 2008, within ten years of the start of the prediction simulation at the end of 1997. A significant gain is obtained in most rivers (an exception being the Rainworth Water), such as the River Meden where approximately 5,500 m^3/d has been gained at the end of the prediction simulation for high and low-flow conditions (Figure 6).

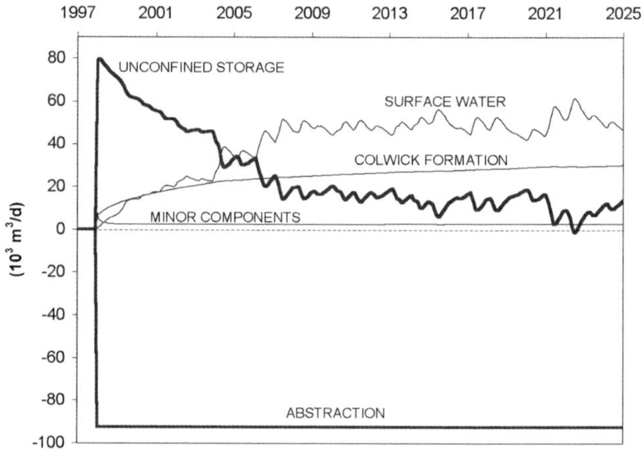

Figure 4. Water balance difference plot. Minor components comprise flows to and from confined storage and the Mercia Mudstone (excluding Colwick Formation).

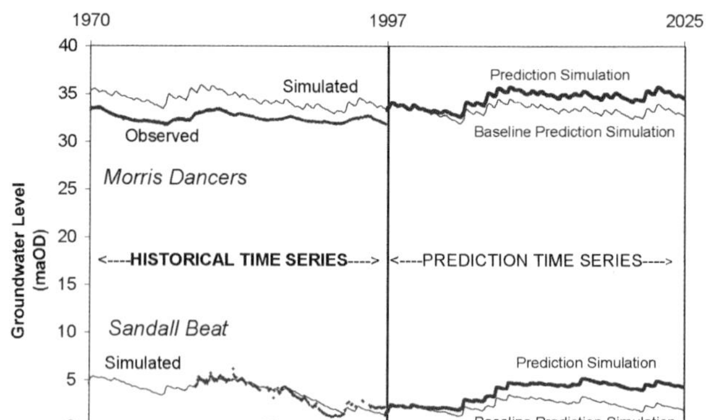

Figure 5. Hydrographs of observed heads, and simulated heads for the historical time series, the baseline prediction simulation and a prediction simulation for Sandall Beat and Morris Dancers observation boreholes (see Figure 1).

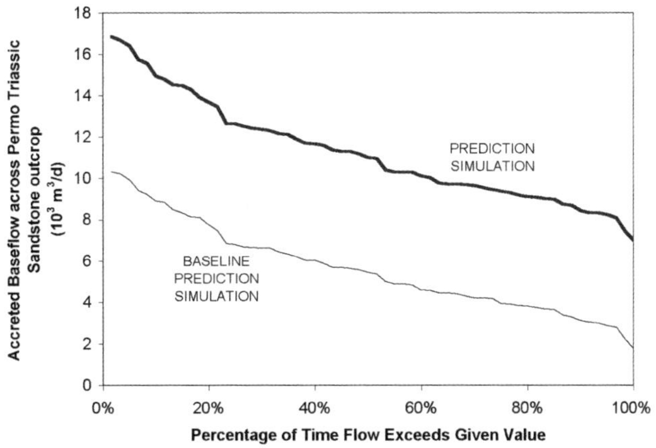

Figure 6. Simulated baseflow duration curve of the River Meden for prediction simulation and the baseline prediction simulation, calculated for 2020-2025.

An interesting feature of this prediction simulation is that it has not reached a dynamic equilibrium and heads are still rising together with the flows to the Colwick Formation, which appears to act as a buffer on the aquifer system. In fact the Colwick formation receives a significant part of the water that is gained by the aquifer from reduced abstraction and this to a large extents limits the increase in flow to surface water features, which only accounts for 55% of the abstraction reduction (approximately 50,000 m^3/d).

6 DISCUSSION AND CONCLUSIONS

There are two main issues that suggest some caution is needed for implementing abstraction changes based on these prediction simulations. Firstly, the possible importance of the Colwick Formation influencing the rebound of the groundwater levels was not evident until this prediction exercise. As the emphasis of the monitoring network has been on the EMPTS aquifer, no systematic groundwater level monitoring has occurred in the Colwick Formation. When abstraction is reduced, it would be highly desirable to have a monitoring network in place in the Colwick Formation to understand if the response in groundwater levels in the Colwick Formation are as expected. Secondly, the surface water features represented in the groundwater model represent the understanding of the hydrological system from the 1960s onwards. As the abstraction is reduced to levels of the 1950s there is a possibility that spring flows may occur that are not represented in the model and that these spring flows are in places that are not desirable.

Given the uncertainty in the prediction simulation a more staged reduction in abstraction would be advisable to give more time to monitor the effects and so gain increased understanding in the aquifer behaviour and more confidence in the modelled results. As the EMPTS has a relatively low diffusivity, it will respond slowly to applied abstraction reduction, as demonstrated by the calculations described in this paper. The prediction simulation suggests that substantial gains in surface water flows can be obtained in most rivers at a timescale that is just about compatible with that of the Water Framework Directive. However, this would require a dramatic cut in abstraction within the next three years. A staged approach would give far more certainty that the size of reduction is not too great, or too small for the environmental benefit in surface water flows desired and would not result in unwanted surface water flows. This would almost certainly require a longer time-scale than allowed for by the Water Framework Directive.

ACKNOWLEDGEMENTS

I am grateful for the valuable help provided by several of my colleagues at the Environment

Agency, particularly Elfyn Parry and Gordon Smith who helped with the collation of the historical information and Steve Fletcher for reviewing the paper. The views expressed in this paper are those of the author and not necessarily those of the Environment Agency, UK.

REFERENCES

Allen, D.J.; Brewerton, L.J.; Coleby, L.M.; Gibbs, B.R.; Lewis, M.A.; MacDonald, A.M.; Wagstaff, S.J. and Williams, A.T. (1997). *The physical properties of major aquifers in England and Wales*, British Geological Survey Technical Report WD/97/34, Environment Agency R&D Publication 8.

Council of European Communities (2000). Directive on establishing a framework for Community action in the field of water policy (2000/60/EC), Official Journal L 327.

Downing, R.A.; Land, D.H.; Allender, R.; Lovelock, P.E.R. and Bridge, L.R. (1966). *The hydrogeology of the River Trent basin*, Trent River Authority report.

Downing, R.A.; Land, D.H.; Allender, R.; Lovelock, P.E.R. and Bridge, L.R. (1970). *The hydrogeology of the Trent river basin*, Water Supply Papers of the Institute of Geological Sciences, Hydrogeological Report No. 6, Institute of Geological Sciences, London.

Hulme, P.; Fletcher, S. and Brown, L. (2002). *Incorporation of groundwater modelling in the sustainable management of groundwater resources*, in: Hiacock, K.M., Rivett, M.O. and Davison, R.M. (eds), Sustainable Groundwater Development, Geological Society of London Special Publications 193, 83.

Lamplugh, G.W. and Smith, B. (1914). *The water supply of Nottinghamshire from underground sources*, Memoirs of the Geological Survey of England and Wales.

Land, D.H. (1966). *Hydrogeology of the Bunter Sandstone in Nottinghamshire*, Water Supply Papers of the Geological Survey of Great Britain, Hydrogeology Report No. 1, Institute of Geological Sciences, London.

Rushton, K.R. and Redshaw, S.C. (1979). *Seepage and Groundwater Flow*, Wiley, Chichester,

Rushton, K.R. and Tomlinson, L.M. (1995). *Interaction between the rivers and the Nottinghamshire Sherwood Sandstone aquifer*, in: Modelling River-Aquifer Interactions, Younger, P.L. (ed.), BHS Occasional Paper, 6, 101.

Rushton, K.R.; Fletcher, S.W. and Bishop, T.J. (1996). *Surface water and groundwater components of the Nottinghamshire Sherwood Sandstone Aquifer System*, BHS 5th National Hydrology Symposium, Edinburgh.

Shepley, M.G. and Taylor, A. (2003). *Exploration of Aquifer Management Options using a Groundwater Model*, Journal of the Chartered Institution of Water and Environmental Management, 17, 176–180.

Importance of hydrogeological modeling in the management of groundwater, a case study: the coast of Hermosillo Aquifer, Sonora, México

Miguel Rangel-Medina[(1)]
mrangelm@marina.geologia.uson.mx

Rogelio Monreal[(1)], Mariano Morales[(1)], José Castillo[(2)] and Herman Valenzuela[(1)]
[(1)] *Dpto. de Geología. Universidad de Sonora, Hermosillo, Sonora, México*
[(2)] *Dpto. de Agricultura. Universidad de Sonora, Hermosillo, Sonora, México*

ABSTRACT

The seawater intrusion into the aquifer of Hermosillo's coast is the result of a physical disequilibrium caused by overpumping in absence of an adequate groundwater administration. In the first 20 years of agricultural operation (1947–1967) the groundwater extraction reached up to 1,200 Mm^3/yr, in consequence the advance of the salinity of seawater intrusion had a strong diffusion, developing a wide cone of depression with a progressive dropping of the piezometric levels until 65 m under the sea level in 2001, nowadays with 550 Mm^3/yr of extraction. Hydrogeological investigations show actually that seawater is acting as an artificial recharge with 98 Mm^3/yr. The policy of groundwater extraction and management were based on an erroneous hidrogeologic model, which induced the aquifer towards a continuous degradation in quality and a loss of fresh water volume. This study of migration of the saline intrusion reveals the way in which severely the resource was exhausted in the last 50 years. Although we offer with a new hidrogeologic modeling a prevention measurement, our results can be taken in advantage to feed a simulation model for the management and restoration to help users and planners who must take decisions for a system, because the aquifer will continue pumping. Unfortunately to revert the generated environmental impact will require of hard decisions and political ability since it will have a high social and economic cost and several decades to give back its natural condition partially.

Keywords: *saline intrusion, overpumping, coastal aquifers, Costa de Hermosillo, groundwater management.*

1 INTRODUCTION

When dealing with exploitation, restoration and management of fresh groundwater in costal aquifers the key issue is saltwater intrusion (Van Dam, 1999). The shape and degree of seawater front intrusion depend on many factors, the type of aquifer, its geology, water table and/or piezometric head, sea water concentration, density, rainfall intensities and frequencies, amount and duration of withdrawal or recharge, natural rate of inflow, physical and geometrical characteristics of the aquifers, land use, geometric and hydraulic boundaries, tidal effects, variations in barometric pressure, earth tides, earthquakes and water wave actions (Sherif, 2003). Precisely the most important factors are intrinsic or naturals and related to the hydraulic and geometric characteristics of the hydrogeological system, the others are artificial related to human activities. This kind of physical assumptions happens in the study area of the coast of Hermosillo aquifer and unfortunately the saline intrusion was growing along the last 34 years creating serious consequences to the future availability of fresh water.

2 STUDY AREA

The coast of Hermosillo aquifer is located to the southwest of Hermosillo city between the coordinates 28°14' and 28°57' of latitude North and 111°15' and 111°45' of length West of Greenwich, includes an approximate surface of 3,200 km^2, it is an exorreic basin located on the northwestern of Mexico and its superficial waters drains towards the Gulf of California (Figure 1).

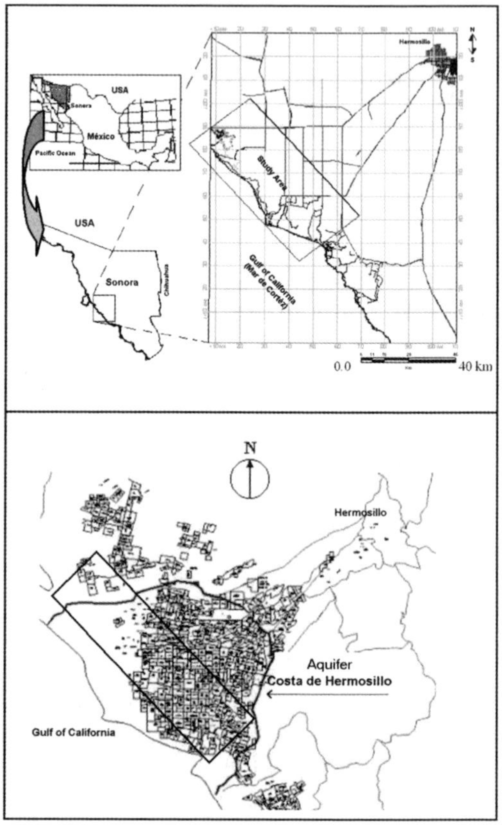

Figure 1. Location of the study area.

3 THE PROBLEM

The aquifer of coast of Hermosillo initiated its exploitation in 1945 with 17 wells, in the year 1965 it reached its maximum volume of extraction, with 800 Mm^3/yr and about 1,100 Mm^3/yr (Arreguin *et al.*, 1968). Since then there has been diminishing the extraction of ground water, up to stooping at present to a value estimated in 550 Mm^3/yr (CPCH, 1999) (Figure 2). Since 1949 this loss of hydraulic head in the aquifer originated a depletion cone and inverted the flow from the littoral towards the center of the plain from (Figures 3 and 4). As consequence, the government started a program to diminish and forbid the exploitation near the coast. The first statement was published in 1951, subsequent amplifications in 1954 and in the agricultural cycle 1963–1964 the Law of Regulation was imposed for the exploitation of the groundwater in the forbidden zone and later amplifications were given due to the strong discouragement in 1967 (Oroz, 2001; CNA, 2000).

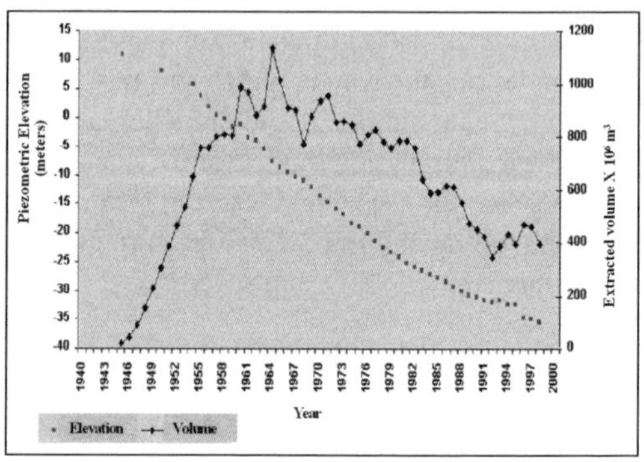

Figure 2. Groundwater surface and pumping rate evolutions in the Coast of Hermosillo aquifer (1945–2000). Modified from Comisión Nacional del Agua, 2000 (CNA/GRN).

Diverse unconnected studies analyzed the critical conditions of the aquifer (Matlock *et al.*, 1966), but the evidences were no attended by the farmers. Arreguin *et al.* (1968) realize a study of

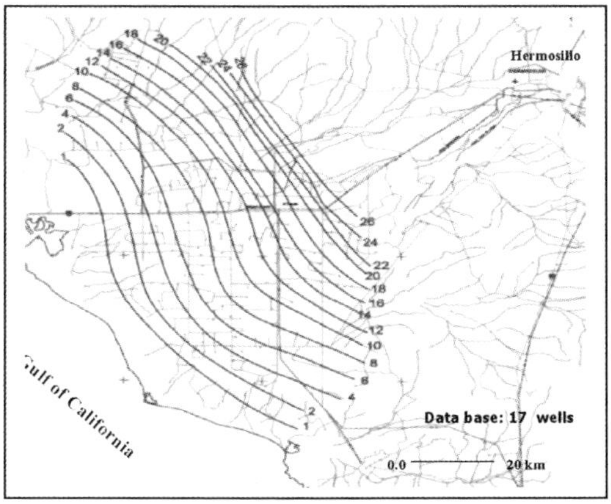

Figure 3. Groundwater surface level 1945.

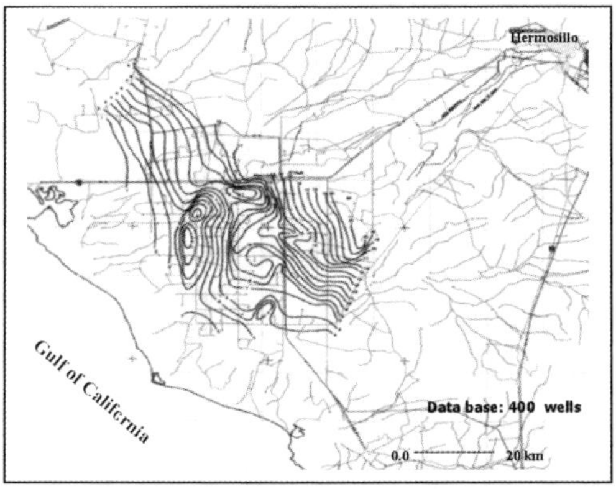

Figure 4. Groundwater surface level 1949.

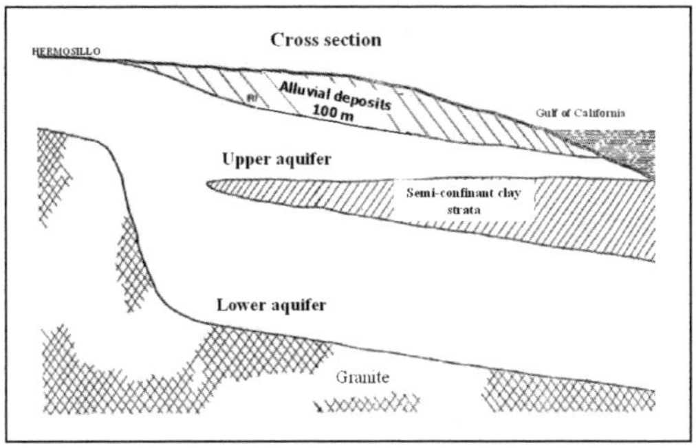

Figure 5. Former conceptual hidrogeological model from Ariel Construcciones, 1968.

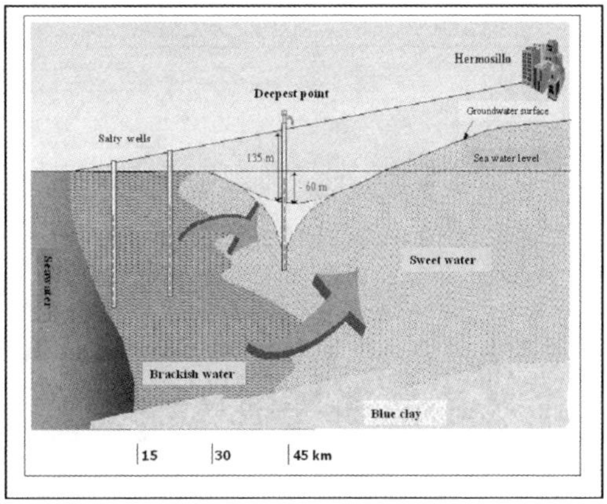

Figure 6. Schematic Groundwater table position. Modified from Comisión Nacional del Agua, 2000 (CNA/GRN).

higher transcendence for the administration of the resource in the following 35 years. In this study, a hydrogeologic model was defined with a system of two aquifers; lower (confined) and upper (unconfined) and the total recharge was calculated in 350 Mm^3/yr; 280 Mm^3/yr of vertical rising infiltration, coming from a low aquifer and 70 Mm^3/yr from superficial infiltration. With base in this scheme, there were relocated 105 wells affected by salinity near the coast. This decision was implying also, to reduce gradually the extraction of groundwater, up to a volume near to its recharge. This hydrogeologic model remained for 34 years (1968–2001) and the dynamical level continued being deepened up to 135 m in the year 2000 (65 m under the sea level) (Figures 5 and 6).

4 METHODOLOGY

A variety of techniques were used to define a new hydrogeological model and the current position of the interface of the saline intrusion in the aquifer. We characterize the geochemistry of the saline and fresh waters and the process of migration using geochemical and geophysical methods to explain how and where the marine water penetrates preferentially (Rangel, 2000 and 2001).

5 MORFOTECTONIC

With the analysis of the image LANDSAT and the map of Schellhorn *et al.* (1991) and the gravity anomalies from ESSA (1971), we located the diverse structures that limit the low basin of the Rio Sonora and its neotectonic relations, also we obtained with gravimetric surveying a tridimensional map of the crystalline basement that shows the depth and the development of an alternation of tectonic basins (grabens) and elevated blocks (horsts) sensitively orientated NW-SE that reach top depths to 1,500 m. Therefore we can say that the morphology of the coastal zone is a product of two neotectonic events (distensive) called the Province of *Basin and Range Sonorense* and to the opening of the Gulf of California. As a result there exist a series of tectonic sedimentary basins formed by fallen blocks of crystalline regional basement constituted of granitic rocks composition, which they form a part of the so called *Batolito Larámide of Sonora*. This rocks present numerous *roof pendants* of ancient rocks, covered in discordance for post-batolitic rocks and recent deposits (SARH, 1978 and 1982; Morales *et al.*, 2000). These basins represent channels of high permeability that facilitate the process of marine intrusion towards the continent and that's why we consider they are the bigger hydrogeologic relevance and control in this process (Figure 7).

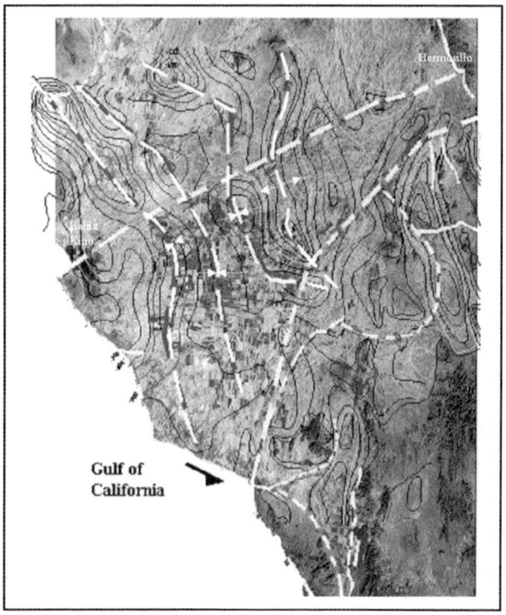

Figure 7. Structural Lineaments with the horst and graven of basement (Basin and Range Sonorense).

6 ESTRATIGRAPHY

The neotectonic and structural scenary of the Basin and Range Sonorense is of supreme importance in the distribution, storage and movement of the groundwater flow, because there exists deltaic and alluvial deposits which constitute isolated aquifers in each basin. In some of this basins the sediments have strong presence of impervious interstrata that restrict its lateral and vertical movement, forming local semiconfinements. These quaternary deposits can rest on the miocenic and mesozoic volcanic rocks, or on the crystalline basement (Figure 8). Based on the geophysical information and litology of shallow and deep wells we propose the presence of three principal units and the basement: 1) Top Quaternary Unit (alluvial deposits); 2) Impervious Miocenic Unit (marine sediments; *blue clay*); 3) Low Miocenic Unit (gravels and sands packed in clays); crystalline Basement (granites and volcanic rocks).

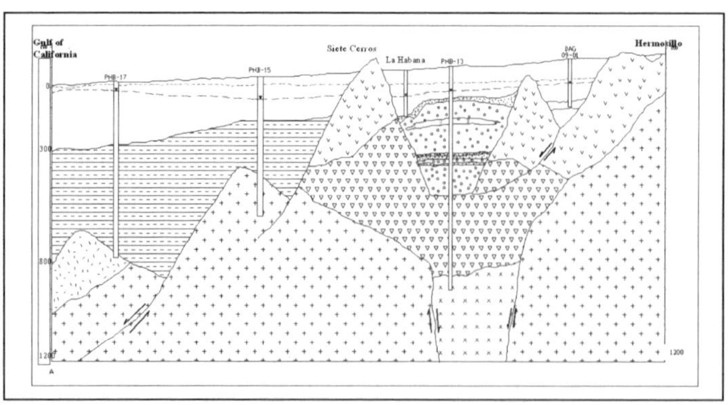

Figure 8. Current conceptual model, showing the estratigrafic correlation in a perpendicular section to the coast, SW-NE.

7 HYDROGEOLOGY

The geometry of the basement is irregular and the deposits in the basin too, therefore in some areas the sediments have a thickness of 150 m and in others of up to 800 m. This materials constitute a multilayer aquifer which works as an unconfined unit. The pumping rate from the aquifer has been evaluated in 527 Mm^3/yr, and the total recharge in 150 Mm^3/yr. We did not find evidence of any lower aquifer. This disequilibrium have got a depression cone with a maximum deep of 65 m under the sea level located 45 km inland (Figure 9). We calculate the intrusion of seawater in 98 Mm^3/yr, and the front of the saline water is currently inland up to a distance of 32 km (Figure 10).

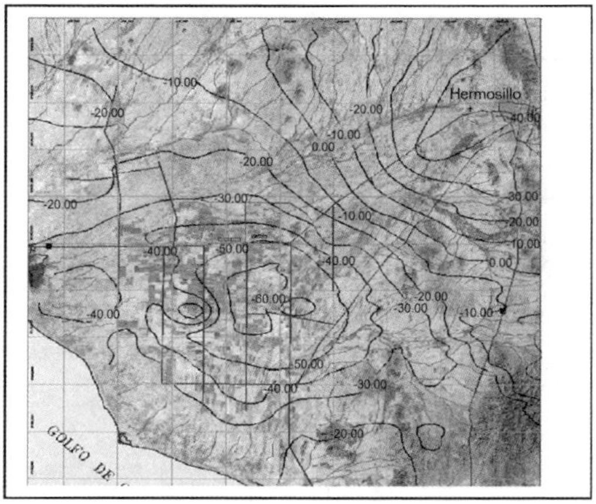

Figure 9. Groundwater surface, year 2000.

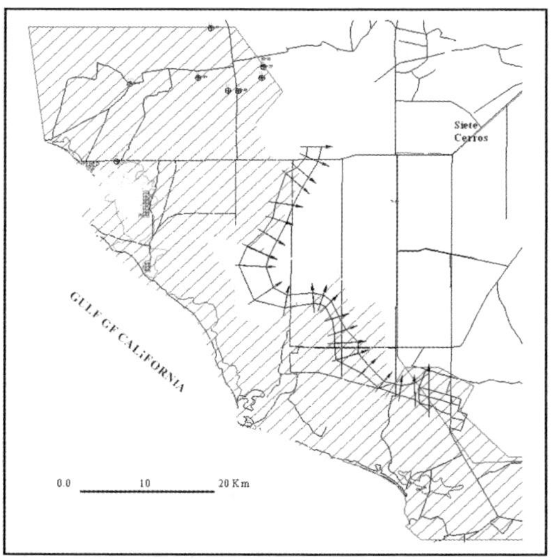

Figure 10. Front of saline intrusión.

8 HIDROGEOCHEMISTRY

Taking the electrical conductance (EC) in 30 profiles along the coast in a net of deep wells we configure the salinity to 40 m, 70 m and 100 m depth and it shows us the vertical distribution of the saline intrusion (Figures 11 and 12). With the spatial distribution of physical and chemical parameters, using a monitoring net of 107 wells we got the consistency to show the representative quality of water, however we could distinguish three main zones; a) saline intrusion 10,000 μS/cm to 40,000 μS/cm; b) interface zone 2,000 μS/cm to 10,000 μS/cm (mixing water) and; c) fresh water zone <2,000 μS/cm (Figure 13). Additionally we use the surveying of 408 electromagnetic soundings (TEM's) to correlate the chemical quality and the fluid resistivity and they all gave us the final spatial and tridimensional distribution of the saline intrusion, and the current position of the plume of saline water (Figure 14).

9 CONCLUSIONS

The excessive pumping rate in the aquifer exceeded the natural recharge since 50 years ago, since then the front of the saline intrusion have been migrating 650 m/yr, however the current position is of 32 km inland. There exist two preferential areas for the saline intrusion related with the tectonic settings.

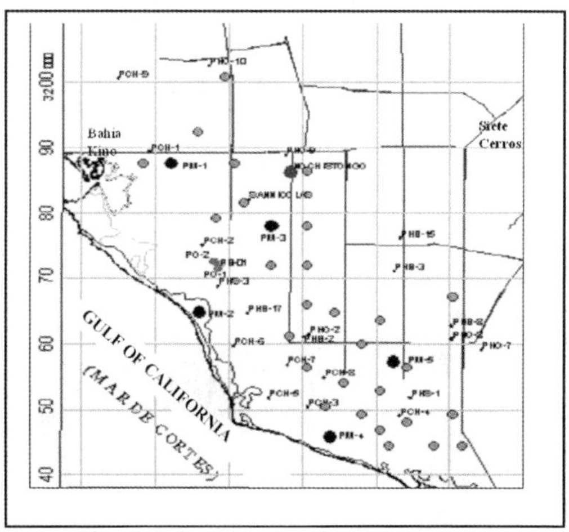

Figure 11. Monitoring wells net and profiles of electrical conductance.

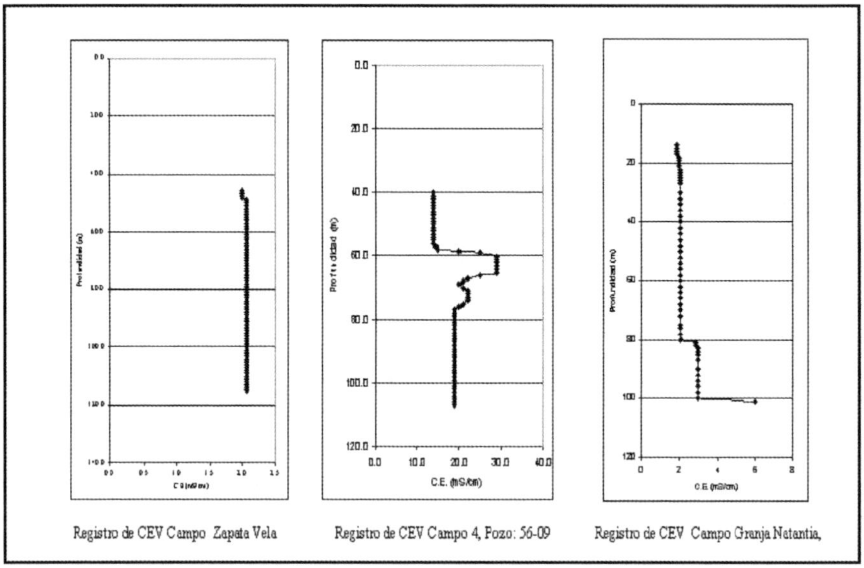

Figure 12. Typical profiles of electrical conductance (µS/cm) in the interface zone.

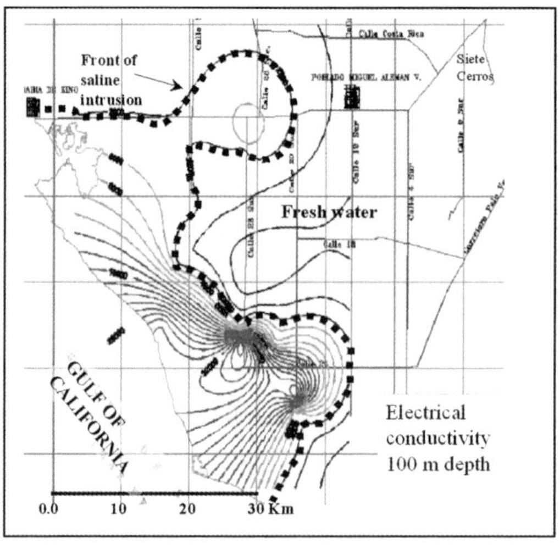

Figure 13. Electrical conductivity distribution at 100 m depth. The dashed line is the 2000 µS/cm.

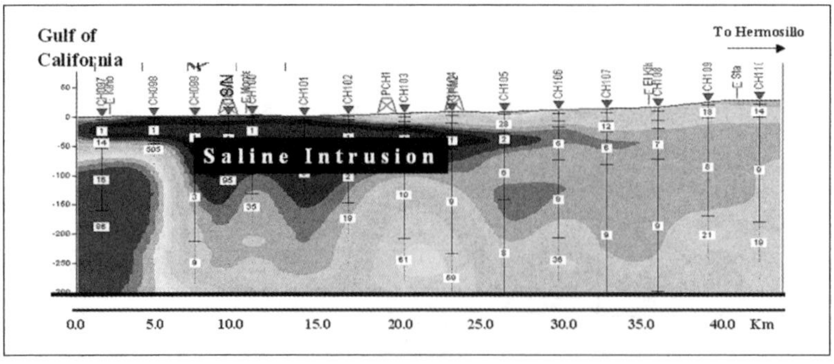

Figure 14. Current position of the saline intrusion, showing the pollution plume (January 2002).

The EC average of fresh groundwater quality is 250 μS/cm. In the interface zone has been getting poorer and goes from 2,000 to 10,000 μS/cm, this dispersion zone has at least a depth of 80 m and 5 km to 10 km width. The saline intrusion zone has 10 km to 20 km width and we found electrical conductances up to 40,000 μS/cm. The aquifer of Costa de Hermosillo needs an integrated water management plan for future administration and control.

REFERENCES

Arreguin, J.; Figueroa, G. and Peña, S. (Ariel Construcciones, S.A. de C.V.) (1968). Estudio hidrogeológico completo de los acuíferos de la Costa de Hermosillo, Sonora, Mex. Secretaria de Recursos Hidráulicos, 220 pp.

CNA (Comisión Nacional del Agua) (2000). Pozos de estudio perforados por la Dirección de Aguas Subterráneas en el periodo 1967–1969 en el acuífero de la Costa de Hermosillo, Sonora. Gerencia Regional Noroeste.

CPCH (Colegio de Postgraduados de Chapingo) (1999). Estudio para la estimación de volúmenes de agua subterránea extraídos para uso agrícola en las zonas de: Costa de Hermosillo, Sonora y Janos Chihuahua, aplicando técnicas de percepción remota. Instituto de Recursos Naturales. Especialidad de Hidrociencias. Montecillos, México.

ESSA (Exploraciones del Subsuelo S. A.) (1971). Informe del levantamiento gravimétrico e interpretación cuantitativa del mismo. Costa de Hermosillo, Sonora. Secretaría de Recursos Hidráulicos. México. 19 pp. Not edited.

Matlock, W.G.; Fogel, M. and Bush, C.D. (1966). Utilization of water resources in a coastal ground water basin. J. *Soil and Water Conserv*. V. 21. No. 5.

Morales, M.; Rangel, J.; Castillo, R. and Monreal, R. (2000). Hidrogeofísica del acuífero de la Costa de Hermosillo, *II Foro del Agua, UNISON-AIMMGM*. October, 2000. Hermosillo, Sonora.

Oroz, L.A. (2001). Modelo Conceptual Hidrogeológico e Hidrogeoquímico de la Costa de Hermosillo, Doctoral thesis.

Rangel, M.M. (2000). Nuevas aportaciones para el entendimiento del modelo hidrogeologico de la Costa de Hermosillo, utilizando isótopos ambientales (O-18; D) y fechamiento de agua (C-14). II Reunión Nal. UGM, Pto.Vallarta, Jal.

Rangel, M.M.; Castillo, G.J.; Monreal, S.R. and Morales, M. (2001). Determinación de la Vulnerabilidad del Acuífero Costero Costa De Hermosillo, Son. Méx. a la Intrusión Salina. XI *Congreso Nacional de Geoquimica*, Ensenada, B.C.

SARH (Secretaría de Agricultura y Recursos Hidráulicos), 1978, 1982. Reportes de perforación de Pozos de Observación y Centinelas en la Costa de Hermosillo, Sonora. Not edited.

Schellhorn, R.W.; Aiken, C.L.V. and de la Fuente, M. (1991). Bouguer gravity anomalies and crustal structure in northwestern Mexico. *AAPG Mem. 47; 197–215.*

Sherif, M.M. (2003). *Seawater intrusión in the Nile delta aquifer: an overview*. Pub. IGME, Spain, Serie: Hidrogeología y Aguas Subterráneas No. 8, Tomo II, pp 295–308. Mar. 2003.

Van Dam, J.C. (1999). *Exploitation, Restoration and Management*, Chapter 4, in J. Bear et al., (eds) Seawater Intrusion in Coastal Aquifers, 73–125, Kluwer Academic Publishers. Printed in the Netherlands.

The analysis of the intensive use of groundwater in the Upper Guadiana Basin (Spain) using a numerical model

Luis Martínez Cortina
Fundación Marcelino Botín (FMB), Madrid, Spain
l.martinez@cospa.es

Joaquín Cruces
Universidad de Cantabria, Santander, Spain
crucesj@unican.es

ABSTRACT

The Upper Guadiana Basin has become in the last decades a paradigmatic world-wide case from the hydrological and ecological point of view, because of the existing conflict between the intensive development of groundwater and the sustainability of very important aquatic ecosystems (the most important is the National Park of *Tablas de Daimiel*).

By using a mathematical model of groundwater flow, it has been analysed the role of this intensive use of groundwater in the evolution of the situation and behaviour of the hydrological system. This analysis will be focused on several aspects like piezometric evolution, relationship between aquifers and rivers, impacts on wetlands, water storage variation, etc.

Keywords: *intensive use, numerical models, evapotranspiration, impact on wetlands, aquifer-river relationship.*

1 INTRODUCTION

During the European Union's GRAPES research project (Acreman, 2000), a numerical model was developed to simulate the groundwater flow of all the aquifers in the Upper River Guadiana Basin (Cruces *et al.*, 2000). This model has allowed investigating the interaction between surface water and groundwater, and has also been used as a tool to analyse the influence of some of the anthropic changes in the functioning of the system. The model was calibrated with existing historical data (piezometric levels, river flows) *(hereafter called: historical model).*

Human influence has produced important hydrological, ecological, social and economic changes in the basin (Bromley *et al.*, 2001). In order to analyse the influence of some of these actions (extraction of water from the aquifers, the building of dams), a second model was developed, simulating a fictitious undisturbed situation of the hydrological system, that is, in conditions where supposedly no anthropic actions have disturbed the system *(hereafter called: undisturbed model)*. This would allow a comparison with the aforementioned historical model, thus giving a qualitative and quantitative idea about the influence of these anthropic effects in the evolution of the system. Furthermore, the behaviour of the system affected only by meteorological variations could be also analysed.

2 GENERAL CHARACTERISTICS OF THE BASIN

The Upper Guadiana Basin (Figure 1) covers about 16,000 km^2, with a population of approximately 400,000 people (about 25 inhabitants/km^2). It is located in the Northeast corner of the Southern Castilian subplateau and characterised by a flat relief, a poorly defined drainage network, and a significant presence of aquifers. It is a semiarid region, one of the driest in Spain (approximately 415 Mm^3/yr). If the specific contribution to the streamflow is considered, it is the driest in Spain (less than 30 mm/yr).

In the early 1970s, the conversion of rain-fed agriculture into irrigated farming using groundwater began to become common. The extraction from the main aquifer of the system (Hydrogeological Unit 04.04, H.U. 04.04) continued to increase up until the end of the 1980s. At that time, areas irrigated using groundwater covered about 130,000 ha, and annual abstraction was about 575 Mm^3. This, of course, implied very positive socio-economic consequences, but it also induced very significant hydrological and ecological alterations in some areas of the basin (Cruces and Martínez Cortina, 2000).

3 THE UNDISTURBED MODEL

In a similar way to the historical model previously developed, the undisturbed model implies

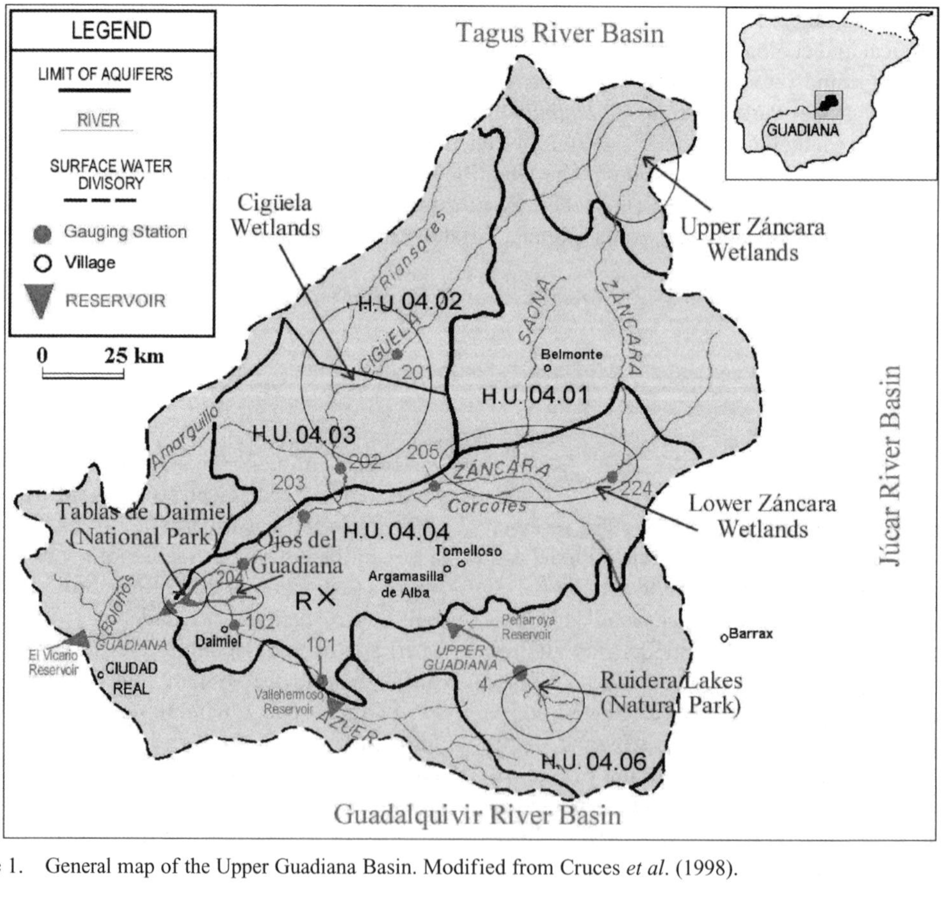

Figure 1. General map of the Upper Guadiana Basin. Modified from Cruces *et al.* (1998).

the simulation of an initial model in a steady state, and three models corresponding to the periods 1959/60–1973/74, 1974/75–1983/84, and 1984/85–1995/96 (transient state).

All these models were developed using the program MODFLOW (McDonald and Harbaugh, 1988), and the pre- and postprocessor PMWIN (Chiang and Kinzelbach, 1993). MODFLOW is a groundwater flow model which doesn't take surface runoff into consideration. So, part of this runoff which infiltrates and incorporates with groundwater flow, is not reproduced in the model. In order to simulate the functioning of the system as close as possible to field conditions, the historical model adopted a scheme of total flows. In this scheme, the streamflows historically measured in the gauging stations were introduced in the model. At these points, all the existing water in the system was considered, being able to infiltrate, evaporate, or be transmitted to the next cells. The difference between the streamflow resulting in the model in the next gauging station and the streamflow actually measured, corresponds theoretically to the surface runoff between two gauging stations.

One of the problems for the undisturbed model is that no data of measured streamflows exist that corresponds to the simulated conditions. This fact makes the use of the total flows scheme difficult. Circulating streamflows in undisturbed conditions can be very different from streamflows measured in disturbed conditions, even with the same meteorological sequence.

The solution adopted was to consider the flows imposed with the total flows used in the historical simulation model as acceptable for the undisturbed model, for gauging stations which, due to their location within the system, were not affected very much by the extraction of groundwater (those situated in the units adjacent to H.U. 04.04). On the other hand, this scheme was not used for gauging stations situated within H.U. 04.04, because the simulated situation in the undisturbed model is quite different from the one where the streamflows and water levels were measured. Therefore, at these points, surface runoff of the corresponding subbasin would not be simulated (scheme of base flows). This seems to be acceptable as the topographic surface in this aquifer is rather flat and surface runoff hardly occurs.

The undisturbed model (steady state and three transient periods), simulates the same historical meteorological sequence of the historical model. The differences with the historical model are: the suppression of all groundwater abstractions; the exclusion of the influence of Peñarroya Reservoir (built in 1959); and the adoption of the aforementioned base flow scheme at gauging stations numbers 202, 203, 205 and 102 (see Figure 1), by replacing the streamflows imposed in the historical model with those coming from upstream.

4 MAIN RESULTS OF THE MODELS

Once the models corresponding to the steady state and the three transient periods in historical and undisturbed conditions were completed, several results were analysed, corresponding to piezometric levels, relation between aquifers and rivers, lateral flows between aquifer systems, different terms of the water budget, etc. (Martínez Cortina, 2001, 2003).

In the following sections some of these results will be summarised. In order to compare the results of both models (historical and undisturbed), and to analyse the temporary evolution of the system, a reference cell in the model has been considered in the main aquifer of the system (H.U. 04.04 of the Western Mancha aquifer). This point (letter R in Figure 1) is located near Manzanares, in the area where the largest piezometric depletions have taken place.

4.1 *Piezometric evolution*

The main difference in the undisturbed model, as was expected, was the small variation of piezometric levels in H.U. 04.04, as compared to the significant decrease registered in the historical model, calibrated with the real situation. Figure 2 shows this difference in the piezometric evolution of the reference cell for H.U. 04.04. In the historical model a decrease of more than 50 m was registered throughout the simulation. It can also be seen that the suppression of the pumping of 60 Mm^3/yr in the steady state of the historical model produces a previous increase of about 2 m in the initial situation, which represents the level corresponding to the undisturbed steady state.

If the scale of the hydrograph corresponding to the piezometric evolution for the undisturbed model is amplified, the small level oscillations in the reference cell can clearly be seen throughout the entire simulation. This has been done in Figure 3, representing the piezometric evolution at the reference point without the influence of pumping or some other anthropic action.

The oscillations of the piezometric level, not affected by anthropic actions, but only through meteorological variations, can be seen in this figure. The low peaks of the piezometric level usually occur after the summer period, when evapotranspiration is high and the rainfall is usually irrelevant, as well as other dry periods. On the other hand, the most relevant level increases are related to periods with an important rainfall event.

The behaviour at a certain moment and point of the hydrological system is controlled

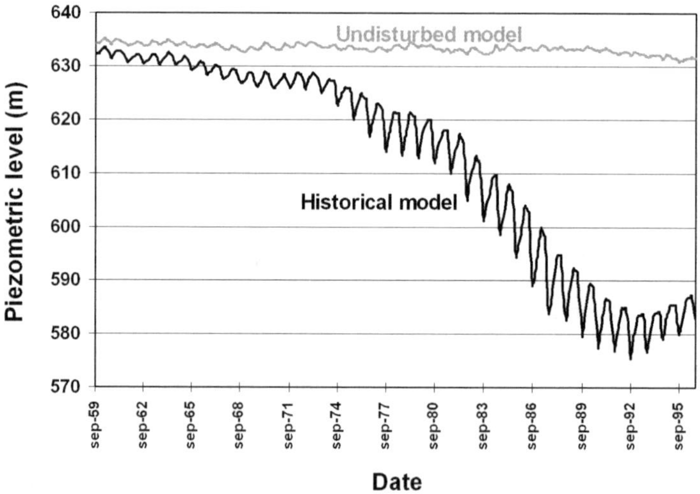

Figure 2. Piezometric evolution in the reference cell. Comparison between historical and undisturbed models.

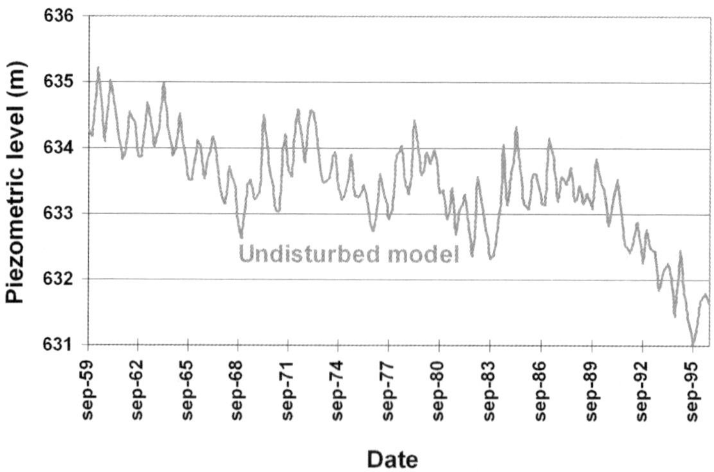

Figure 3. Piezometric evolution in the reference cell. Undisturbed model.

by the intrinsic characteristics of the point, and also by the situation of the system at that moment. For example, Figure 2 shows how the reduction of groundwater extractions in the early 1990s implies a small recovery of the piezometric level in the historical model, in spite of the occurrence of a dry period. In this situation, the great decrease in extraction is more relevant than scarce rainfall. At the same time, the undisturbed model (with no extractions), shows a decrease in the piezometric level, as it can be seen in Figure 3. In this case, the dry meteorological sequence is more relevant in the piezometric evolution. The different situation of the system explains these behaviours.

A situation with significant drawdowns of the piezometric level, with rivers perched over the aquifers, facilitates rainfall infiltration. Part of the rainfall that was rejected by evapotranspiration in an undisturbed situation, is now infiltrated into the aquifer. Water circulating in the rivers also tends to infiltrate and the discharge areas of the aquifer are significantly reduced, thus decreasing the volume of water drained. In some areas, the piezometric changes produce pumping cones, changing the direction of flow, also favouring the recharge to the aquifer.

4.2 *Water budget*

Table 1 contains the main figures of the water budget from the point of view of the aquifers, corresponding to the undisturbed model, for each simulated period. In order to compare both models, the figures corresponding to the historical model can be seen in brackets.

For the steady state, the amount of water pumped in the historical model (60 Mm^3/yr) is practically reallocated – in the undisturbed model – to the discharge of the aquifers to the rivers (from 365 Mm^3/yr to 420 Mm^3/yr). The increase of the evapotranspiration in the undisturbed model is minimum, because in this steady state of the historical simulation the piezometric levels were sufficiently close to the land surface to have the evapotranspiration value near the one obtained in the undisturbed model.

Table 1 shows an important difference between both models in relation to the decrease of the storage volume in the aquifers, especially in the last two transient periods (10 and 12 years). These periods were rather dry (especially the latter), and the extracted volumes, as shown in the table, increased considerably. In the historical calibrated model, the decreases of the storage volumes in these two periods were 1,800 Mm^3 and 2,700 Mm^3, respectively. According to the results of the undisturbed model, about 1,000 Mm^3 of storage reduction, are due to the meteorological dry sequence, and not to anthropic actions. The total difference with the historical model of more than 3,000 Mm^3 is principally due to the intensive use of groundwater.

4.3 *Evapotranspiration*

As shown in Table 1 and also in Figure 4, the value of the evapotranspiration decreases slightly

Table 1. Water budget for the undisturbed model. In brackets, water budget for the historical model. Average figures in Mm^3/yr.

Water budget (aquifers)	Steady state	1959/74	1974/84	1984/96
	IN			
Direct recharge (rainfall)	455 (455)	505 (505)	375 (375)	330 (330)
Indirect recharge (from rivers)	135 (140)	130 (110)	120 (125)	115 (115)
Total IN	590 (595)	635 (615)	495 (500)	445 (445)
	OUT			
Pumping	0 (60)	0 (110)	0 (410)	0 (540)
Evapotranspiration	170 (170)	165 (160)	160 (110)	160 (45)
Discharge to the rivers	420 (365)	455 (360)	375 (160)	335 (85)
Total OUT	590 (595)	620 (630)	535 (680)	495 (670)
Storage variation		15 (–15)	–40 (–180)	–50 (–225)

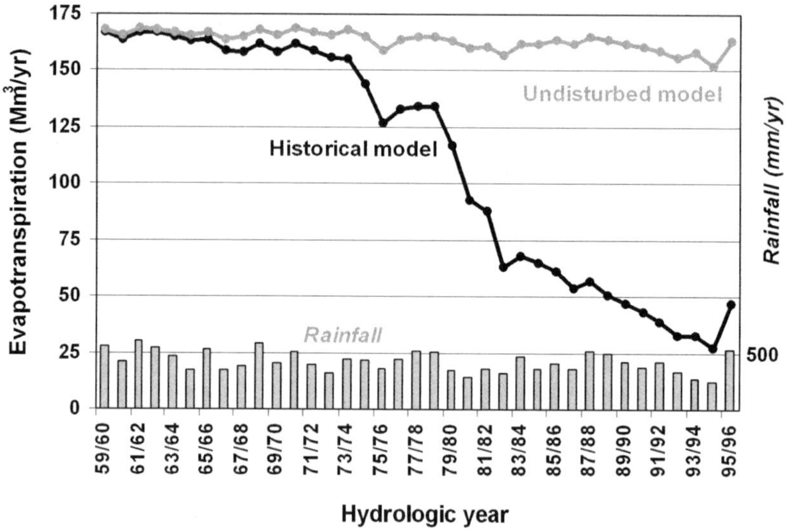

Figure 4. Evolution of evapotranspiration. Comparison between historical and undisturbed models.

throughout the simulation of the undisturbed model. In most of the areas, the piezometric levels are sufficiently high, the reduction of the wetlands surfaces is not very important, and the natural behaviour of the discharge areas of the aquifers do not change, at least qualitatively.

Under these conditions, the evapotranspiration maintains high values in the undisturbed model, with only a little reduction after dry periods when there is a small decrease in piezometric levels. There was a large difference, however, with the average values of evapotranspiration obtained in the historical simulation, especially in the third period, when the situation of the system in this historical model was already completely disturbed.

4.4 *Indirect recharge from the rivers*

Figure 5 shows the annual values of indirect recharge to the aquifers from the rivers, for the historical and the undisturbed simulations.

Average values are similar in both models, in spite of the fact that the situation of the aquifer is quite different. In the historical model, when the system is disturbed, many sections of the river change their behaviour, from gaining rivers that receive the aquifer discharge, to losing rivers recharging the aquifer, making a greater value of the indirect recharge to the aquifer possible. The indirect recharge in the historical model had more extreme values, especially the high values. When the rainfall is scarce, the absence of circulating streamflow limits the value of the indirect recharge. On the other hand, in wet years the infiltration to the aquifer was very high, because of the disturbed situation of the system (as shown for the wet year 1995/96).

In the undisturbed model, there were less recharge areas, but the circulating streamflows were greater, relatively compensating the total infiltrated volume.

4.5 *Discharge from the aquifers to the rivers*

The small oscillations of the piezometric level in an aquifer discharging to a river imply variations in the discharged volume.

As shown in Figure 6, even in the undisturbed model, variations of drained flows are important, in spite of little drawdowns in the piezometric level. This drained volume can decrease to less than a half from a wet year to a dry year after a dry sequence (almost 550 Mm^3 in 1959/60, and about 250 Mm^3 in 1994/95).

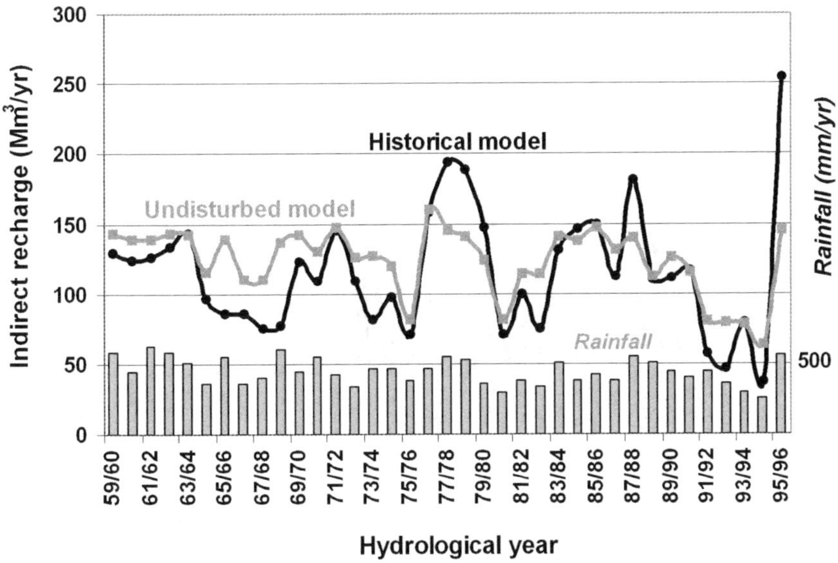

Figure 5. Indirect recharge from rivers. Comparison between historical and undisturbed models.

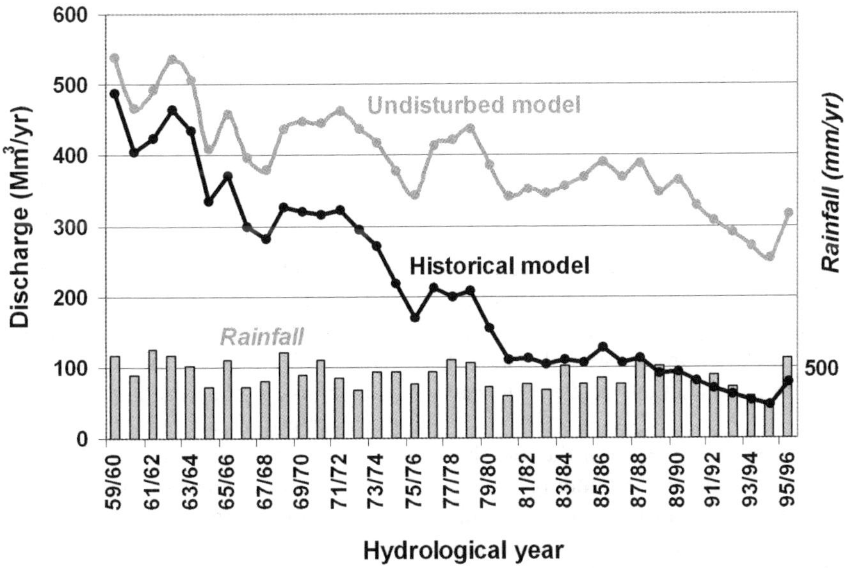

Figure 6. Discharge from aquifers to the rivers. Comparison between historical and undisturbed models.

This figure clearly shows the scheme of natural regulation of the aquifers in this basin. In dry sequences there is a decrease of the discharge from the aquifers to the rivers, and therefore a reduction of the circulating streamflow, but maintaining almost the same water table.

In the historical model, there was a much greater reduction of the drained volume to the

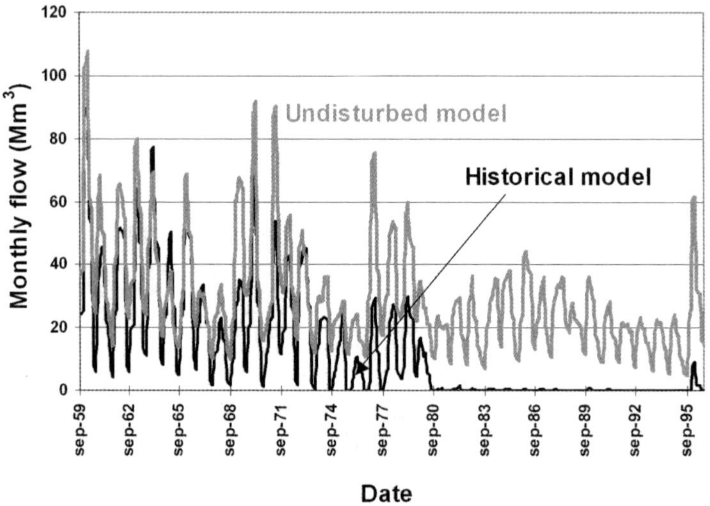

Figure 7. Monthly flows in El Vicario. Comparison between historical and undisturbed models.

rivers. In this case, there aren't moderate decreases of the piezometric levels that reduce the discharged flows, but total changes of behaviour in the aquifer-river relationship, with the disappearance of many discharge areas because of the disconnection between aquifer and rivers.

4.6 *Streamflows*

The simulation of the undisturbed model allows independent analysis of the effect of meteorological variability, with reductions of circulating streamflows, exclusively due to the dry characteristics of the last two transient periods, but not comparable with the reductions of the historical model.

This can be seen in Figure 7 for the monthly flows for both models corresponding to the Guadiana River in El Vicario, at the end of the Upper Guadiana Basin. In the historical model, the streamflows were irrelevant from the early 1980s. In the undisturbed model – with the system in natural conditions–, there was a reduction of streamflow during dry periods, but there was always a significant flow, mainly supported by the contribution of groundwater.

The undisturbed model can be used, by itself or as a previous step to more detailed models, to analyse the effects of certain meteorological sequences (for example, information about the vulnerability of a wetland in relation to the drawdowns caused by a drought). Figure 8 shows, as an example, the streamflows registered by both models (historical and undisturbed), in the most relevant discharge area in the basin, at the end of *Ojos del Guadiana*. This corresponds to the baseflow contribution to *Tablas de Daimiel*, a National Park which is the most important wetland in the basin.

In the undisturbed model, the small oscillations of the piezometric level imply moderate reductions of the discharge flows, but never the disappearance of this contribution, as did occur in the real situation (historical model), with the disconnection between aquifer and rivers. During the three considered transient periods of the undisturbed model, the annual average value of the contribution to *Tablas de Daimiel* was 125, 105 and 95 Mm^3/yr, respectively (120 Mm^3/yr in the steady state). On the other hand, in the historical model this average contribution was 95 and 15 Mm^3/yr in the two first transient periods, and there was no contribution during the third period (in the steady state was 100 Mm^3/yr).

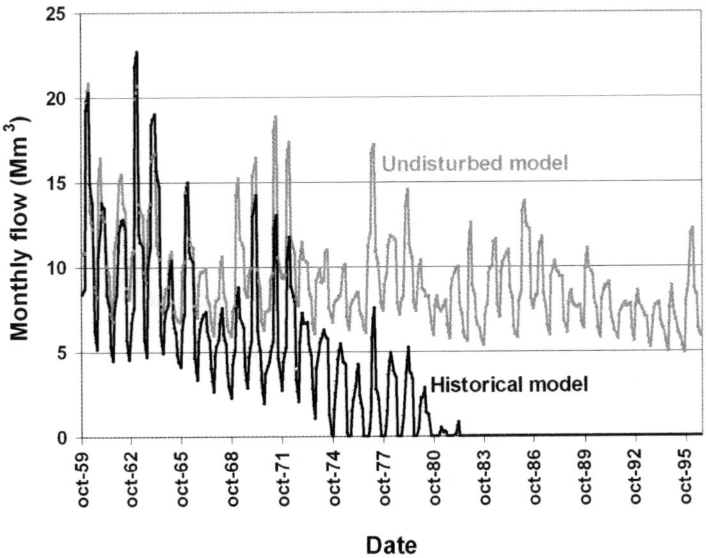

Figure 8. Monthly baseflow from *Ojos del Guadiana to Tablas de Daimiel.* Comparison between historical and undisturbed models.

5 CONCLUSIONS

The simulation of an undisturbed model of the Upper Guadiana Basin, corresponding to a supposed natural functioning of the system, allows an analysis of the behaviour of the system when it is not affected by anthropic actions. By comparing this undisturbed model with the historical model previously developed and calibrated with the real situation, an idea can be given about the relative influence of anthropic actions in the actual situation of the basin.

This analysis of the undisturbed model and its comparison with the historical model has taken different aspects into consideration such as piezometric evolution, relationship between aquifer and rivers, streamflows, evapotranspiration, indirect recharge from the rivers, aquifer discharge, etc.

As an example of the results, the undisturbed model quantifies a decrease in the stored water volume in the aquifers of about 1,000 Mm^3, mainly as a consequence of the dry period which occurred in late 1980s and early 1990s. On the other hand, the reduction of this volume in the historical calibrated model was about 4,500 Mm^3. The difference of about 3,500 Mm^3 can be assigned to the intensive use of groundwater.

Another conclusion inferred from the comparison of both models is the self-regulating behaviour of the system according to its state. While there is a discharge of the aquifer to the river, the regulation of the system consists of a variation in the circulating river flow. The piezometric levels in all the aquifer systems, especially in H.U. 04.04, range slightly around an equilibrium point, and the influence of the meteorological variability is not very important. On the other hand, when the disturbance of the system is important, and the discharge of the main aquifer disappears, the differences between inputs and outputs in the water budget of the aquifer have to be compensated by large variations in the piezometric level. In this situation, the riverbeds only receive the surface runoff, which normally infiltrates before the end of the basin.

REFERENCES

Acreman, M. (ed.) (2000). *Technical Report to the European Union ENV4 – CT95-0186 – Groundwater and River Resources Action Programme on a European Scale (GRAPES)*. Institute of Hydrology, Wallingford, UK, 248 pp.

Bromley, J.; Cruces, J.; Acreman, M.; Martínez Cortina, L. and Llamas, M.R. (2001). Problems of sustainable groundwater management in an area of over-exploitation: the Upper Guadiana catchment, Central Spain. *Water Resources Development*, 17(3): 379–396.

Chiang, W.H. and Kinzelbach, W. (1993). *Processing Modflow (PM), Pre- and postprocessors for the simulation of flow and contaminants transport in groundwater system with MODFLOW, MODPATH and MT3D.*

Cruces, J. and Martínez Cortina, L. (2000). *La Mancha Húmeda. Explotación intensiva de las aguas subterráneas en la cuenca alta del río Guadiana*. Papeles del Proyecto Aguas Subterráneas, Fundación Marcelino Botín, Serie A(3), 66 pp.

Cruces, J.; Fornés, J.; Casado, M.; De la Hera, A.; Llamas, M.R. and Martínez Cortina, L. (1998). El marco natural: agua y ecología. In: J. Cruces; J.M. Hernández; G. López Sanz and J. Rosell (eds), *De la noria a la bomba. Conflictos sociales y ambientales en la cuenca alta del río Guadiana*. Editorial Bakeaz, Bilbao, Spain: 15–130.

Cruces, J.; Bromley, J.; Bradford, R. and Martínez Cortina, L. (2000). Numerical groundwater modelling. In: M. Acreman (ed.), *Technical Report to the European Union ENV4 – CT95-0186 – Groundwater and River Resources Action Programme on a European Scale (GRAPES)*. Chapter 2.3. Institute of Hydrology, Wallingford, UK: 118–145.

Martínez Cortina, L. (2001). *Estimación de la recarga en grandes cuencas sedimentarias mediante modelos numéricos de flujo subterráneo. Aplicación a la cuenca alta del Guadiana*. Doctoral Thesis. Universidad de Cantabria. 418 pp.

Martínez Cortina, L. (2003). Marco hidrológico de la cuenca alta del Guadiana. In: C. Coleto; L. Martínez Cortina and M.R. Llamas (eds), *Conflictos entre el desarrollo de las aguas subterráneas y la conservación de los humedales: la cuenca alta del Guadiana*. Fundación Marcelino Botín and Ediciones Mundi-Prensa. Madrid, Spain: 3–68.

McDonald, M.C. and Harbaugh, A.W. (1988). *MODFLOW, a modular three-dimensional finite difference groundwater flow model*. U.S. Geological Survey. Open-file report 83–875, chapter A1.

Evolution of groundwater intensive development in the coastal aquifer of Telde (Gran Canaria, Canarian archipelago, Spain)

M. Carmen Cabrera
Department of Physics (Geology). University of Las Palmas de Gran Canaria, Spain
mcabrera@dfis.ulpgc.es

Emilio Custodio
Geological Survey of Spain and Technical University of Catalonia, Spain
e.custodio@igme.es; emilio.custodio@upc.es

ABSTRACT

A 75 km^2, intensively developed coastal aquifer in eastern Gran Canaria island, Canarian Archipelago, around the town of Telde, has been surveyed and studied, using data from existing wells. The aquifer consists of Pliocene to Recent volcanic materials, with an intercalated detrital formation called Las Palmas Detritic Formation (LPDF). Groundwater development in the area started late in the 19th century by means of shaft wells fitted with horizontal water galleries and later on with horizontal drainage boreholes. Groundwater development became intensive since the 1950s, mostly for cash crop irrigation and town supply. First surveys are of the 1970s. Groundwater development conditions follow a quickly changing pattern. There is a conspicuous water table drawdown, up to 40 m in 20 years, following a strip parallel to the coast, although inland water table elevation is little affected. There is also a progressive water salinity increase with changes of the hydrochemical water type. Recent basalts were a good aquifer some decades ago but now are mostly drained. In spite of being considered the area as highly *overexploited*, groundwater reserve depletion contributes only about 5% of abstracted water, and more than 60% is transferred from inland areas. Discharge to the sea seems to be still significant, perhaps about 30% of total recharge.

Keywords: *Volcanic terrains, hydrogeology, groundwater flow, intensive use, hydraulic properties, Canary Islands.*

1 INTRODUCTION

The Canary Islands support an intensive exploitation of their groundwater resources, up to 300 Mm^3/yr, for a population close to 2 million inhabitants, especially in Gran Canaria and Tenerife. The general hydrogeological structure of the Canary Islands can be sketched as a *core* of low permeability old volcanic, intrusive bodies, and thermally metamorphosed rocks, with successive covers of younger, more permeable ones (Figure 1) (SPA-15, 1975; Custodio, 2003; Custodio y Cabrera, 2002). Groundwater flows from the central, high altitude areas, towards the coast, where it discharges mostly in diffuse form. A fraction of groundwater may discharge in inland areas, feeding springs and wet strips, and producing flow in some tracts of the gullies *(barrancos)*. These discharges depend on internal rock structure, and relief dissection by the gullies. Due to intensive groundwater exploitation these groundwater manifestations have

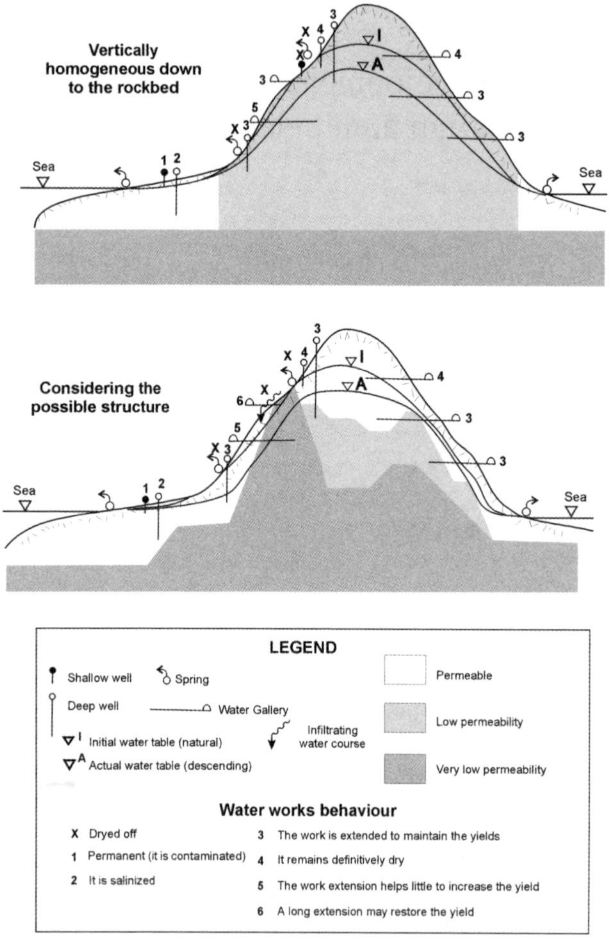

Figure 1. Schematic cross-sections to explain the hydrogeologic behaviour under natural and under intensive exploitation in Canary Islands, as inspired from Gran Canaria Island. The upper figure shows a low permeability core above a very low permeable bedrock, with young volcanic rocks or sediments in the periphery. The lower figure shows a central caldera partially slipped at one of the sides, and a young volcanics cover with intra- and extra-caldera areas. In both cases coastal water table gradients do not change significantly, at least at early times, from natural to intensive exploitation situations if groundwater development is mostly in the middle and high parts of the island (Custodio and Cabrera, 2002).

mostly disappeared and the water flows now through canals and pipes.

This paper deals mainly with the area of Telde (Cabrera *et al.*, 1992; Cabrera 1995; Cabrera and Custodio, 2003), in eastern Gran Canaria Island (Figure 2), which is the oldest of the intensively developed areas for local groundwater in the island. This area has not been studied in detail before, but for preliminary surveys during the SPA-15 (1975) study. Some new data has been added by the unpublished MAC-21 study (about 1980), and the twice-a-year monitoring network of the Geological Survey of Spain (IGME) from 1986 to 1990, on 42 wells, and the 12 wells surveyed by the Water Plan of Gran Canaria (PHGC) in 1991. Actually, a bigger area including Telde is being studying by the IGME in order to elaborate a plan to reorganise the groundwater exploitation in the east of Gran Canaria.

Telde is located in the eastern part of Gran Canaria. The headwaters of the area are outside it and extend up to the top of the island where average rainfall is close to 1,000 mm/yr. The *Barranco Real de Telde* is the main gully, which was one of the island permanent streams in the past. Now it is permanently dry. The study area is the coastal strip, up to an elevation of about 200 m, and extends along 10 km, with a total surface area of 75 km^2. It is a gently slopping lava platform, relatively flat, bounded by east-west trending gullies at the north and the south, and dotted with small recent volcanic cones. Most of the coast consists of 10 m to 30 m high cliffs on recent volcanics.

The Telde area is characterised by an average rainfall of 150 mm/yr with an average yearly temperature of 20°C. It is swept by frequent and relatively intense winds from the north. Besides it is at the boundary of the persistent northern cloud cover, which results in a high sunshine exposure. It can be considered an arid area, at the foot of the relatively rainy highlands.

The area has been traditionally used for irrigated agriculture, with changing dominant crops from the 15th century: sugarcane, vineyards, cereals, cochineal, banana and tomatoes. Banana trees have been cultivated from the end of 19th century to the 1970s. Afterwards, they were progressively replaced, first by pepper and cucumber cultivated under greenhouses in the

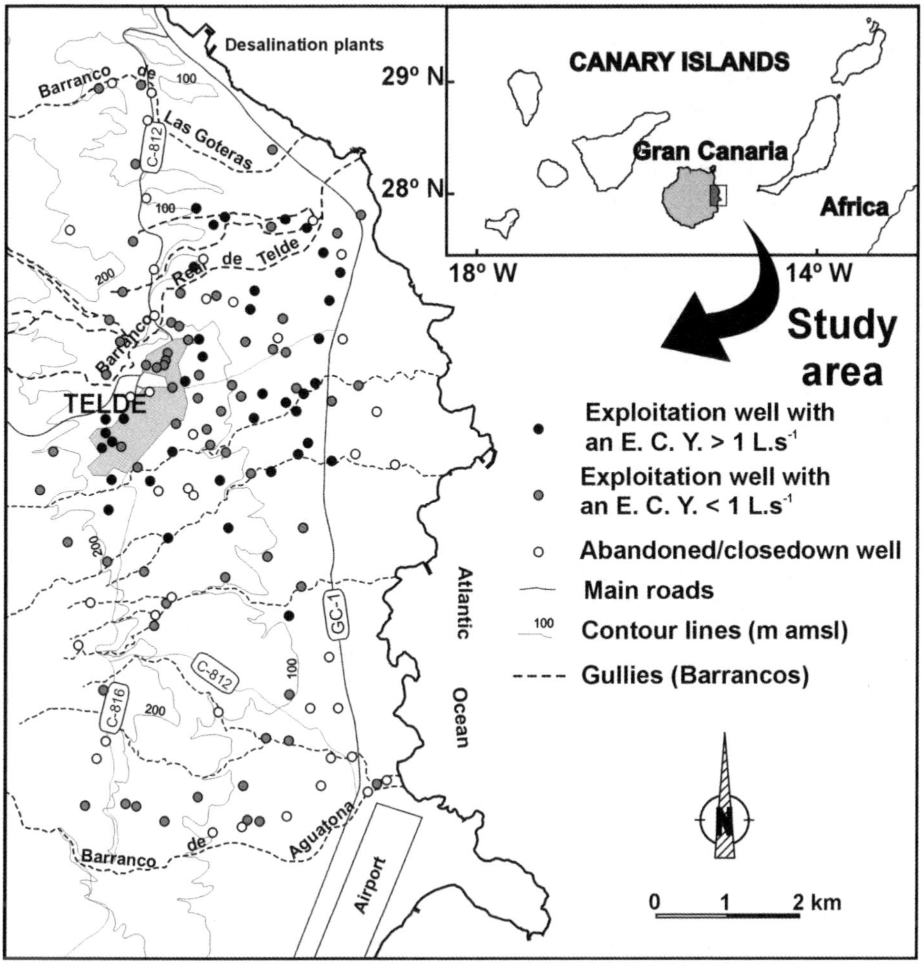

Figure 2. Location of Telde area in Gran Canaria, Canary Islands (Spain). It is shown the topography, main gullies (*barrancos*) and the location of wells with different exploitation regimes.

early 1970s, and later on by orange trees in the late 1970s. From 1980 the area has suffered an important urbanistic occupation, with the development of industrial and commercial parks and residential zones. The cultivated land has been reduced to some potato plots and fruit tree plantations for self-consumption. Currently about 80% of the study area is urban or commercial, or consists of abandoned plots of land that were previously used to cultivate tomatoes.

Water supply for irrigation was initially spring water coming from higher lands, and uptakes from the *Barranco Real de Telde*. After the 1880s the growing irrigation needs fostered the use of local groundwater by means of shaft wells, which are the oldest in the island. In 1980s, half of irrigation water as well as domestic water supply came from upper parts of the island and the other half was from wells in the Telde area. Now, urban and domestic water supply include desalinated sea water. Wells continue to be exploited for local needs, although less intensively, and the water is party mixed with desalinated water. Groundwater exploitation has followed the same scheme: from 1900 to 1950 many wells were excavated; later on water needs for irrigation caused an abstraction increase which was accompanied by the deepening of wells until the 1980s, when water exploitation has decreased due to salinity increase and land use changes.

From bottom to top, the stratigraphic column includes: Cycle I phonolitic lava flows and ash-flows (Miocene); Las Palmas Detritic Formation (LPDF) sediments; Cycle II basanitic and basaltic lava flows and volcanic breccias (Pliocene); and Cycle III basanitic lava flows and fall pyroclastic deposits (Plio-Quaternary). The sedimentary deposits correspond to the LPDF, that is divided in Lower, Middle and Upper members, according to their characteristics and sedimentary environments. The spatial disposition of these materials in the subsoil is deduced from visits to the shaft wells of the area, mainly descending into them, but also by studying the voluminous tips. Materials at about sea level change from north to south: they are mostly lava flows and pyroclastic deposits of phonolitic composition at the northern part of the area, and LPDF in the central and southern area. Recent basalts with Roque Nublo Formation represent the non-saturated zone in most of the area; they are thicker in the south-west.

There are 145 wells in the area, 30 of which were excavated before 1924. The oldest wells have a diameter of 6 m in the first metres to hold the old, bulky, animal powered pumping machinery (water wheel). The younger traditional wells of the area, most of them built on the 1940s and the 1950s, are hand and explosive dug shafts wells in the rock, 2.5–3 m of diameter. Some parts of the shaft are lined with concrete or holds a reinforced annulus when rock is unstable; 4 of them have small diameter vertical boreholes at the bottom, built in the 1980s. They use to have secondary works like galleries at different depths and horizontal, small-diameter drains (*catas*), that have been drilled in different stages as well water yield decreased due to the general water table level drawdown. Then, it is very common that these old secondary galleries are now useless because they are above the water table. Well depths range from 15 m to 313 m, with an average of 102 m, without significant differences between large diameter wells and wells with bottom drains. A total of 12 (350–400 mm diameter) boreholes were drilled in 1980s. They are deeper than the shaft wells, with an average depth of 165 m, and generally uncased or with only a slotted tube to protect the pump from falling stones.

The exploitation regime is very variable, from automatized to manually operated wells. Most irrigation wells work several hours a day, resting one or two months a year. Pumping rates range from 3 L/s to 50 L/s, with an average value of 12 L/s. Since the pumps do not work continuously, an equivalent continuous yield is calculated after interpreting the data obtained during field surveys (Cabrera, 1995; Cabrera and Custodio, 2001).

2 GROUNDWATER EXISTENCE AND EXPLOITATION

Groundwater in the Telde area originates mostly in inland areas. Under natural conditions this water discharged in diffuse form along the coast. There are neither references about discrete coastal and submarine freshwater outflows nor they were not identified in an airborne thermographic survey (SPA-15, 1975). They were not observed during the detailed surveys along the coast carried out for recent studies (Cabrera, 1995).

The time evolution of potentiometric levels in the study area comes from data from different times and origins: 1970–1971 for the SPA-15 study, 1980–1981 for the MAC-21 study, 1986–1989 for the operation of the IGME network, and 1991–1992 for the technical reports for the Island's Water Register, plus the 1988 data from the exhaustive inventory carried out by the first author. The analysis has been made considering potentiometric levels, well depth and exploitation data expressed as equivalent continuous yield. Levels go down in 38 wells out of the 50 wells with data, even in the case of abandoned wells. The water table drawdown range from 1 m to 30 m in wells that have not changed their depth, most of them located in the central part of the study area. In wells that have been extended, groundwater levels use to go down in parallel with well depth as a result of increased abstraction. This makes deeper the local drawdown cones around wells. In spite of abstraction decrease, only in four wells a water level stabilization or recovery has been observed.

Figure 3 shows an example of the evolution of three wells with the typical behaviour of different zones. Well A, located at the north of the area, is out of use and show a level drawdown close to 14 m. Well B is located at the central part of the area, where the exploitation is more important. This well has been deepened about 90 m by drilling a borehole in the bottom of the shaft well; this allows an increase of abstraction (until almost 20 L/s of equivalent continuous yield) and has produced a piezometric level decrease of 10 m. Well C shows the behaviour of the wells at the southern part of the area, where there is a progressive closedown of wells due to groundwater salinization.

The water table elevation has been drawn in three different times: 1970–1971 (SPA-15 data), 1980–1981 (MAC-21 data) and 1988–1992

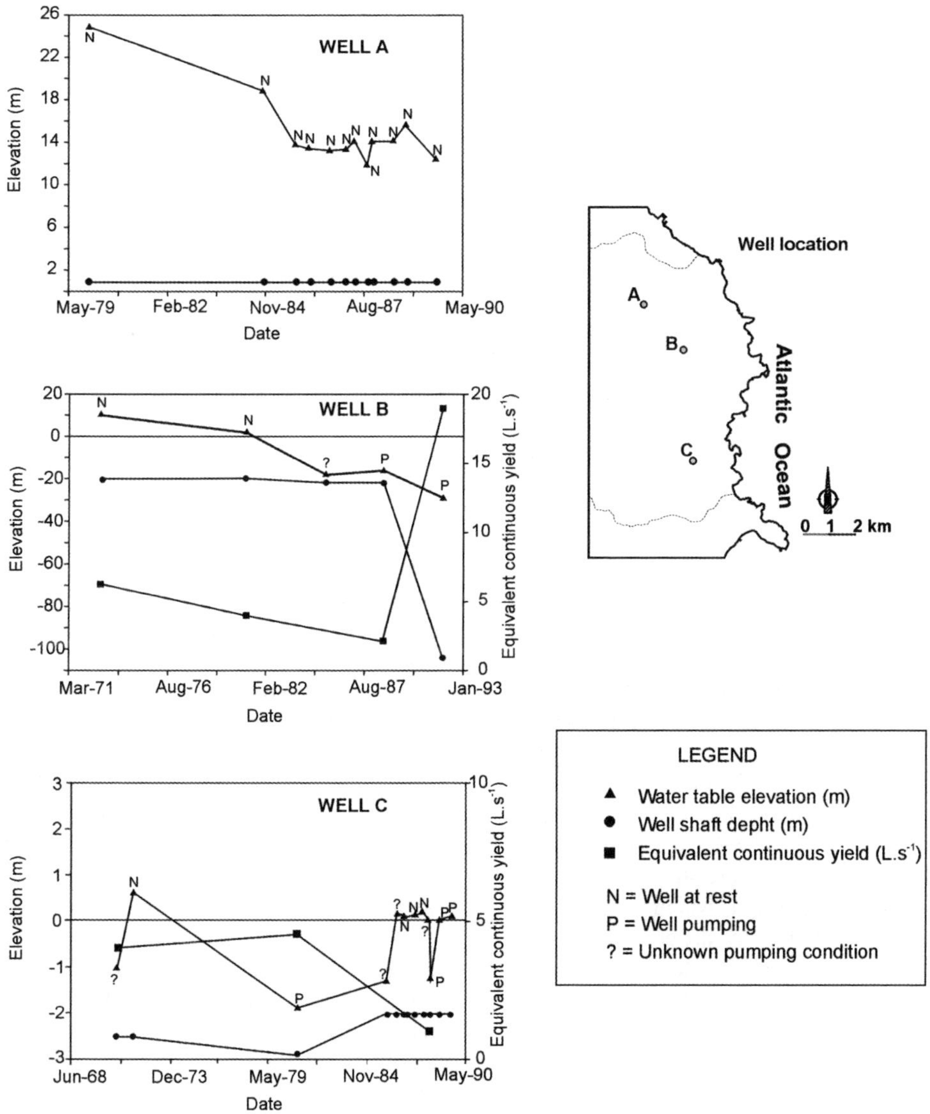

Figure 3. Time evolution of water table elevation, well shaft depth and abstraction (expressed in equivalent continuous yield) in three wells of the study area. Well A is out of use. Well B is located in the central part of the area, where exploitation is more intensive. Well C presents a slight depth decrease, possibly due to up filling with sediments or rock falls.

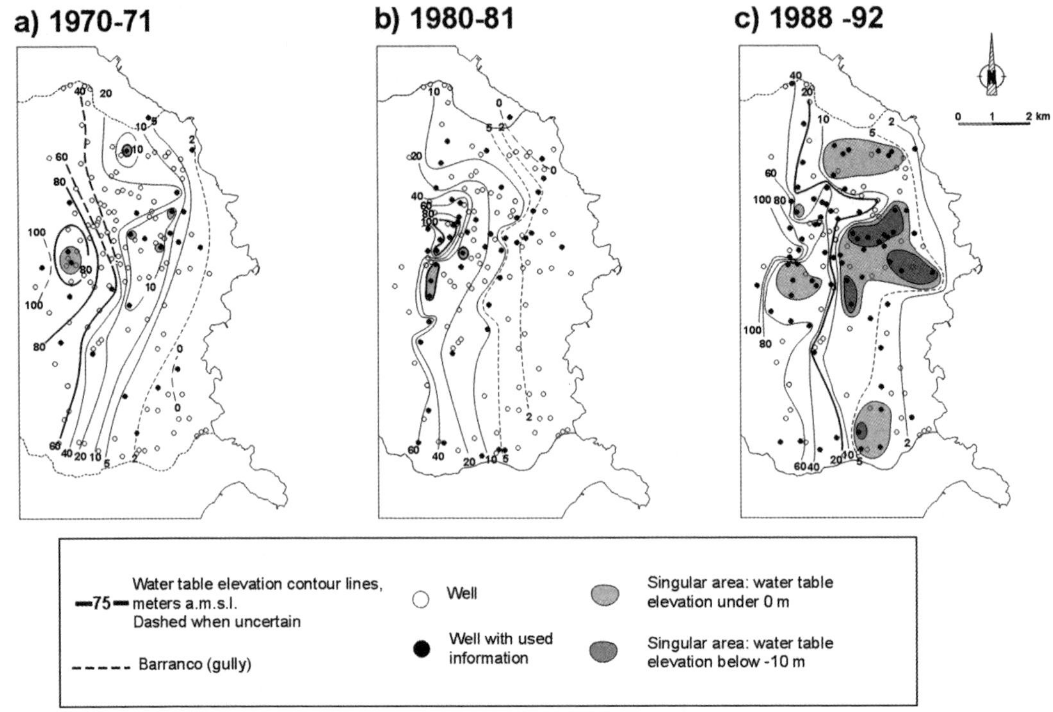

Figure 4. Water table elevation contour lines in three different times: a) 1970–1971; b) 1980–1981 and c) 1988–1992. Singular areas with coalescent drawdown pumping cones have been identified (Cabrera and Custodio, 2003).

(authors' and island water register data) as shown in Figure 4. The area is not uniformly covered with information. There are several zones where there are no wells, especially near the coast. In some cases altitude errors may be up to 5 m and water level depths may reflect residual or active dynamic effects. In the figure, low values that represent the local effect of a well or a close-by cluster of wells have been deleted, but the more extensive depressed areas are shown.

The materials through which groundwater flows and is abstracted are different from what shows the surface geology due to the depth of the water table (often more than 100 m) and the horizontal and vertical complexity of the area. The possibility to directly observe the penetrated materials by descending into the wells is a rare opportunity when a trustable, detailed geological description is not available. The emplacement of wells and the geological formation exploited by each one are shown in Figure 5. Wells in the southern zone exploiting the dominant Roque Nublo Group volcanic breccias and Recent Basalts formation have a low yield. The other formations may yield more than 1 L/s per well in areas that cluster several wells. The correspondence between well yield and the geological formations is not clear; there are several factors which intervene, that include well penetration and the effective functioning of the drainage horizontal boreholes at the study time. The most productive area is at the east of the town of Telde, where the LPDF and/or the Phonolitic Formation are developed, although most of the well productivity is probably due to the LPDF contribution, directly or by leaking water into fractures in the phonolites below. The coastward extension is poorly known.

Of the different formations involved only the LPDF is a typical porous one. The other formations are rocks with fissures and porous-like blocks, as happens to most subaerial volcanic

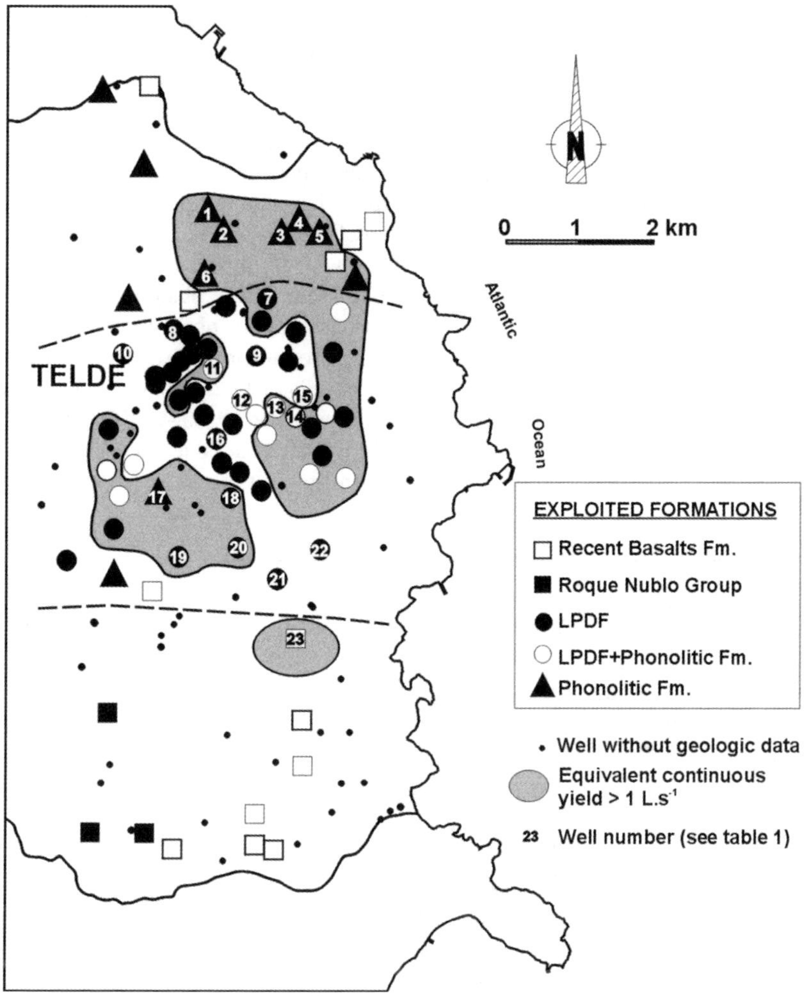

Figure 5. Wells with data on the exploited formation. Three main zones are distinguished. Wells with an equivalent continuous yield higher than 1 L/s are indicated (modified from Cabrera and Custodio, 2003).

formations (Custodio, 2003). Calculated aquifer hydraulic properties refer to bulk rock properties (Cabrera and Custodio, 2003).

3 GROUNDWATER QUALITY

Groundwater in the area is divided in several water families depending on the water-rock interaction with the different materials. Interaction is enhanced by volcanic CO_2 contributions and mixing with other waters. Typical groundwater from the zone is of the Mg-CO_3H type when basaltic and basanitic materials from Cycles II and III are exploited. They are located at the south of the study area, above 100 m elevation, in an area where deep CO_2 contributions are important. Water coming from Cycle I phonolitic lava flows and from sedimentary LPDF is of the Na-Cl type and are located in the north and centre of the area. Na-Cl-SO_4 type water is found in the coastal strip, pointing to the mixing with sea water which is intruding as a result of freshwater abstraction. This is accompanied by cation exchange processes in the mixing front, even with possible local gypsum precipitation due to water hardening, and later redissolution by the intruding seawater. There are no enough monitoring wells to check this behaviour, but it shows up more to the south of the study area (unpublished). Mg-Cl type water at the south indicates the existence of saline recharge and mineralization from the rocks in a CO_2 rich media. Groundwater in the central part of the zone is of the Na-SO_4 or Na-CO_3H-Cl types, with high nitrate contents (up to 300 mg/L), that points to the mixing with irrigation return flows, or urban waste water, which have already penetrated the thick unsaturated zone.

The analysis of the hydrochemical evolution from 1970s to 1992 shows a generalised salinity increase (Figure 6), which follows the groundwater level drawdown. In the coastal-central part of the area the pumped water temperature also increases as well depth increases, but other wells located at the southern and northern part can yield high temperature water associated with deep CO_2 contributions. The salinity and temperature increase in the coastal-central part of the area goes along with a change of the water type when the wells are deepened, from the Na-Cl-SO_4 or Na-SO_4 types to the Na-CO_3H-Cl type, with an important nitrate content decrease. This is due to the abstraction from deeper formations (mainly phonolitic lava flows), with a lower proportion of irrigation return flows. In other wells that have not been deepened, groundwater evolves to Na-Cl-SO_4 type, showing a salinity increase due to increasing irrigation return flows effect.

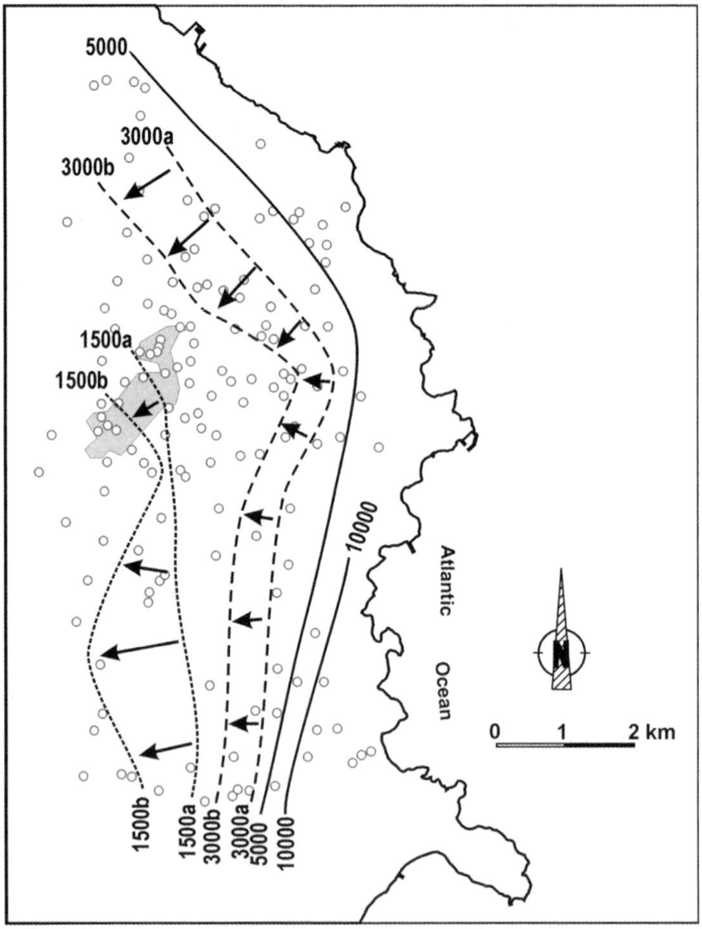

Figure 6. Spatial distribution and time evolution of electrical conductivity of groundwater from 1972 (a) to 1992 (b) in the study area. Arrows show the movement of isolines.

4 RESULTS OF INTENSIVE EXPLOITATION

Information about the aquifer in the Telde area starts when it was already intensively exploited, so the natural situation cannot be known, but for some scattered data. It seems that the *Barranco Real de Telde*, and perhaps some of the other gullies, were draining groundwater from the aquifer in the headwater areas, but not in the lower reaches in the study area. Here the water-table was below the gully channel, and now it is very deep below it. There were recharge areas in which possible surface flows infiltrated, and perhaps continue to infiltrate today when sporadic storm runoff is produced. In practice it seems that this recharge can be currently considered a small groundwater balance term, and little affects the water table. It can be safely assumed that in 1970, the date of the first water-table map, there were already some clear drawdown in the area.

Figure 4 shows that in the intensively groundwater developed area of Telde:

- along the coast there are no water-table elevations below sea level, except locally.
- below sea level water-table elevations have been developed along a central strip and they have progressed toward the coastal area.
- the central strip roughly corresponds to the most permeable areas shown in Figure 5, where groundwater abstraction concentrates.
- at the western boundary the water table elevation is little unchanged.
- groundwater flow presents an average slope of about 20‰ to 40‰ at the inner boundary, and flattens to 3‰ to 5‰ near the coastal line, but there are local disturbances created by the exploitation well clusters.
- an area of small drawdown and high water table goes W-E through the town of Telde; it may be the result of preferential recharge by leakage from the urban water supply network, sewage system and irrigated areas at the time.
- no clear relationship shows up between water table position and lithology.

Figure 7 is an attempt to reconstruct the little disturbed water table before 1970; it also shows a representative cross-section though Telde, in which the water table was inside the Recent Basalt Formation; it is currently mostly drained.

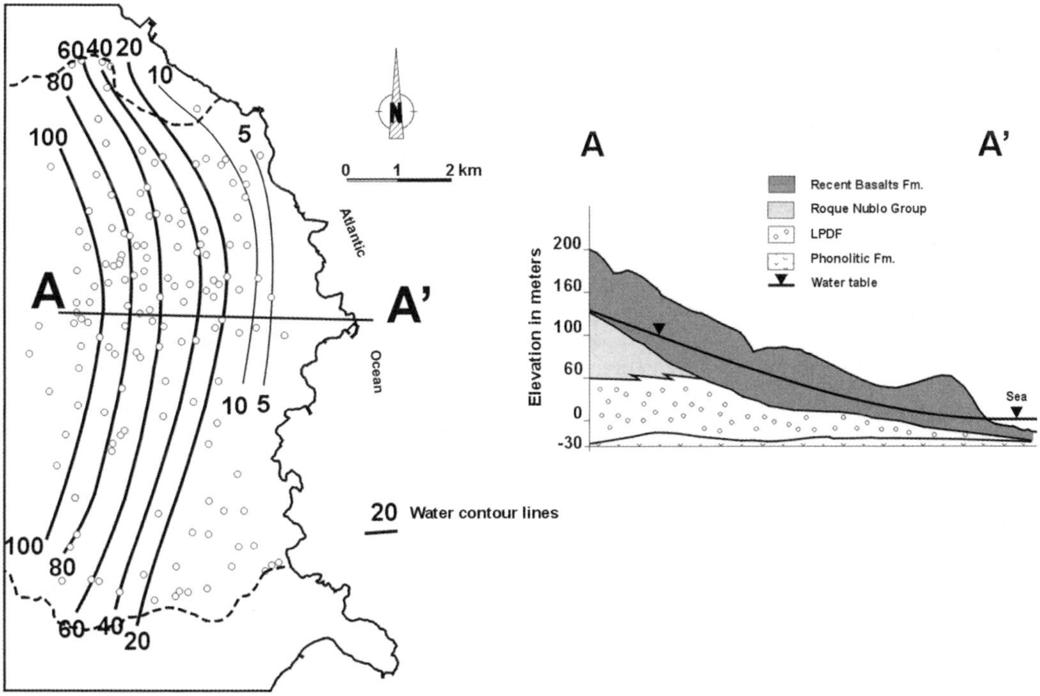

Figure 7. Assumed water table map and hydrogeologic cross-section for Telde area before 1970, when groundwater abstraction was small (Cabrera and Custodio, 2003).

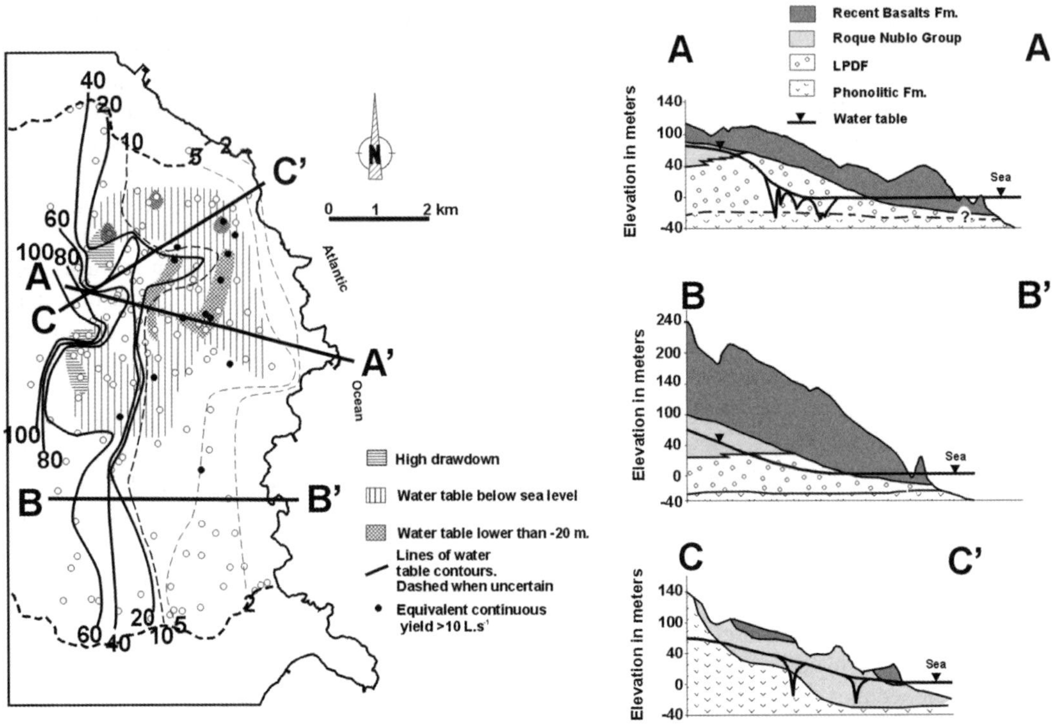

Figure 8. Smoothed water table map and hydrogeologic cross-section of the 1988 situation in Telde area. Local pumping cones are not considered in the map. In the cross-sections they are depleted only for illustration, and do not necessarily correspond to real wells (Cabrera and Custodio, 2003).

Figure 8 is an attempt to show the current situation by extrapolating the presented results with new available information, which shows the depletion of groundwater in a central N-S strip, except around the town of Telde, where there are the recharge circumstances explained before. The three cross-sections show that now the water table is mostly inside the LPDF except to the north of the town of Telde, where static levels are in the Roque Nublo Group breccias and depleted water levels are in the Phonolitic Formation below.

Drawdown has produced the evolution from a situation where presumably the Recent Basalts Formation was the most productive unit, sometimes draining water from the LPDF emplaced below, to the actual situation. The evolution was that water came first from LPDF, and when the water table went down it was the Phonolitic Formation which started to be exploited, compensating its low permeability with the greater penetration of the recent boreholes.

As a secondary effects of drawdown, the groundwater chemistry shows a generalized salinity increase (Figure 6) and changes in the chemistry water type. The abstraction of water from the Recent Basalts Formation and FDLP produces groundwater of the Na-Cl-SO_4 or Na-SO_4 types, with a high content in nitrates due to irrigation return flows. When wells are deepened and groundwater come from the phonolitic lava flows, water changes to the Na-CO_3H-Cl type, with an important nitrate content decrease. This can be due to the dilution of the irrigation return flows and/or to a decrease of the agriculture in the area from 1980s. Nevertheless, certain wells that have not been deepened show an increase in the salinity, which may be due to the increase in the recharge from irrigation return flows.

Recharge mechanisms have not been studied in detail. The Recent Basalts and the thin soil cover in some areas allow the downward vertical transfer or recharge water without the formation

Table 1. Indicative groundwater balance of the 75 km² of the Telde area for the situation in the 1990s. Average values are given. Uncertainty may be large.

Inflow terms	Mm^3/yr	*Comments*
Lateral, from the W boundary	7.4	Applying Darcy's law
Infiltration from Bco. Real Telde	0.5	Data from SPA–15 (1975)
Recharge from local rainfall	1.8	Chloride balance
Return flows from irrigation	1.7	Uncertain
Leakage from supply network	0.7	Municipal data
Depletion of groundwater storage	0.7	$25\ km^2$; 0.06 drainable porosity
TOTAL	12.8	Uncertainty $\pm$ 5 Mm^3/yr
Outflow terms	Mm^3/yr	*Comments*
Abstraction from wells	9.0	After the survey
Discharge to the sea	3.8	Applying Darcy's law; uncertain
Outflow to water courses	0.0	No drainage
Direct evaporation	0.0	Deep water table
TOTAL	12.8	Uncertainty $\pm$ 3 Mm^3/yr

of permanent or temporal perched aquifers, or at least they have not been seen during the descents into the wells. The only falling in water that was observed was very shallow and related to leakages and irrigation.

A groundwater balance in the study area of Telde is a crude exercise but shows the relative values of the different terms. Balance terms of table 1 are average values in the 1990s, in million m³ per year (Mm^3/yr).

In such an intensively exploited area, with conspicuous water table depletion, lateral inflow plus local recharge and other sources of recharge may still exceed actual groundwater exploitation. The recent evolution of the area to transform the cultivated areas into urbanised land causes the progressive decrease of local recharge and irrigation return flows. Also the leakage from the water supply network has been greatly reduced after major investments and repairs to save water. So, the inflow terms are decreasing, even if lateral inflow is maintained. But also well abstraction is decreasing, according to new data on the area.

Groundwater storage depletion is a minor term and may mostly reflect the transient evolution of the system, but it may continue to increase if well abstraction does not sharply decrease. This seems to be happening as expected if irrigation land becomes urban. A major drawback is progressive groundwater quality impairment by saline intrusion, but this is still a too poorly known process in the area, especially due to the lack of observation points in the critical areas.

5 CONCLUSIONS

The study area has a complex disposition of materials and a relatively deep (40 m to 100 m) water table. Groundwater flow is variable areally and also along time, as the water table position changes due to exploitation. Exploitation causes also an increment in water salinity and changes in the hydrochemical water type, when water flows through deeper formations. The existence of an intercalated detritic formation (LPDF) plays a major hydrogeological role as a preferential pathway for groundwater. The area shows a central strip of groundwater reserve depletion due to abstraction, up to 40 m of drawdown between 1970 and 1992, where salinity increase is higher than in other areas.

Groundwater reserve depletion contributes only about 5% to input balance terms while groundwater transfer from inland areas is more than 60%. Recharge was almost entirely discharged into the sea in dispersed form along the shoreline; currently about 30% may be still discharged, much more than reserve depletion, in spite of persisting the drawdown trend, and the existence of local seawater intrusion.

In spite of the intensive groundwater exploitation of the area of Telde and the continuous drawdown, the area may be managed to get a sustainable use if abstraction is reduced and the spatial distribution of wells is corrected to avoid saline water problems. This means a groundwater users' association, which must include inland stakeholders and a monitoring network.

ACKNOWLEDGEMENTS

The authors thanks the General Directorate for Water of the Canarian Government and the Gran Canaria Hydrologic Plain office for the support during the works and the supply of data. The Geological Survey of Spain (ITGE, IGME) has contributed data, comments and support through the local office in Las Palmas de Gran Canaria and directly from the headquarters in Madrid. The Technical University of Catalonia, Barcelona has provided tuition and supported some of the works. Special thanks to all the well owners and operators who have spent time to supply and comment the data. The starting part of this work was carried out with the support of Project CCA 8510/10 of the Spain-USA Joint Committee for Science and Technology.

REFERENCES

Cabrera, M.C. (1995). Caracterización y funcionamiento hidrogeológico del acuífero costero de Telde (Gran Canaria) [Characterization and hydrogeologic functioning of Telde coastal aquifer]. Doctoral Thesis, University of Salamanca, Spain.

Cabrera, M.C. and Custodio, E. (2001). Análisis de la explotación por formaciones en el acuífero de Telde (Gran Canaria) [Exploitation analysis according to formations in Telde aquifer]. Hidrogeología y Recursos Hidráulicos, XXIII, Madrid: 269–279.

Cabrera, M.C. and Custodio, E. (2003). Groundwater flow in a volcanic sedimentary coastal aquifer: Telde area, Gran Canaria, Canary Islands, Spain. Accepted by Hydrogeology Journal.

Cabrera, M.C.; Núñez, J.A. and Custodio, E. (1992). Contribución al conocimiento geológico del subsuelo de Telde (Gran Canaria) [Contribution to the geological knowledge of Telde deep formations]. Proc. III Cong. Geol. España and VIII Congr. Latinoam. Geol., Salamanca. 2: 256–260.

Custodio, E. (2003). Hydrogeology of volcanic rocks. In Groundwater Studies. UNESCO. Paris. Chap. 16 (in press).

Custodio, E. and Cabrera, M.C. (2002): ¿Cómo convivir con la escasez de agua?. El caso de las Islas Canarias. [How to survive with water scarcity?. The case of Canary Islands]. Boletín Geológico y Minero de España, Vol. 113-3: 243–258.

MAC-21 (1980-81). Proyecto de Planificación y Explotación de los Recursos de agua en el Archipiélago Canario. Com. Interminist. Coord. Est. Mat. Aguas Canarias. Internal reports, not published.

PHGC (1991). Actualización del conocimiento hidrogeológico de Gran Canaria [Update of Gran Canaria hydrogeologic knowledge]. Gran Canaria Hydrologic Plan Office. Las Palmas de Gran Canaria, 3 vol.

SPA-15 (1975). Estudio científico de los recursos de agua en las Islas Canarias [Scientific study of water resources of the Canary Islands]. Project SPA/69/515 MOP (DGOH)–PNUD (UNESCO). Ed.: Dirección General de Obras Hidráulicas-UNESCO. Madrid. 4 vols.

MANAGEMENT

Water management issues related to aquifer intensive use

Eduardo Usunoff
Instituto de Hidrología de Llanuras, Azul, Buenos Aires, Argentina
eusunoff@faa.unicen.edu.ar

ABSTRACT

Born out of the increasing needs for food and water supply, there exist many regions in the world where regional groundwater resources have been and are intensively exploited. The consequences are sadly and widely known – rising cost of pumping, decreasing crop yield, water quality worsening, disappearance of ecologically-sound environments, falling groundwater-levels, land subsidence, so that innovative management approaches are eagerly sought. On the positive side, however, groundwater is so widely spread and has so many attractive characteristics that it has become the central element for development in both developed and developing countries.

Tied to the concept of system sustainability in the context of integrated water resources management, this paper presents some ideas and examples of conjunctive use of surface water and groundwater, management of groundwater demand in heavily stressed aquifers, and stakeholder participation in the decision-making process.

Keywords: *groundwater overdraft, groundwater management, stakeholder participation.*

1 INTRODUCTION

The ever-increasing rate of population growth throughout the 20th century and the inherent demand for food and water supply has led to intensive groundwater development in vast regions of the world (chiefly, North China, North America, Central America, West and South Asia, South Europe). Groundwater levels decline, groundwater and surface water quality deterioration, decreasing crop yields, loss of water-dependent natural environments, unbearable pumping costs, are among the most frequently cited effects of such an intensive use. To be fear, though, groundwater offers overwhelmingly more benefits than faults of outmost importance in poor regions: easy access, great areal distribution, progressive development, low capital investment, relative low cost, ease of available technology, widespread use by a large number of users, relative resilience to droughts, general good chemical and bacteriological quality (Llamas and Custodio, 2003).

Intensive groundwater use is a general purpose term, that embraces many other concepts/facts widely found in the literature, such as *aquifer overexploitation*, *aquifer depletion*, *groundwater overdraft*, and even *groundwater mining* (see Custodio, 2002, for a lucid review on the subject). Any of those terms is usually attached to a rather negative view, disregarding physical and socio-economic issues. As a matter of fact, the intensive use of groundwater has been the major drive behind the well-being in many regions, particularly in those located in arid or semi-arid climates and devoted primarily to agricultural activities.

On the other hand, and given the scenario for the 21st century in terms of population growth and the coupled demands for food and access to water, it is also true that significant efforts have to be made in order to improve several water management issues that have been downplayed, overlooked or disregarded in the past (system sustainability, supply and demand management, conjunctive use, stakeholders participation).

This paper aims to bring about the essential facts behind the issues already mentioned, highlighting the keypoints of successful experiences as well as the major drawbacks detected.

2 NEW PARADIGMS OF WATER RESOURCES MANAGEMENT

Although the intensive use of groundwater can be traced back several decades, its effects on the hydrological and the associated socio-economic systems have been made evident more recently. Indeed, in the early period of development the prevalent action was water use, and the management attitude pointed almost exclusively at meeting the water demand. Entering the new millennium, however, the actual and future consequences of water scarcity due to the population rate of growth have imposed a major change in the way that water and other related natural resources are to be managed. The unplanned use of water is not longer acceptable, and water resources have to be managed for long-term sustainability. Perhaps the key phrase in that respect is integrated water resources management (IWRM). In general terms (Mariño and Simonovic, 2001), IWRM can be described as "a form of coordinated management of land and water resources within a region, with the objective of preventing land degradation, protecting the quality of the freshwater resource, protecting biodiversity, and continuing sustainable use; within a context which includes genuine community/government partnerships and recognition of socio-economic objectives".

The main principles of IWRM, according to García (1998), include all actions and projects aimed at increasing water use efficiency and water conservation. In doing so, it should count on: 1) seeking for all complementary ways of water usage, 2) minimizing conflicts among competitive uses (in quantity and quality), 3) incorporating the demand-oriented approach along with the conventional supply-oriented management, 4) having an adequate set of rules (laws, political decisions, strategies, plans, and norms), and 5) assuring that enough human and economic resources are available to reach the goals.

It may readily be realized that such principles are of direct and much-needed application at regions with intensive groundwater use. The scope can be much comprehensive and long-reaching, but taking precaution in balancing extreme positions (deep ecology, economics above all).

3 CONJUNCTIVE USE

At the global and even national level, groundwater availability is far in excess of the current use (Shah *et al.*, 2000). However, the spatial imbalances in the occurrence of groundwater and the all-too varying demand pattern of it brings about the existence of many regions with problems associated to intensive groundwater use. One of the strategies proposed to ameliorate such a situation is conjunctive management of surface water and groundwater.

Mariño (2001) indicates that conjunctive or joint management "refers to the coordinated operation of a groundwater basin and the overlying surface water system to increase total supplies and enhance total water supply reliability". The idea is to use as much surface water is needed in wet years, preserving the groundwater for dry years, and enhancing infiltration to the aquifer of excess surface water, precipitation, return flow from irrigation and sewage.

One of the leading examples of conjunctive use is that of Southern California. Large volumes of groundwater had been pumped since the late 1940s for agricultural, industrial, and urban purposes, which led to regionally declining groundwater levels and water quality degradation from land-use practices and other causes. Lately, the United States Geological Survey has joined efforts with the Water Replenishment District of Southern California and other minor water management agencies in order to bring surface water from Northern California and the Colorado River and recharge it into the local aquifers, mostly by surface spreading. Mariño (2001) points out that, although people in California agree that conjunctive use is an economic way to manage better the total water supply, questions remain as to who decides when to store water, when to take the water out, and who is going to own and operate the facilities. So far, the involved agencies concentrate their efforts in applying mathematical optimization techniques largely based on decision-support systems (ASCE, 1998). Nonetheless, the Californian experience is probably the best known and long-lasting in the world, and may serve as an example of at least how to avoid certain management problems (Smith, 2003). Indeed, as correctly pointed out by Smith (2003), the so-called dominant social paradigm in USA includes predominantly free market economics, individualism, faith in science and technology, and a growth orientation. Groundwater pumping in California and Texas accounted for 50% of groundwater use in USA, a figure that nowadays is down to about 25%. In spite of that, neither state has a law that establishes individual pumping rates. Yet, Smith (2003) recognizes that "competition for groundwater resources within the states of the USA has often been managed reasonable well".

Conjunctive use and artificial recharge has been used for many years at the Llobregat River, Catalonia, Spain. Induced recharge to the alluvial aquifer is achieved through scrapping the river bed. Deep recharge wells to inject treated river water have also been set up (Candela, 2002). The so-recharged aquifer stores water for emergency supply of Barcelona City.

Several spots of intensive groundwater use are found along the USA-Mexican border, which brings about the issue of transboundary aquifers and its constraints in terms of integrated management. For the Santa Cruz and San Pedro Rivers (Arizona, USA; and Sonora, México), Liverman *et al.* (1997) have shown that differences in socio-economical and cultural status do have a profound effect on the quantity and quality of available water, and that an exhaustive assessment is in order if integrated water resources management in the region is ever to be achieved. The enactment of the North American Free Trade Agreement (NAFTA) may result in unforeseen changes in agricultural, municipal, and industrial demand for water, which have to be properly addressed.

Many regions in Spain, where the common features are water scarcity and high water cost, have adopted and put into action the conjunctive use principle (North, Duero, Tajo, Guadiana, Guadalquivir, South, Júcar, and Ebro basins). It is generally deemed that the potential of the whole Spaniard water resources to adopt the conjunctive use of surface water and groundwater is well above what has been achieved in the mentioned basins. Moreover, the current successful experiences were born from the needs and engagement of stakeholders and backing-up agencies more than from a truly national strategy for management of water resources (MIMAM, 1997).

Another example is the Dan River Project in Israel, which incorporates collection, treatment, groundwater recharge, and reuse of municipal waste from Tel Aviv, to use it later for irrigation purposes. It is essentially a soil-aquifer treatment, using the natural cleansing properties of soil to treat the water. The primary system relies on biological treatment in oxidation ponds, with recirculation and lime magnesium treatment, partial free ammonia stripping and natural recarbonation. Since 1989, recovery wells have pumped about 360 Mm^3 reclaimed water for irrigation, with complete removal of BOD, suspended solids, and reduced nitrogen and phosphorous. The system ensures that as long as there is waste water from Tel Aviv, there will be water for crops.

4 MANAGEMENT OF DEMAND AT MASSIVE USED AQUIFERS

Some years back the exploitation of groundwater had little or no restrains. Aquifers were taken as common pool resources, so that the use of water was on the *first come, first serve* basis. Those using the resource had no motivations for preserving the water quantity and quality, just because anyone conserving it for future use was simply leaving it for others to use. In other words, there was no regard of the effect of current pumping on future water availability or future pumping cost. On top of that, the external costs (unwanted damages on third parties) were also disregarded. Time has shown the adverse effects of such proto-management attitude, which is superbly described in the classical manuscript by Hardin (1968). Interesting enough, Hardin (1968) noted that "the morality of an act is a function of the state of the system at the time it is performed". In the early days, our ancestors had at their disposal as much water as they wanted, and could not foresee the consequences of intensive groundwater use; they are not to blame.

Nowadays, however, the increase in population (i.e., more water demanded for domestic, industrial, agricultural, and other purposes) as well as the increase in the average standards of living, cry for innovative approaches to groundwater demand management. Right to the point, Young (1991) warned about the need to address two types of collective decisions: "managing the water" (decisions on the appropriate annual rate of pumping, on the geographic distribution of pumping, on whether to augment water supplies, and/or whether to artificially recharge the aquifer), and "managing the people" (to determine the institutions and policies which divide the extraction rate among potential individual users and user classes, the monitoring and enforcing rules for limiting pumping, the nature and extent of education and research programs, and the conflict resolution mechanisms).

Both aspects include a large number of subtopics that this paper does not attempt to cover in detail. The general impression is that, even in the most developed societies, the degree of knowledge on the hydrological systems and the commitment of the society at large (politicians, decision-makers, stakeholders) is not enough to build sound, well-coordinated plans to deal with the complex interactions between the physical environment and the human society. Some examples below will help substantiate such an statement:

- According to a study by the International Water Management Institute, 24 nations (mainly in the sub-Saharan Africa) are classified as suffering from *economic water scarcity* (Radford, 2000). There might be enough water to meet demands in 2025, but to do so they would have to double their efforts to extract water. They are caught in the poverty trap: they cannot afford dams and irrigation systems, badly needed for their survival, let alone their rational management.
- 87% of all freshwater resources in the Middle East and Africa region are used in mainly low value agriculture (Charrier *et al.*, 1999). In the same report is reads that experts on different aspects of Middle East water issues have agreed that (partial list): 1) there is a large potential for water markets in the region as a way of promoting greater water use efficiency, although the lack of experience by agricultural workers and the large economic discrepancy among countries may hinder the initiative; 2) integrated water resources management is more easily discussed in the abstract than implemented, especially given the political situation in the Middle East; and 3) saving water for environmental purposes may be problematic as the local population increases and greater demands are made on already scarce water resources.
- Davis and Lund (2000) found that, in spite of the wealthy economy in Chile and the broad private-sector involvement, the Chilean system is incomplete in that is based on an economic formulation that does not take into account for recognized market failures and does not reconcile economic, environmental, and social objectives. Simple market economics are in conflict with IWRM, as stated bluntly by Harris (2000): "The market economy works well while the environmental damage is being done".
- Referring to the Southeastern Australia experience with water trade, Bjornlund and McKay (2000) found that water entitlements and policy instruments need to be more sophisticated, market operations and water allocation mechanisms must be transparent and predictable and the factors affecting supply and demand for water as

well as the level of annual allocations must be communicated to users.

- In South Africa, mining and industrial uses represent 8% of the water availability, whereas irrigation and afforestation take 62%. The former contributes 37% of the Gross Domestic Product (GDP), as opposed to only 6% from the latter. Along the same line, Llamas (2001) indicated that 80-90% or water consumptive use in Spain is attributed to agriculture, with a return of 4% in terms of GDP.
- In Africa, the cost of a borehole drilled by a truck-mounted rig can be extremely high in relative terms (10–20 times the cost of the pump), and many times the cost of well drilling in Asia, for example. Thus, the groundwater potential remains almost untapped (Shah *et al.*, 2000).
- Commitment by locals is a must for integrated water resources management, but it not always comes for free. A report by the World Bank (2000) indicates that people participates in environmental conservation activities only because they expect benefits from the entry activity (or from labor payments involved). Once these benefits have been achieved, people's interest in long-term management of the environmental conservation activity may easily fade.
- Active management of overstressed aquifers may be even more complicated when two or more countries share the resource. The Israeli-Palestinian conflict is an example of it, although the hope is that water may be the instrument to bring peace into the region.

In spite of the rather discouraging examples above, a fact (not always recognized) is that the role of groundwater as a tool for human development is far beyond the perception of people, in general, and politicians, in particular. Evidence in India suggests that the crop yield/m^3 on groundwater-irrigated farms tends to be 1.2–2 times higher than on surface-water-irrigated farms (Shah *et al.*, 2000). Hernández-Mora *et al.* (2001) found just about the same in Andalucía, Spain. Groundwater development is much more keen to poverty targeting, as it can be seen mostly in developing countries of Asia and Africa. Surface water development entails large investments, long execution times, and is prone to misuse of allocated funds (overbudget, corruption). On the contrary, groundwater development is rather individual, and apt to the scale economies of most regions in the world. Therefore, and backing up to the central matter of this paper, there are several ways of mitigating groundwater overdraft (examples taken from Shah *et al.*, 2000):

- Aquifer recharge with imported water: excessive pumping at the San Joaquín Valley in California (USA) led to 30–100 m drawdown of groundwater levels, changed the direction of water flow in the confined units, pumping lift increased to US$ 250 million, and land subsidence stood out as a severe problem. Such economic looses drove the decision to import surface water from the California Aqueduct. The recovery in potentiometric surface from 1967 to 1984 was nearly one-half the drawdown that occurred from the predevelopment years to 1967. The Azraq Oasis (Jordania), a Ramsar wetland that covers 7,500 ha, dried out as a result of groundwater overdraft upstream for irrigation and water supply to the city of Aman. The whole ecosystem collapsed, the salinity of local groundwater increased by a factor of three, and revenues from tourism felt down. Cutting water supply to Aman and/or banning irrigation were politically unfeasible. A UNDP-supported project brought water to the oasis from a distant well-surplus well field, the flora and fauna returned, and with them, the tourists.
- Recharge with rainwater: this approach seems particularly important in monsoonal regions, where most of the annual rainfall concentrates in a few hours, thereby providing little time for recharging the aquifers. Asia has a long tradition and ancient structures for water harvesting, which have been put back into use. China may have a great potential, as the country holds 7 million ponds for water harvesting and recharge. In Western India, a region

where groundwater depletion is critical, locals created spontaneously (i.e., not waiting for governmental actions) a massive well-recharge movement. Some 300,000 wells are being used to divert rainwater to them, along with constructing thousands of ponds, check dams and other minor structures to prevent the rainwater from flowing into the Arabian Sea.
- Vegetative treatment of recharge areas: There is a delicate balance between the benefits and the drawbacks of a vegetation cover in the recharge areas of an aquifer. Some species consume too much water and their rooting effect is not that important. In the last years there is a growing worldwide movement that promotes the use of vetiver grass (details at http://www.vetiver.com) for soil and moisture conservation. This grass is able to grow in a wide range of soil-pH conditions (4 to 11), and its roots may reach 3–4 m depth, which enhances groundwater recharge by creating downward preferential flowpaths. In addition to that, vegetative cover help decreasing the effects of otherwise fast and eroding overland flow.
- Domestic rainwater harvesting: aside from major rainwater harvesting structures, many regions in Asia have in the past practiced rainwater harvesting at the household level, and the ancient techniques are coming back. Roofs are built in such a way that the rainwater is channeled to a family cistern or a tank. In the coastal desert of northern Chile, a fog collection project has provided an average of 11,000 L/d of water to a community of 330 people.

5 STAKEHOLDERS PARTICIPATION IN WATER POLICY DEVELOPMENT

Admitting that each country is unique as far as its political, hydrological, economic and social features, still there are a number of characteristics of the process of policy development and sector reform which are common: vision, political endorsement, technical expertise, stakeholder engagement, and realism. No matter how sound a given policy may be, people's attitude ultimately define its successful application. In that respect, stakeholders participation is a crucial point. Optimally, it is to be seen as the government participating in people's programs, and not the other way around.

Community involvement, however, is not an easy task. It does require some abilities and skills of local people, in what may be termed *capacity building*. Abrams (2001) defines it as "the process whereby a community equips itself to undertake the necessary functions of governance and service provision in a sustainable fashion". In as much as many countries may not have efficient administrative structures nor properly-trained professionals, capacity building should include officials, technicians, and the general awareness of the local population regarding their services and development in general. The amount of capacity building required for a given sustainable effort will depend on the level of capacity already within the community, and the level of service and the technology chosen. Given so, Abrams (2001) indicates that there are minimum requirements to be met, or *capacity thresholds*. Items to be evaluated for checking such thresholds are: a) technical skills, b) administrative skills, c) governance skills, d) conflict resolution ability, e) public health/ecological awareness, f) project support/awareness, g) financing systems, h) community wealth/poverty, and i) support infrastructure. It is essential that the capacity of the community remains at the threshold levels during the lifespan of the project, as if one area of capacity falls below the threshold there is likely to be a *domino effect*.

Even in countries with rich history in water management issues, citizen participation is complex. Ubbels and Verhallen (2001) indicate that stakeholders opinions are legally formalized in the Dutch system of decision-making. Yet, the existing structure of citizen participation does not lead to a basis of agreement, but rather to conflicts. As a result, plans are delayed for long times. In fact, the traditional approach to decision making in IWRM relied on scientific, technical, and economic information produced and delivered by experts, whereas the new methods point towards making use of the knowledge, experience and opinions that are present in society. However, technical experts are not used to working with people from different backgrounds and varying degree of knowledge. It is then imperative to come up with tools that can support the required integration of scientific-technical knowledge and the dialogue among stakeholders. A generic name for such tools is decision support systems (DSS). There is an ample variety of DSS, and the interested reader is referred to ASCE (1998) for specifics on the subject. Any DSS uses mathematical calculations to produce information that is comprehensible and helpful to decision-makers and stakeholders, facilitates public accountability in the way decisions are reached, and helps winning acceptance for plans (Jamieson and Fedra, 1996).

Stakeholder participation, in the framework of collaborative planning, has to take into account the different activities embodied in the planning process. Ubbels and Verhallen (2001) have identified six phases: 1) design of the process, 2) problem definition, 3) search for solutions, 4) analysis of the alternatives and modeling, 5) presentation and discussion of results, and 6) the policy choice. Although such phases may not occur sequentially, it is important to preserve the division in order to appreciate the different activities involved, as shown in the table below.

Table 1. Activities in a collaborative planning process (after Ubbels and Verhallen, 2001).

Activities/Phases	1	2	3	4	5(*)
Communication	X	X	X	X	X
Storing generated knowledge	X	X	X	X	X
Collective problem definition	X				
Incorporating divergent views	X	X			
Identifying objectives/goals	X	X			
Consensus building	X	X		X	
Identifying evaluation criteria	X	X		X	
Seeking possible solutions		X			
Initial ranking of possible solutions		X			
Translation of solutions into alternatives		X	X		
Estimating effects of alternatives			X		
Visualization of effects				X	
Comparing alternatives				X	
Choosing an alternative					X

(*) Decision-making is in fact a matter of choosing between different alternatives. The decision-maker is responsible for selecting the final alternative (phase 6). That is why there are a limited number of activities during phase 5. It is the responsibility of the decision-makers to ensure that the final decision is communicated to all stakeholders.

More on the subject, the consensus building activity may be a little elusive because has no units of measure. As it relates to sustainability, the following definition may be offered (Simonovic, 1997): "Consensus is an equitable compromise which is robust with regard to: a) resource management uncertainties, and b) stakeholders perspectives". In the same paper, Simonovic (1997) presents a consensus sustainability approach that, more than looking for correct answers, provides sources of feedback to assist in: a) shortening up the number of appropriate alternatives, b) identifying sources of disagreement, c) tracking the progress of negotiations, and d) adding additional insight to our perceived degree of robustness.

The USA experience (Smith, 2003), with primary emphasis on property rights and money, shows that rigid, state-handled regulatory pumping measures do not work. Instead, voluntary management systems administered with effective input by local stakeholders, may reach the level of flexibility required at each place given the local hydrological, social, and economic conditions. Education, in the widest meaning of the term, may play an important role in building the necessary trust among the interested parties (McClurg and Sudman, 2003).

Community involvement in groundwater management has been at the heart of long-lasting debates in Spain. There are, however, some fruitful experiences which demonstrate that the *bottom-up* attitude (Llamas, 1996) proved to be the right track in establishing the Groundwater Users Councils or CUAS (*Comunidades de Usuarios de Aguas Subterráneas*). One of the leading examples (Valdés, 1996) is that of the CUAS for the Lower Llobregat Basin (about 110 km^2). The *Baix Llobregat* CUAS was launched in the late 1970s, facing the evidence of deteriorating groundwater quality in the local aquifers by seawater intrusion due to groundwater levels decline from excessive pumping. The overall goals were to decrease groundwater extraction and to make better use of existing surface water surplus. As of 1996, the *Baix Llobregat* CUAS supervised directly the operation of 600 wells (251 for agriculture irrigation, 292 for industrial uses, and 57 for freshwater supply). The *Baix Llobregat* CUAS is self-sustainable, in the sense that it does not receive federal funds. Notably, its annual budget increased 10-fold in the period 1987–1996.

Latin America holds a great portion of the world's freshwater resources. Yet, many places in the region (North and Central America, and along the Andes in South America) rely almost exclusively on groundwater resources for water supply and irrigated agriculture. Mexico offers a wide range of examples, from the intensive use of groundwater to supply the Mexico DF megacity, to the growing demand for crop irrigation and industrial uses (along the USA-Mexican border). Water management agencies are pushing towards the establishment of basin councils, seeking for active stakeholder engagement. Current developments to consolidate the institutional and participatory structures are undertaken at the Lerma-Chapala River basin, the Bravo (Grande) River basin, and the Valley of Mexico groundwater system. The IMTA (*Instituto Mexicano de Tecnología del Agua*) reported some successful initiatives on water use efficiency and increased product yield in small family orchards. Likewise, a special program has been set out to save water in official buildings (details may be retrieved at http://www.imta.mx).

The Mendoza Province (Western Argentina) counts on a relative highly evolved water law, tailored to the economy of a large number of farms (winery industry oriented, mostly) who rely on a complex gravity-driven irrigation system. The instituted rules were drafted by provincial water manager agencies in consultation with the farmers and, up to date, no major conflicts have existed. Although the official water management agencies determine the water quotas according to the available resources for a given year, the day-by-day operations are virtually in the hands of farmers.

6 FINAL REMARKS

The actual effects of massive groundwater use and the increasing need for water in the same places where such effects are felt claim for immediate actions. Business as usual is no longer an option, as too many vital issues related to life are at stake. Although many scientific and technological pieces of knowledge are still missing, one may say that there is enough know-how to face the *managing the water* aspect of the problem. On the contrary, the *managing the people* side is still pretty much in its infancy stage. Politicians, water managers, technicians, and end-users are to show their will and profound commitment if peace, food and water for all is to have a chance in the near future.

REFERENCES

Abrams, L. (2001). *Capacity building for water supply and sanitation development al local level.* Paper delivered at the 2nd Symposium on Water Sector Capacity Building. Found at: http://www.thewaterpage/capacity_building.html.

ASCE (1998). *Sustainability criteria for water resource systems.* Task Committee on Sustainability Criteria, Water Resources Planning and Management Division, American Society of Civil Engineers and Working Group of UNESCO/IHP IV Project M–4.3, 253 pp.

Bjornlund, H. and McKay, J. (2000). Are water markets achieving a more sustainable water use? *Proceedings of IWRA's 10th World Water Congress*, Melbourne, Australia.

Candela, L. (2002). *Integrated management of conventional and non-conventional water resources in arid countries.* 1st ASEM Workshop on Water Management, Changsha, China, 11 pp.

Charrier, B.; Dinar, S. and Curtin, F. (1999). *Water, conflict resolution and environmental sustainability in the Middle East.* Green Cross International, at: http://www.greencrossinternational.net/greencrossprograms/waterres/water/waterconflictresolution.

Custodio, E. (2002). Aquifer overexploitation: What does it mean? *Hydrogeology Journal* 10(2), 254–277.

Davis, M. and Lund, J. (2000). Reconciling economic, environmental, and social objectives in Chilean water resources management. *Proceedings of IWRA's 10th World Water Congress*, Melbourne, Australia.

García, L. (1998). *Strategy for integrated water resources management.* Technical Study No. ENV–125, Inter-American Development Bank, Washington, D.C., 36 pp.

Hardin, G. (1968). The tragedy of the commons. *Science*, 162: 1243–1248.

Harris, G. (2000). Water, science and ecology in an age of cynicism. *Proceedings of IWRA's 10th World Water Congress*, Melbourne, Australia.

Hernández-Mora, N.; Llamas, M.R. and Martínez Cortina, L. (2001). Misconceptions in aquifer overexploitation: implications for water policy in Southern Europe. In: C. Dosi (ed.), *Agricultural use of groundwater: towards integration between agricultural police and water resources management*. Kluwer, Dordrecht: 107–125.

Jamieson, D. and Fedra, K. (1996). The "WaterWare" decision-support system for river-basin planning. Conceptual design. *Journal of Hydrology*, 177: 163–175.

Liverman, D.; Merideth, R.; Holdsworth, A.; Cervera, L. and Lara, F. (1997). *Assessment of the water resources in the San Pedro River and Santa Cruz River Basins, Arizona and Sonora*. Latin American Area Center and Udall Center for Studies in Public Policy, University of Arizona, Tucson, 75 pp.

Llamas, M.R. (1996). *Importancia básica de las Comunidades de Usuarios de Aguas Subterráneas*. Jornadas Técnicas, 30 Aniversario Curso Internacional Hidrología Subterránea, Barcelona, Spain, 7 pp.

Llamas, M.R. (2001). *Cuestiones éticas en relación con la gestión del agua en España*. Discurso de Ingreso, 4 April 2001. Real Academia de Doctores. Madrid, Spain. 102 pp.

Llamas, M.R. and Custodio, E. (2003). Intensive use of groundwater: a new situation which demands proactive action. In: M.R. Llamas and E. Custodio (eds.), *Intensive use of groundwater. Challenges and opportunities*. Balkema Publishers. The Netherlands: 13–31.

Mariño, M. (2001). Conjunctive management of surface water and groundwater. *IAHS Publication 268*: 165–173.

Mariño, M. and Simonovic, S. (2001). *Integrated Water Resources Management. IAHS Publication 272*. Preface.

McClurg, S. and Sudman, R. (2003). Public and stakeholder education to improve groundwater management. In: M.R. Llamas and E. Custodio (eds.), *Intensive use of groundwater. Challenges and opportunities*. Balkema Publishers. The Netherlands: 271–283.

MIMAM (*Ministerio de Medio Ambiente*) (1997). *Integración de los acuíferos en los sistemas de explotación*. Ministerio de Medio Ambiente de España, Secretaría de Estado de Aguas y Costas, 78 pp.

Radford, T. (2000). High and dry. *The Guardian Weekly*, 4 June, page 26.

Shah, T.; Molden, R.; Sakthivadivel, D. and Seckler, D. (2000). *The global groundwater situation: overview of opportunities and challenges*. International Water Management Institute, 21 pp.

Simonovic, S. (1997). Consensus as the measure of sustainability. *Hydrological Sciences Journal*, 42(4): 493–500.

Smith, Z.A. (2003). Groundwater collective management systems: the United States experience. In: M.R. Llamas and E. Custodio (eds.), *Intensive use of groundwater. Challenges and opportunities*. Balkema Publishers. The Netherlands: 257–269.

Ubbels, A. and Verhallen, A. (2001). Collaborative planning in integrated water resources management: the use of decision support tools. Integrated Water Resources Management. *IAHS Publication 272*: 37–43.

Valdés, L. (1996). *Aspectos técnicos y económicos de la Comunidad de Usuarios de Aguas del Delta del río Llobregat. Experiencias*. Jornadas Técnicas, 30 Aniversario Curso Internacional Hidrología Subterránea, Barcelona, Spain, 10 pp.

World Bank (2000). *Participation in the Himalayan foothills: Lessons from watershed development in India*. Social Development Papers, Paper 28, 53 pp.

Young, R. (1991). Managing aquifer overexploitation: An economic and policy analysis. *Proceedings of the 23rd IAH International Congress*, Canary Islands, Spain, April 1991, 261–262.

Groundwater depletion and its socio-ecological consequences in Sabarmati river basin, India

M. Dinesh Kumar[(1)] and Om Prakash Singh[(2)]
International Water Management Institute, India Project Office, Anand, Gujarat, India
[(1)] *d.kumar@cgiar.org*
[(2)] *o.singh@cgiar.org*

Katar Singh
INREM Foundation, Anand, Gujarat, India
inrem@earth.planet.net.in

ABSTRACT

Sabarmati river basin is a trans-boundary river basin in Gujarat, India. Nearly 63% of the water requirements in the basin are met by groundwater; also nearly 80% of the irrigation demand. Uncontrolled abstraction of groundwater in the basin is causing several adverse consequences. This paper discusses the complex social, economic, hydrologic, hydrodynamic and ethical considerations involved in assessing the degree of *groundwater over-development*. It assesses the nature, extent and degree of groundwater problems in Sabarmati Basin using some of the most commonly used definitions of *groundwater over-development*, and analyzes their socio-ecological consequences.

The symptoms of an impending groundwater crisis in the basin include declining groundwater levels; poor economics of groundwater irrigation; unethical water use practices; drinking water scarcity; and ecological problems such as land degradation. The socio-ecological consequences are inequity in access to groundwater for irrigation; increased dependence of farmers on irrigation water; and threats to human biological and physiological health due to high levels of fluorides. The social, ecological and environmental costs associated with groundwater depletion in the basin are huge.

Keywords: *Groundwater depletion, Sabarmati river basin, groundwater overexploitation, water market.*

1 INTRODUCTION

The Sabarmati Basin is a trans-boundary river basin in Gujarat, India. With a total drainage area of 21,674 km^2, it accommodates a human population of 10 million persons and animal population of 2.84 million. The basin has a total of 282.2 km^3 of static groundwater (GOI, 1999), and 2.56 km^3 of dynamic groundwater (GOG, 1994). Nearly 63% of the water required to meet the various needs in the basin come from groundwater. Groundwater accounts for nearly 75% of the gross irrigated area, and is the main source for rural and urban domestic requirements (Kumar and Singh, 2001).

Uncontrolled exploitation of groundwater has led to several problems. So far, debates on groundwater depletion have focused on the manifestations of the problems rather than the nature, degree, extent and the root causes thereof. Very little clarity exists about the ways to define the problem. Several misconceptions exist about the causes of the problems among the users, mainly farmers. There is also very little understanding of the socio-ecological consequences of groundwater problems, while a lot of empirical evidence is available on the economic impacts such as rising cost of extraction of water, and emergence of markets.

There are several complex considerations involved in assessing groundwater overexploitation. The purpose of this paper is to demystify these complex considerations; discuss different definitions of groundwater overexploitation, which integrate these concerns; provide a realistic assessment of the nature, extent and degree of groundwater problems in Sabarmati basin using these concerns and definitions; and assess the economic, social and ecological consequences of the problems.

2 ASSESSMENT OF GROUNDWATER OVEREXPLOITATION IN SABARMATI BASIN

The concept of groundwater overexploitation predominantly deals with negative aspects of groundwater development (IGME, 1991; Custodio, 1992; Delgado, 1992; Margat, 1993; Custodio, 2000). Such consequences may include: large and continuous drops in groundwater levels over long time periods; large seasonal drops in water levels in wells and the drying up of wells in summer season; increase in groundwater salinity; land subsidence; enormous increase in cost of abstraction; and reduction of groundwater dependent vegetation and springs and seepage.

Custodio (2000), however, argues that though undesirable consequences appear when abstraction exceeds recharge, often there is no clear proof of the same being the cause of these undesirable consequences. Thus, the concept of *groundwater over-development* or aquifer overexploitation does not appear to be simple, merely linked to recharge and extraction balance, but is rather complex linked to various *undesirable consequences* which are physical, social, economic, ecological, environmental, and ethical in nature. Therefore, an assessment of groundwater over-development involves complex considerations/concerns such as fundamental rights, basic survival needs, health, and economic, ecological and ethical issues and hence it is not possible to capture its essence with simple definitions.

2.1 *Definition and assessment of groundwater overexploitation*

Several researchers have tried to define groundwater overexploitation and evolve criteria for assessing degrees of over-development by integrating some of the concerns discussed early. The 1986 Regulations of the Public Water Domain of the Spanish Water Act (1985), define overexploitation by its effects: an aquifer is considered over-exploited or in the risk of overexploitation, when the sustainability of existing uses is threatened as a consequence of abstraction being greater than or close to, the annual mean volume of renewable resources, or when they may produce a serious water quality deterioration problem (Custodio, 2000). Young (1992) defined overexploitation from an economic point of view, de-linking pumping rates from mean recharge values, as the non-optimal exploitation. Llamas (1992) introduced the notion of strict overexploitation leaving room for definitions with broader scope as a groundwater abstraction producing effects whose final balance is negative for present and future generations, taking into account physical, chemical, economic, ecological and social aspects.

The concept of sustainability used in the context of natural resource development by the World Commission on Environment and Development headed by Brundtland, Gro Harlem (WCED, 1987) based on the principle of inter-generational equity is also used to define groundwater overexploitation (Custodio, 2000). However, Georgescu-Reogen (1971) and Custodio (2000) have argued that the concept is too broad and cannot be applied to local specific situations, as it does not take into account the impossibility of complete recycling of matter. Another point of contention of Custodio is that if one strictly follows the principle of sustainable development, as proposed by the Commission, the non-renewable resources like the large and deep confined aquifers of arid regions yield no benefit to anyone.

Finally, the way overexploitation is perceived depends on points of view of different stakeholders involved such as farmers, water development administrators, ecologists, conservationists, mass media, naturists, and citizens and professionals such as engineers, scientists, economists, management specialists, environmentalists, lawyers, sociologists and politicians (Custodio, 2000). For instance, one of the dominant perceptions of the farmers about falling water levels and drying up of wells etc., is that they happen due to frequent failure of monsoons and the long term sharp declines in annual rainfalls. In fact, *declining rainfalls* is a *hydro myth* existing among millions of farmers in the region, and droughts are perceived as a major cause of depletion. Nevertheless, the farmers do not seem to recognise increased groundwater draft as the major cause of over-development.

On the other hand, the official agencies claim with the support of their data of recharge and extraction that there are no reductions in the quantum of recharge over time. However, here we do not rule out the chances of *bias* in the estimates as they are often influenced by strong political interests. The direction of such a *bias* could change depending on the kind of vested interest (for details see Kumar *et al.*, 2001). Custodio (2000) has also talked about this *bias* and manipulation as an important factor influencing the perception of overexploitation. The official perceptions of over-development are driven by *aggregate views*, and miss out several hidden phenomena such as excessive drawdowns in water levels due to large well-fields, groundwater pollution, and excessive rise in water levels causing water-logging, which are often localised.

In sum, defining and assessing of groundwater over-development are both difficult and complex and not amenable to simple formulations (Custodio, 2000).

2.2 *Characteristics of aquifers in Sabarmati basin*

The Sabarmati river originates in the metamorphic rocks of the Ajabgarh series of the Delhi system consisting mainly of calcschist, calcgneiss and limestones, which have a regional northeast-southwest strike direction dipping towards the northwest as well as the southeast. The basin is underlain mainly by the rocks belonging to Precambrian age in its northern and eastern parts and recent alluvial deposits in the western and southern parts. The alluvial deposits underlie approximately two-thirds of the basin and are themselves underlain by rocks of Precambrian age. The maximum thickness of the sediments in the alluvial zone is 2,600 m. There are two distinct aquifer systems in Sabarmati basin area. First is a multi-aquifer system formed by alluvial sediments and the second an unconfined hard rock aquifer.

The alluvial aquifer system in the basin extends over parts of Kheda, Ahmedabad and Gandhinagar district, a part of Kheralu taluka[1] in Mehsana district, and parts of Himmatnagar, Idar, Prantij and Dehgam in Sabarkantha district. It is quite complex: it has a shallow unconfined aquifer as the upper layer, followed by alternate layers of clay and sand forming semi-confined layers and aquifers. The thickness of the aquifer increases in the northeast to southwest direction. The shallow alluvial aquifer has a vertical extent of over 30 m. According to a demarcation made for a United Nation Development Programme study carried out in 1976, apart from the upper unconfined aquifer, there are seven aquifers below; but are not completely confined. The alluvial aquifers have very high porosity and permeability and have high storage coefficient and transmissivity. The yield of the deep confined aquifers is as high as 90 m^3/h. The yield of the unconfined aquifer ranges from 36.32 m^3/h to 45.4 m^3/h in shallow water table areas.

Unconfined aquifers in the northeastern parts of the basin are comprised mainly of igneous and metamorphic hard rocks with low yields indicated by transmissivity values ranging from

[1] Taluka is the second largest administrative unit in Gujarat state after *District*.

20–100 m²/d. This is also the case with Deccan Traps (basalt) occurring in the eastern parts of the basin under Sabarkantha district, and with igneous and metamorphic rocks in the north-northwestern part of the basin under parts of Kheralu taluka of Mehsana district, and Danta taluka of Banaskantha district. Unconfined aquifers in the sedimentary sandstone formations (mainly occurring around Himmatnagar in Sabarkantha district) have medium yield characteristics with transmissivity values of around 250 m²/d. Open wells are still the most common groundwater extraction structures found in the hard rock areas. Though farmers drill tube wells to tap the water occurring in deep fractures, yield is very poor.

2.3 *Dimensions of groundwater over-development in Sabarmati basin*

2.3.1 *Hydrological aspects*

The official estimates of groundwater recharge and extraction available for the talukas falling in Sabarmati basin's drainage area available for the year 1992 and 1997 are analysed here to see the overall changes in groundwater balance during 1992–1997. In our analysis, we have considered the gross withdrawal figures provided in the report of a High Level Committee as the *groundwater draft* for comparison against the utilisable recharge, which includes the return flows from irrigation.

Based on the stage of groundwater development, the talukas in Gujarat state are classified into four different categories as "white", "grey", "dark" and "over-exploited". "White" talukas are those in which the average annual groundwater draft is less than 70% of the average annual utilizable recharge. "Grey" talukas are those in which the draft is between 70% and 90% of the utilizable recharge, "dark" talukas are those in which the draft is above 90% of the utilizable recharge, and "over-exploited" are those in which the annual draft is more than 100% of the utilizable recharge. Going by the 1991 official figures of utilisable recharge and gross draft, 15 out of the 28 talukas falling (fully and partly) are "over-exploited". Also, there were five "dark" and four "grey" talukas. But, according to the estimates based on their 1997 official figures, the number of "over-exploited" talukas and dark talukas are 11 and 3 respectively. But, the number of "grey" talukas had gone up to eight from five in 1991. At the aggregate level, the stage of groundwater development is 112% according to our estimates using official figures of 1991, while it is reduced to 102% according to our estimates using the 1997 official figures (Table 1).

The above estimates mean that the degree and extent of overexploitation has reduced during 1991–1997 in comparison to the previous 7 years from 1984–1991. This is caused by increase in recharge and reduction in abstraction owing to better annual precipitation during 1991–1997 as compared to 1984–1991, which experienced three consecutive years of drought during 1985–88. But, change in the level of groundwater development is not sufficient to reverse the negative water head evolutions caused by overdraft in those blocks that were earlier categorised as "over-exploited", the reason being that the cumulative deficit in groundwater recharge-extraction balance had actually increased. If we consider the official figures of groundwater recharge and abstraction as reliable, in those talukas, which have shifted from the "over-exploited" category to "dark" category, slight improvements in geo-hydrological conditions is expected to have occurred resulting from the

Table 1. Recharge, abstraction and stage of groundwater development in Sabarmati basin (1991 and 1997).

Utilisable GW recharge		Total Groundwater draft		Stage of development (%)	
1991	1997	1991	1997	1991	1997
1,939.8	2,046.5	2,175.2	2,087.3	112	102

Source: GOG (1992, 1999).

reduction in the overall deficit in groundwater recharge-abstraction balance.

2.3.2 *Hydrodynamics*

The geo-hydrological condition of the basin had undergone dramatic changes during the past forty years, though significant variations are found in the pace at which change have taken place across different parts of the basin (for details, see Kumar *et al.*, 2001). In the early sixties, abundant groundwater was available in the upper phreatic aquifers. During 1970–1985, farmers shifted gradually from open wells to dug-cum-bore wells and tube wells. While energised extraction started with diesel engines in the late 1960s, electric motors and submersible pumps were introduced by mid 1970s. The tube wells drilled during mid 1980s tapped the deep confined aquifers. Analysis of primary data collected for 10 wells from two villages, namely, Ode and Jetalpur, showed a rise in depth to water level from 18 m in 1970 to 1979 m in 2000. The average annual drop is two metres.

The analyses of data on piezometric levels in the deep confined aquifers recorded by automatic water level recorders of Central Ground Water Board (CGWB) available for the months of January, May, August and November showed declining trends in all the three stations. In the case of Bhavela, the level declined from 12.95 m to 16.02 m during May 1996-January 2000. The maximum drawdown was observed during November 1997 to January 2000, and was of the order of 5.52 m. In the case of Ingoli, the drawdown during the same period was 4.55 m. For shallow aquifer in Danta, the decline in water level was of the order of 4.19 m. Such heavy drawdowns are primarily due to the heavy withdrawals that have taken place during the three consecutive years of drought from 1998 to 2000.

Though the water level trends in the first confined aquifers during the years 1995–2000 are moderate, they are mostly from those areas which receive canal irrigation. According to CGWB (2000), "large declines even up to more than 70 m have taken place in the confined aquifers over the period of nearly one and a half decades. The hydrographs of several other piezometers also indicate declines in the range of 30 m to more than 50 m during the last two decades, especially in Ahmedabad, Gandhinagar and Mehsana districts". Again, in the first confined aquifer in the central parts of north Gujarat, especially Ahmedabad and Gandhinagar, the water levels have undergone average annual declines in the range of 4–6 m during 1996-2000 (CGWB, 2000).

The analyses of water level data (1996–2000) of five open wells in hard rock areas obtained from the CGWB show declining trends in water levels from year to year (for details, see Kumar *et al.*, 2001). In one of the observation wells (village: Gamdi), the depth to water level consistently increased from 3.39 m in May 1996 to 10.5 m in May 1999 with a total drop of 7.11 m in four years. Now, the groundwater balance for Vijayanagar, Bhiloda and Meghraj shows only 80%, 86% and 94% development respectively during 1992–1997. Contrary to this, May–May water levels in Vijaynagar, Samlaji (Bhiloda), and Meghraj show declining trends at the average rates of 2.82 m, 1.33 m and 3.31 m respectively during 1995–1997. The seasonal fluctuation in water levels was much higher than annual fluctuation. For instance, in the case of Vijayanagar well, the water level dropped by 7.71 m from August 1995 to May 1996 owing to pumping; but recovered by 7.27 m during May 1996-August 1996 owning to the recharge from rainfall. The water level dropped by 6.2 m during August 1996-May 1997; then rose by 6.8 m during the monsoon. On the other hand, the average annual water level drop over a four-year period was found to be only 1.27 m.

The water level trends in the open wells of alluvial areas are different from those in the hard rock areas. In Ghatlodia village of Ahmedabad, the total drop in water level (from 1995–2000) was 4.25 m in six years, though in this area, groundwater withdrawal and recharge are precariously balanced (abstraction 123% of the recharge). This is because almost all the abstraction is from the lower aquifers, while the upper unconfined aquifer receives the direct recharge from top. Then the level drop could be attributed to the interaction between the upper aquifer and the tapped underlying aquifer wherein the water from upper aquifer could be contributing to the lower one. Unlike in the case of hard rock areas, the seasonal drops are not very significant. A maximum rise in water level of the order of 3.74 m was observed during 1999 when there were floods in Ahmedabad.

2.3.3 *Economics of groundwater overexploitation*

Economics of groundwater development helps in decision-making with regard to investment priorities in water and in the selection of competitive alternatives (Custodio and Gurgui, 1989; Howitt, 1993). In the areas where large-scale exploitation of groundwater had led to undesirable consequences, such analysis provides a strong indication of whether groundwater is depleted to an extent that it is no more economically accessible. In the economic analysis, the variable costs should not only include the actual water cost, but also the indirect costs such as the social and ecological costs of using fossil fuel and water, where ever applicable (Custodio, 2000). In the real sense, farmers in Gujarat are not paying the actual cost of generation of electricity used for groundwater pumping, i.e. the use of electricity for pumping water for irrigation is subsidised (IRMA/UNICEF, 2001). Therefore, if we consider the price farmers are paying for electricity used for groundwater pumping, the benefits are likely to be inflated.

The economics of groundwater irrigation depends on the fixed cost of the irrigation system and the life of the system, the variable cost of water, and the returns from irrigated farming. The values of these parameters had increased remarkably over a period of time, owning to the drastic changes in the hydrological setting groundwater, resulting from continued exploitation. We will now examine what has been the net effect of these changes on the economics of well irrigation.

First: The fixed cost of the irrigation system has increased dramatically over the last 25 years from US$ 2,173.91 in 1975 (1 US$ = 46 Indian Rupees) to nearly 8,695.65 US$ today (2000), which is mainly attributed to the increase in the depth of tube wells.

Second: the variable cost of water includes the direct cost of energy used for water extraction and the indirect cost (social and environmental) of mining water. The direct cost has increased over the 25-year period from 0.48 US cent per cubic metre in 1975 to 0.72 US cent per cubic metre in 1985 to 0.89 US cent per cubic metre in 1990 to 1.28 US cent per cubic metre in 2000. The social and environmental costs of water abstraction increase as one goes deeper due to the increased use of fossil fuel and fossil water. One of the social imperatives is that with increased mining of water, the drought proneness of the region increases. Secondly, the farmers have to spend more time in irrigating their crops as water table declines due to several factors that are linked to depletion: increase in frequency of watering, reduction in well yield etc. While it used to take an average of only 22 hours in 1970 to irrigate one hectare of paddy, it took 110 hours in 2000 for a paddy field of same size. In the case of fodder jowar, the total hours of irrigation went up from 66 per ha in 1970 to 122 per ha in 2000. Farmers in the region practice flood irrigation for paddy and therefore have to irrigate excessively to create pounding of water. The opportunity costs, both social and economic, associated with spending extra hours in the field for irrigation need to be taken into consideration while evaluating the costs.

Third: the economic returns from irrigated farming (exclusive of irrigation cost) reduced over time due to: [a] reduction in area under irrigation, which is the result of both reduction in well yields and the increase in irrigation water rate for different crops; and [b] increase in cost of inputs. The mean yield of the tube wells has gone down from 157 m^3/h to 78 m^3/h, though the average pump horsepower had increased from 23.3 HP to 31 HP. The average area under irrigation for the tube wells had come down from 6.55 ha to 4.32 ha.

Taking these variables into consideration, we carried out a comprehensive analysis of economics of well irrigation. The economic value of the environmental cost of water (per cubic metre) is tentatively taken as 0.76 US cents. The analysis shows that the net economic return from tube well irrigation would be negative (B/C ratio is 0.80), even if one does not includes the economic value of the environmental costs associated with exploiting the non-renewable groundwater. The net economic loss will be more (B/C ratio is 1.72), when we include the environmental cost of the pumped water. This means, as time passes, tube well irrigation will become totally unviable in the deep alluvial areas inside the Sabarmati basin.

2.3.4 *Social aspects*

The social impact of groundwater exploitation would include: the changes in groundwater availability for drinking and domestic purposes in terms of quantity and quality is due to over-exploitation; changes in livelihood patterns of the farming community; and patterns of migration caused by water scarcity.

Groundwater over-draft has severely affected the sustainability of drinking water sources in a large part of Sabarmati basin area due to several factors. *First*: in the hard rock areas, the overall availability of groundwater is limited. There is severe competition for water for irrigation and drinking purposes. *Second:* the unique nature of the drinking water demand places it in a less advantageous position as compared to irrigation demand. While demand for drinking water is more or less uniform across the year, the irrigation demand is highest during winter. A major share of the available water in the shallow aquifers gets drawn up through the hundreds of thousands of farm wells. Due to large seasonal drops in water levels (in the order of 6–7 m), by the end of the winter, the open wells and hand pumps get dried up. While the rich farmers manage to drill vertical bores to provide protective irrigation to their summer crops, the poor farmers have to leave their lands fallow. The poorer sections of the village communities face drinking water shortages.

The shortage of water becomes acute during droughts as water levels drop further in the wells during the monsoon season itself. In such situations of absolute water scarcity as encountered during the past two years (1999–2000), several hundreds of villages in Sabarkantha district which fall within the basin area were supplied drinking water by tankers. The poor farmers in this region have to migrate to other perennially irrigated areas in search of employment. In the alluvial areas, excessive withdrawal of groundwater from the deep aquifers had resulted in contamination of freshwater aquifers with fluorides and salts. This also had severe adverse impact on the availability of good quality groundwater for drinking.

2.3.5 *Ethics of water use*

Groundwater use involves certain ethical considerations (Custodio, 2000). The phenomenon of groundwater overexploitation and mining, therefore, needs to be assessed from the point of view of water use ethics. If water from a regional aquifer in an arid or semi-arid region is extracted using unethical practices, then the aquifer is most likely to be mined. The question is: how does one determine whether ethical practices are followed in water extraction and use or not? How does one differentiate ethical from unethical practices? Answers to such questions are beyond the scope of natural sciences, engineering and economics.

The fact is that the groundwater crisis in the form of overall scarcity of water, and inequitable access emerges as a result of lack of solidarity among farmers and other water users, unsustainable use and pollution. Therefore, any use of water, which jeopardises this solidarity among water users who are sharing a *common good*, threatens the sustainability of water use, or causes pollution of freshwater resources can be called *unethical.* The ethical considerations concerning water use mainly revolve around the distribution of benefits and costs of water use and risks associated with it (Llamas and Priscoli, 2000).

Following are our observations in the context of Sabarmati basin.

First: farmers are engaged in wasteful irrigation practices. The water used by farmers for irrigation (5,832.6 Mm^3) is much higher than the actual irrigation requirement (4,350.7 Mm^3) (Kumar and Singh, 2001). Owing to lack of adoption of appropriate soil management and on-farm water management practices, farmers apply water more frequently and in larger depth to take care of the excessive percolation, runoff and evaporation losses from the fields.

Second: farmers are engaged in a competition to abstract water from the aquifers. Owing to limited hours of power supply, they start their pumps as soon as the supply is restored. Simultaneous pumping by several farmers leads to excessive drawdowns within a short span of time at the regional/local level in the hard rock areas. This leads to temporary drying of shallow wells, mainly affecting the poor farmers who then have to wait for their wells to recuperate.

Third: a large number of industries, which generate hazardous effluents and which do not want to invest in treatment systems, are resorting to pumping their effluents into the underground formations. This leads to contamination of the aquifers. This practice has been going on for quite some time in the industrial area of Odhav in Ahmedabad.

Finally, the official perceptions of groundwater overexploitation are *absolute* and not *relative*. According to the UNESCO Working Group on Ethics in Freshwater Use, as quoted in Llamas and Priscoli (2000), to view water crisis in absolute terms is ethically insufficient and one must also view it in relative terms as the two different perceptions lead us to different ethical norms. The same concept can be applied to groundwater overexploitation also. Apart from describing the *absolute consequences* of overexploitation problems such as drops in water levels and quality problems, the official agencies must also describe the *distributional/relative consequences*. Some of the distributional consequences are denial of access to groundwater by the poor, and inequitable access to water resulting from disproportionate distribution of the benefits of electricity subsidy.

2.3.6 *Ecological consequences*

2.3.6.1 *Breaking of soil structure, loss of soil retention capacity and loss of organic matter and nutrients*

One of the consequences of groundwater level drops is the introduction of tube well technology. Though tube well irrigation was costly, the technology turned out to be a blessing in disguise for farmers in the initial years as it provided assured higher well yields. The yields from deep confined aquifers were not affected by abstraction and seasonal changes in natural recharge. The outcome was the expansion in area under irrigated crops and intensification of land use. Excessive irrigation resulted in leaching of minerals and organic matter into the soil. As gross irrigated area increased, the availability of organic manure per unit area of cultivated area got reduced substantially. Chemical fertilisers had to be used in greater quantities. This, in a way, substituted for the organic, bio-fertilisers. The chemical fertilisers enhanced the secondary productivity of the soils, which the primary productivity of soils declined. Increased fertiliser use also became necessary for modern farming using green revolution, hybrid varieties.

Analyses of historical changes in the intensity of fertiliser application for three major crops in Daskroi taluka of Ahmedabad show rising trend in the use of fertilisers (urea and di-ammonium phosphate) per ha of land. The highest increase in the rate of fertiliser application was observed in the case of paddy. It went up from a mere 137 kg/ha in 1970 to 404 kg/ha in 2000. The intensive use of fertiliser was accompanied by increased rate of application of irrigation water. This resulted in breaking of soil structure. In sum, three major changes in land use occurred and had serious implications for land productivity; increase in cropping and irrigation intensity; increased rate of water application for each of the irrigated crops; and increased rate of application of fertilisers.

2.3.6.2 *Increase in soil salinity resulting from irrigation with high TDS water*

Irrigation with saline groundwater is leading to soil degradation in the groundwater irrigated areas in the basin. A study carried out by Singh and others (2000) cites groundwater overexploitation as one of the major causes of inland salinity in Gujarat, like increased use of fertilisers, and lack of soil nutrient management practices. The water from deep aquifers generally has high levels of total dissolved solids (TDS). The analysis of pre-monsoon water samples collected from selected piezometer stations shows high TDS in alluvial areas. For instance, in the case of Baola in Dholka, the TDS of water samples during May 1995, 1997 and May 1999 were 4,080 ppm, 3,300 ppm and 3,156 ppm respectively. In Kheda, TDS values of 2,580 ppm and 2,460 ppm were observed. All these values are far above the permissible TDS level of 1,500 ppm for use in irrigation.

Farmers are continuing irrigation with the saline groundwater in these areas. This is leading to increase in soil salinity causing hardening of soil surface and lump formation. In order to break the soil lumps to enable better growth of crops, the farmers had to increase the water application rates. Thus, over a period of time, more salts get accumulated on the soil surface and thus the soils become saline. Excessive irrigation to leach the salts causes faster loss of organic matter and nutrients present in the soils. All these ultimately result in soil degradation. According to a study by Dubey *et al.* (1999), six talukas falling in Sabarmati basin account for 44,286 ha of salt-affected land in the State. This includes areas affected by both alkalinity (12,505 ha) and salinity (31,781 ha) in the districts of Mehsana, Ahmedabad and Kheda. Of the total salinity-affected area, 9,300 ha falls in the two talukas of Kheda that are also affected by the twin problem of water logging and salinity, namely, Matar and Khambhat.

ORG (1994) estimated a figure of 10,757 ha as the total areas affected by salinity due to water logging in these two talukas (as cited in ORG, 2000), slightly more than the figure (9,300 ha) provided by Dubey *et al.* (1999). Therefore, we assume that the entire problem of soil salinity in the two talukas of Kheda district, namely, Matar and Khambhat, is due to water logging alone. This leaves us with a figure of 34,986 ha as the land affected by salts in the talukas falling in Sabarmati basin area. The salt-affected soils are degraded soils and suffer from low primary productivity. The salt-affected soils are concentrated in the alluvial areas having high TDS groundwater. This further strengthens the argument that groundwater degradation is the major cause of salts in soils.

3 SOCIO-ECOLOGICAL CONSEQUENCES OF OVEREXPLOITATION

3.1 *Declining water levels and emergence of water markets*

One of the most striking impacts of depletion in Sabarmati basin has been on the ability of farmers to access groundwater. In the 1960s and 1970s, shallow open wells served as the cheapest and the simplest source of instantaneous irrigation. As water table started declining with proliferation and energisation of wells, the farmers responded by deepening their wells. This did not create much financial burden on the farmers, as they used family labour for deepening wells.

But, with the drying up of shallow aquifer, the resources and technology available with the poor farmers to extract groundwater became inadequate. The farmers had to dig up to 107 m to get adequate water from the next lower layer. Open well construction was not feasible for such a large depth and tube wells had to be drilled. Tube well construction required expensive drilling. Submersible pumps were needed to lift water from large depths. While the progressive, wealthy farmers aggressively adopted tube well, the poor farmers, who could not mobilise the necessary funds, were deprived of direct access to groundwater. They had to depend on the well owners, who found this as the most convenient way of recovering the large investments made in tube wells.

Water markets form a significant portion of the groundwater irrigation in the alluvial areas of Sabarmati basin. In Daskroi taluka, the 10 well owners surveyed, together sell water for a total area of 29.3 ha, after irrigating 75 ha of own land. This was the scenario in 2000, after two consecutive years of drought, when the sellers' own demand was very high and all of them had made sharp cuts in the sale of water. In Kalol taluka, out of the 20 tube well owners surveyed, eight were found to be selling water to their neighbouring farmers. The total area irrigated with purchased water was 49.15 ha, against a gross area of 132.5 ha irrigated.

The well owners charge for water on hourly basis. Analyses of primary data collected from 30 farmers in Daskroi taluka of Ahmedabad district showed that the price at which water is sold ranges from US cent 0.65–1.28/m^3. The implicit unit cost of water (found to be ranging from US cent 0.39–1.78/m^3), which is primarily decided by the hours of operation, does not have any bearing on the price at which water is sold. For instance, a farmer who incurred the lowest implicit cost of US cent 0.28/m^3 for pumping water charged US cent 0.83/m^3. At the same time, a farmer who incurs US cent 1.78/m^3 for pumping water charges only US cent 1.00/m^3.

Thus, when groundwater resources deplete and the cost of well construction and pumping increases, the system of trading water provides greater economic opportunities to well owners having large holdings, and lesser opportunities to well owners having smaller holdings and water buyers. This is due to the fact that for a large farmer, the implicit unit cost of water is much lower as compared to small farmers. At the same time, a small farmer will not be able to raise the water charges to match with the implicit cost of pumping.

3.2 *Increased dependence on irrigation and fertilizers*

In Sabarmati basin, soil degradation has caused in increased irrigation requirement for several crops. Similarly, the fertiliser input requirement has also increased due to loss of organic matter content in the soils. In order to compensate for the losses in crop production due to declining primary productivity of soils, farmers are forced to apply more water and fertilisers to their crops. Due to increase in soil salinity, the farmers have to apply extra water to flush out the salts in the soils. Often, extra water is needed to break the soil pan formed as a result of salinity. The deterioration of water retention capacity of the soil leads to higher percolation losses from irrigation, which further calls for increasing the frequency of irrigation.

Analyses of historical changes in the irrigation water application based on primary data collected from the field on the number of waterings from a sample of 25 farmers in Daskroi taluka shows that the number of waterings applied to all the four major crops grown in the region has increased consistently over the period, 1970–2000. For instance, the average number of waterings for paddy (kharif) had gone up steadily from 2.5 in 1970 to nearly 3.5 in 1985 to 5.5 in 2000. Similar differences were found in the case of wheat and summer jowar (fodder) also. The number of waterings for jowar had gone up steadily from 3.3 in 1970 to 4.75 in 1985 to 6.2 in 2000. For alfalfa, the number of watering significantly increased from 9.5 in 1970 to 13.4 in 1985 to 14 in 2000. A similar trend was found in Kalol taluka of Gandhinagar (for details, see Kumar *et al.*, 2001).

3.3 *Health impacts of groundwater quality deterioration*

The rising level of fluoride content in pumped groundwater is one of the most alarming trends observed in Sabarmati basin. This is posing a serious threat to drinking water supplies and human and animal health in the State. According to GOG (1994), the basin has 532 villages affected by fluorides. Incidences of high fluoride are most common in the deep confined aquifers of the alluvial parts. Over the years, the degree and extent of fluoride contamination have been increasing and so also the number of villages affected. There are enough empirical evidences to suggest that in north Gujarat, excessive fluoride in pumped groundwater is directly associated with aquifer overexploitation and mining.

Due to absence of any surface water based schemes, the rural people in the region are fully dependent on decentralised rural water supply schemes based on groundwater for drinking purpose. While the schemes supply fluoride free groundwater in the initial stage of their operation, over a period of time, fluorides start building up. The Gujarat Water Supply and Sewerage Board (GWSSB) does not have any system for regular monitoring and analysis of the chemical quality of the water being supplied for drinking in the rural areas. The decentralised schemes also lack water treatment facilities. In addition to the decentralised public schemes, the rural communities also use water from their own private irrigation wells for drinking and domestic purposes also. The problem of fluoride in drinking water is not limited to rural areas alone. Widespread occurrence of fluoride in groundwater has been reported from many pumping stations of Ahmedabad Municipal Corporation (AMC). According to newspaper reports, water from AMC tube wells has fluoride levels as high as 6 ppm against the permissible level of 1.5 ppm.

Presence of fluorides in water cannot be detected without the help of water quality testing equipment. The presence of high fluoride levels in groundwater is often detected from its manifestations in such symptoms as yellowing of teeth, and joint pain, which occur from long years of exposure to contaminated water. Due to these reasons, the communities in the region have been exposed to drinking water containing high levels of fluoride for a long period of time without being much aware of it. By the time the communities realised the *menace* and its long-term effects on health (crippling effect and hunchback), a large section of the population had already been affected.

The biological and toxicological effects associated with the use of water containing fluoride for drinking and cooking are yet to be fully explored. But if the available scientific knowledge is any indication, the potential effects are dangerous. Studies on fluorotic human populations of north Gujarat carried by Sheth *et al.* (1994) revealed an increase in frequency of sister chromatic exchange in fluorotic individuals as compared to the control indicating that fluoride might have genotoxic effect. Fluoride had been reported to cause depressions in DNA and RNA synthesis in cultured cells (Strochkova *et al.*, 1984). Several human conditions including ageing, cancer, and arteriosclerosis have been associated with DNA damage and its disrepair.

4 CONCLUSIONS

The study revealed several symptoms of an impending groundwater crisis including: declining groundwater levels; poor economics of groundwater irrigation; unethical water use practices; and land degradation due to breaking of soil structure, loss of water retention capacity, organic matter and nutrients from soils, and increase in soil salinity. Overexploitation of groundwater was found to have socio-ecological consequences such as emergence of water markets, with access inequity impacts; increased dependence of farmers on irrigation water; and increasing levels of fluoride in groundwater, which pose serious threats to human biological and physiological health. The social, economic, ecological and environmental costs associated with groundwater degradation are huge. As a result, the investment decisions for groundwater management projects should be based on all economic, environmental, ecological, and social considerations.

REFERENCES

CGWB (Central Ground Water Board) (2000). *Declining Groundwater Levels in the Deep Aquifers of Gujarat.* Central Ground Water Board, West Central Region, Ministry of Water Resources, Ahmedabad.

Custodio, E. (1992). Hydrological and Hydrochemical Aspects of Aquifer Overexploitation. Selected Papers on Aquifer Overexploitation. Intern. Assoc. of Hydrogeologists, Heise, Hannover, 3: 3–28.

Custodio, E. (2000). *The Complex Concept of Overexploited Aquifer.* Uso Intensivo de las Aguas Subterráneas, Madrid.

Custodio, E. and A. Gurguí (1989). *Groundwater Economics.* Elsevier. Amsterdam.

Delgado, S. (1992). *Sobreexplotación de acuíferos: una aproximación conceptual.* Hidrogeología y Recursos Hidráulicos. Asoc. Española de Hidrología Subterránea. 25: 469–476.

Dubey, D.D.; Sharma, O.P.; Sethi, M. and Gupta, R.K. (1999). *Salt Affected Soils of Gujarat: Extent, Nature and Management*, CSSRI, Karnal.

Georgescu-Reogen, N. (1971). *The Entropy Law and the Economic Process.* Harvard University Press, Cambridge.

GOG (Government of Gujarat) (1992). *Report of the Committee on Estimation of Groundwater Resources and Irrigation Potential in Gujarat State (1991).* Narmada and Water Resources Department, Gandhinagar.

GOG (Government of Gujarat) (1994). *Integrated Plan of Sabarmati River Basin.* Draft Report, Gujarat State Water Resources Planning Group, Gandhinagar, Gujarat, India.

GOG (Government of Gujarat) (1999). *Report of the Committee on Estimation of Groundwater Resources and Irrigation Potential in Gujarat State (1997).* Narmada and Water Resources Department, Gandhinagar.

GOI (Government of India) (1999). *Integrated Water Resources Development: A Plan of Action*, Report of the National Commission on Integrated Water Resources Development, Volume I, Ministry of Water Resources, New Delhi: Government of India.

Howitt, R.E. (1993). *Resolving Conflicting Water Demands: A Market Approach.* The Water Economy. Soc. General de Aguas de Barcelona. Barcelona: 151–164.

IGME (1991). *Sobreexplotación de acuíferos: Análisis Conceptual.* Instituto Tecnológico Geominero de España, Madrid.

IRMA/UNICEF (2001). White Paper on Water for Gujarat, Institute of Rural Management, Anand.

Kumar, M.D. and Singh, O.P. (2001). Rationed Supplies and Irrational Use: Analyzing the Water Accounts of Sabarmati River Basin. Monograph 1, INREM Foundation, Anand.

Kumar, M.D.; Singh, O.P. and Singh, K. (2001). Groundwater Depletion and its Socioeconomic and Ecological Consequences in Sabarmati River Basin. Monograph 2, INREM Foundation, Anand.

Llamas, M.R. (1992). La surexploitatión des aquiféres: Aspects Techniques et Institutionnels. Hydrogéologie, Orleans. 4: 139–144.

Llamas, M.R. and Priscoli, J.D. (2000). *Report of the UNESCO Working Group on Ethics of Freshwater Use, Water and Ethics, Special Issue*. Papeles del Proyecto Aguas Subterraneas, Madrid.

Margat, J. (1993). *The Overexploitation of Aquifers*. Selected Papers on Aquifer Overexploitation. Intern. Assoc. of Hydro-geologists. Heise. Hannover. 3: 29–40.

ORG (Operations Research Group) (2000). *State Environmental Action Plan, Draft Report on Hydrologic Regimes Sub-Component*, Submitted to Gujarat Ecology Commission, Government of Gujarat, Vadodara.

Sheth, F.J.; Multhani, A.S. and Chinoy, N.J. (1994). Sister Chromatic Exchange: A Study of Fluorotic Individuals of north Gujarat. *Fluorosis*. 27 (4) 215–219.

Singh, Y.D.; Jeevan, C.P. and Dixit, A.M. (2000). *State Environmental Action Plan Phase-1 Report on Ranns and Desertification*, submitted to Gujarat Ecology Commission, Government of Gujarat, Vadodara.

Strochkova *et al.* (1984). Effects of Fluoride on Morphological Modifications in Hela Cell Culture. *Tsitologiya*. 26: 299–206.

World Commission on Environment and Development (1987). *Our Common Future*. Oxford University Press.

Young, R.A. (1992). *Managing Aquifer Overexploitation: Economics and Policies*. Selected papers on Aquifer Overexploitation, International Association of Hydrogeologists, Heise, Hannover.3: 199–222.

Intensive use of groundwater in North Gujarat, India. Challenges and opportunities

A.K. Sinha
Department of Geology, University of Rajasthan, Jaipur, India
sinhaaa@sancharnet.in

Ratanchand Jain
Central Groundwater Board, West Central Region, Ahmedabad, Gujarat, India

ABSTRACT

Gujarat is the western most state of India with dry climate having erratic and unreliable rainfall. In absence of any perennial source groundwater is the most dependable source of water in this part of the country. However this region has witnessed intensive use of groundwater in recent past mainly owing to an enhanced development process coupled with exponential growth in population. The frequent drought has also abated the groundwater depletion process endangering the sustainability of the life process. This paper discusses the causes and consequences of the intensive use of groundwater and identifies the measures towards addressing the challenges posed by it.

Keywords: *groundwater, aquifer overexploitation, declining water table, sustainability.*

1 INTRODUCTION

Geographically, the North Gujarat region includes the area covered by Mahesana, Banaskantha and Sabarkantha districts of the Gujarat State of India (Figure 1). This region, though, has multi-aquifer system, with prolific yields but the excessive groundwater pumping over last many years has resulted in de-watering of a substantial part of aquifer and reduction in pressure heads. The overexploitation of groundwater has resulted in conspicuous decline in water levels seriously affecting agriculture production in this part of the state. The region is

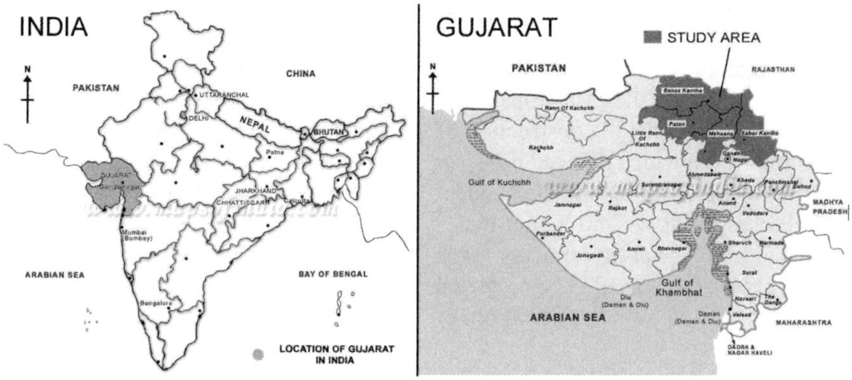

Figure 1. Location of North Gujarat region, India.

highly vulnerable to such type of adverse effects due to low reliability of rainfall and frequent recurrence of droughts.

The detrimental effects of over exploitation of groundwater i.e., decline in water levels have not only added to the capital cost of construction of the tube wells and the energy bill, but have also added to annual cost of maintenance of tube wells. Conventional dug wells and dug-cum-bored wells have already, gone out of operation as in many areas the dug wells have dried up due to lowering of groundwater. Further, during drought years in the absence of adequate replenishment and sustained groundwater pumping, the pumped groundwater far exceeds that of normal years, resulting in enhanced de-watering of aquifers and further lowering of water table.

No land subsidence is reported in this area, but the same cannot be completely ruled out under the conditions of continuous decline in water levels. If suitable artificial recharge measures are not taken up timely, land subsidence would be highly hazardous as the region is densely populated.

Owing to continuous development during the last three decades for meeting the increasing demands for irrigation, drinking water supplies and industrial needs, the dependence on groundwater has tremendously increased. During the first and second five-year plan periods (1950–1960) the tube wells were drilled between 60 m and 100 m and their water levels ranged between 10 m and 15 m bellow ground level (bgl).

These tube wells have also gone out of use. The tube wells being drilled now are between 250 m and 300 m depth with the water levels ranging between 80 m to more than 120 m bgl (Jain *et al.,* 2000).

In order to maximize the economic gain, farmers are largely growing the cash crops in the area and are taking high risks in construction of the deep tube wells even with a reduced life span due to lowering in water levels and consequent reduction in discharge, higher and higher maintenance cost. Large financial investments made in construction of tube wells need to be protected from implied hazards of over exploitation of groundwater.

The adoption of high yielding crops for boosting the agricultural production and the impetus in rural electrification has resulted in substantial increase in groundwater pumping. Earlier, an area of 0.4 ha to 0.6 ha could be irrigated through conventional open well with discharge ranging 4–6 m^3/h (NWRD, 1986). Now with energisation of open wells and construction of deep tube wells the average groundwater abstraction rates have gone as high as 90 m^3/h to 112 m^3/h, with irrigation areas ranging from 40 ha to 60 ha (NWRD, 1999).

2 GEOLOGICAL AND STRUCTURAL FRAMEWORK

The North Gujarat region is formed by two main structural elements: the Aravalli hills, built up by Precambrian basement rocks in the northeast, and the Cambay sedimentary basin in the rest of the area (Merh, 1995). This basin was formed as a tectonic graben at the end of the Mesozoic era, through development of tensional faults accompanied by large scale volcanic activity. The main fault system trends NNW-SSE, approximately parallel to the main trend of Aravalli hills. The basin was subsequently filled up with Tertiary and Quaternary sediments. It is assumed that subsidence ceased in late Miocene times. Major part of north Gujarat is underlain by thick Quaternary and Tertiary sediments deposited in the Cambay basin. These sediments rest in the central part of the area upon Deccan Traps and on Himatnagar Sandstone in the eastern and northern part.

3 HYDROGEOLOGICAL SCENARIO

Pre-Cambrian hard rocks, semi-consolidated Mesozoic's, Tertiary formations, and unconsolidated alluvial deposits, form aquifers in the area. Groundwater occurs both under phreatic and confined conditions, however, its development is restricted depending upon the aquifer geometry and yield characteristic of individual aquifer and/or the salinity of formation water. Thick alluvial deposits (Post-Miocene), which occupy more than 75% of the area, form most prolific multiaquifer system (UNDP, 1976). The hard rocks exposed in the northeast form aquifer with low yield prospects. Based on the studies carried out so far, various rock formations have been grouped under two broad hydrogeological units namely fissured formation and porous formation.

3.1 *Groundwater in fissured formations (hard rocks)*

The meta-sediments of Achaean age belonging to the Delhi supergroup and granites and basic intrusive of Post-Delhi period constitute the hard rock aquifers in the area are grouped under fissured formations from the hydrogeological point of view. These aquifers are generally limited to shallow depths not exceeding 100 m.

Primarily the thickness and extent of weathered zones and size and interconnection of fissures and joints, which provide secondary porosity, govern occurrence and movement of groundwater in hard rocks. The high hills, in general, act as run off zone because of steep gradients and impervious nature of formations. Only in low lying terrain and intermountain valleys groundwater occurs in shallow weathered and fractured zones under water table to semi-confined conditions. These formations, in general, do not form good repository of groundwater. The depth of wells ranges from 8 m to 18.5 m bgl and depth to water level in open wells varies from 5 m to 14 m bgl. The water table is shallow near streams and in topographically low areas. Yield of wells ranges from 30 m^3/d to 120 m^3/d with an average of 75 m^3/d. Open wells generally sustain intermittent pumping during summer season.

In general the quality of groundwater is fresh and suitable for domestic and irrigation use with EC less than 2,000 μs/cm at 25°C. However, in select areas of Sabarkantha district, fluoride contents above 1.5 ppm are observed.

3.2 *Groundwater in porous formations (sedimentaries)*

A major part of the area is underlain by post-Miocene alluvium and older sedimentary formations. Thickness of alluvium gradually increases from piedmont zone in the northeast towards the west and southwest. Maximum thickness of alluvium in the area is estimated to be about 550–600 m in the central part. Groundwater occurs both under phreatic and confined conditions in arenaceous horizons within sedimentaries. The occurrence and movement of groundwater is mainly controlled by inter-granular pore spaces. Groundwater is extensively developed by dug, dug-cum-bored in the shallow aquifer while tube wells form the main groundwater withdrawal structures in the deeper aquifer in the alluvium.

Two major aquifer units have been identified within the explored depth of 600 m in the area underlain by sedimentaries. The upper unit is mostly phreatic but becomes semi confined to confined in some parts. It is designated as aquifer *A* and consists of relatively coarse-grained sediments.

The lower unit, comprising a few hundred meters of alternating sandy and argillaceous bed, forms confined aquifer system. It is subdivided into aquifer *B, C, D and E* within post-Miocene sediments and aquifer *F and G* in the Miocene sediments. The post Miocene aquifers are mainly coarse to fine sands, with occasional gravel beds (B) whereas the Miocene aquifers are mainly fine to medium grained sand and sandstone, interbedded with clay, claystone and siltstone.

The Himatnagar Sandstone (Cretaceous) is a medium to fine grained poorly cemented sandstone with shale and conglomerate beds and constitute a local aquifer in the north-eastern part. It has been designated as aquifer *H*.

Towards the north and northeast, the nature of sediments is more uniform and one single, phreatic aquifer exists. This is considered the main recharge zone for the confined aquifer system. Towards the centre of the area, the argillaceous intercalations act as confining layers of varying efficiency and are leaky in nature at places. In such areas the upper phreatic aquifer (A) and six deep confined aquifers (B to H) have been identified. General properties of various aquifers are given in Table 1 and their disposition is shown in Figure 2.

Aquifer characteristics are generally less favourable in the deeper aquifers as compared to the shallower ones. The shallow aquifer (A) shows the most favourable hydraulic parameters and contains the best quality of groundwater in the vicinity of the recharge zone in the northeast. The quality of groundwater deteriorates towards the southwest. The same trends are noticed in the confined aquifers also.

Phreatic aquifer – Aquifer "A"

Groundwater occurs under phreatic condition in the major part, however, semi-confined to confined condition occur in some parts in the central and south-western area. Groundwater flows from the main recharge area in northeast towards discharge area in southwest. Dug and dug cum bored wells and shallow tube wells tap this aquifer. The groundwater quality is generally fresh in the central and north-eastern parts with EC generally <3,000 µs. Towards the west and south-west the quality of groundwater is brackish to saline. Higher fluoride values are also observed in central, northern and southern parts.

Confined aquifers – Deep aquifers

Aquifer *B*: This is the upper most confined aquifer and extends over major part of the area. This aquifer is tapped by a large number of dug cum bored wells and tube wells. This aquifer was over-exploited aquifer during early seventies. Depth to water level during mid seventies was about 20 m bgl in central and southern part

Table 1. General description of Aquifer Properties.

Stratigraphy	Formation Group	Lithology	Depth to top of aquifer (m)	Thickness		Remarks
				Range (m)	Average (m)	
Recent to Post - Miocene	Unconfined Aquifer A	Coarse sand, gravel pebbles, medium and fine sands with clayey sand.	5-71	35-125	62	Variable water quality
	Confined Aquifer B	Medium to coarse sand and gravel inter bedded with sandy clay.	78-162	10-80	45	Generally good water quality
	Confined Aquifer C	Medium to coarse sand in north-east, and fine to medium in central part inter bedded with sandy clay and clay	154-274	13-62	34	Generally good water quality
	Confined Aquifer D	Medium sand interbedded with sandy clay	229-402	11-105	52	Variable water quality
	Confined Aquifer E	Fine to medium sand and sandy clay	300-542	15-57	24	Developed in central part, water quality good
Miocene	Confined Aquifer F	Fine to medium sand, sandstone interbedded with siltstone	200-574	7-68	39	Variable water quality
	Confined Aquifer G	Fine to medium sand inter bedded with silt stone	264-513	9-124	48	Water quality generally saline
Cretaceous	Confined Aquifer H	Himatnagar sandstone	214-547	98-145	121	Variable water quality

Source: Groundwater surveys in Rajasthan and Gujarat, UNDP-CGWB, 1976.

and the direction of groundwater flow was to the southwest. The piezometric surface at present rests between 68 m to 125 m bgl and ranges from −40 m to more than 60 m amsl. This aquifer is heavily exploited and at places water levels are within or close to the aquifer top.

Aquifer *C*: Because of its great depth only a few wells existed during early seventies, however, large number of tube wells at present tap this aquifer and it is highly exploited.

Aquifer *D* and *E*: These aquifers are also being developed at present in the central part.

Aquifer, *F, G* and *H* are generally not exploited so far.

The tube wells are the main groundwater withdrawal structures and range in depth from 60 m to 350 m. Shallow tube wells (<100 m) is restricted to the alluvial area in the north eastern part. In the central, south-western and southern parts, deep tube wells tap one or more aquifers. The depth to piezometric surface of deeper confined aquifers ranges from near surface in the south-western part to more than 120 m bgl in the central part. The discharges of tubewells vary from 20 L/s to 60 L/s for 8 m to 13 m of drawdowns. The average yield of a 250 m deep tube well is around 20 L/s. The transmissivity of deeper aquifer varies from 300 to more than 1,200 m^2/d (GWRDC, 1997).

4 MONITORING NETWORK

Groundwater levels are being monitored since 1969 by the CGWB (Central Groundwater Board), when few hydrograph network stations were established. Subsequently, the number of stations was increased from time to time. During the Hydrology Project (1996–2002) a number of purpose built piezometers have been added to the network for better appreciation of aquifer-wise groundwater scenario especially in respect of deeper aquifers. The water levels are monitored four times a year, during January, May, August and November by CGWB since mid-seventies.

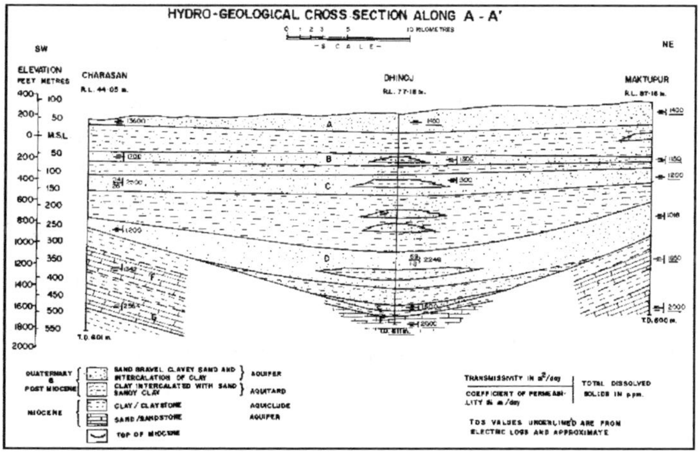

Figure 2. Multi-aquifer system in North Gujarat Region, India.

Table 2. Present network of CGWB hydrograph stations in North Gujarat.

Aquifer	Dug Wells	Piezometers	Total Network
Phreatic	80	50	130
Confined I (Aquifer -B)	-	34	34
Confined II (Aquifer -C)	-	22	22
Total	80	106	186

5 IMPACT OF INTENSIVE USE ON GROUNDWATER REGIME

For understanding the impact of intensive groundwater use on different aquifers the available data of water levels of phreatic and confined aquifers I and II have been analyzed and it is found that the secular declines in the water levels in both aquifers have set in (Table 3).

5.1 *Phreatic Aquifer*

In this aquifer the declines ranged from less than 1 m to about 21 m over a five-year period (1996–2001). The summarized data of analysis of the absolute change in water levels over this are presented in Table 3 and depicted in Figure 3. Declines of more than 15 m are concentrated in the central, northern and southern part of the area. The largest decline of about 21 m is seen around Kamboi in southern part of Banaskantha district.

The declines also became significant in the Himatnagar sandstone aquifer during the drought period of 1999–2001 recording declines of up to 5 m during this period. This is ascribed to the deficient rainfall during the two consecutive drought years (1999–2001) resulting into higher groundwater withdrawals for sustaining the irrigated agriculture.

The analysis of trends of groundwater level for the period 1996–2001 reveals that declines of 1–2 m/yr and >2 m/yr are confined to central part of North Gujarat (Table 3, Figure 4).

It is also seen that declines during 1999–2001 are of twice the magnitude of average yearly decline during 1996–2001. This is ascribed to the deficient rainfall in the region. The highest rate of decline is observed at Kamboi (4.13 m/yr) in southern part of Banaskantha district

The analysis of the long-term trends (1985–2001) reveals that the entire region has been experiencing general decline ranging from <1 m/yr to more than 3 m/yr. Declining trends of higher magnitude >2 to 3 m/yr and more are seen in Banaskantha district (Table 3, Figure 5).

5.2 *Confined Aquifer*

In first confined aquifer the declines range from less than 5 m to about 30 m over a five-year period (1996-2001). Declines of more than 15 m are concentrated in the central part of the area. The data of analysis of the absolute change in water levels over this period are presented in Table 3 and Figure 6. The largest decline of about 29 m is seen around Matarwadi-II in northwestern Mahesana district during the two consecutive drought years (1999–2001) resulting into higher groundwater withdrawals for sustaining the irrigated agriculture.

In this aquifer the trend of short-term decline ranges from less than 1 m/yr to about 8 m/yr (Table 3, Figure 7). The highest rate of decline

Table 3. District-wise summary of declines in Water levels in different aquifers in North Gujarat.

DISTRICT	Range of decline during 1985-2001		Range of decline during 1996-2001		Range of decline during 1999-2001		Remarks
	Absolute value (m)	Avg. m/yr	Absolute value (m)	Avg. m/yr	Absolute value (m)	Avg. m/yr.	Name of station with max. decline (short term)
Unconfined aquifer							
Banaskantha	-3 to -15	0.3 to 1.5	-7 to -21	1.4–4.2	-5 to -15	2.5–7.5	Kamboi -21.46 m
Mehsana	- 1 to - 6	0.1 to 0.8	-1 to -8	0.2–1.6	-1 to -13	0.5–6.5	Mehsana-V -12.96 m
Sabarkantha	-1 to -12	0.1 to 0.9	-1 to -2	0.1–0.4	-1 to -5	0.1–2.5	Himmatnagar - 4.83 m
First confined aquifer							
Banaskantha	-10 to -80	1 to 5	-1 to -22	0.2–4.5	-1 to -20	0.2–4.0	Jerda-II -20.45m
Mehsana	-12 to -22	1 to 2	-2 to -30	0.4 –6.0	-1 to -22	0.5–11.0	Matarwadi-II -29.65m
Second confined aquifer							
Banaskantha	-10 to -24	>1 to > 3	-10 to -12	2 –2.5	-6 to -13	3.0–6.5	Miyal-I -12.60m
Mehsana	-15 to -66	>1 to >6	-2 to -16	0.4–3.3	-1 to -18	0.5–9.0	Malekpur -18.12m

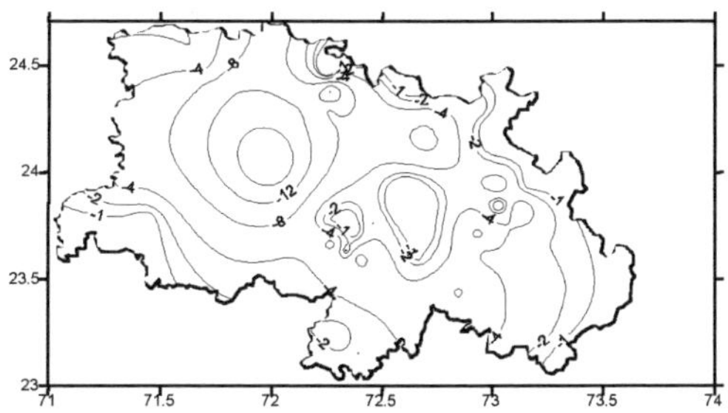

Figure 3. Decline of Water Level in unconfined aquifer (1996–2001) in meters.

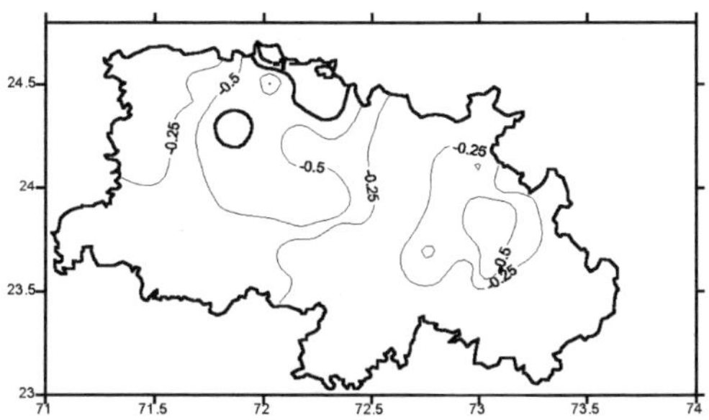

Figure 4. Long term trend of decline of Water Level in unconfined aquifer (1985–2001) in m/yr.

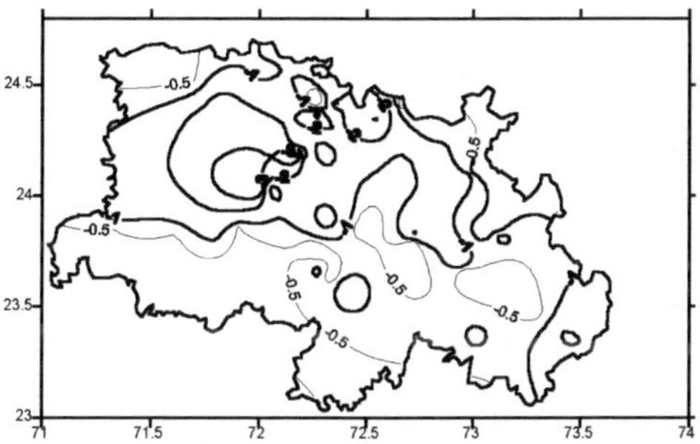

Figure 5. Short term trend of decline of Water Level in unconfined aquifer (1996–2001) in m/yr.

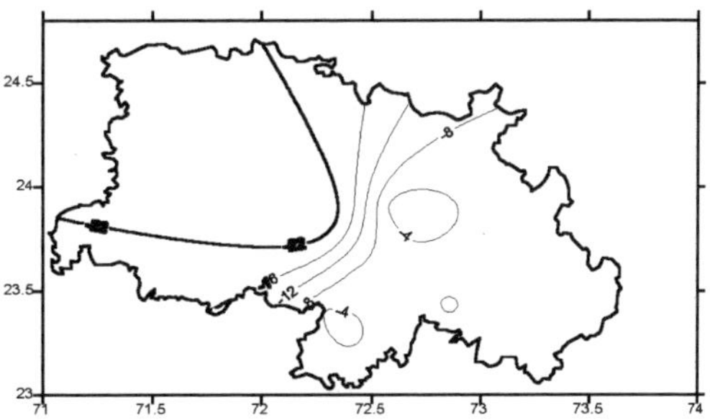

Figure 6. Decline of Piezometric surface in 1st Confined Aquifer (1996–2001) in meters.

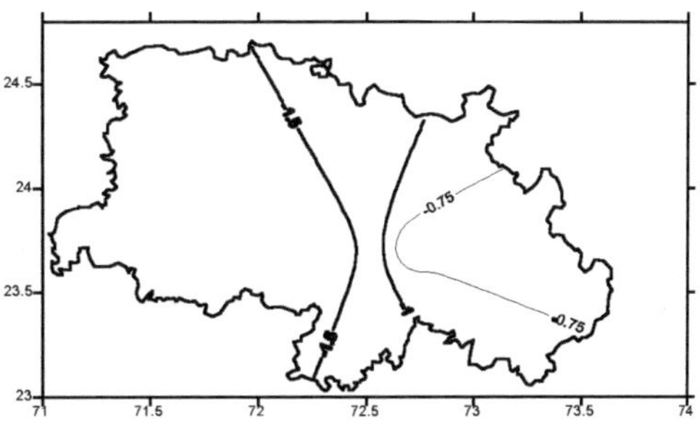

Figure 7. Long term trend of decline of Piezometric surface in 1st Confined Aquifer (1985–2001) in m/yr.

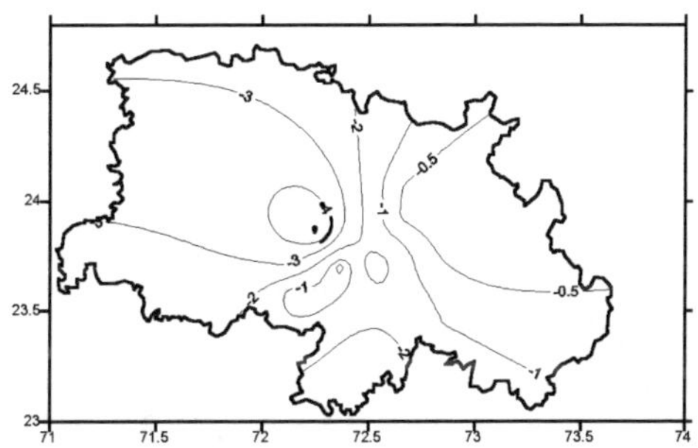

Figure 8. Short term trend of decline of Piezometric surface in 1st Confined Aquifer (1996–2001) in m/yr.

being 7.97 m/yr at Chanasma Pz in Mahesana dist. Declines of more than one m/yr are seen in the elongated area extending from Dhanera in the north to beyond south of Mahesana. The Matarwadi-II piezometer has recorded total absolute decline of 29.65 m during the period 1996–2001 of which 22.44 m is contributed by the years 1999–2001 only. Similarly the Jerda-II piezometer has recorded a decline of 20.45 m over the period 1999–2001, which works to about 10 m/yr.

Large declines even up to more than 60 m have taken place over the period of nearly two decades as revealed by the Mahi-tw III and Bhandu III PZ. The long term declines in this aquifer are depicted in Table 3 and Figure 8. The hydrographs of several other piezometers also indicate declines in the range of 10 m to 30 m during the last two decades especially in Banaskantha and Mahesana districts.

In the second aquifer declines ranged from less than 1 m to more than 20 m over the five year period (1996–2001). The details of the absolute change in water levels are given in Table 3 and presented in Figure 9. At some of the piezometers large part of the total decline is contributed by the years 1999–2001. The pockets of large declines (more than 15 m) are concentrated in the northern part of Mahesana district.

In general there is a declining trend all over north Gujarat (Sinha, 2001). The trends of decline in water levels (1996–2001) in this aquifer have been tabulated in Table 3 and presented in Figure 10. The annual declines range from less than a meter to more than 4 m/yr for the period 1996–2001. The declines in the higher range of 4 m/yr to 6 m/yr are located in the central part of north Gujarat spread over Mahesana district.

Long term decline in this aquifer range from 1.4 m/yr to 6.6 m/yr (Table 3, Figure 11). The central part of north Gujarat has recorded declines of more than 2 m/yr. This aquifer has recorded declines of even up to more than 60 m during past two decades in the central part of the area.

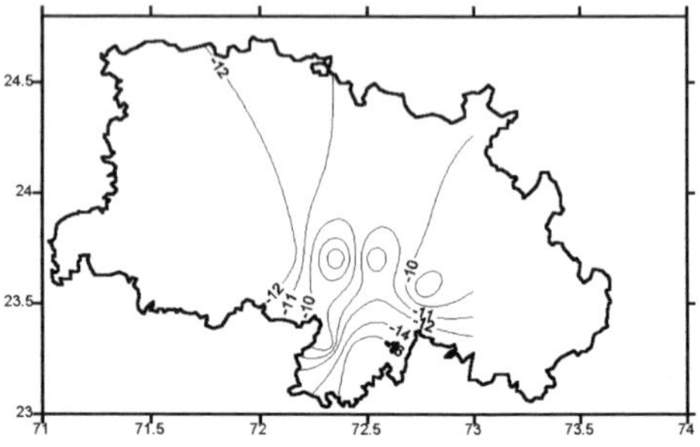

Figure 9. Decline of Piezometric surface in 2nd Confined Aquifer (1996–2001) in m.

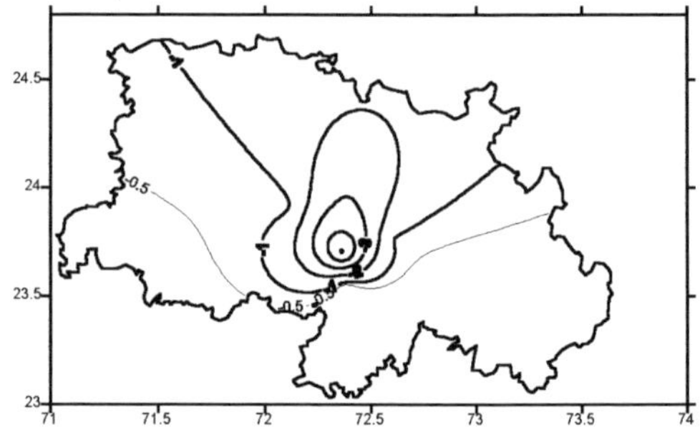

Figure 10. Long term trend of decline of Piezometric surface in 2nd Confined Aquifer (1985–2001) in m/yr.

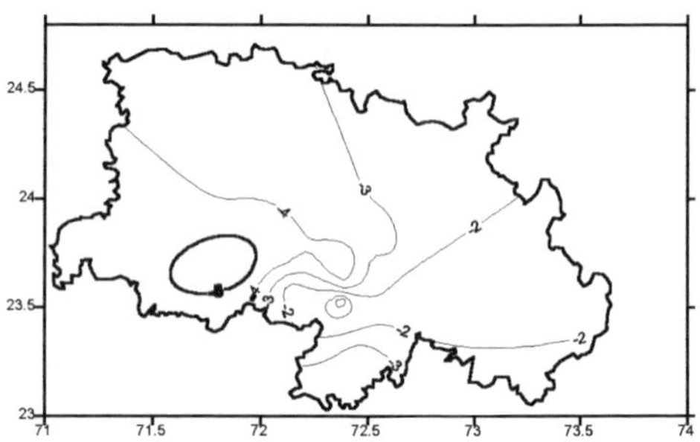

Figure 11. Short term trend of decline of Piezometric surface in 2nd Confined Aquifer (1996–2001) in m/yr.

6 GROUNDWATER RESOURCES AND STATUS OF DEVELOPMENT

The ground resource estimation exercise was taken up in the state of Gujarat for the first time in 1986 as per the recommendations of the Groundwater Estimation Methodology 1984, published by the government of India. The resource assessments have been carried out subsequently in 1997. It is seen that the level of groundwater development has shown a consistent rise from 67% in 1984 to 121.63% in 1997 (NWRD, 1986, 1997). The gross draft has increased by about 1,000 Mm^3 from 1990 Mm^3/yr in 1984 to 2,990 Mm^3/yr in 1997.

Table 4. Groundwater resources North Gujarat.

District	Utilizable Groundwater Resources in Mm^3/yr	Gross Draft - 1997 in Mm^3/yr	Balance - 1997 in Mm^3/yr	Level of Groundwater Development (%) 1997
BANSKANTHA	795.84	887.29	-91.45	111.49
MAHESANA	862.22	1,419.64	-557.42	164.65
SABARKANTHA	769.47	682.92	86.55	88.75
TOTAL	2,427.53	2,989.85	-562.32	121.63

The increase in draft has been more pronounced in the districts of Mahesana and Banaskantha. The extent of over-exploitation being of the order of 560 Mm^3/yr. The number of talukas where the gross draft has exceeded the annual recharge has increased from 4 in 1984 to 14 in 1997. Table 4 clearly brings out the situation of recharge *vs.* draft for 1984 and 1997.

7 CHALLENGES AND OPPORTUNITIES

The Challenges posed by the declining water levels as discussed above has assumed alarming proportions in Mahesana, Banaskantha and parts of Sabarkantha districts. However these challenges may be surmounted through adoption of following multi-disciplinary approaches which provides opportunities in itself towards the addressing the challenges posed as above:

- Enactment and implementation of effective groundwater legislation with the need to develop clear and enforceable right to groundwater since such rights are not defined in India.
- The framework of legal, administrative and regulatory measures in groundwater legislation should be encouraged in integrated water management by paying attention to both quantity and quality aspects of groundwater.
- Introduction of license system for abstraction of water from aquifers and the artificial recharge of groundwater. Notification of certain area as protected aquifers and allocation of high quality groundwater only for human consumption.
- Setting up of Groundwater Management Centre for dissemination of technical and scientific information to public for effective enforcement of environmental rules.
- Implementation of economic measures to provide for an effective incentive to use groundwater rationally in proportion to the volume abstracted in relation to total available resources.
- Management strategy should aim at sustainable use of groundwater, preservation of its quality and should respond to the changing conditions. Protection measures should include intensive micro level monitoring of groundwater, preparation of aquifer vulnerability maps and zoning of groundwater protection areas.
- Provision of recharge tube wells in the upstream areas of centers of heavy pumping specially the regional water supply schemes for speedy percolation into deep aquifers.
- Introduction of statutory authority for grant of permits for groundwater exploration and abstraction. Drilling and sinking of tube wells and boreholes should as a rule be carried out by qualified and skilled personnel with appropriate equipment.
- For protecting aquifers around present and future abstraction sites, establishment of groundwater protection zones over and above the general protection of groundwater through legal/statutory provisions needs to be considered as preventive measures
- Encourage farmers through incentives like switching over to micro-irrigation systems such as sprinkler and drip irrigation for demand-side management that will induce sustainable extraction and use of groundwater.
- Periodic impact assessment surveys to review any adverse impacts on groundwater resources due to anthropogenic interventions. Results of such assessment should be incorporated in decision-making.
- Creation of legally empowered water associations having competence to arbitrate among various competing demands and diverging interests regarding groundwater abstraction and use. Such bodies should interact with other authorities competent for land use planning, soil management, wastewater management, etc.
- Research should be undertaken to improve knowledge of dynamics of aquifers, detailing of guidelines and technologies with regard to development and calibration of groundwater models, which allow predictive analysis of groundwater levels and quality under various pumping alternatives.
- Promoting greater awareness of the inherent groundwater problems at all levels, raising the level of knowledge of public in

general and of water users in particular with regard to behaviour and vulnerability of groundwater resources.

8 CONCLUSION

Intensive use of groundwater is one of the manifestations of the competitive use of the resources towards maximizing the economic benefit without caring for the consequences of such use. Debate has started whether such use be allowed or not. Conservation of the natural resources is widely accepted way of living but instead we find more intense competition. Challenges posed by intensive use of groundwater resources may only be addressed if the community become more aware of the importance of the conservation in day to day life and indulge in participatory management of the groundwater resources. Community must get skilled in harvesting each and every drop of the rainfall to augment the groundwater resources. Optional sources of water may also be looked into to reduce the stress on groundwater.

ACKNOWLEDGEMENTS

The authors acknowledge the Chairman, Central Groundwater Board, Govt. of India, and the Regional Director, WCR, CGWB, Ahmedabad for utilizing part of the data and information generated by CGWB.

REFERENCES

GWRDC Ltd. (1997). *Hydrogeological data of Gujarat State*. Unpublished Technical Rep., Govt. Of Gujarat.

Jain R.C., Nagar A., Jain P.K. and Krishna V.S.R. (2000). *Declining groundwater levels in deep aquifers of Gujarat*. Unpublished Technical Rep., Central Groundwater. BoardWCR, Ahmedabad.

Merh S.S. (1995). *Geology of Gujarat*, 1st Edition Geological Society Of India, Bangalore.

NWRD (Narmada and Water Resources Dept) (1986). *Groundwater resources and irrigation potential of Gujarat State*. Technical report by State Department.

NWRD (Narmada and Water Resources Dept) (1999). *Groundwater resources and irrigation potential of Gujarat State*. Technical report by State Department.

Sinha A.K. (2001). Sustainability of India's Water Resources: Challenges and Perspectives. *Journal of Sustainable development* International Spring 2001, pp. 129–133.

UNDP (1976). *Groundwater Surveys in Rajasthan and Gujarat*. A technical report prepared under UNDP project.

The study on sustainable groundwater resources and environmental management for the Langat Basin in Malaysia

Saim Suratman
Minerals and Geoscience Department Malaysia, Kuala Lumpur, Malaysia
saim@jmg.gov.my

ABSTRACT

Rapid and ultra-large scale development is progressing in the national capital of Kuala Lumpur and the adjoining areas in Selangor State, Malaysia. To cope with the water deficit in Selangor State, the groundwater is considered as a possible supplementary water source. Uncontrolled use of groundwater, however, might induce serious environmental problems, e.g. land subsidence, drying-up of wetlands and saltwater intrusion.

The Langat Basin includes Kuala Langat, Sepang and Hulu Langat Districts of Selangor State, and the western part of Seremban District of Negeri Sembilan State. The industrial sector will play an important role in the development of the Kuala Langat District. Three development corridors are proposed in the development plan of Kuala Langat District; namely, Telok Panglima Garang–Banting-Olak Lempit; Morib-Tg, Gabang-Tg, Tumbuk; and Pulau Carey. The Multimedia Super Corridor (MSC) is the largest and the only urban development plan in Sepang District. Its establishment is expected to greatly influence development not only of the district but the State and the whole nation as well. The MSC will also influence the land use development pattern in Hulu Langat District.

Major surface water sources in the Langat Basin are the Langat and Semenyih treatment plants with 93% of the total water treatment capacity of 1,003.8 ML/d for the Langat Basin. The National Water Resources Study was carried out for planning, development and management up to the year 2050 of the overall water resources of Peninsular Malaysia. Water demand in the Langat Basin is increasing steadily. The future demands of the three districts of Selangor State in the Langat Basin would also be supplied by the Southern Interstate Transfer from Pahang State from 2007. Usage of groundwater during 1999 was estimated as 2,845 m^3/h. Industrial activities utilised 1,385 m^3/h, followed by domestic 1,341 m^3/h.

Groundwater simulation with the objectives of estimating groundwater resources potential in the Langat Basin area and to predict environmental impacts caused by groundwater abstraction was also carried out. In the modelling work, a number of application scenarios have been carried out. The results of nine selected scenarios (Simulation Variants), characterise the groundwater conditions in the study area and potentially enable to support technical decisions and policy making.

Periodical monitoring is being carried out for the following items: groundwater abstraction volume, static groundwater table, groundwater quality, land subsidence, and surface water level.

The establishment of a balanced, multi-sector and integrated groundwater resources and environmental management plan is deemed urgent to attain a sustainable groundwater resources use and to maintain a favourable groundwater quality in the Langat Basin.

Keywords: *aquifer, Langat Basin, Malaysia, modelling, Visual MODFLOW.*

1 BACKGROUND

The importance of groundwater resources in the Langat Basin (Figure 1) has been increasingly recognised in helping to cope with the water deficit in Selangor State. Thick Quaternary layers are deposited in the flat lowlands spreading the downstream of the Langat Basin. The aquifer in the study area consists of a sand and gravel layer, distributed continuously around 15–20 m below the ground with a depth of 20 m to more than 100 m in the lowlands. Groundwater can be developed economically in this area.

Groundwater abstraction through wells construction and dewatering activities in the Basin is estimated to be 45,000 m^3/d, which is nearly equivalent to the sustainable groundwater yield of the Basin. Groundwater monitoring and modelling reveals that the groundwater abstraction results in lowering of the groundwater level around the pumping area.

Groundwater quality in the basin is good. However, monitoring especially for heavy metals, such as lead and arsenic, and organic compounds by using integrated isotope, hydrogeology and modelling will be of great importance. In addition, seawater intrusion and land subsidence that may affect the environment in the Basin, as well as water level in Paya Indah lakes should also be monitored closely.

Rapid and ultra-large scale development is progressing in Kuala Lumpur, the national capital, as well as the adjoining areas in Selangor State. As a result, emerging are social problems, namely, urban environmental deterioration including water deficit and water quality deterioration. To cope with the water deficit in Selangor State, the groundwater is considered as a possible alternative supplementary source of water, and some factories have already started shifting their water source from surface water to groundwater. The establishment of a balanced,

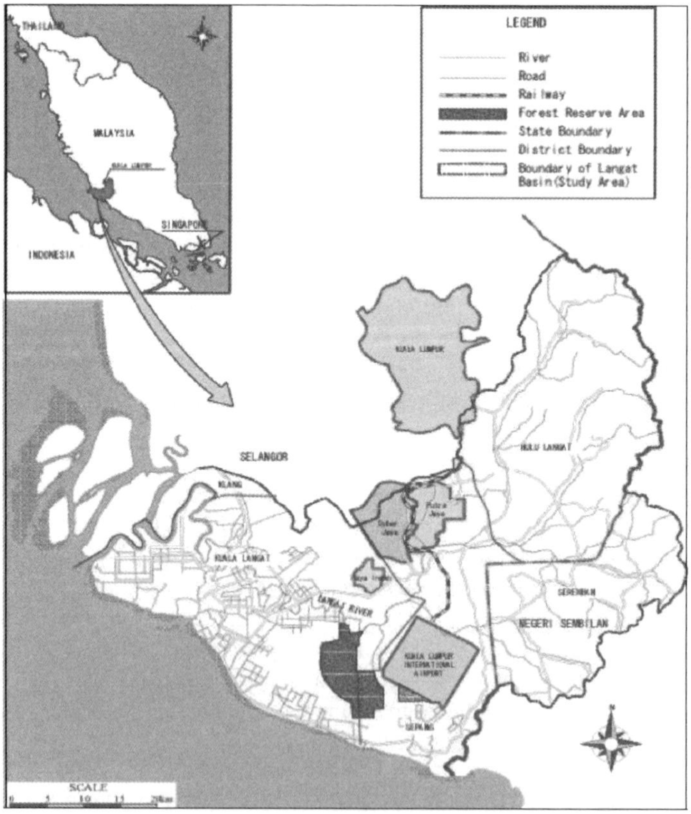

Figure 1. Location map of the study area.

multi-sector and integrated groundwater resources and environmental management plan is thus deemed necessary.

This study commenced in April 1999 and ended in March 2001. Due to a drought that hit the Klang Valley about one and a half years before this study, The Minerals and Geoscience Department Malaysia was directed to explore the possibility of groundwater exploitation in the Basin. Groundwater abstraction at the Megasteel complex is about 10,500 m^3/d (Geological Survey Department, 1998) while another 23,600 m^3/d of groundwater is removed by sand mining operation.

It is thought that the abstraction of groundwater at Megasteel complex and the proposed additional wellfields may affect the groundwater table and may lead to drying up of the Paya Indah Wetlands.

2 OBJECTIVES OF THE STUDY

The objectives of the Study include:

i. To formulate a sustainable groundwater resources and environmental management plan for the Langat Basin;
ii. To establish a monitoring system and Geographic Information System to support the Management Plan; and
iii. To formulate human resources and institutional development plan for the implementation of the Management Plan, and to be able to utilise the Management Plan for other basins.

3 TOPOGRAPHY, RIVER SYSTEM, HYDROLOGY AND GENERAL GEOLOGY

3.1 *Topography and river system*

The Langat Basin can be broadly divided into three; namely, mountainous areas, hilly areas and flat lowlands, from the upstream to the downstream (Figure 2). Groundwater recharging areas are in the upstream mountainous and hilly areas. The aquifers are distributed widely in the flat alluvial lowlands. The Langat River and its left bank tributary, the Semenyih River, originate in the western slope of the Main Range. Both rivers flow southwest in the mountainous terrain and gradually change their direction to the west. The two rivers then join, enter the flat land near Kg. Dengkil and flow westward and finally into the Straits of Melaka.

3.2 *Hydrology*

The Malaysia Peninsula receives rainfall throughout the year. There are two distinct monsoon

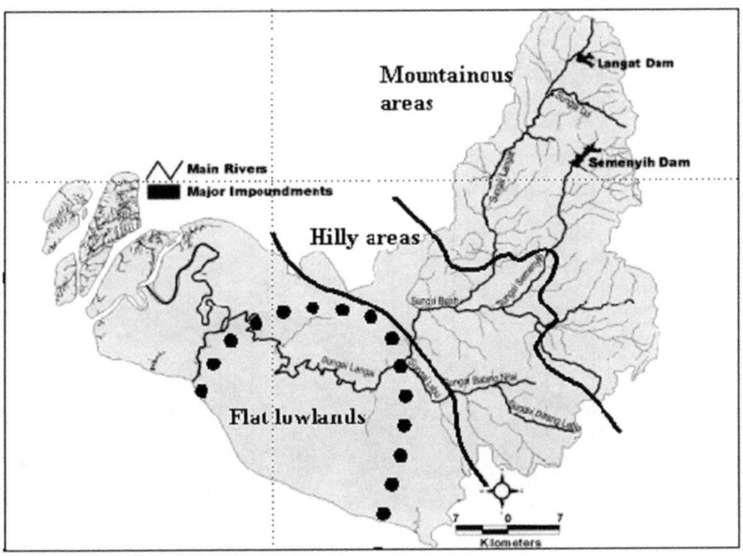

Figure 2. Topography and river system in the study area.

periods: the northeast monsoon (October to January) and the southwest monsoon (May to September) with a *dry* period in between. The average annual rainfall is 2,238 mm and the average annual pan evaporation is 1,606 mm.

3.3 *Geology*

The area is made up of 60 m to 130 m (east to west) Quaternary alluvium emplaced on top of sandstone and shale of the Kenny Hill Formation bedrock and Kajang Formation at the eastern fringe. Four units of Quaternary alluvium are recognised; namely the Beruas Formation, Gula Formation, Kempadang Formation and Simpang Formation (Perunding Utama Sdn. Bhd., 1998). Simpang Formation of terrestrial Pleistocene deposits made up the lowermost alluvium unit and made up of largely coarse to very coarse sand and gravel (main aquifer) with minor clay content and overlying the Kenny Hill Formation bedrock.

4 LAND USE AND WATER DEMAND

4.1 *Land use*

Comparison made on the land cover between 1995 and 1998 for the Langat Basin (2,750 km^2) revealed increases of built-up area and plantation by 255.5 km^2 and 31.7 km^2 respectively, and decreases of oil palm, swamps, rubber and grassland by 102.5 km^2, 92.7 km^2, 75.5 km^2 and 36.6 km^2 respectively.

4.2 *Water demand and supply*

Major surface water sources in the Langat Basin are the Langat and Semenyih treatment plants with 93% of the total water treatment capacity of 1,003.8 ML/d for the Langat Basin. The National Water Resources Study (GOM, 2000) which was carried out for planning, development and management up to 2050 of the overall water resources of Peninsular Malaysia indicated that water demand in the Langat Basin is increasing steadily (Figure 3). The future demands of the Langat Basin is to be enhanced by the Interstate Transfer from Pahang State from 2007. The construction of Kelau Dam is very critical for the demand-supply balance in the Langat Basin.

The Mid-term Review of the Seventh Malaysia Plan 1996–2000 describes the necessity of sustainable development of groundwater as an alternative source. Most of the groundwater is used for industrial purposes.

5 GROUNDWATER STATUS

5.1 *Groundwater quality*

Groundwater quality from 34 wells was analysed. The surface water quality was also measured at 10 locations. Low pH values ranging from 3.9

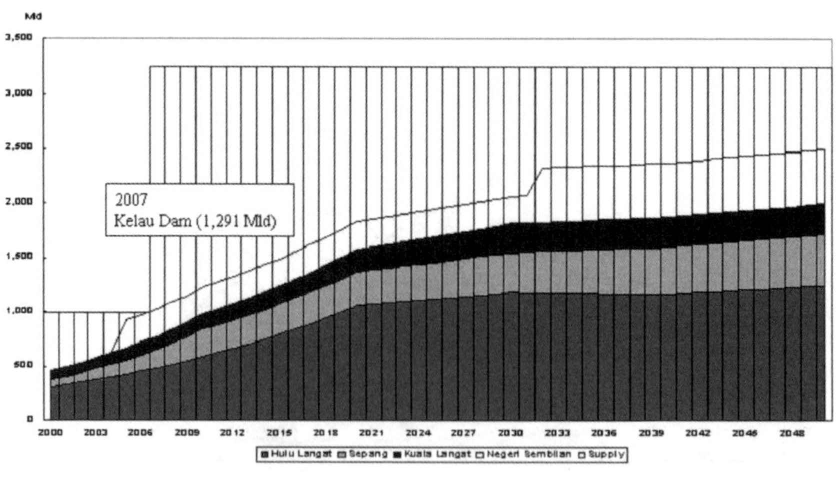

Figure 3. Water demand in Langat Basin (after GOM, 2000).

to 4.7 were observed mainly in surface water near Paya Indah area. The highly acidic surface water has significantly reduced the pH values of the groundwater to the range of 4.5–4.7.

Total iron values mostly exceed the mandatory level of iron values specified in the Malaysian National Drinking Water Guidelines, i.e., 0.3 mg/L, generally ranging from 1 mg/L to 10 mg/L. Arsenic is widely detected although almost all were below the drinking standard of 0.05 mg/L and the highest concentration of arsenic are found in shallower wells. Organic compounds (m & p-xylene, o-xylene, 1,2,4-trimethylbenzene, napthalene and 2-methyl-napthaline) are observed in four wells in only one sample.

The groundwater can be classified as Na_2SO_4 or NaCl Type. The chloride water indicates the influence of the seawater and marine origin sediments.

Potential sources of groundwater pollution are from a total of 30 industrial estates in the Basin. Other potential source of pollution comes from the agricultural activities dominated by palm oil and rubber plantation. Other potential pollution sources include mines, solid waste landfill site, wastewater treatment plants, and petroleum storage tanks.

5.2 *Groundwater usage*

The existing wells in the Langat Basin, are divided into 5 different categories; namely, domestic, industrial, observation, test well and unknown. Usage of groundwater during 1999 was estimated as 2,845 m^3/h. Industrial activities utilised 1,385 m^3/h, followed by domestic 1,341 m^3/h.

The main wellfields in the basin is located in the Megasteel property at Brooklands Estate. The smaller capacity wellfields are in Brooklands Estate and in Olak Lempit. The natural groundwater flow is affected by abstraction of water from the Imuda Tin Mine.

5.3 *Land subsidence status*

Elevations of twenty shallow benchmarks were precisely measured in July and November 2000, and March and August 2001 to monitor the land subsidence. Records of 15 and 11 out of 20 benchmarks show a tendency to subside in the second and third measurement, respectively, while 19 benchmarks sank in the fourth one. A total of eight benchmarks indicate continuous settlement. Benchmarks situated near the Langat River have settled from around 11 mm to 16 mm for one year period since July 2000. These benchmarks are located in the area where much groundwater is extracted presently.

6 GROUNDWATER MODELLING

6.1 *Modelling approach*

The overall objectives of groundwater simulation are to estimate groundwater resources potential and to predict environmental impacts caused by groundwater abstraction. A three-dimensional groundwater flow and contaminant transport modelling software (Visual MODFLOW) was used.

Langat basin is a typical coastal confined aquifer with geologically defined boundaries and recharge areas. The aquifer is replenished from precipitation. Recharge areas are distributed according to geological conditions (e.g. thin or absent aquitard, bedrock outcrop). The aquifer recharge is simulated as aerially distributed recharge to the groundwater system. Aquifer can be also replenished (or drained) to some small degree from rivers (stretches with permeable riverbed).

The model, which simplifies the actual geological conditions, consists of the following four layers:

- Layer 1 – peat and peaty soil (Beruas Formation)
- Layer 2 – clayey soil (Gula Formation and Kempadang Formation, if exist)
- Layer 3 – the Aquifer, alternatives of sandy soils/gravelly soil and clayey soil (Lower Member of Simpang Formation)
- Layer 4 – bedrock (this layer is not explicitly modelled)

6.2 *Model discretization*

The modelling domain was selected according to natural hydrogeological boundaries, over an area with *x* coordinate 378,000–418,000 m, and *y* coordinate 285,000–330,000 m (area $40 \times 45\,km^2$)

as shown by Figure 4. Inside the modelling domain, rectangular model grid with different lengths and widths of cells is set-up (37,240 cells). The model has a 70 × 76 cell grid in horizontal direction; with minimum cell size 250 × 250 m^2 (well field area at Megasteel, Kajibumi WF 2, Paya Indah area) and maximum cell size 2,000 × 2,000 m^2 (edges of the model). Generally, node spacing is finer where the hydraulic gradient and flux are subject to greater change (e.g. the areas near extraction wells).

6.3 *Initial and boundary conditions*

Initial conditions represent the initial state of the considered model parameters and variables for start of computation. Boundary conditions are constraints imposed on the model grid that express the nature of the physical boundaries of the aquifer being modelled (Figure 4). Along the coast is specified as general head boundary while river boundary is specified along the Langat River. Pumping wells represent the discharge of groundwater out of the model.

6.4 *Model parameters*

Hydraulic parameters of the 3D groundwater flow model were based on data interpreted from pumping tests at Paya Indah, Kajibumi WF 2, and Kanchong Darat as shown by Table 1. For all three localities a local pumping test model was set up, and this model was calibrated and verified against groundwater levels measured during pumping and recovery tests. The 3D model was calibrated against groundwater levels in the model area measured on March 9 to 13, 2001. For model calibration and verification, two other measurements of groundwater levels and time series from long-term groundwater level monitoring at existing wells were analysed. Average rainfall in the area northwards from Lower Langat was calculated as 2,238 mm/yr and evaporation as 1,284 mm/yr. Thus, at least 954 mm/yr could be infiltrating into the ground.

6.5 *Model calibration and verification*

In order to demonstrate the applicability of the established modelling tool, a number of application scenarios were carried out. As a results, 9 selected scenarios (simulation variants) characterising the groundwater conditions in the study area were modelled.

Simulation variants 0, 1, 2, 3, 6, and 7 represent steady-state flow simulation while simulation variants 4, 5, and 8 were performed as

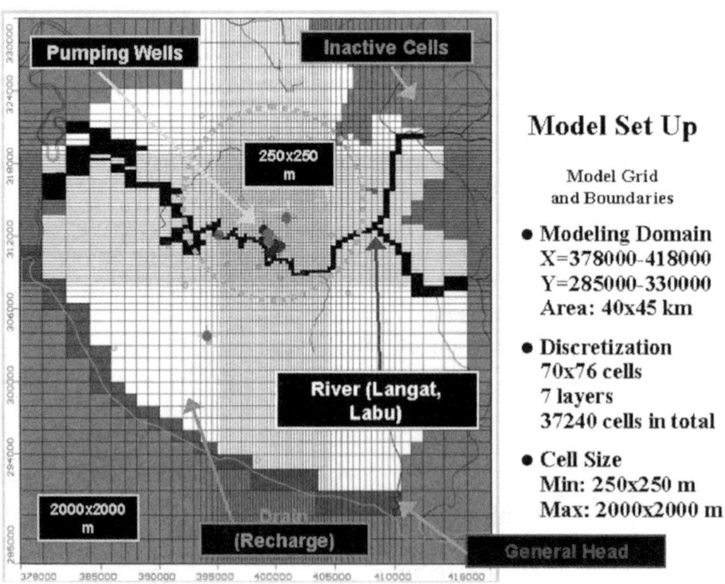

Figure 4. Model discretization.

transient flow simulation (with duration of 3 years, or 20 years, respectively). Table 2 provides an overview of simulated daily pumping rates in each model scenario.

6.6 *Model simulation result*

6.6.1 *Natural groundwater flow*

Simulation Variant 1 (SV1) represents recent status and calibration variant. The steady-state simulation was performed with pumping rate scenario, which corresponds to conditions during field investigation in the Study period (see file *Variant 1* in Table 3). The modelling result for Solution Variant 1 of Head Equipotentials in Model Layer 3c is presented in Figure 5 which is indicating the current conditions in the study area.

6.6.2 *Pollution transport*

Result of three-dimensional particle tracking calculation is presented in the form of map of pathlines (Figure 6). These show the movement of groundwater towards the pumping wells.

Table 1. Hydraulic parameters of regional 3D groundwater flow model before calibration.

Layer N°	Zone	Prop. ID	Conduct. K_H (m/s)	Conduct. K_V (m/s)	Anisotr. K_H/K_V	S_S (m^{-1})	S_Y (−)	Soil description
1		18	1.00E-04	1.00E-05	10	1.00E-03	0.3	Peat- laterite soil
2		8	1.20E-08	9.20E-09	1.3	6.00E-06	0.01	Silty clay
3a	A	21	6.00E-05	6.00E-06	10	3.70E-05	0.15	Clayey sand with gravel
	B	1	1.00E-04	1.00E-05	10	6.00E-06	0.15	
	C	28	1.60E-04	4.00E-06	40	6.00E-06	0.15	
3b		9	1.00E-08	7.00E-09	1.4	6.00E-06	0.01	Clayey silt
3c	A	23	4.00E-04	9.60E-06	42	4.50E-05	0.2	Silty sand and gravel, sandy clay
	B	12	1.20E-03	4.80E-05	25	3.00E-05	0.2	
	C	25	1.20E-03	2.40E-05	50	4.10E-05	0.18	
3d	A	38	1.00E-04	5.00E-06	20	4.50E-05	0.2	Silty sand and gravel
	B	15	2.00E-04	1.00E-05	20	3.00E-05	0.2	
	C	40	8.00E-04	3.20E-05	25	4.50E-05	0.2	
4	A	20	3.20E-06	3.20E-06	1	6.00E-06	0.02	Bedrock
	B	24	1.00E-08	1.00E-08	1	6.00E-06	0.02	
Dredge Ponds		2	5.80E-07	5.80E-07	1	4.50E-05	1	Mining area

Table 2. Simulated daily pumping rates from well fields and dewatering pits.

Well field	Sc 0 (m^3/d)	Sc 1 (m^3/d)	Sc 2 (m^3/d)	Sc 3 (m^3/d)	Sc 4 (m^3/d)	Sc 5 (m^3/d)	Sc 6 (m^3/d)	Sc 7 (m^3/d)	Sc 8 (m^3/d)
PWM-1	0	6,450	6,450	13,440	6,450	13,440	0	6,450	6,450
Well A (new)	0	0	0	13,730	0	13,730	0	0	0
Well B (new)	0	0	0	13,730	0	13,730	0	0	0
PWM-2 (JBA-1)	0	2,640	2,640	2,640	2,640	2,640	0	2,640	2,640
PWM-3	0	4,440	4,440	4,440	4,440	4,440	0	4,440	4,440
PWM-4	0	4,440	4,440	4,440	4,440	4,440	0	4,440	4,440
PWS-1	0	1,200	1,200	1,200	1,200	1,200	0	1,200	1,200
TW-4	0	1,600	1,600	1,600	1,600	1,600	0	1,600	1,600
Kajibumi WF 1	0	0	9,200	9,200	0	9,200	0	9,200	0
Kajibumi WF 2	0	0	6,900	6,900	0	6,900	0	6,900	0
Imuda Tin Mine	27,028	24,884	23,756	18,933	20,619	20,931	0	0	24,432
Total	27,028	45,654	60,626	90,253	41,389	92,251	0	36,870	45,202

In brief the characterisation of scenarios is in Table 3:

Table 3. Characterisation of modelling scenarios.

Variant	Conditions
Variant 0	The natural (original) conditions in the past, before putting existing well fields into operation, with active Imuda Tin Mine.
Variant 1	Present conditions (during field investigation in the Study), with abstraction of groundwater from existing well fields and dewatering pits.
Variant 2	Near future conditions, with increased abstraction of groundwater, after putting Kajibumi WF 1 and Kajibumi WF 2 into operation.
Variant 3 & Variant 5	Future conditions, with increased abstraction of groundwater in the Megasteel/Amsteel II area, with Kajibumi WF1 and Kajibumi WF 2 in operation.
Variant 4	Present conditions (during field investigation in the Study), with abstraction of groundwater from existing well fields and dewatering pits, similar to Variant 1 but without recharge from rainfall.
Variant 6	The natural conditions in the future, similar to Variant 0 but without any abstraction of water from well fields and dewatering pit at Imuda Tin Mine.
Variant 7	The conditions with increased abstraction of groundwater, after putting Kajibumi WF 1 and Kajibumi WF 2 into operation, similar to Variant 2 but without abstraction of water from existing dewatering pit at Imuda Tin Mine.
Variant 8	Present conditions (during field investigation in the Study), with abstraction of groundwater from existing well fields and dewatering pits, similar to Variant 1 but with reduced effective recharge from rainfall in the Multimedia Super Corridor area.

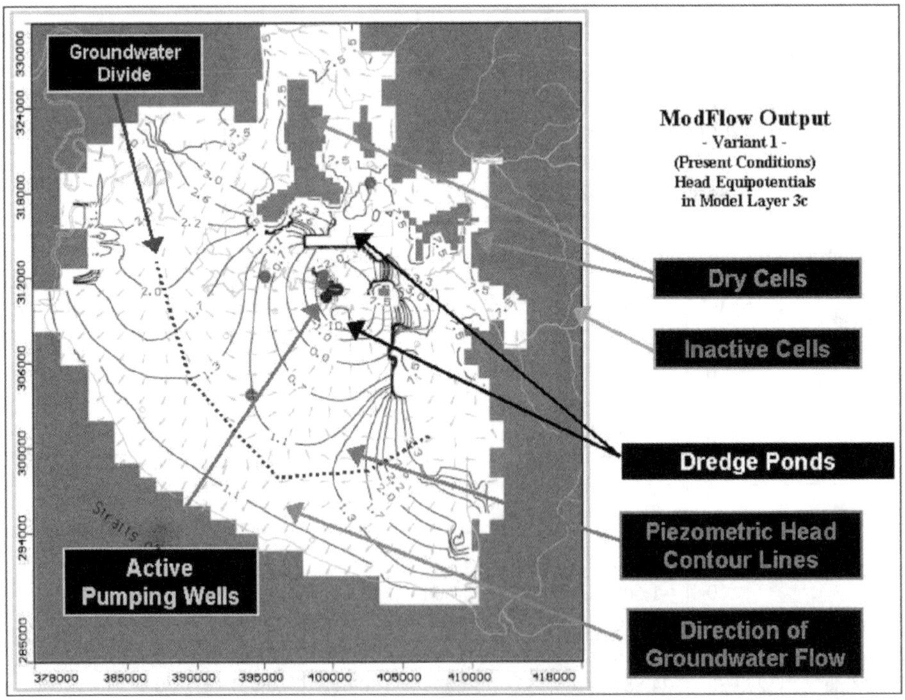

Figure 5. Head equipotential in model layer 3c.

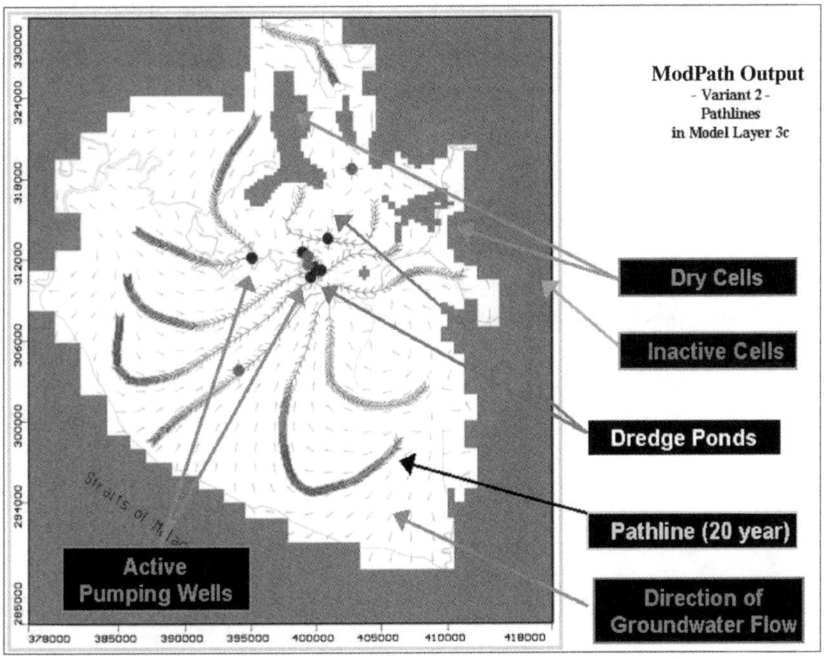

Figure 6. Pathline in model layer 3c.

7 INFORMATION MANAGEMENT

The Internet Web-application by using ArcIMS (Arc Internet Map Server) has been developed for the Management Information Systems. Topographic and thematic maps as well as the other relevant maps have been prepared. The available Well data are converted into Oracle database. ArcIMS Server is introduced and PCs are utilised as Data Server and ArcSDE Server. The system employs a web-based three-tier architecture, namely, Client Front-End (User Interface), ArcIMS Middle Tier (Process Management), and Database Back-End (Database Management).

8 CONCLUSION

The results of groundwater flow modelling in the form of piezometric heads or drawdown create a base for ecological predictions and evaluation of environmental impacts caused by groundwater abstraction. The decrease of groundwater level can be calculated for any combination of modelling scenarios; the piezometric heads are simulated in all calculation nodes of discretised model layers. The results of pumping tests and groundwater flow modelling indicate that groundwater abstraction in the main aquifer will cause only negligible drawdown in the wetlands and shallow peat layer.

The model would also be used as a technical scientific based management tool for planning of the sustainable groundwater resources development. The model would enable the authorities to formulate optimal management strategies leading to a protection of the water resources and development of the area.

The shallow layer is affected more by lowering of surface recharge (rainfall). Influence of 3 year long dry season or increased abstraction of groundwater was studied in simulation variant 4 and 5. Model is based on assumption that the old dredge ponds consist of low-permeable backfill. Land subsidence is an important soil-engineering problem mainly in sedimentary basins. It can be ascribed to several causes. In the Study Area, compaction of the soft clay layer following the decline of groundwater levels is the most important cause of land subsidence.

For the establishment of the Sustainable Groundwater Resources and Environment Management Plan in the present study, the Management Information System is used as a tool for the execution of the Management Plan.

REFERENCES

Geological Survey Department (1998). Potensi Air Tanah di Kawasan Telok Datuk – Olak Lempit Daerah Kuala Langat, Selangor, Internal Report No. Laporan (Hidro) 01/98.

Government of Malaysia (GOM) (2000). The National Water Resources Study 2000–2050, Vol. 1. (Study undertaken by SMHB, Ranhill and JPZ).

Perunding Utama Sdn. Bhd. (1998). Environment Impact Assessment Report for Proposed Groundwater Abstraction on Lots 1632, 2319, 2320, 2321 & 2323, Kawasan Perindustrian Olak Lempit, Mukim of Tanjung Dua Belas, Kuala Langat District, Selangor Darul Ehsan. Subang jaya, Malaysia, unpublished.

CONJUNCTIVE USE

Effective management of water resources in a region of intensive groundwater use: a lesson from Arizona's Valley of the Sun

Mario R. Lluria
Salt River Project, Phoenix, Arizona, USA
mrlluria@srpnet.com

ABSTRACT

The Phoenix metropolitan area, a group of more than twenty municipalities, is the fastest growing population center and the sixth largest urban area in the United States. Located in a semiarid region of North America, it has very limited surface water resources. Surface water comes from the Salt River and its tributaries as runoff from snow accumulated during the winter in distant mountains to the north during wet climatological cycles. In prehistoric times and until the 1940s agriculture predominated in this *Valley of the Sun*. The large alluvial aquifer system that underlies this region provided sufficient groundwater to boost agricultural products after the advent of the turbine pump in the late 1920s. This intensive use of groundwater resulted in several undesirable impacts among them groundwater level declines, land subsidence and groundwater quality deterioration. In 1980 a strict groundwater code was adopted for the state and areas of closely regulated groundwater abstraction were established. Increased water conservation was mandated from the municipal and industrial water users and better management practices were required in agriculture. Artificial groundwater recharge and water reuse were promoted and extensively adopted. A long aqueduct brought water from the Colorado River and added a new water resource. Public information and education in schools provide guidance and encourage conservation. This has created a public water consciousness. Water exchanges with neighboring states will soon augment the groundwater stored in the local aquifers using large scale recharge facilities. The coordinated use of all these water management techniques will continue to provide the necessary resources to support the population growth.

Keywords: *Artificial recharge, conjunctive use, water management, Phoenix, Arizona.*

1 INTRODUCTION

The concern for a worldwide water shortage, especially of potable water, which may result from the present rapid population growth, has been increasing during the last two decades. The subject has and continues to be addressed by a broad spectrum of individuals and organizations that range in all continents from top scientists and research institutions to the ordinary layman. This water shortage issue has been considered by many as a real *crisis* and by some as a part of an on-going worldwide environmental disaster coupled with such other factors as global warming and ocean and atmospheric pollution. The uneven geographic distribution of water resources and its relation to population density is perhaps the principal cause for the concern of its present and future supply. Highly water stressed populations, where people use more than 40 percent of available renewable water, are numerous and irregularly distributed throughout our planet. These are presently estimated at approximately two billion inhabitants. Many regions of high water-stressed populations on earth have arid to semiarid climate. The countries bordering the

Mediterranean Sea and the Persian Gulf are good examples of highly water-stressed populations. Most of the western United States and large areas of South Africa are also in this category. Certain areas in the state of Florida in the United States and western India have highly water-stressed populations but enjoy relative humid subtropical weather. They are in this condition because of other factors. In the state of Florida for example, unfavorable topographic conditions do not allow economically feasible surface water reservoirs devoid also of severe environmental impacts. The solution of the water scarcity in any region, regardless of prevailing climatological, geologic or demographic conditions, is appropriate water resources management. This management has to be applicable to the conditions of the region, and to be successful it should be technically efficient, cost-effective and be accepted by the population.

1.1 *Effective water management and its essential components*

The major components of water management are listed below. Each of these components varies from region to region and their composition, as well as the degree of their importance will depend on the needs of the local population, in addition to the prevailing environmental conditions. Of these conditions, climate, physiography and geology are among the most important ones because they cannot be modified by human endeavors. The principal components of effective water management are:

- Technical
- Political
- Environmental
- Economic

Commonly there is overlapping between these components and one determines the composition of the other. The high cost of certain water management facilities, such as dams, may eliminate them as the preferred alternative and may favor a lower cost option such as an aquifer storage project which may not be as technically efficient for the prevailing local environmental conditions.

This article will show how water management has been effective in providing sufficient water for urban, industrial and agricultural activities in a region with very limited water resources and experiencing very rapid population growth. It is hoped that the methods and procedures employed for this region may guide water managers in other areas, especially in those with similar environmental conditions, to improve their water management practices.

2 ARIZONA'S VALLEY OF THE SUN

2.1 *Physical setting*

The area under consideration is located in the southwestern part of the United States of America (Figure 1). It is situated in the semiarid region of North America known as the Desert Southwest, which includes the states of Texas, New Mexico, Arizona, California, Nevada and Utah, and the states of Chihuahua, Sonora, and Baja California in Mexico. Mean annual precipitation in the Valley of the Sun is 180 millimeters with most of it occurring in the winter season as a result of northwest Pacific Ocean storms. High intensity, very short duration storms prevail in the summer. The major drainages of the area are the Salt, Verde and the Agua Fria Rivers, all with ephemeral regimes. All these rivers have surface water storage reservoirs in the upper reaches in the surrounding mountains. All are tributaries of the Gila River, itself a tributary of the Colorado River. The Valley of the Sun is an intermountain trough formed during the Mid-Tertiary Orogeny commencing approximately 25 million years ago during a transition from a compressional tectonic regime to a tensional tectonic regime. The compressional regime was the result of the collision of the Pacific tectonic plate and the North American tectonic plate and the subduction of the Farallon plate under the continental mass. The tensional regime that followed was dominated by the San Andreas transform fault system. This was followed by a metamorphic core complex event, detachment faulting and a late high angle fault Basin and Range episode. These last two episodes generated the physiographic setting of the Valley of the Sun and created the deep alluvial basin that host the local aquifer system. The Valley of the Sun grades gently to the west-southwest between 400 meters and 275 meters above mean sea level. It is

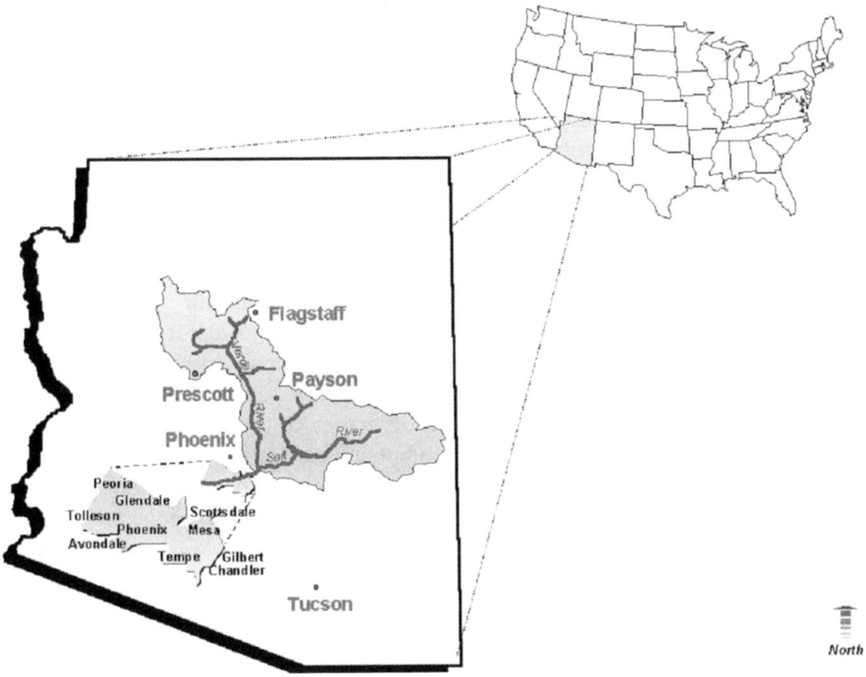

Figure 1. Location of the Phoenix metropolitan area and the watershed of the Salt and Verde Rivers.

surrounded by mountains which attain elevations that range between 900 meters and 1,800 meters above mean sea level. Native vegetation consists of mesquite and Palo Verde trees and cacti.

2.2 *Demographic setting*

In prehistoric times, between 300 B.C. and 1400 A.D., the valley was settled by the Hohokam tribe a community that developed extensive farming using the limited runoff of the Salt, Verde and Agua Fria Rivers. They constructed diversion dams and an extensive network of irrigation canals. A drought caused the disappearance of this civilization. During the Spanish colonial time this area was uninhabited. In the 1860s, after the Civil War, some pioneers moved into the area and commenced farming near the Salt River. In the 1880s Phoenix started taking shape as a small urban community to support the needs of the farmers of the Salt River Valley. Before 1900 several irrigation and canal companies were established to supply irrigation water to the local farmers. However, the unpredictable supply of water from the Salt and the Verde Rivers encouraged the farmers of the area to incorporate for the purpose of obtaining a more reliable source of water. They asked the federal government to construct a large dam in the Salt River and pledge their lands to guarantee its construction. This resulted in the establishment of the Salt River Valley Water Users Association (SRVWUA) in 1903. The SRVWUA was the first serious effort to effectively manage the limited water resources of the Valley of the Sun. It was the first reclamation of land project with the U.S. Reclamation Service, a section of the U.S. Department of the Interior. The SRVWUA is now a part of the Salt River Project (SRP), Phoenix's largest water purveyor and one of the nation's largest public electric power utilities.

The Phoenix metropolitan area has a population which exceeds three million and continues to have the fastest growth rate in the nation. It has been transformed from a large agriculture area to one of the USA's major centers for high technology. Its warm climate attracts people with a desire for outdoor activities and has also become a major tourist attraction. It is this large population and the expanding industrial activities that require effective management of the limited water resources in this semiarid region to support its continued development. Although agriculture has been reduced more than 80% of its maximum production in the 1940s and 1950s, it still requires a reliable water supply and has to be included in regional water resource management plans.

3 WATER SOURCES OF THE PHOENIX METROPOLITAN AREA

Water resources to supply the needs of this large urban center and its peripheral agriculture are:

- Salt and Verde Rivers water (SRP water)
- Groundwater
- Colorado River water (CAP water)
- Reclaimed water (municipal and industrial)
- Local runoff
- Agriculture tail water

The volume and availability of SRP water and CAP water are dependent on weather conditions in the watersheds of the Salt, the Verde and the Colorado Rivers (Figure 1). The supply of CAP water is also dependent on the priority of water allocation of the various states that use the Colorado River. This allocation is regulated by a series of laws and decrees, which is referred to as the Law of the River. Of the 1,850 Mm^3 of CAP water delivered in 1997, 835 Mm^3 (45%) were delivered to the Phoenix area. SRP water deliveries to Phoenix average 1,230 Mm^3. The balance of the water supply comes mostly from groundwater and a progressively increasing municipal and industrial reclaimed water volume. Local runoff, particularly urban and agriculture tail waters are at present considered unusable because of potential health hazards and are usually delivered to blend with runoff or percolate in detention basins.

4 WATER DEMAND OF THE PHOENIX METROPOLITAN AREA

The present municipal annual demand for Phoenix is 1,120 Mm^3. This represents 40% of the total demand. Agriculture, although reduced still consumes 55% of the total demand. Industry only uses 5% of the total demand. The average municipal water use per person for the Phoenix area averages 1 m^3/d.

5 WATER MANAGEMENT IN THE PHOENIX METROPOLITAN AREA

To effectively manage the volumes of water from the different sources described in Section 3 and fulfill the demands mentioned in Section 4 a complex water management system is in place in the Valley of the Sun. This system has many components which are owned and operated by several entities. The operation of this system requires planning and cooperation among the various water entities to optimize the efficiency of its operation.

5.1 *Water resources management system*

The principal components of the water resources management system of the Phoenix metropolitan area are:

- SRP water system
- CAP water system
- Water system of the municipalities
- Agricultural irrigation system

5.1.1 *Salt River Project water system*

The largest volume of water for the Phoenix area is supplied by the SRP using its large water system. Both surface and groundwater is delivered for potable, agricultural and industrial uses. The SRP has the water rights for watersheds of the Salt and the Verde River (Figure 1). These rights were acquired by the farmers of the Valley of the Sun in the early 1900s for irrigation. These rights are inseparable from the land it irrigates. The SRP manages these rights for the landowners. Since these agricultural lands are now predominantly part of a municipality, the water can be delivered for potable or industrial uses and is conveyed to water treatment plants. The SRP water system consists of the following components (Figure 2):

- Surface water storage: four reservoirs in the Salt River and two in the Verde River

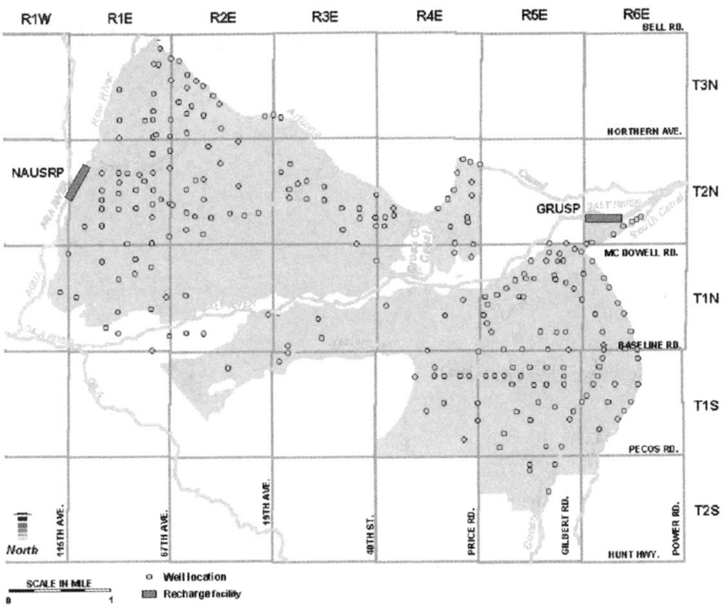

Figure 2. The groundwater supply system of the Salt River Project in the Phoenix metropolitan area.

- A surface water delivery unit also connected to the CAP water system
- A groundwater production unit
- An aquifer storage unit

The total surface storage capacity is 2,880 Mm^3. There is also additional storage space of 1,850 Mm^3 in Roosevelt Lake for floodwaters. The water delivery unit consists of a diversion dam and an extensive network of two major canals and branching laterals, which cover most of the metropolitan area. All water delivery is by gravity. The groundwater unit consists of 250 high capacity wells (160 L/s to 315 L/s) that range in depth from 150 m to 900 m. The aquifer storage unit consists of two large water-spreading recharge projects each with a storage capacity of 2,500 Mm^3 and an annual recharge capacity of 125 Mm^3. This unit also has 15 high rate recharge wells. The wells can pump approximately 620 Mm^3/yr. The SRP has been successfully using conjunctive water management since the 1930s when the entity also became an electric utility in order to provide inexpensive power to the irrigation wells.

5.1.2 *Central Arizona Project water system*

The Central Arizona Water Conservation District (CAWCD) operates the Central Arizona Project water system (Figure 3). It consists of the following units:

- An aqueduct: 540 km long with turnout structures and a capacity of 1,850 m^3/yr
- Pumping stations
- One surface reservoir
- Four groundwater recharge facilities with an annual capacity of approximately 250 Mm^3

The CAP system handles a single water source: Colorado River water (CAP water), which is pumped from a reservoir in this river and delivered to agricultural districts and to the cities of Phoenix and Tucson.

5.1.3 *Water system of the municipalities*

The Phoenix metropolitan area is a conglomerate of more than 20 municipalities. Each municipality is responsible for the treatment of raw water to potable. They also hold ownership of the effluent and are responsible for its conveyance. Most use both SRP and CAP water and supplement the demands with groundwater from their own wells or from SRP wells. Their water systems vary in capacity according to the population they serve but usually include the following components:

- Water treatment plants (WTP): filtration and disinfection
- Water reclamation plants (WRP)
- Potable water and sewage delivery units
- Wells connected directly to the potable water delivery system. Water is chlorinated at the wellhead

There are two large WRPs which treat the effluence of several municipalities. In addition, there are smaller plants which deliver the reclaimed water for direct use (turf irrigation in golf courses, parks, recreational lakes, …) or recharge the highly treated water for credits. The Water Campus Project of the City of Scottsdale is the nation's most technologically advanced WRP and recharge facility.

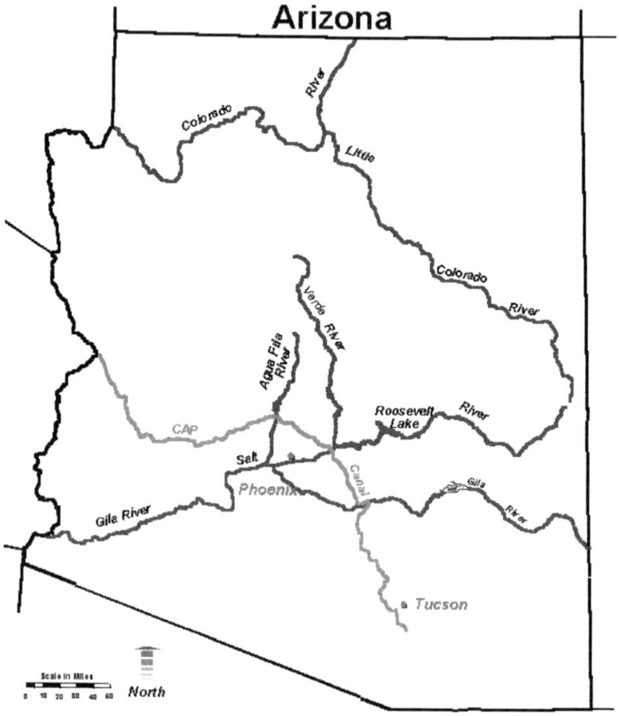

Figure 3. The Central Arizona Project Aqueduct. Imported water from the Colorado River to Southern Arizona.

5.1.4 *The irrigation system for agriculture*

The remaining farms in the periphery of the Phoenix metropolitan area receive SRP and CAP water, which is delivered by gravity to the cultivated fields via the SRP delivery system. Water for the industrial facilities of the area is delivered from the water system of the municipalities or the SRP system.

6 MANAGEMENT OF THE WATER SYSTEM OF THE PHOENIX METROPOLITAN AREA

The technical aspects of the water system described in Section 5 require appropriate management to operate in an efficient and cost-effective manner. This is done by the contribution, coordination, supervision and direction of several entities following an operation plan that follows general guidelines that are controlled by rules and regulations contained in local, state and federal laws. In the USA, the regulation of surface and groundwater are the domain of each state. In the state of Arizona this group of regulations is contained in Title 45 of the Arizona Revised Statutes. However, the federal government has jurisdiction over navigable waters in any state and over water quality. The applicable federal laws are:

- The Federal Water Pollution Control Act or Clean Water Act (CWA)
- The Safe Drinking Water Act

The state of Arizona follows the doctrine of prior appropriation for its surface water and regulates the use of the groundwater. The applicable laws are:

- The Groundwater Management Act (GMA) of 1980
- The Recharge and Underground Storage and Recovery Act of 1986

- The Environmental Quality Act (EQA) of 1986

The Arizona Department of Water Resources (ADWR) regulates all water resources in the state with assistance from the Department of Environmental Quality (ADEQ) in water quality matters. The local regulators are the municipalities and such specialized entities as flood control districts. In the Phoenix area the water system of each entity operates independently and for the benefit of its users following the local, state and federal regulations. There is also close cooperation among all water entities driven by organizations like the Arizona Municipal Water Users Association.

Three other entities that contribute to the management of water resources in the Phoenix area are:

- The Arizona Water Banking Authority (AWBA)
- The Central Arizona Water Replenishment District (CAWRD)
- The Maricopa Association of Governments (MAG)

The AWBA was created in 1996 to fully utilize the allocation of the Colorado River water assigned to Arizona by the U.S. Supreme Court in 1966, resulting from the Arizona *vs.* California lawsuit. The CAWRD assists the smaller and economically weaker communities to obtain water. The MAG promotes the cooperation among municipalities for an effective management of the region's limited water resources. For the Phoenix Active Management Area, one of four areas in Arizona of groundwater intensive use, the ADWR prepares a 10-year plan to protect this valuable resource. The various components of the plan, such as water conservation, are then promoted and enforced by this agency. This has resulted in considerable decrease in wasteful use of both surface and groundwater during the last 20 years and has also mitigated the effects of aquifer stressing by unregulated groundwater pumping. Two of the benefits from this control has been reduction of land subsidence and groundwater quality deterioration.

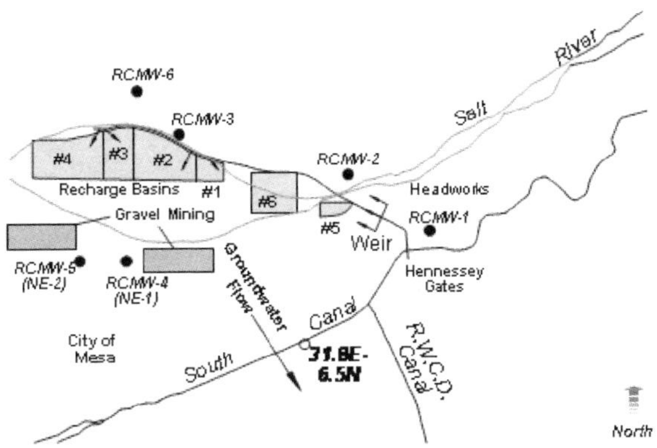

Figure 4. Plan view of the GRUSP recharge facility showing its various components.

7 FACTORS THAT DETERMINE THE EFFECTIVENESS OF THE WATER RESOURCES MANAGEMENT OF THE PHOENIX METROPOLITAN AREA

There are several key factors in the management of water resources to provide a safe and reliable supply of water to the large and rapidly expanding population in the Valley of the Sun. These are:

- Water conservation
- Water importation
- Water recycling/reuse
- Aquifer water storage
- Water transmission

For water conservation both public education and creating a fair but adequate user cost rate have been successful. The importation of CAP water has provided a supplemental and reliable supply, which has considerably reduced the use of groundwater. Direct aquifer recharge has augmented the volume of groundwater for use during drought periods and has served as an effective tool to mitigate demand peaks. The Granite Reef Underground Storage Project (GRUSP) recharge facility (Figure 4) has stored 925 Mm^3 of water in a stressed aquifer during its seven years of operation (Lluria, 1995). Indirect recharge programs have reduced the use of groundwater in the high water demand agricultural areas. The continued maintenance and rehabilitation of the water delivery and distribution units of all the water systems is a high priority for all water purveyors of the Phoenix area and has avoided unnecessary loss of water.

8 CONCLUSIONS

The success of providing a safe and reliable supply of water to the sixth largest urban center in the USA is the result of well planned, adequately coordinated and appropriately regulated water resources management operation. Up to the 1970s, Phoenix depended on groundwater for more than 40% of its water demand. By conjunctively managing all water supplies, the use of groundwater has decreased. From 1990 to 1995 the mining of groundwater in the Valley of the Sun has been reduced by 45% (ADWR, 1999). Although the decrease of agriculture has influenced this reduction, the principal factor has been a considerable improvement of the regional water management. Further improvement will be needed for the condition of safe yield to be attained by 2025 as mandated by the GMA. A tri-state water agreement is presently being negotiated between Arizona, California and Nevada that will assist in reaching that goal.

REFERENCES

ADWR (1999). *Third Management Plan for the Phoenix Active Management Area. 2000–2010.* Arizona Department of Water Resources: 453 pp.

Lluria, M.R. (1995). A larger aquifer storage facility for the Phoenix area. *In Proceedings of the Second International Symposium on Artificial Recharge of Groundwater*: American Society of Engineers: 129–138.

Water resources management of an overexploited aquifer. The Adra-Campo de Dalías system, Almería, Spain

A. Sahuquillo[(1)], J. Andreu[(2)], M. Pulido[(3)], J. Paredes
Dept. Ing. Hidráulica y M. A. Politechnical Univ. of Valencia, Spain
[(1)] *asahuq@hma.upv.es*
[(2)] *ximoand@upvnet.upv.es*
[(3)] *mapuve@hma.upv.es*

A. Sánchez
Gestión Dominio Público Hidráulico. M. Medio Ambiente, Spain

V. Pinilla
PROINTEC S.A.

J. Capilla
School of Civil Eng., Politechnical Univ. of Valencia, Spain

ABSTRACT

The Campo de Dalías aquifer area in western Almeria (Spain) has faced a spectacular increase in cultivated land and population, becoming the main factor of economic growth in the province. Nevertheless, like in all arid areas, water resources are very scarce and aquifers became overexploited producing undesirable effects including seawater intrusion, important local drawdowns, and a general decay in water quality. Currently the piezometric level in almost half of the surface of the highly transmisive limestone aquifers, which are the most productive in Campo de Dalías, have their piezometric levels between 10 m and 20 m below sea level. Several conjunctive use strategies are analyzed in this work with the goal of minimizing overexploitation effects and stopping seawater intrusion by importing water from the new Benínar dam in the nearby Adra river, minimizing the impacts and maintaining the water rights in the donating basin, maximizing reliability of water supply, and relocating water pumping between the different aquifers. The Benínar reservoir has important losses by infiltration. In order to develop an analytical tool, the whole water resources system is modeled with the simulation model SIMGES, a component of the Decision Support System known as AQUATOOL developed at the Technical University of Valencia for complex water systems planning and management. SIMGES allows the inclusion of several elements of regulation and storage, either surface or subsurface ones, water intakes, channels, consumption units, pumping from aquifers, etc., and is able to consider conjunctive use of surface and groundwater. The Adra delta aquifer and the aquifers of the Campo de Dalías are integrated in the model using the Eigenvalues Method, a distributed approach, also developed at the UPV and based on previously calibrated groundwater flow models, that solves explicitly the space discretised groundwater flow equation allowing a very efficient integration of an aquifer distributed model in the simulation of complex systems with conjunctive use, without losing accuracy. Simulations shows that the evolution of the aquifers is not sustainable with current pumping. Therefore, the decision support system has been used in order to devise adequate measures to ensure a sustainable exploitation. Several reductions for the demand to be satisfied have been considered, the safe abstraction from the Adra basin have been determined, as well as the operating rules of the system to minimize the environmental impacts and relocation of groundwater abstractions to improve water quality and reduce seawater intrusion.

Keywords: *conjunctive use, seawater intrusion, dam leakage, spring regulation, eigenvalues.*

1 INTRODUCTION

The high value of the crops produced in the coastal plain of Campo de Dalías (Almeria) led to an spectacular increase of cultivated land and population that started four decades ago. In fact, this development became the main factor of economic growth in the province. Currently, the area of cultivated land under greenhouses is over 20,000 ha.

Associated to this quick development there has been an important increment of water demand, mostly supplied by groundwater from the Campo de Dalías aquifers. These aquifers also supply the urban water demand of a population about 250,000 inhabitants, including Almeria city. In the period from 1984/85 to 1994/95 the average annual water volume abstracted from the aquifers was 127 Mm^3. These withdrawals have caused drawdowns in piezometric heads of the most productive limestone aquifers leading to water table levels below the sea level. This has caused seawater intrusion problems and progressive water quality degradation in sectors of eastern and especially western area of the Campo de Dalías. In the last 16 years more than 300 pumping wells have been abandoned (because of water quality or quantity problems) and about 125 new wells have been drilled (Dominguez *et al.*, 2001). Since 1987/88 this system receives water imported from the Benínar reservoir, located in the contiguous Adra River basin, by means of the Benínar-Aguadulce channel.

In the research described in this paper, several conjunctive use strategies are analyzed with the goal of determining optimal management operational rules for the available resources. This is done with the objective of minimizing overexploitation effects by importing water from the Adra River (limited by stakeholders rights and maximizing reliability of water supply) and relocating pumping wells. A simulation model reproduces all the relevant aspects of the system, allowing the evaluation of different management alternatives.

2 DESCRIPTION OF THE SYSTEM

The system includes the following subsytems: Campo de Dalías, Sierra de Gádor and Adra River basin.

The Campo de Dalías is a coastal plain of 330 km^2, delimited by the Mediterranean Sea in the South and the Sierra de Gádor in the North. It constitutes the main water supply source in the western Almería, either for irrigation (main consumption destination) or for urban and industrial supply.

From an hydrogeological point of view, the Campo de Dalías is made up by several interconnected aquifer units in the plain and by the southern part of Sierra de Gádor. The main recharge area of this complex system is Sierra de Gádor, where infiltration takes place in an area of 160 km^2, made up by high permeable limestones and dolomites. The aquifer subunits are interconnected, with a complex spatial disposition. In some areas there are up to three units in a vertical cross sections (IGME, 1989, 1995). Depending on their structural disposition we can distinguish the Upper and Medium aquifers, mainly detritic, and the Lower aquifers, very fracturated and fragmented in blocks, belonging to the carbonated triassic series of the Gádor formation. There are six main subaquifers in the Campo de Dalías: in the western, the Inferior Occidental Aquifer (AIO) and the Escama of Balsanueva aquifer (AEBN), in the central area the Central Upper Aquifer (ASC) and in the north-eastern the Inferior North-easthern Aquifer (AIN), Intermediated North-East (AMN) and North-East Upper (ASN). These subunits are assimilated to large blocks of a spatial discretization. The contact surfaces among them define the possible areas of flow. The pumping wells and recharge are assigned to the subunit or block to which they physically belong.

The Adra River basin is located in the southern slope of Sierra Nevada, limiting with the Llanos de Berja to the East, eastern limit of Sierra de Gádor, and with the Sierra la Contraviesa to the West.

The Benínar reservoir, located in the Adra River, has an active storage capacity of 60 Mm^3 and important losses by infiltration. In this basin there are several aquifers, corresponding to different lithologies and hydrodynamic properties of the existing rock formations. However, the main aquifer is the Turón-Peñarrodada, made up by limestone and dolomites of the Lujar formation. It reaches the surface in three outcrop areas: Peñarrodada, Turón and Cerrón-Álvarez. Downstream the reservoir, the Adra River flows through the calcareous area of Turón, where it has carved a deep canyon at the end of which several springs are found, the Fuentes de Marbella springs. Before the construction of Benínar dam, the springs were fed by infiltration in the river bed and through the limestone outcrop of Turón, approximately in the same proportion. Currently water rarely flows through the canyon and the inflows of the springs are explained by the rainfall infiltration and the reservoir losses. Several wells have been drilled to capture part of the reservoir infiltration to the limestone aquifer. Pumped water can be incorporated to the Benínar-Aguadulce channel.

In the lower Adra River basin the Neogenous and Quaternaries materials make up an aquifer system called delta of the Adra River, in a general sense, the limits of which are the impermeable Alpujarrides materials and the sea. Surface water from the Adra River is used to irrigate crops on the right bank of the delta. On the left bank, applied water is pumped from the aquifer.

3 MATHEMATICAL SIMULATION MODEL

The conceptual model of the system is shown in Figure 1. The system is represented by a mathematical model based on the simulation model SIMGES. This is part of the Decision Support System known as AQUATOOL (Andreu *et al.*, 1996), developed at the Technical University of, Valencia for complex water systems planning and management. SIMGES is a general model for

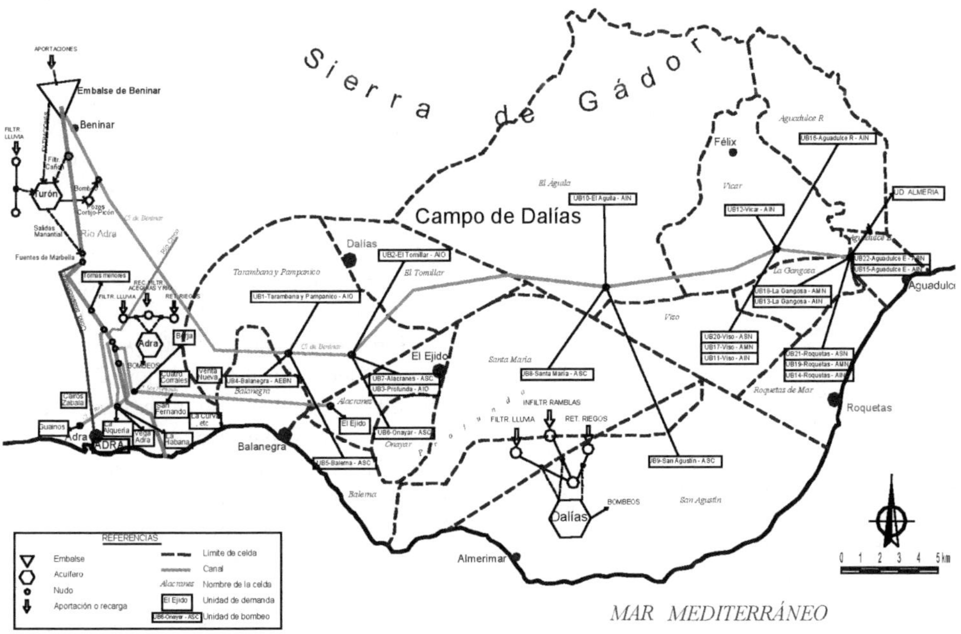

Figure 1. Scheme of the conceptual model of the system.

the simulation of water resources systems with several elements of regulation and storage, either surface or subsurface ones, water intakes, channels, consumption units, pumping from aquifers, etc., and is able to consider conjunctive use of surface and groundwater. The design of a scheme representing the system, introduction of data and analysis of results are facilitated by graphical user interfaces and postprocessors. The simulation is made at a monthly time step and reproduces the water flow through the system following the operation rules defined by the model user. The results of the model simulation comprise the evolution of all the interesting variables: inflows, water stored in reservoirs, releases, streamflows, control variables of aquifers, etc., monthly and annually. Average values for the simulation horizon are also computed, as well as deficits and reliability for the different demands. The simulation has been made for the historical hydrological series from 1945/46 to 1994/95.

3.1 *Modelization of the Adra subsystem*

The scheme includes only one reservoir, Benínar. The reservoir operation simulation is based on a simple mass balance, adding the inflow to the reservoir for every month to the initial storage, and subtracting the outflow, infiltration and evaporation losses. Part of the water stored in Benínar reservoir is transferred, by infiltration to the Turón-Peñarrodada aquifer, process that has been incorporated to the model by an equation that relates water stored and infiltration. The coefficients in the equation are calibrated using data of daily balance in the reservoir.

The Turón aquifer receives the recharge of rainfall infiltration, reservoir seepage and water infiltration in the calcareous canyon of Turón. The outflow in the Fuentes de Marbella springs depends on the state of the aquifer and the recharge of the period. The simulation model of the aquifer has been integrated in the global model using a semi-distributed Model (Sahuquillo 1983a; Pulido and Sahuquillo, 2001) comprised by four virtual blocks that are simulated as aggregated models. Two main considerations justify the use of this type of model:

- Given the goals of this study, the model of the Turón aquifer is only a minor piece of the system model. An exhaustive knowledge of the spatial and temporal evolution of the aquifer is not needed. This type of model allows a simple and mathematically sound way to model the relationship between recharge and discharge in karstic aquifers.
- This semi-distributed model requires the calibration of few parameters that can be estimated directly from the input and output data of the system. Thus, the limited knowledge of the formation is properly accounted for avoiding the unnecessary construction of a large distributed model where a great number of parameters would have to be estimated.

Wells at Cortijo de Picón allow pumping water from the Turón aquifer to the Benínar channel, which transfers water from the Benínar reservoir to the Campo de Dalías.

The first intake ditches along the river (zone close to Fuentes de Marbella springs) have been grouped under the name of Tomas Menores demand unit. The groundwater abstraction in the alluvial of the Adra River, downstream the Fuentes de Marbella springs and in the deltas is made by means of subsurface draining galleries, wells and boreholes.

The two existing galleries are related with the Antonio Ruiz and San Fernando canals, which conveys the water to the irrigation districts of Guainos and San Fernando canal.

The surface water still flowing downstream the above diversions, is used for urban water supply and partially diverted through minor ditches that irrigate mainly the eastern area of the delta. The urban water demands are considered in the mathematical model as an aggregated unit identified as UDU Adra. They are satisfied by an intake in the spring of the Fuentes de Marbella springs and two wells in the delta aquifer. Generally the river flow does not reach the deltaic plain because of the existence of two large diversions that feed a complex ditches network. The most important irrigation districts in the municipality of Adra are: Adra community, San Fernando canal community, and the Cairos-Zabala one. The Adra Community manages all the surface water resources of the river, and it is divided in three sectors: Vega de la Alqueria, Vega del Adra and Vega de la Habana. The Otros

Riegos comprises the irrigation demands of the remaining irrigation districts of the Adra delta.

The delta aquifer is integrated in the simulation scheme of the system by a distributed model using the Eigenvalue Method (Sahuquillo, 1983b), based on the aquifer calibration by a finite difference model. The Eigenvalue Method allows a more efficient integration of an aquifer distributed model in the simulation of complex systems with conjunctive use, without losing accuracy (Andreu and Sahuquillo, 1987). The model (CHSE, 2004) comprises 139 active cells, of 400 m in the coastal direction by 253 m in the perpendicular direction, and a thickness of 100 m (the simulation has been done with a constant transmissivity layer for each cell). The AQUIVAL module (Capilla and Andreu, 1996) has been used for the preprocess of the aquifer by the Eigenvalue Method. The recharge stresses considered over the aquifer are: rainfall infiltration, infiltration loses in the Alqueria, Vega Adra – La Habana ditches, infiltration in the river and irrigation and urban return flows. Pumping is made to satisfy the demands of La Habana, Cairos-Zabala, San Fernando in Adra, Otros Riegos and Adra Urban Demand. There is also a borehole close to the river that pumps water to the San Fernando canal.

3.2 *Modelization of Campo de Dalías subsystem.*

The aquifer system of Campo de Dalías is integrated in the simulation scheme of the system by a model based on the Eigenvalue Method, using the calibration of the aquifer by a integrated finite differences model (Pinilla and Sahuquillo, 2001). The spatial discretization used in the model consists on 22 element of large size or macrocells. The delimitation of the macrocells coincides, with some variations, with the zones in which the IGME (Spanish Geological Survey) has traditionally grouped the exploitation. These zones were established following mainly hydrogeological criteria. In spite of their size, the representation of the system by means of these macrocells is considered appropriate because those that are more important, regarding the abstraction volume and lateral exchange with others, have very high transmissivities that makes them to be well represented by a single average level. The degree of knowledge for the hydrodynamic properties of the aquifers, their interconnection, the exploitation and the piezometric head evolution during the calibration period, are very different. In general, the data of exploitation and piezometric levels are more reliable for the most recent periods. The best quality data corresponds to 1976/77 to 1994/95.

In the Campo de Dalías aquifers, recharge is mainly due to rainfall and runoff infiltration in the ravines of the south slope of the Sierra de Gádor (main contribution to recharge), to rain infiltration in the Campo de Dalías plain and to infiltration in irrigated areas. Based on the volume balance at a daily time step, the infiltration of precipitation and runoff that takes place in the Sierra de Gádor has been evaluated for each macrocell of the aquifer model using an aggregated rainfall-runoff model of few parameters, Temez model (Témez, 1997). This has been done for each element of a grid in which the whole area has been previously discretized. The infiltration values for irrigation flow returns, increased by rain and losses in urban supply network, depends on the irrigation surface at each moment. The loses in the supply network imply a low value, especially considering the uncertainty to which their evaluation is subjected. The values of irrigation returns obtained for the 1994/95 period were extended to the whole simulation period. Regarding the effect of precipitation in the humid years, we consider that is partly compensated by the consequent reduction of irrigation, and it can be considered in average values. For each cell of the model the recharge by irrigation return values, rain and runoff infiltration in the Sierra de Gádor were added in a monthly recharge series.

One of the objectives of the present study is the determination of alternative abstraction distributions in the aquifers of the Campo de Dalías with the goal of reducing the effects of overexploitation on it. This objective has conditioned the criteria of modeling pumping and water uses in the Campo de Dalías. It has been considered that it was not reasonable to go into the detail of water distribution for its use in the Campo de Dalías, assuming that, given a global quantity of available resources in the Campo de Dalías, this is applied in the field with a similar distribution to which was applied in an *standard* campaign (such as 1994/95). It is important to reflect appropriately the abstraction in each macrocell of the model in order to be able to analyze the evolution of the overexploitation. Thus, 22 *units of pumping* have been defined in the conceptual model. Each unit of pumping has been associated to a macrocell of the aquifers system of the Campo de Dalías and all of them with possibilities of receiving contributions of the Benínar channel. The following conceptualization of the operation has been made:

- For each alternative, a global demand in the Campo de Dalías, DT, is defined. It should be supplied by water from Benínar channel and aquifer withdrawals.
- For each alternative, a coefficient of pumping distribution, α_i, is defined for each macrocell i, so that .

$$\sum_{i=1}^{22} \alpha_i = 1$$

- As a consequence of the simulation of Adra subsystem, the quantity of transferred water from Benínar for every period, TB_t, is obtained.
- The global quantity to bc withdrawn from the Campo de Dalías aquifers is: $BT^t = DT - TB_t$
- The quantity to be withdrawn from each macrocell is: $B_{t,i} = \alpha_i * BT^t$

Therefore, all the demands that at each moment are supplied from the aquifers of the Campo de Dalías and the water imported by the Benínar channel are represented by the 22 units of pumping (UB).

4 MANAGEMENT STRATEGIES

Water management of the system is made every month in the following way. The urban demand UDU Adra (highest priority) makes use of spring outflow. If the outflow is enough to satisfy the monthly demand, the excess flows downstream. If it is insufficient, the water released from the reservoir is increased to satisfy the UDU Adra demand considering the infiltration losses to the aquifer when water flows through the canyon of Turón. For each intake of irrigation demand in the river with the right to surface water supply, the water available is obtained by adding up the springs outflows and the water released from the reservoir, and subtracting the water diverted to higher priority demands. If this water is insufficient, the release is increased in the precise quantity.

When there is not enough water in the reservoir the different demands are satisfied in order of priority until reaching the minimum storage volume of the reservoir. The less prioritized demands will suffer the corresponding shortage. If the flow in the downstream intakes of the Adra River is lower that the minimum flow required to recharge the delta aquifer, additional release from the Benínar reservoir will occur. The resources of the reservoir are released so that, in case of existing shortage, UDU Adra is the first demand to be supplied. Second priority corresponds to the demands of Vega de Adra, Alquería and Tomas Menores. They are followed by the demands of La Habana, UDA San Fernando and Otros Riegos C.S.F., which can compensate shortage by means of pumping from the aquifer of the delta. Finally, the remaining available stored water is transferred to supply the demands of the Campo de Dalías. The Otros Riegos and Cairos-Zabala demands are supplied exclusively with groundwater. In the demands that can obtain surface and groundwater supply, the latter is used as a complement to the former when the surface water supply is not enough.

In the different simulations, we have considered the possible use of the Pozo-Picón wells to transfer water to the Benínar channel. This possibility allows to take advantage of the infiltration from the reservoir to Turón aquifer and the capacity of regulation that the latter offers. The effect of using the wells to pump water to the Adra River has also been studied as a means of completing the guarantees in satisfaction of the demands associated with the river.

Using the model with the scheme of the system corresponding to the situation previous to the construction of the Benínar dam and channel, the historical guarantees of satisfaction of the Adra traditional demands have been determined. These guarantees of satisfaction turn out to be relatively high, and the objective is that the management with the new infrastructure does not worsen them.

By means of the use of the model with the complete system scheme, some preliminary simulations have yielded a first estimation of the average volumes to be transferred to the Campo de Dalías. The repercussions that this derivation would have on the guarantees of the demands of the delta of Adra and on the state of the delta aquifer have also been analyzed. The main conclusion of these preliminary simulations is that, in order to avoid damaging the demands of the Adra River and the delta aquifer, it is necessary to determine and establish a reservation curve for the Benínar reservoir. When the stored volume is located below this curve, transfer to the Campo de Dalías will not be allowed. Likewise, it shows the convenience of establishing a minimum flow at the beginning of the delta with the purpose of assuring enough recharge to the Adra aquifer.

By means of multiple simulations and using a critical biennium of inflows to Benínar reservoir, a reservation curve has been determined to maintain the indicators of demand satisfaction and delta aquifer levels under acceptable conditions. Once the Adra demands have been satisfied, if there is still water in the reservoir above the reservation curve, either the Campo de Dalías demand for this month or the available volume up to the reservation curve (if lower than the Campo de Dalías demand) will be transferred by the Benínar channel. The water of the channel is distributed among the Campo de Dalías demands. If there is a water shortage, the Campo de Dalías demands can obtain the water by pumping from the corresponding aquifers until the demands are satisfied.

Finally, a sensibility analysis of the results of the simulations versus variations in the reservation volumes adopted for the Benínar reservoir has been carried out. Transferred volumes through the Benínar channel, levels of satisfaction of the demands, and balances in the aquifers and surface outflows into the sea have been compared.

5 ALTERNATIVES CONSIDERED

Using the reservation curve previously determined, we have proceeded to the simulation of alternatives that affect directly the aquifers of the Campo de Dalías. The first simulation (Simulation 130) considers the Campo de Dalías demand as equal in magnitude and distribution among cells to the historical one registered in the period 1994/1995. This demand was 130 Mm^3/yr, with 0.5 Mm^3 of them satisfied with water from Benínar channel, and the rest, 129.5 Mm^3, with water abstracted from the aquifers. The objective of this simulation is to evaluate the improvements achieved in the system management with the use of the reservation curve, maintaining the value of the demands of the last period from which reliable data is available. An average annual transfer of 22.5 Mm^3/yr from Benínar reservoir to the Campo de Dalías is obtained. In spite of it, the evolution of levels in the Campo de Dalías is not sustainable. Most of the surface of the inferior carbonate aquifers (AIN and AIO) remains below sea level, sometimes more than 20 meters, during most of the simulation period. It was necessary then to look for alternatives for the system management. The two bases that have been used to define these alternatives were:

- The total volume of demand to be satisfied jointly by surface and groundwater in the Campo de Dalías. The simulations show that the evolution of the aquifers is not sustainable if the pumping volume corresponding to the demand of the campaign 1994/95 is maintained. The possibility of obtaining additional resources has been analyzed, so that the total water volume to be satisfied from the Adra and the aquifers of the Campo de Dalías were lower. Concretely, reductions to 100.5 Mm^3/yr, 90.5 Mm^3/yr and 80.5 Mm^3/yr of the demand to be satisfied by means of transfer and pumping have been considered. These decreases would be compensated by the improvement of the efficiency of the uses and for the use of alternative sources, as desalination or transfer from other systems.
- The reorganization of groundwater abstractions from the Campo de Dalías aquifers. Starting from the distribution of historical abstractions corresponding to the year 1994/95, the simulations demonstrate that there are aquifer subsystems of the Campo de Dalías that do not evolve in a sustainable way,

even when reduction of abstractions is considered. Therefore, it is necessary to study the consequences of reorganizing the abstractions. Diverse alternatives with different distribution of abstractions among the aquifers of the Campo de Dalías have been simulated for the different hypotheses of global demand (Table 1) and the evolution of the resulting levels has been studied.

The different alternatives have been simulated and the results have been compared and analyzed. These results have been used to establish conclusions on the most advisable alternatives and, in general, on the management of the whole system.

For the study of the effect of the different alternatives on the system of aquifers of the Campo de Dalías, the results have been analyzed for different control variables along the simulation. For each alternative the inflows and outflows to the sea have been plotted in aggregated balance for the whole system, as well as individually for each cell in connection with the sea.

6 CONCLUSIONS

The main conclusions on the different alternatives of conjunctive management of the Adra-Campo de Dalías system are synthesized in the following aspects.

Regarding the Adra basin we conclude that:

- To preserve the satisfaction of the Adra basin demands and the situation of the delta aquifer it is necessary to establish the reservation curve in the Benínar reservoir.
- If the resources of the Adra basin are properly managed, the average annual volume that is transferable to the Campo de Dalías ranges from 21 Mm^3/yr to 22 Mm^3/yr. The average annual water volume in Benínar reservoir released toward the delta of Adra would range from 4.2 Mm^3/yr (Simulation 80.5) to 4.8 Mm^3/yr (Simulation 130). Maintaining the reservation curve and the minimum flows defined in the beginning of the delta of Adra (0.15 Mm^3/month), the guarantees of the demands and the delta aquifer state are acceptable for these values of transfer.

If the increase of levels in the occidental part of the delta caused drainage and salinity problems, it would be convenient to reorder the groundwater abstraction in the delta by transferring part of the pumping from the eastern to western part. Although it does not seem probable according to the obtained results, if seawater intrusion problems were presented in the oriental riverbank during a drought period it would be possible to supply the community of La Habana temporarily with surface water.

The Cortijo-Picón wells should be used to increment the total flow at the Benínar channel (therefore, for the Campo de Dalías supply) whenever the outflows from the Fuentes de Marbella springs are greater than 2 Mm^3/month. The average annual volume of pumping ranges from 0.7 Mm^3/yr to 1.7 Mm^3/yr, depending on the alternative, although the use of the wells is sporadic. Moreover, it is possible to guarantee 100% reliability for those demands of Adra with surface supply. This can be achieved by using the Cortijo-Picón wells to supply the Adra River in those few occasions in which the water volumes of Benínar reservoir plus the Fuentes' outflows are insufficient.

Regarding the Campo de Dalías we conclude that:

In order to achieve a sustainable evolution of the Campo de Dalías system of aquifers, it is necessary to decrease the total demand supplied by resources from the Campo de Dalías aquifers plus the transfer of 40–50 Mm^3/yr. It can be attained either by improving the efficiency of the uses and/or resorting to alternative sources, as desalination or importing water from other systems. Given the uncertainties associated to the behavior of the system, the most prudent option would be a reduction of 50 Mm^3/yr.

Likewise, it is necessary to reorganize groundwater abstractions according to table 1 in order to accomplish a sustainable evolution of the carbonate inferior aquifers of the Campo de Dalías. This reordering produces and increase groundwater abstraction in the upper aquifers (ASC and ASN) and a decrease in the lower carbonate aquifers (AIO and AIN) that at the moment bear 85% of the total water pumped. Since the ASC presents a high salinity, it will be

Table 1. Annual demand in each pumping unit for each simulation.

	SIMUL.130	SIMUL. 100.5	SIMUL. 90.5	SIMUL. 80.5
AIO	Annual demand (Mm^3)			
Taramb.-Pampánico	31.13	14.64	14.18	11.46
Tomillar	27.16	12.77	12.43	10.05
Profunda	7.05	3.31	3.32	2.68
TOTAL	65.33	30.71	29.93	24.19
AEBN	0.74	0.00	0.00	0.00
ASC	Annual demand (Mm^3)			
Balerma	0.00	0.66	0.70	0.70
Onáyar	0.23	1.76	1.34	1.34
Alacranes	0.00	1.16	0.50	0.50
Sta. María	7.42	20.23	16.07	16.07
San Agustín	1.81	6.50	4.80	4.80
TOTAL	9.46	30.31	23.41	23.41
AIN	Annual demand (Mm^3)			
El Águila	25.16	15.39	15.45	12.49
El Viso	10.39	6.35	6.38	5.16
Vicar	0.26	0.21	0.21	0.17
La Gangosa	0.00	0.00	0.00	0.00
Roquetas	0.00	0.00	0.00	0.00
Aguadulce E	9.66	0.21	0.21	0.17
Aguadulce R	0.00	0.00	0.00	0.00
TOTAL	45.46	22.16	22.25	17.99
AMN	Annual demand (Mm^3)			
El Viso	1.53	2.35	2.35	2.35
La Gangosa	2.46	1.76	1.76	1.76
Roquetas	0.00	0.00	0.00	0.00
TOTAL	3.99	4.11	4.11	4.11
A.S.N.	Annual demand (Mm^3)			
El Viso	4.25	11.77	9.47	9.47
Roquetas	0.29	0.55	0.65	0.65
Aguadulce E	0.47	0.88	0.68	0.68
TOTAL	5.01	13.20	10.80	10.80
TOTAL	**130.00**	**100.50**	**90.50**	**80.50**

necessary to treat the water to obtain an acceptable salinity for irrigation.

In the proposed solutions it is observed that, in the sequences of humid years in the aquifers of the AIN, certain levels that can produce important outflows of fresh water to the sea are reached. To determine the most appropriate form of withdrawing part of those flows it would be necessary to have a more accurate modeling of the carbonate aquifers of the northeast part of the area.

Nonetheless, it seems highly advisable to begin diminishing the exploitation of the inferior carbonate aquifers and taking strict control of exploitation and evolution of levels and seawater intrusion in the affected aquifers. In spite of the achieved advances (Pulido Bosch *et al*., 2000; Domínguez, 2000) there are still uncertainties about the real situation of the seawater intrusion in the aquifers and its possible evolution. To reduce these uncertainties a detailed analysis of the necessary control and the most appropriate methods is required, as well as the elaboration of annual reports on the evolution of the control variables (abstraction, levels and water quality).

ACKNOWLEDGEMENTS

The model that has been described is integrated in the *Estudio para la determinación del Plan de Ordenación del Campo de Dalías* (Study for the determination of the Campo de Dalías Water Plan), carried out for the *Confederación Hidrográfica del Sur de España* (Water Agency of Southern Spain), under the technical direction of the General Subbureau of Hydraulic Public Domain Management of the Ministry of Environment of Spain. This study has been commissioned to the consultant company PROINTEC, Corp. The Department of Hydraulic Engineering and Environment of the Technical University of Valencia has collaborated in its realization. We appreciate the *Confederación Hidrográfica del Sur* authorization for the presentation of these results and the information provided. We also thank the IGME for their collaboration and information supplied.

REFERENCES

Andreu, J. and Sahuquillo, A. (1987). Efficient aquifer simulation in complex system. *Jour. Water Resources Planning and Management*, ASCE, 113 (1), pp. 110–129.

Andreu, J.; Capilla, J. and Sanchís, E. (1996). AQUATOOL, a generalized decision-support system for water-resources planning and operational management. *Journal of Hydrology,* No. 177, pp. 269–291.

Capilla, J. and Andreu, J. (1996). AQUIVAL: A GUI for groundwater modeling incorporated into the simulation of complex water resources systems. *Hydraulic Engineering Software*. Computational Mechanics Publications (ed. W.R. Blain). pp.71–80.

CHSE (2001). *Estudio para la determinación del Plan de Ordenación del Campo de Dalías.* Confederación Hidrográfica del Sur de España. Ministerio de Medio Ambiente.

Domínguez, P. (2000). *Funcionamiento hidrogeológico y mecanismos de intrusión marina en sistemas carbonatados de estructura compleja: aplicación al Acuífero Inferior Noreste (AIN) del Campo de Dalías (Almería).* Doctoral thesis . DET-UPC–IGME.

Domínguez, P.; González, A. and Franqueza, P.A. (2001). Situación actual de los acuíferos del Campo de Dalías. Un ejemplo de la necesidad de conocer el estado actualizado del funcionamiento de un sistema complejo intensamente explotado. *VII Simposio de Hidrogeología,* Murcia. Ed. IGME. Tomo XXIII. p. 211–225.

IGME (1989). Síntesis hidrogeológica del Campo de Dalías (Almería). *Propuesta de primeras actuaciones de investigación y gestión.* Ministerio de Industria y Energía, Madrid, 169 pp.

IGME (1995). Situación de los acuíferos del Campo de Dalías en relación con su declaración de sobreexplotación. *VI Simposio de Hidrogeología,* Sevilla, Tomo XXI, pp. 443–467.1995.

Pinilla, V. and Sahuquillo, A. (2001). Modelo matemático para la evaluación del flujo de agua subterránea en los acuíferos del Campo de Dalías. *Conferencia Internacional Las caras del Agua Subterránea,* UPC, Barcelona, IGME.

Pulido, M. and Sahuquillo, A. (2001). Modelación de las relaciones río-acuífero. Modelo Pluricelular Englobado. *VII Simposio de Hidrogeología,* Murcia, IGME, XXIII, pp. 151–163.

Pulido Bosch, A.; Vallejos, A.; Molina, L. and Pulido Leboeuf, P. (2000). Problemática hidrogeológica del Campo de Dalías-Sierra de Gádor (Almería). *El Campo de Dalías, Paradigma de uso intensivo.* Papeles de Proyecto Aguas Subterráneas, Serie A, 4: 5–36. Fundación Botín.

Sahuquillo, A. (1983a). Modelos pluricelulares englobados. *Utilización conjunta de aguas superficiales y subterráneas.* nº 43 del SGOP, Serv. Publ. del MOPU, Madrid.

Sahuquillo, A. (1983b). An eigenvalue numerical technique for solving unsteady groundwater continously in time. *Water Resources Research*, 19 (1), pp. 87–93.

Témez, J.R. (1977). *Modelo matemático de transformación precipitación-aportación.* ASINEL, España.

Opportunities of conjunctive use of groundwater and surface water

Alfonso Rivera
Geological Survey of Canada, Quebec, Canada
arivera@nrcan.gc.ca

Andrés Sahuquillo [1] and Joaquín Andreu [2]
Politechnical University of Valencia, Spain
[1] *asahuq@hma.upv.es*
[2] *ximoand@upvnet.upv.es*

Aditi Mukherji
International Water Management Institute, Anand, Gujarat, India
a.mukherjee@cgiar.org

ABSTRACT

At the occasion of the *International Symposium on Intensive Use of Groundwater: Challenges and Opportunities* (SINEX; Llamas and Custodio, 2003), a roundtable was held to analyze current practices of the conjunctive use of groundwater and surface water in the world, and to discuss challenges, benefits and opportunities. The roundtable consisted of world-class groundwater specialists from 5 countries in the Americas, Europe and Asia. Selected cases of key countries, having practices and experience in the conjunctive use of surface water and groundwater resources, were presented: India, Pakistan, Spain, Mexico and Canada. The discussions included definitions and misconceptions on the practices of conjunctive use of water resources as well as technical aspects of their effectiveness and applications.

Results of the discussion indicated that emphasis should be given to the integrated management of water resources, via the conjunctive use of surface water in natural or artificial basins, and aquifers. The joint assessment at the regional scale was considered the most adequate approach to effectively manage both resources. Available ad hoc tools and models are vital to the assessment and operation of conjunctive systems. Some of the main conclusions dealt with the strong political will, needed to implement such systems, and the enormous economical and environmental benefits that can be obtained.

Keywords: *Conjunctive use, artificial recharge, alternate use, droughts, efficiency.*

1 INTRODUCTION

In a very basic definition, one would see the conjunctive use of surface water and groundwater as being a mechanism through which the use of available water resources are optimized, and the benefits of doing so are greater than if both sources were managed in an un-coordinated manner. It has to be clear that un-coordinated simultaneous use of surface water and groundwater should not be considered as conjunctive use (although this is a frequent misconception).

Conjunctive use involves at least decisions on when, where and in which amount to use each one of the sources of water. It has been demonstrated (Sahuquillo and Lluria, 2003) that such a coordinated use of both resources may help to solve, or at least attenuate, water quality and water quantity problems. At times, conjunctive use can prove to be a cheaper solution than either sole dependence on surface water or groundwater.

Among the advantages of the conjunctive use of available water resources, are the economic, operational and strategic benefits, or improvements, a society may obtain when optimizing both resources. Although not very obvious at the start of a project, the *economic* advantages become clear when new investments for water supply sources (construction of large dams) decrease and the operational costs of integrated systems are lowered. The *operational* advantages include the increase of available water resources for water supply without necessarily increasing the storage in the basins. Furthermore, some problems due to the over-exploitation of either one of the surface water or groundwater resources, may be reduced or completely solved, such as the drainage and salinization of soils in irrigated lands in arid and semi-arid regions, land subsidence due to excessive pumping, etc.

The rationale behind adopting an approach of conjunctive use of water resources are mainly, although not exclusively, to take advantage of the storage capacity of aquifers, the hydrological inter-linkages between groundwater and surface water, and the differences in the timing of water circulation between these water bodies. The main basic schemes for conjunctive use include: artificial recharge and alternate conjunctive use (ACU). The last includes stream aquifer systems and integrated management of complex groundwater and surface water systems. Artificial recharge makes more emphasis on the storage capacity of aquifers, whereas the emphasis on stream-aquifers systems is on groundwater-surface water relationship. Finally, integrated management explores all of the above-mentioned properties.

A frequent misconception among hydrologists and water planners is to identify conjunctive use mainly with artificial recharge practices. Hence, to demonstrate that ACU is in most cases much more cheaper and easier to implement than artificial recharge, particularly in developing countries, was one of the motivations of the round table in SINEX seminar.

The use of subsurface storage in *artificial recharge* is clear and very intuitive. The objectives of artificial recharge can be the storage of surplus water, reduction of aquifer drawdown, in some cases the recovery of seawater intrusion, water treatment and water banking. The usual practices of artificial recharge are through injection wells and infiltration ponds. In arid regions, artificial recharge is an appropriate option, but this practice may also be applied in other areas and for other purposes. Artificial recharge requires adequate technical operation and monitoring and permanent supervision. In less economically and technically developed semi-arid regions, the influence of operation and maintenance in final water cost could be high for most irrigation needs.

Alternate *conjunctive use* is a simple type of conjunctive use, whereby surface water is used preferentially in wet periods and groundwater is used preferentially in dry periods. However, pure surface water demands, pure groundwater demands and alternate water demands usually coexist. The use of subsurface storage is achieved by differences in storage between the higher levels at the end of several wet years with important groundwater recharge and less pumping, and the lower levels at the end of a dry period with less recharge and considerable abstractions from the aquifer. The concept is less intuitive than artificial recharge, but in no way less effective and in most cases much cheaper. Alternate conjunctive use is currently applied in coastal aquifers, large interior aquifers, alluvial aquifers, and in the *drought supplemental wells* approach. In less developed semi-arid regions, it could be a better option than artificial recharge because in general it is more economic, has less technical problems and is more suitable to developing countries. Moreover, in addition to being more costly and complex in operation, artificial recharge needs a clear identification of payers and beneficiaries, and it needs a complex technical and institutional development. These conditions are infrequent in developing countries.

Nevertheless, that does not preclude the convenience in many cases of the enhancement of natural recharge or development of methods to lower the cost of artificial recharge.

Cases discussed during the round table, with details on the practices of conjunctive use of five countries, are presented in the following sections.

2 EXAMPLES DISCUSSED

2.1 *South Asia (India and Pakistan)*

In South Asia, there are three main kinds of conjunctive use, firstly, total dependence on groundwater during drought years and lesser dependence on it during good rainfall years in non-canal command areas, second, conjunctive use in canal command areas and third, artificial recharge of aquifers. The history of first type of conjunctive use during drought years is well documented (Bhatia, 1992). However, conjunctive use in modern canal commands are much widespread, but are neither planned nor organized. In recent years, in view of severe water shortages, artificial recharge has acquired importance, especially in water-scarce parts of this region.

The most important characteristic of conjunctive use in India is that it is done more through the lack of options than through intention, e.g. farmers in tail end of canal commands do not receive enough water and resort to tube-well water as an alternative irrigation source. The complementary use of both sources is more common than pure substitution, e.g. due to groundwater quality problems as is the case in the Pakistan Punjab. In Pakistan Punjab, groundwater is often saline and so farmers mix groundwater with canal water in order to get desirable water quality. Decisions to have conjunctive use are taken by individual farmers rather than as the systematic application of a public policy.

Some examples of conjunctive use in South Asia are in the canal commands in Pakistan and parts of India. In Pakistan, when canals were built almost a century ago, conjunctive use was not a part of the principal design. As such, water was allocated through *warabandi*.[1] Public tube-wells were eventually built in order to cope with problems of water logging and also to supplement canal supply through direct discharge into wells. But over the years, due to mismanagement of these publicly managed tube wells, there is now a scheme to hand them over to the public for operation and maintenance. As a direct consequence of not so good performance of canals, during the late 1970s and early 1980s a phenomenal increase in private shallow tube-wells was observed in all canal command areas in Pakistan Punjab. Generally, two factors determine farmers' decision to make conjunctive use of groundwater and surface water. These are quality of groundwater and reliability of surface water. In order to maintain hydraulic equilibrium, common sense demands that there should be more groundwater use in the head reaches where water tables is close to the surface, thereby encouraging vertical drainage. However, because canal water is cheaper than groundwater, the farmers in the head reach rarely rely on groundwater irrigation. As a result, it is seen that the percentage of groundwater used as proportion of total water use, is more towards the tail than the head reaches of the canal command. This is because, at the tail end, canal water supply deteriorates necessitating farmers to use more and more groundwater. In certain parts of Pakistan Punjab, it is seen that groundwater quality deteriorates with distance from head. Therefore, conjunctive use of marginal quality groundwater actually leads to further deterioration of water quality and also affects soil pH adversely (Murray-Rust and Vander Velde, 1994).

In India, conjunctive use is found in almost all canal commands. One well-documented example of conjunctive use is found in the Mahi Canal Command, in the western state of Gujarat. In order to maintain hydraulic equilibrium, conventional wisdom demands that there be more groundwater pumping in the head reaches to reduce water logging and ensure that enough water reaches tail ender. Empirical evidence in

[1] Warabandi is a system whereby each plot gets water at a pre-specified time every week depending upon that size of the plot. This system does not take into account the nature of the crop grown.

Mahi command shows that pumping fails to stabilize water tables because it tends to be low in areas with very high or very low water tables (Shah, 1993).

In south Asia, planned conjunctive use of canal and groundwater gives an excellent opportunity of preventing water logging in the head reaches and allowing more and more water to flow towards the tail end- thereby resulting in a win-win situation. Empirical evidence, however, shows that no planned conjunctive use takes place in order to maintain hydrological equilibrium. The main reason behind this is the unequal pricing of canal and groundwater. While canal water is highly subsidized, extracting groundwater entails incremental costs, which are 5 to 10 times larger than cost of canal water. This price anomaly leads to lop sided conjunctive use, with deleterious effects.

Artificial recharge is seen as a means of conjunctive use in India. Artificial recharge and water harvesting are increasingly becoming important given that artificial recharge helps to maintain or supplement groundwater, in order to ameliorate conditions of coastal salt water intrusion and to store underground water imported from other basins (e.g. Narmada water). In areas that are particularly hit hard by water scarcity (such as western Indian provinces of Rajasthan and Gujarat), there has been a groundswell of popular/NGO/religious movements to popularize artificial recharge and water harvesting. There is as of yet no firm quantitative estimates about its efficacy, but some analysis shows that there has been a decline in TDS and that water in wells is available for longer duration. The very fact that village after village continues investing their efforts and money into building rain water-harvesting structures amply speaks for its efficiency, even if there might not be any scientific data to vouch for the same at this very moment.

In South Asia (Pakistan and India), it is clear that conjunctive use offers numerous benefits, but only if private interest of individual groundwater users coincide with societyís interest. This is, unfortunately, seldom the case. Thus, there is neither concerted public policy to do so, nor enough incentives for the farmers to practice conjunctive management as a planned choice; they do it on their own because they have no choice.

2.2 *Spain*

Except for the humid one third northern region, most of Spain can be considered as semi-arid. Three and half million hectares are irrigated. One million hectares are irrigated with groundwater and the rest with surface water. Current population is around 40 million persons, as a permanent population, in addition to around 60 million tourists, that introduce an important peak on the summer water demand mainly in the Mediterranean coastal areas. Two thirds of the cities larger than 20,000 inhabitants rely mainly on surface water, while two thirds of the cities smaller than 20,000 inhabitants rely on groundwater. Development of Spanish basins started many centuries ago, but it is in the 20th century when a big pressure was put on water resource development. Currently, most of the two thirds driest basins face a ratio of water demand to renewable resource higher than 60%; many of them are very close to 100%; and some of them more than 100%. These figures mean that, in many basins, aquifer overexploitation is taking place, reliability of the surface supply is not too high and water quality as well as environmental problems are severe, among other inconveniences. They also mean that a big effort has been done in the past, and must be done in the future, to render the situation more sustainable. Alternative resources, as treated wastewater direct reuse (230 Mm^3/yr, currently), and desalination (280 Mm^3/yr, currently) have been developed, and will be further increased. But an important element in the solution of those problems is provided by the design of optimal (or at least better) integrated water resources management policies for the basins. And conjunctive use of surface and ground waters is the main one.

Although a very interesting application of artificial recharge for urban use exists in the Llobregat delta near Barcelona, in Spain conjunctive use is mainly performed trough ACU, mainly in the Mediterranean slope. The territory of the Júcar Water Agency has a good number of basins where ACU is applied in a very efficient way. The name of the agency is taken from the Júcar, the biggest river in their territory. The agency territory covers 42,988 km^2, population is above 4 million, and irrigated land is around 370,000 ha. Of this irrigated land, 33% relies on surface water, 37% relies on groundwater, and 30% has availability to surface and ground waters trough ACU. A medium size basin in the agency is the Mijares one that is presented as a representative example of ACU. It has three storage reservoirs, one upstream in the Mijares River with 100 Mm^3 of capacity, the second, downstream in the main river and the third in a non permanent tributary with 50 Mm^3 and 28 Mm^3 of storage respectively. Those last two reservoirs in karstified limestones undergo important water loses, in the order of 45 Mm^3/yr that recharge the detritic aquifer of La Plana de Castellón. The main river, being perched above the aquifer, also feeds it with a similar yearly volume. When more surface water is available, aquifer recharge increases, less groundwater is pumped, and in dry years, recharge is lower and pumping increases. The difference between high and low volumes of water in storage in the aquifer can be as high as 700 Mm^3 (Figure 1), more than four times the existing surface storage. ACU schemes are used all over the basin (Figure 2) and a very efficient use of existing water resources is obtained; this has allowed maintaining most water demands during droughts.

Figure 1. La Plana de Castellón aquifer. Change in storage for different use alternatives.

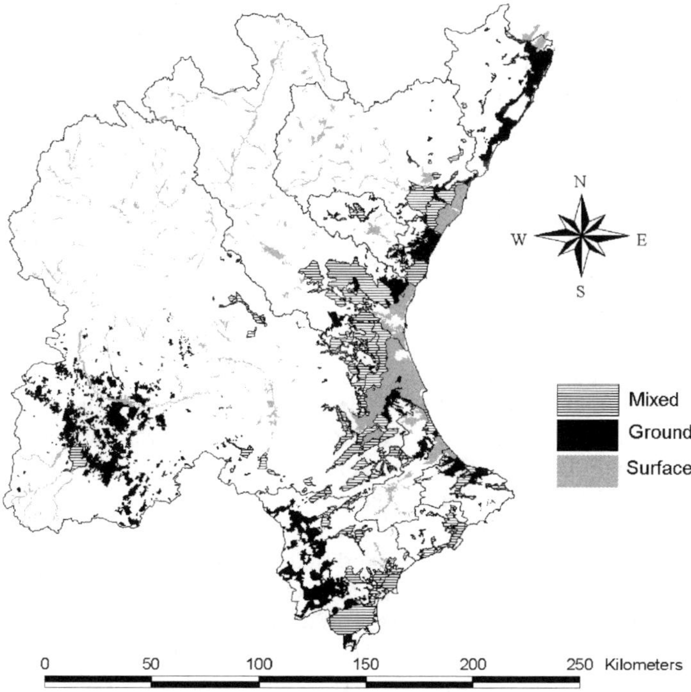

Figure 2. Conjunctive use in the Júcar basin, Spain.

In Spain there are many cases of regulating karstic springs with wells near the spring outlets in the vicinity of existing canals or aqueducts used to transport the spring flow. Pumping is used to increase water availability and starts operating when spring flow is below the existing water demand. So pumping effect over spring flow dries it and all required flow has to be pumped. Operating in this way, the water supply can be increased well over natural spring flow during irrigation season, or for urban or industrial needs. Very large flows have been obtained in several cases, with the consequent low cost. For example, up to 1,200 L/s were obtained with two wells in Los Santos river spring in Valencia, and more than 2 m^3/s with five 100 m deep wells near Deifontes, Granada, for irrigation; as well as 800 L/s in two pumping wells for the ACU scheme of La Marina Baja in the Júcar basin for urban water supply.

Most ACU in Spain have been promoted by users. In many Mediterranean basins in Spain, besides the fields traditionally irrigated with prior rights, additional areas seem to have been irrigated with surface water in humid years. After the rapid increase of aquifer exploitation in the 1960s, they were integrated smoothly into the existing systems. So more surface water was used during wet periods and more groundwater was pumped during drought periods. In all cases, schemes were developed and handled by users, and only in cases when they wanted technical or economic help, they submitted the operating schemes to the water authority. They are efficient, and in most cases there are practically no legal or social problems. The lack of problems is probably due to the fact that the proponents are the same water use associations located in the existing irrigated areas with surface water. This is a very well established tradition in Spain.

As a semi-arid country, Spain is concerned with the use of surface water and groundwater resources and has acquired experienced in the analysis and management of conjunctive use systems by applying advanced Decision Support Systems (DSS). AQUATOOL is a generalized DSS developed during the last twenty years at the Universidad Politécnica de Valencia (UPV), to optimise and simulate complex systems including conjunctive use. This method, which has been applied in many Spanish basins, can handle several tens of dams, aquifers and demand areas including rivers, canal and aqueducts and aquifer-river interaction, and it can tackle most common non-linear situations. It has been designed to help decision makers to analyze complex systems in order to answer specific questions facilitating the use of a set of models and data bases in an interactive way in a user-friendly control framework. One of its capabilities is the possibility of using a methodology that solves the space discretized groundwater flow equation allowing a very efficient integration of an aquifer model in the simulation of complex systems with conjunctive use.

In Spain, some of the applications of these systems were performed in three large basins in central and southeastern regions. The Tajo river basin (56,000 km^2), the Júcar river basin (43,000 km^2) and the Segura river basin (19,000 km^2), were studied to analyze water transfers, regional conflicts and the evaluation of alternatives in their conjunctive use of water resources management. The Segura basin has a series of 152 natural and artificial canal reaches and more than 18 stream reaches connected to aquifers; 18 reservoirs and 26 identified aquifers complement the basin and there have been included more than 90 consumptive demands. The application of the DSS in the Júcar basin led to a detailed analysis of a system management under drought conditions. The decision tools developed during the study of this basin were applied in all decision boards for water allocation during the campaign of 2001 and 2002.

2.3 *Canada*

Canada is a water-rich country. The availability of groundwater and surface water varies considerably in Canada. Canada holds about 20% of the world's freshwater; three major rivers of Canada rank among the top 10 rivers of the word, and the country has over 32,000 lakes with areas greater than 10 km^2. Close to 1 million km^2 of space is covered by Canada's freshwater lakes, ponds and rivers (excluding the Great Lakes). These numbers, however, do not tell the whole story. Approximately 60% of Canada's freshwater drain north, whereas 90% of the population lives in the South in a small fringe along the Canadian-USA border. In many cases, and in particular in rural areas, groundwater is the main source for domestic and agricultural uses. This poses some problems to manage the abundant high-quality water demands, having the second highest consumption in the world with 350 L/d, per capita, after the USA.

Contrary to its two southern neighbors and NAFTA partners, the USA and Mexico, the groundwater resources of Canada are *under*-exploited. Groundwater use in Canada varies from 0% in the North to 100% depending on the provinces and territories. However, several looming issues are currently raising awareness of Canada's groundwater resources:

- increase in water demands: groundwater use increased from 10% in the 1970s to 30% in the 1990s;
- instances of aquifer contamination;
- recognition of large knowledge gaps in the country's groundwater resources;
- bulk water exports. Exports of water to the USA under the North American Free Trade agreement (NAFTA) and to other countries;
- climate change impact and adaptations.

On the last item, recent federal-provincial discussions have identified potential effects of climate changes onto groundwater, among which, the effects that surface water would have on land and the possible increased demands of groundwater as a backup source of water supply. Furthermore, it is evaluated that groundwater development may affect low flow in streams, or that minimum levels in lakes could reduce groundwater discharge to coastal areas during droughts causing seawater intrusion. Finally, the combined effects of groundwater development and climate change could lead to less dilution of contaminants in streams.

As a country, Canada does not have a national strategy to manage water resources, these are managed independently, and sometimes, differently, by the 10 Canadian provinces. Thus, there are no specific plans at the national level. However, a general sense is slowly emerging in that groundwater is to be kept as a strategic resource against climate changes, severe droughts, and/or water exports. A couple of examples from two provinces show how direct measurements relative to droughts are already being implemented.

2.3.1 *Example in the Province of Alberta*

In 2001, the government of Alberta declared a drought disaster. Extreme dry conditions and lack of significant precipitation prompted the province to declare a drought disaster and responded to the immediate needs of livestock producers. Dugouts that usually store water for livestock were empty due to the lack of rainfall that summer, fall and spring, and the lack of snow the previous winter.

To assist the most affected people in the province, compensation was offered to farmers and ranchers. The program offered money to drill wells on their property to develop new water resources, namely from the ground, to fight for drought disaster. Additional funding was made available to help farmers develop long-term drought solutions for their land. Eligible projects to help develop permanent on-farm water supplies may include: farm wells, stock watering dams, and other projects that reduce the risk of drought due to water.

A comprehensive package of $93 million (Canadian dollars) was announced as drought relief assistance in an effort to salvage Alberta's $7.3-billion agricultural industry from the worst dry conditions to hit the province in more than 130 years.

The Alberta Farm Water Program (AFWP), however, is looking beyond the 2001 drought. In order to ensure farmers have sufficient and sustainable water sources, they will submit a long-term water supply management plan to the provincial ministry.

The province is currently evaluating a comprehensive water law and regulations to manage the provincial surface water and groundwater resources in an integrated manner. The study of possibilities to use the resources conjunctively in the future is in the province agenda. Having four large rivers (e.g. Peace, Athabasca, Saskatchewan and Bow) and more than ten regional aquifers (i.e. Paskapoo, Cold Lake, McMurray), it is likely that an integrated, conjunctive use, is the solution in the province to future drought scenarios due to climate changes.

2.3.2 *Example in the province of Quebec*

Even tough the province of Quebec has higher yearly precipitation than Alberta, the drought has also affected this province in the last few years. In 2002 the alarm was rang in the city of Montreal and its suburbs because the water supplies, whose sources are practically all from rivers, were at their lowest in decades. The severe drought (Canadian standards) observed in Quebec in 2002, prompted municipal authorities North of Montreal, to prohibit the use of water during several hours of the day.

On the other hand, recent hydrogeologic studies indicated that the St-Lawrence Lowlands Regional Aquifer, North of Montreal is composed of high quality groundwater with large quantities in storage in the sedimentary fractured rocks. Modeling of the regional aquifer showed that current pumping conditions could be tripled and their exploitation could still be sustainable (Nastev *et al.*, 2003). The model showed that the very low levels observed in the rivers during the drought, were actually due to base flow whose source was the regional aquifer.

Quebec has many large rivers, and approximately 36 large regional aquifers in the areas where most of the population lives. Approximately 26% of all water use comes from groundwater resources, yet there are large gaps in knowledge of the province's groundwater resources.

This province has recently created its water policy (Gazette Officielle du Québec, 2002) whereby the water resources of the province will adopt an integrated watershed management practices. Although not written in the policy, these measurements will eventually lead to a conjunctive use of the provincial water resources.

2.4 *Mexico and Mexico City*

Mexico, a fast-growing country, heavily relies on groundwater with approximately 70% of the population depending on groundwater. Total groundwater extraction is close to 30 km^3, from which 19 km^3 are for irrigation, approximately 7 km^3 is the supply to close to 70 millions inhabitants, and the rest is used by industry and other users. Failure or success of Mexican's economy and well being clearly depends on its water resources. History has shown the importance of water in all levels of society including the revolution and thereafter. With more than half the country being arid or semi-arid, water is a day-to-day struggle; groundwater resources have been the backbone of Mexican agriculture and urban development. Clearly, water has helped shaped this nation since the Spanish conquest until today. Alternate use has been a common practice in Mexico for centuries. But a coherent long-term plan for conjunctive use has not.

Mexico has been recently divided in 13 Water Administrative Regions (Figure 3). The inventory of Mexican groundwater resources indicates

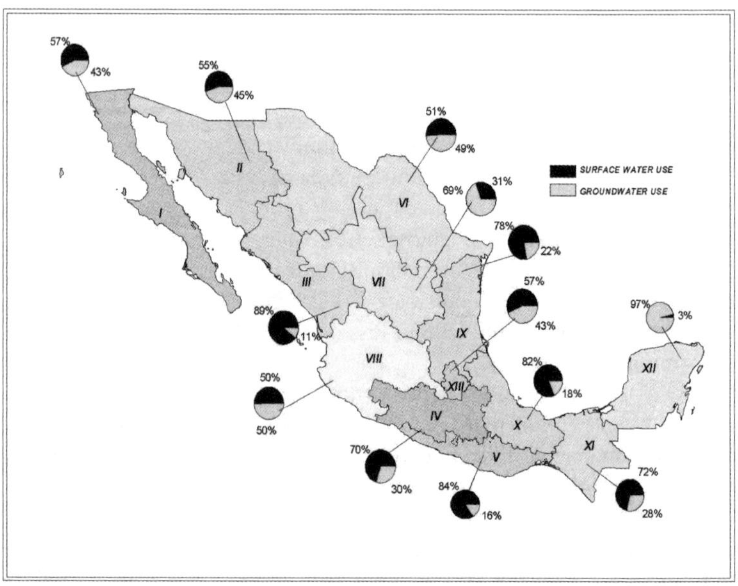

Figure 3. Water administrative regions of Mexico.

667 aquifers identified of which 258 are overexploited. It is estimated that of these aquifers, recharge is about $10\,km^3$, and pumping is about $15\,km^3$. Most overexploited aquifers are located in arid and semi-arid zones.

The effects that have been observed across the country, due to the overexploitation of groundwater resources are: increase in pumping costs; well abandonment; reduction of base flow in streams and springs; decrease in water quality; saltwater intrusion in coastal aquifers; and land subsidence.

Perhaps the clearest example of how man can completely modify a whole city because of its water use, is the City of Mexico. The magnificent Aztec city of *Great Tenochtitlan*, as it was know at the time of the arrival of Spanish conquistador Hernán Cortés, followed a quick metamorphosis in the last five centuries due to its eternal water problem of having too much or too little.

For centuries after the conquest, the city of Mexico had abundant surface water and its problems were essentially related to flooding. As the city developed and grew, surface water in the form of rivers and lakes slowly began to dry as tunnels and canals were dug to solve the flooding problem. Historically, the whole valley floor has been the discharge zone for the Mexico City Valley. Dewatering of the Mexico City has been a troublesome engineering challenge since the time of the Aztecs. Currently, the untreated wastewaters are disposed of on the surface and in an underground drainage system. These waters are used for irrigation north of Mexico City.

Until the end of the last century, the supply of drinking water for Mexico City was provided by springs located to the west and south of the city. Between 1900 and approximately 1930, when the city's population increased but still remained below one million, water-supply sources shifted progressively from springs to artesian wells. With time, these wells, and other new wells, were drilled deeper and deeper and were equipped with pumps, thereby rapidly modifying the regional groundwater head.

Due to the thick clay layers that are present throughout the Valley of Mexico, land subsidence became a major problem once groundwater extraction began on a regular basis in the middle of the 20th century. This problem (which continues today) became more acute in the 1940s when major groundwater withdrawals from the regional aquifer started. In the late 1950s, for example, the subsidence rate in the center of the Valley was on the order of 40 cm per year. Land subsidence greater than nine meters was reported in the Valley of Mexico as a result of groundwater withdrawals.

Mexico City, with a population of close to 10 million inhabitants (within the city), and approximately 20 million including the surrounding areas, obtains approximately 55% of its drinking water from groundwater (on the order of $19\,m^3/s$). As the population of Mexico City has continued to increase, so has the demand for groundwater. For example, as of 1988 it was estimated that more than $33\,Mm^3$ of groundwater were being withdrawn from storage annually, and that this volume was in excess of the recharge to the aquifer system.

In order to provide the larger amounts of water needed for economic growth, city authorities created a very ambitious program of groundwater exploitation. Local groundwater remained the only source for the city's water supply until the beginning of the sixties, when the city authorities started to import both surface and groundwater from other basins in neighbouring states. Figure 4 is a histogram of the pumping data in Mexico City for the period of 1934–1986.

During the same period, more than 6 meters of land subsidence was observed at some locations (Figure 5), constituting one of the most remarkable cases of subsidence in the world because of its magnitude and its extent. Since the forties, this phenomenon, observed at a regional scale, has been ascribed principally to groundwater exploitation.

The key for survival of this great metropolis is no doubt water. The city authorities have mentioned that "water will be the limiting factor for future development". Current working hypothesis estimate a 30% to 35% increase in water needs for the city by 2010, when the city could reach 30 millions inhabitants. To cope with this growing problem many solutions are envisaged, among which: restrictions in water use; water

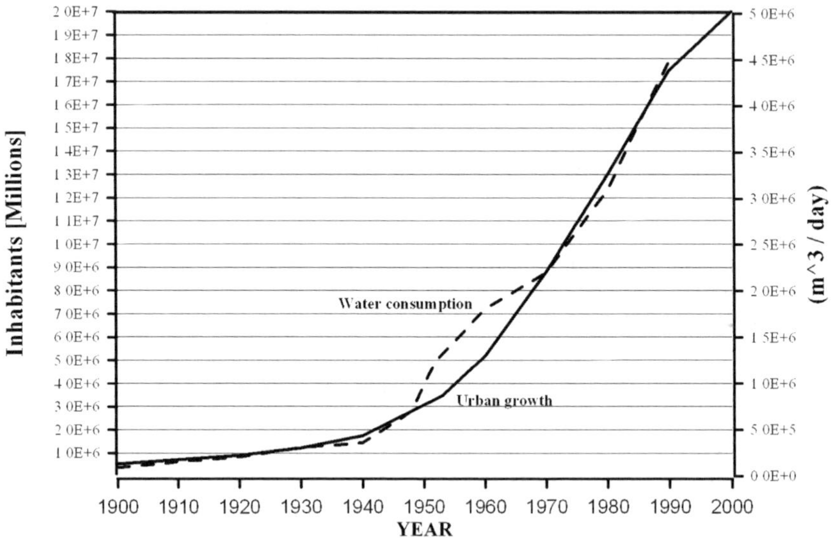

Figure 4. Urban growth and water consumption in Mexico City (Rivera *et al.*, 1991).

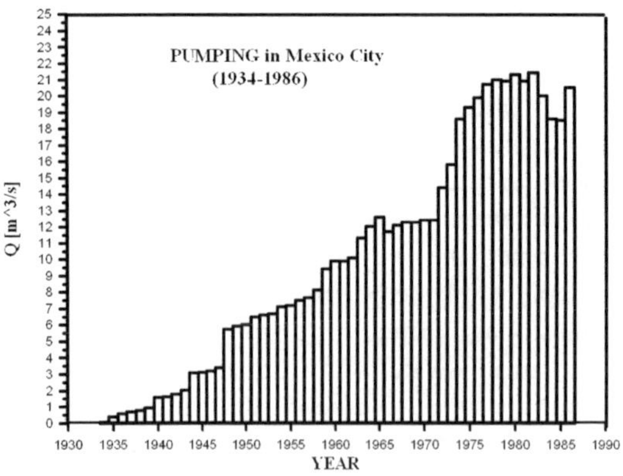

Figure 5. Groundwater pumping in Mexico City for the period of 1934 to 1986.

transfers from other basins outside the city (plans are to bring water from as far as 300 km away); artificial recharge; extraction of other groundwater resources believed to be located in deep aquifers at an elevation of –2,000 m.a.s.l. The first solution is not sufficient; the second is extremely expensive and poses many problems; and the fourth is yet to be confirmed. In the 1990s, the artificial recharge was believed to be a good solution, or at least a partial solution, to the overall problem of water.

A three-dimensional numerical model was used to evaluate the water constraints in the city and to reproduce the observed subsidence (Rivera *et al.*, 1991). The model was intended to reproduce the land subsidence coupled to the excess pumping and to be used for groundwater management practices. Following the model simulations, it was suggested to use surface water from the lake Texcoco with artificial recharge into the sandy confined aquifers in order to recover groundwater pressures and to stop land subsidence.

Artificial recharge would have as a main objective to restore groundwater pressures in the valley in order to decrease and stop land subsidence. However, it was also hoped that by recharging the aquifers artificially with re-used water, these could be used as conduits for storing water for later use and help alleviate the problem of over-exploitation.

The model simulations indicated an increase in water pressures but at a very modest rate and with modest consequences, it would slightly decrease the land subsidence rate, but it would not stop it. These concepts and plans were not fully realized given technical and economical problems. Water transfers from other basins were strongly increased and it is the current partial solution to the problem.

The Mexico City basin is globally overstressed both in the groundwater and surface water resources. Few gains are possible even if a comprehensive conjunctive use plan is applied; in

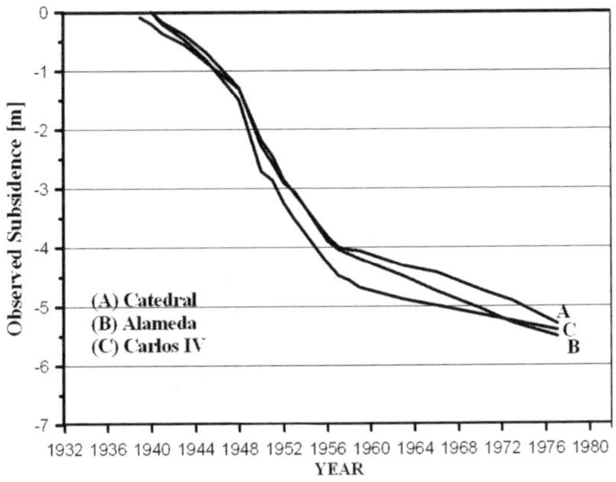

Figure 6. Subsidence observed in Mexico City for the period of 1934 to 1977.

this case only reallocation of resources is feasible. This practice, however, has other environmental consequences, for it disturbs other basins around the valley and posses all type of social, economical and political problems.

3 DISCUSSION

To be effective, the application of conjunctive use of groundwater and surface water requires at the very minimum: a good organization of stakeholders and good management. As is the case in other issues associated with the management of groundwater resources involvement of direct users is highly recommendable at the early stages of a project. An important issue in conjunctive use is the legal and sociologic aspects. In many cases, these may become important obstacles to develop actions that are acceptable hydrologically and economically. Otherwise if they are considered carefully, they may facilitate their realization.

Adequate tools and models are important to evaluate gains trough different operational alternatives to convince governments regulators and user of the different possibilities and restrictions. Decision-support systems are one of the most promising tools to improve operational aspects of complex water resources systems. The regular management practices for a conjunctive use requires tools for risk estimation in order to adapt the general operating guidelines to hydrological circumstances. The anticipation of measures (i.e. droughts) should become a regular practice in management. A set of operating rules are necessary for a better management and ad-hoc tools and models may be used for decision-support systems.

3.1 *Analysis of conjunctive use systems*

It is important for the design and operation of conjunctive surface water and groundwater resources systems to adequately evaluate their performance. Performance that is also required to convince all stakeholders in water related problems: governments, water agencies and other public administrations and users.

As discussed above, the design and management of conjunctive surface water and groundwater resources systems have a higher degree of complexity than systems of surface water alone or groundwater alone. In any case, a good system analysis practice is recommended to obtain good results, in conjunctive use this is a necessity. Thus, some particular aspects of the analysis have to be carefully considered:

- The assessment of surface and groundwater resources has to be done jointly. It is very common to make a separate assessment of the resources by building and calibrating more or less sophisticated models for surface hydrology and more or less sophisticated models for groundwater hydrology. Both types of models are usually calibrated in order to best reproduce the observed values at some locations of surface flow in the first case, and of piezometric levels in the second case. Due to the great number of parameters used in the models, this separate calibration does not guarantee that the interactions between the surface system and the aquifer are well captured. In fact, according to the Spanish experience, coupling of such separately calibrated models often produces incoherence's in the resulting joint models. The situation is much worse when the natural surface and ground hydrological regimes are disturbed by man activities. In such cases, the duplication of natural flows needed for the calibration of surface hydrology models require the duplication of groundwater interactions, which require a groundwater model, which in turn might need the values of the surface flows as inputs. Consequently, joint modeling and calibration is needed, with models oriented to capture in the best possible way the interactions between both subsystems. For conjunctive use modeling, this is more important than achieve a better tuning to other responses.

- The analysis of conjunctive use alternatives has to include streams, reservoirs, canals, aquifers, and flow interchanges between groundwater and surface water, in addition to water supply facilities for different uses. Consequently, it has to be conducted at regional scale. The basin scale being the most adequate in many cases.
- Conjunctive use is a matter of management. Therefore, operating rules are important components of the alternatives. The same set of structural facilities can produce very different yields depending on the operation of the system, so they have to be explicitly incorporated in the analysis, and be realistic enough to be applied in real life. There are many political, historical, sociological and cultural factors that may impede the application to real world of otherwise perfect operating rules.
- Modeling of groundwater components must be as detailed as needed for the purpose, but emphasizing surface water-groundwater interactions. The use of the tools must be facilitated through the possibilities provided by the modern concept of computerized decision support systems. State-of-the-art models and methodologies should be put in the hands of the real world practitioners and decision makers in order to study such complex systems for a large number of alternatives.

If these conditions are met, the more convenient conjunctive use strategies can be devised, including both components, the design of the infrastructure needed, and the operating general guidelines. On the latest steps, it is advisable to achieve the effective application of the conjunctive use strategies and continuous monitoring of the water bodies. This is better accomplished trough the users because the users associations can prevent individual objectives from community interests.

It is also necessary to have tools that help in the decisions in regular basis management of the conjunctive use, in order to adapt the general operating guidelines to the existing hydrological circumstances. From the information provided by the continuous monitoring of the water resources system, and the information on the water requirements, future scenarios for the short to medium term (e.g. some months) can be analyzed, and the risks of affording shortages can be evaluated. Then, anticipated measures can be adopted to mitigate the effects of such an operational drought.

4 CONCLUSIONS AND RECOMMENDATIONS

Due to the usually very high investments associated with dams and canal building and the present trend towards groundwater development, there is a great potential for conjunctive use in many developing countries in arid and semiarid regions. Increase of groundwater pumping during droughts has been a common practice all over the world for decades; and it will continue. In many cases groundwater pumping complements surface water availability but usually to a limited extend. It is a limited stage of *conjunctive use.* In such cases surface water surplus can occur often during wet years while groundwater resources are extensively exploited. The logical extension is not only to mitigate droughts by augmenting groundwater pumping, but also to try to use as much surplus surface water in wet years as possible and proceed to an integrated *alternate conjunctive use.* The rationale behind this strategy is: that advantages can be obtained for the aquifer whose stress diminishes and achieve a higher use of surface water resources during wet years. Both gains are obtained without augmenting surface storage and without the need of artificial recharge.

In arid countries, due to surface water irrigation, return flow increases aquifer's recharge increasing groundwater levels. Drainage and salinity problems often arise (e.g. India, Pakistan,

China, Egypt, Asian countries of the former USSR, Argentina) and sometimes millions of hectares are affected and abandoned. Groundwater pumping can solve or attenuate drainage and salinity problems, but it is only practiced in a few cases and in a limited way. In some cases, more dams have been built thereby exacerbating drainage problems. Conjunctive use could be used both to increase water availability and to treat drainage problem. The case of the Indus irrigation scheme in Pakistan discussed before is one of the most enlightening and interesting one.

As a general rule, conjunctive use can help whenever an aquifer and a river (with or without a dam) coexist.

Improvement of many schemes can be achieved rather cheap and quickly through ACU, but adequate institutional and social changes would be needed in most cases.

The roundtable discussions concluded that a conjunctive use is an essential aspect of integrated water resources. Among the conjunctive use schemes, *alternate conjunctive use* is very attractive for semi-arid regions of developing countries. On the other hand, the analysis for implementing conjunctive use has to be carefully performed; management is perhaps the most crucial single factor, and a strong political will is needed to implement such systems. It was obvious from the experiences in various countries that some level of organization must be provided for an effective application of conjunctive use. One important point to stress is that every improvement should be made according the users needs and cultural behaviour. We are confident that enhancement of the roll of water users can work in most cases and the instauration of Water User Associations will work to make irrigation systems more equitable (Shah *et al.,* 2000; Rao, 2000).

Generalized decisions-support systems can help in the planning and management of conjunctive use.

Some of the most benefic aspects we can expect form the conjunctive use of surface water and groundwater are:

- alleviation of drainage and salinity problems;
- alleviation of aquifer over-exploitation;
- alleviation of sea-water intrusion;
- higher reliability;
- smaller infrastructures;
- increase in economic optimization.

However, it was observed that in globally over-stressed basins (e.g. Mexico City, Segura basin in Spain) few gains are possible, only reallocation of resources is feasible.

REFERENCES

Bhatia, B. (1992). Lush fields and parched throats: political economy of groundwater irrigation in Gujarat. *Economic and Political Weekly*, Vol. 19(26): A142–170.

Gazette Officielle du Québec (2002). Regulations and Other Acts. June 14, 2002, vol. 134, N° 24A.

Llamas M.R. and Custodio, E. (2003, eds). Intensive use of groundwater: challenges and opportunities. A.A. Balkema Publishers, the Netherlands: 478 pp.

Murray-Rust, D.H. and E.J. Vander Velde (1994). Conjunctive use of canal and groundwater in Punjab, Pakistan: Management and policy options. *Irrigation and Drainage Systems*, Vol. 8(4):201–231.

Nastev, M., Rivera, A., Lefebvre, R., and Savard, M. (in press) Hydrogeology and numerical simulation of the regional groundwater flow of the St. Lawrence Lowlands of South-western Quebec. *Hydrogeology Journal.*

Rao, P.K. (2000). A tale of two developments of irrigation: India and USA. *Int. J. Water, Vol N° 1, pp.41–60.*

Rivera, A., Ledoux, E. and de Marsily, G. (1991). Nonlinear modelling of groundwater flow and total subsidence of the Mexico City aquifer-aquitard system. *Proceedings of the IV International Symposium on Land Subsidence*. IAHS Publ. No 200. p 45–58.

Sahuquillo, A. and Lluria, M. (2003). Conjunctive use as potential solution for stressed aquifers: social constraints. In Llamas, M.R. and Custodio, E. (eds). Intensive use of groundwater: challenges and opportunities. A.A. Balkema Publishers, the Netherlands: 478 pp.

Shah, T. (1993), Groundwater Markets and Irrigation Development: Political economy and practical policy, *Oxford University Press*, Bombay, India.

Shah, T., Hussain, I and Saed-ur-Rehman (2000). Irrigation management in Pakistan and India: Comparing notes on institutions and politics. Colombo, Sri Lanka: International Water Management Institute (IWMI). 21p. (Working paper 4).

VALENCIA DECLARATION

*The first draft of this Declaration was prepared by the organisers of the *International Symposium on Groundwater Intensive Use* (Valencia, Spain, 14–16 December 2002) and submitted to the participants in order to receive their suggestions. A good number of suggestions were received and most of them accepted. Nevertheless, the final responsibility of the final text belongs to the organisers of the Symposium. This Declaration has similarities with the Mar del Plata Declaration done during the AIH Congress in Mar del Plata on October 2002, but it emphasizes those aspects more directly related to the intensive use of groundwater, a phenomenon that mainly occurs in arid and semiarid regions.

1. There is intensive development of groundwater when a significant proportion of the average annual renewable resource is withdrawn from the aquifers, which in turn noticeably modifies their hydrogeological functioning, causes significant ecological, political or socio-economic impacts, or important changes are produced to river-aquifer relationships.
2. The intensive use of groundwater, mostly but not exclusively developed in the last few decades in arid and semiarid countries, has been a driving force to produce a large number of benefits to society. These include the affordable supply of drinking water and the development of irrigated land, which have contributed to health improvement and famine alleviation of hundreds of millions of people in developing countries.
3. The large water storage capacity of aquifers provides a more reliable means for adapting to interannual precipitation variability. Aquifers, either as a sole source of supply or as part of a conjunctive use management program, become an efficient solution for overcoming or mitigating drought impacts.
4. The guarantee in supply, coupled with the low cost of extraction of groundwater facilitated by the scientific and technological advances, have led to a spectacular increase in groundwater use, especially for irrigation, in numerous arid and semiarid regions and in many coastal areas.
5. Groundwater is typically developed near the designated place of use. The full direct costs of groundwater development are generally born by local users. This situation has contributed to a better and more responsible general use of groundwater compared to cases of surface water in which subsidies dominate.
6. Due to previous socio-economical factors, a large amount of the irrigation developments that use groundwater have been implemented by the final users, with almost no planning or control by public or governmental agencies. As a result, there is a lack of the necessary human and economic means, and technical knowledge for coping with this new situation.
7. Numerous problems and potential long-term impacts have arisen due to intensive use of groundwater when a local institution for control and management does not exist. These problems include excessive lowering of groundwater levels, groundwater storage depletion, land subsidence, impacts to other users (e.g., dewatering of shallow wells and increased costs of abstraction), decreased discharge to – and effects on base flow of – rivers and springs, potential mobilisation of contaminants,

and impacts on aquatic ecosystems. Most of these problems can be avoided, corrected, or at least mitigated with a comprehensive groundwater management program.

8. Problems associated with a relatively recent increase in groundwater development are generally of a form that are short-term and medium-term and can be mitigated, particularly if there is a management plan in place where data are gathered and there is early recognition of a problem. Efforts to overcome development issues should not deviate the attention of the water policy decision-makers from serious longer-term problems, particularly those relating to groundwater quality. Some groundwater quality impacts (i.e., the potential effects of natural contaminants or saline water intrusion) are linked to the quantity of the available potable water supply, but others (i.e., the effect of man-made contaminants) depend more on land use than on intensive use of groundwater.

In order to maintain the benefits derived from groundwater as a source of supply, and to avoid the already mentioned negative aspects of unmanaged intensive groundwater development, the organisers present the following

PROPOSALS

First:

Public Administrations must play a key role in the planning and integrated management of water resources. Among other things, this implies developing a better understanding of the intrinsic relationship between surface and groundwater resources and coordinating water resources management with land use planning. To achieve these goals, water administrations of most countries need to improve their hydrogeological capabilities much beyond what has been the case in recent years.

Second:

In parallel with the above large-scale planning, immediate groundwater decision-making has to be performed by local groundwater management institutions. Active and democratic participation of all the users or stakeholders is important for the successful implementation of the groundwater management program. Therefore, to have effective stakeholder participation, groundwater management program development should be initiated with training that conveys hydrogeological principles and also a clear understanding of the need for a management program to the stakeholders.

Third:

International organisations should recognise the great diversity of the world's hydrogeological and socio-economic situations. While overall groundwater management principles are broadly applicable, unique regional and local conditions ultimately determine the management objectives, the groundwater management tools that are appropriate, and how the tools will be applied to achieve the objectives.

* This version (29 January 2003) contains some minors grammatical changes from the version presented at the end of the Symposium.

SUBJECT INDEX

AUTHOR INDEX